Arboriculture

ARBORICULTURE

Care of Trees, Shrubs, and Vines in the Landscape

RICHARD W. HARRIS

Department of Environmental Horticulture
University of California at Davis

Illustrations by Vera M. Harris

PRENTICE-HALL, INC., *Englewood Cliffs, New Jersey 07632*

Library of Congress Cataloging in Publication Data

HARRIS, RICHARD WILSON
  Arboriculture: care of trees, shrubs, and
vines in the landscape.

  Bibliography: p.
  Includes index.
  1. Arboriculture. 2. Ornamental woody
plants. I. Title.
SB435.H315   635.9'7   82-3867
ISBN 0-13-043935-5   AACR2

Printed in the United States of America

10

Editorial/production supervision and interior design by Joan L. Stone
Cover design and page layout by Judith Winthrop
Manufacturing buyer: John Hall

All photographs, unless otherwise credited, are the property of the author.

Most of the nutrient deficiency symptoms of conifers from previously unpublished material
of R. F. Powers, U.S. Forest Service, are from *Principles of Silviculture*, 2nd ed. by T. W.
Daniel, J. A. Helms, and F. S. Baker. Copyright © 1979 McGraw-Hill, Inc. Used with the
permission of McGraw Hill Book Company.

Table 21-1 from DOULL, J., C. D. CLAASSEN, and M. O. AMDUR. Eds. *Casarett and Doull's
Toxicology: The Basic Science of Poisons*. 2nd ed. Copyright © 1980 by Macmillan
Publishing Co. Inc.

PRENTICE-HALL INTERNATIONAL, INC., *London*
PRENTICE-HALL OF AUSTRALIA PTY. LIMITED, *Sydney*
PRENTICE-HALL CANADA, INC., *Toronto*
PRENTICE-HALL OF INDIA PRIVATE LIMITED, *New Delhi*
PRENTICE-HALL OF JAPAN, INC., *Tokyo*
PRENTICE- HALL OF SOUTHEAST ASIA PTE. LTD., *Singapore*
WHITEHALL BOOKS LIMITED, *Wellington, New Zealand*

Contents

CHAPTER FOUR

Planting Site: Climate, *80*

CHAPTER FIVE

Modifying Climatic Influences, *117*

CHAPTER SIX

Planting Site: Soil, *138*

CHAPTER SEVEN

Planting Site: Preparation, *166*

CHAPTER EIGHT

Planting, *200*

CHAPTER NINE

Transplanting Large Plants, *227*

Preface

This book is written and illustrated for those who are concerned about the preservation and care of woody plants for human well-being and pleasure. It is to serve as text and reference for students, professionals, and those interested in landscape horticulture, urban forestry, landscape architecture, nursery production, landscape construction, and the management of parks and gardens. These areas represent a wide range of skills and concerns, but a primary focus of all is the wise and effective landscape use of woody plants.

Research, observation, and professional ingenuity during the last 20 years have greatly increased our understanding of landscape plants and their environment. By bringing much of this information together, this book attempts to help the reader more completely understand plant growth and response and more effectively use and care for woody plants in the landscape.

New findings have usually verified the wisdom of early practitioners. More recent research and experience in new situations, however, suggest changes in a number of commonly accepted practices. This book describes the course of such changes, as well as some current practices, but it also describes earlier practices and their sources, so that the reader may have a broader exposure. The bases for favored practices are presented so that they may be evaluated on their merits. Research and observations that postdate this book or that have escaped my attention may warrant some modification of the recommendations or explanations that appear here. I would appreciate receiving such information so that all may benefit from the most current advances in the field.

The main emphases of the book are on the plant environment,

plant growth and response, considerations for selecting plants, cultural practices, diagnosis of problems, and nonparasitic problems. Examples of diseases, insects, and pest management complete the coverage on the use and care of woody plants in the landscape. More comprehensive coverage of insects, diseases, and their control, however, are presented in several fine references. Federal and state publications provide up-to-date information on specific pests and the most recent control measures.

Topics that have not been adequately covered or brought together in previous literature are discussed in more detail than are more familiar topics. In the citations, publications giving the more traditional presentations are named. The appendices include lists of plants that are organized according to certain characteristics, such as tolerance to landfill conditions. The more common lists of plant characteristics are available elsewhere and are cited in a final appendix.

Measurements are given in metric units, followed by their English equivalent in parentheses. In many situations, approximate values are accurate enough, so conversions between metric and English systems are rounded for simplicity. To minimize possible confusion, length measurements are given in millimeters (mm), meters (m), or kilometers (km); similarly mg, g, and kg, and ml and l are most commonly used for weight and volume measurements, respectively.

This book has been possible only with the encouragement and support of many people. First I am deeply indebted to my wife Vera for her patience and understanding, for her help in proof editing and index compiling, and for her skill as an illustrator. I drew on the research, experience, teaching, and writings of many people for my understanding of plant growth and arboricultural practices. We are all the richer and wiser for the writings of arborists like A. Bernatzky, T. T. Kozlowski, P. J. Kramer, P. P. Pirone, and A. L. Shigo.

My graduate and undergraduate students sharpened my thinking by asking perceptive questions and making helpful suggestions on earlier versions of portions of this book. Peggy Sears McLaughlin and Nelda Matheny were particularly helpful.

Many colleagues generously assisted by sharing their expertise and reviewing portions of the manuscript. These included Spencer H. Davis, Jr. of Rutgers University; Eugene B. Himelick of the Illinois Natural History Survey, at Urbana; Subhash C. Dormir of the USDA Agriculture Research Service, in Delaware, Ohio; Dan Neely of the Illinois Natural History Survey; Robert F. Powers of the USDA Pacific Southwest Forest and Range Experiment Station, in Redding, California; and Alex L. Shigo of the USDA Northeastern Forestry Experiment Station, in Durham, New Hampshire.

Among the many colleagues of the University of California who also helped are Leland R. Brown and Henry Z. Hield at Riverside; Carlton S. Koehler and Robert D. Raabe at Berkeley; Clyde L. Elmore, Jerry L. Hatfield, Harry C. Kohl, Jr., William B. Davis, Andrew T.

Leiser, Jack L. Paul, and Roy M. Sachs at Davis. Although I was helped by many, I take full responsibility for the ideas presented here, for application and their accuracy.

I appreciate the generous help of those who supplied illustrations and photographs used in this book. They are specifically acknowledged in the captions.

Many of my thoughts and most of my writings were clarified by the late Kelvin Deming, who edited about half of the initial manuscript. The final editing is the fine work of Linda Gregerson. Anita Deming and Paula Deming skillfully completed the initial editing. Gayle Bacon masterfully turned a jumble of edited pages into finished manuscript. Although not directly involved, Judy Martin, Winifred Floyd, Patricia Romejko, and Marsha Dallas of the office staff of the Department of Environmental Horticulture lightened many university responsibilities while I was preparing the manuscript.

Joan L. Stone, Judith Winthrop, Dudley R. Kay, and Paul J. Feyen of the Prentice-Hall staff were extremely patient and helpful throughout the preparation of the book.

When information was originally obtained the following sources, now included on text pages 339 and 274, respectively, were unavailable to the author:

HEERMANN, D. F., and R. A. KOHL. 1980. Fluid Dynamics of Sprinkler Systems. In *Design and Operation of Farm Irrigation Systems*, ed. M. E. Jensen. St. Joseph, Mich.: Amer. Soc. Agr. Engr., pp. 583–618.

LEONARD, O. A., W. B. McHENRY, and L. A. LIDER. 1974. *Herbicide Residues in Soil of the Vine Row 21 Months Following 9 Successive Annual Applications.* Proc. California Weed Conf., Sacramento: 26:115–22.

CHAPTER ONE

Landscape Trees, Shrubs, and Vines

While plants contribute to our pleasure, comfort, and well-being, they also help conserve energy and the quality of air, water, and soil. Plants, particularly trees, are an important part of our lives—around homes, schools, shopping centers, and places of work, along streets and highways, in the central city, parks, and other landscaped areas.

Trees have been held in high esteem since earliest times. More than 4000 years ago, Egyptians wrote of trees being transplanted with a ball of soil around the roots of each (Chadwick 1971); some were moved 2400 km (1500 miles) by boat. In Greece, Theophrastus (370–285 B.C.) and Pliny (A.D. 23–79) gave rather complete directions for tree planting and care. Many books on the care of plants, including trees, have been written since these times.

In the Middle Ages, botanical gardens primarily contained plants that had medicinal potential. Later, the gardens of private estates boasted exotic plants brought in through trade and travel. Many of these gardens have since become public and are great sources of information and pleasure.

By the 1700s, trees were being planted with some frequency in the cities and estates of Europe. In the early settlements of North America, trees were cut to make room for farms and towns, but by the late 1700s, trees were being planted in town squares. After the plantings, however, few of these trees received much care, except on large estates. As settlers moved west onto the prairies, they planted the seeds of fruit and shelter trees around their homes.

In the early 1900s, national research agencies in Europe and North America and state agricultural experiment stations began to study fruit and forest trees. By the 1950s, state and national research stations had begun working specifically on landscape tree problems. The deva-

1

station caused by Dutch elm disease, phloem necrosis, Gypsy moths (*Porthetria dispar*), and Japanese beetles (*Popillia japonica*) in the Northeastern and Midwestern United States was largely responsible for the increased interest in tree research, though this research consequently focused on disease and insect control. Experiment stations, botanical gardens, arboreta, and some large nurseries have long been involved in landscape plant introduction; increasing efforts have sought trees that will be better able to withstand the rigors of the urban environment.

Governmental agencies sponsor extensive parks and landscape tree plantings. Many cities require street-tree plantings in new residential and commercial developments. The Federal Home Administration has long required that at least one tree be planted in the yard of each home that receives an FHA-insured loan. People are more aware of trees and their value. Many cities in the United States have ordinances controlling the removal of trees, even on private property. One of the stipulations of England's Civic Amenities Act of 1967 addresses the preservation and planting of trees on private property (Chadwick 1971).

Several professional organizations are concerned with tree planting and care. The National Shade Tree Conference, organized in 1924 in Connecticut, later became the International Society of Arboriculture, with current headquarters in Urbana, Illinois. The Society has members on every continent except Antarctica. Other United States organizations include the National Arborists Association (commercial arborists), the Society of Municipal Arborists, and the American Society of Consulting Arborists. The Arboricultural Association in Great Britain is similar to the International Society of Arboriculture. Many other organizations are concerned with landscape trees; among them are the American Association of Botanic Gardens and Arboreta, the American Society for Horticultural Science, the American Entomological Society, the American Society of Landscape Architects, the Society of American Foresters, and the Phytopathological Society.

Many technical schools, colleges, and universities offer courses in arboriculture, landscape horticulture, and urban forestry, with supporting courses in botany, entomology, plant pathology, landscape design, and soils.

ARBORICULTURE

Arboriculture, as herein defined, is primarily concerned with the planting and care of trees and more peripherally concerned with shrubs, woody vines, and ground-cover plants. Arboriculture is commonly defined as the cultivation of trees and shrubs only (Bailey 1976, Webster 1976), but woody plants that are called *vines* in the United States are called *wall shrubs* in England (Brown 1972, Halliwell, Turpin, and

Wright 1979) and *climbing shrubs* in Australia (Mullins 1979). It seems reasonable to assume that the common definition of arboriculture includes woody vines. All of the plants mentioned are woody, perennial plants that have many common needs and characteristics and differ primarily in their form and training requirements. A wisteria vine, however, can be trained into a tree, and the mature form of English ivy is shrub-like (see Fig. 2-16). *Hortus Third* (Bailey and others 1976) further defines arboriculture as the cultivation of plants as individuals rather than as elements in a forest or orchard.

Arboriculture is one of the branches of horticulture within the plant sciences. Following are the fields within plant science, their definitions, and their relationships to one another.

Plant Science

Agronomy and *Range Science* deal primarily with the cultivation of field crops and range and pasture plants.

Forestry concerns the commercial production and utilization of timber.

Silviculture is the practice of raising forests.

Urban Forestry is the management of trees in urban areas on larger than an individual basis.

Horticulture concerns plants that are intensively grown for food and aesthetics.

Pomology is the cultivation of perennial fruiting plants, primarily woody trees and vines.

Vegetable Crops (Olericulture) is the growing of herbaceous plants for human consumption.

Environmental Horticulture is the cultivation of plants to enhance our surroundings.

Floriculture is the production of cut flowers and potted plants.

Nursery Production (Ornamental Horticulture) is the production of primarily woody plants for landscape plantings and fruit production.

Landscape Horticulture is the care of plants in the landscape.

Arboriculture concerns the cultivation of woody plants, particularly trees.

Landscape Construction involves the installation of structural and plant materials according to a landscape plan.

Landscape Maintenance (*Gardening* or *Grounds Maintenance*) specializes in the planting and care of a wide variety of plants used in the landscape.

Turfgrass Culture concerns the growing of turf for landscape and sports use.

Landscape Architecture concerns the planning and design of outdoor space for human use and enjoyment.

Park Management concerns the total responsibility of planning, developing, and managing public and private landscaped areas, from housing developments and city parks to heavily used national parks.

Even though these areas of specialization are often thought to be mutually exclusive, the lines of demarcation between them have blurred. A company or agency may engage in landscape design, nursery production, and all phases of landscape horticulture. Another may specialize in one or two. Many organizations find that by engaging in more than one specialization they are able to maintain a more stable labor force and to spread financial risks.

PLANTS
IN THE LANDSCAPE

As public and private plantings become more common and more expensive to maintain, people responsible for such plantings are becoming more concerned about the proper selection and care of landscape plants. The landscape comprises a palette of plant growth habits: trees, shrubs, vines, grass, and herbaceous flowering and foliage plants. This book focuses on the selection and care of woody perennial plants, primarily trees, but also devotes some attention to shrubs, vines, and ground-cover plants (see Fig. 2-1). Perennial plants with these habits of growth have much in common as regards selection and care in the landscape. Because of the ultimate size and long life of trees, their proper selection and care is of greater importance. Woody plants respond much as trees do to most environmental conditions and cultural practices.

Plants not only provide food and fiber but enhance our surroundings in a variety of ways: physical, aesthetic, economic, and psychological. Some influences are important to our immediate surroundings; others are significant only on a more extensive scale. Cited most often when the value of landscape plantings is extolled are *physical* attributes: influence on climate, air purification, noise reduction, and erosion control. The microclimate can be greatly enhanced by plants, particularly trees, when they are properly selected, placed, and maintained.

PHYSICAL BENEFITS

Microclimate Enhancement

Plants absorb heat as they transpire, provide shade that reduces solar radiation and reflection (Fig. 1-1), can reduce or increase wind speed, and can increase fog precipitation and snow deposition. Trees are frequently called nature's air conditioners. A hectare (2.47 acres) of vegetation transpires about 17,000 liters (4000 gal) of water on a sunny summer day. Oke (1972) estimates that a 30 percent cover of

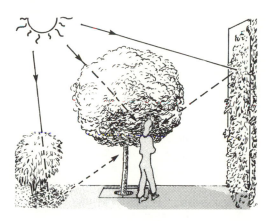

FIGURE 1-1
In summer, plants provide comfortable surroundings by intercepting direct solar radiation and reradiation from heated surfaces. (Adapted from Robinette 1972)

vegetation will give two-thirds as much cooling as will a complete plant cover. Thus, in a 0.08 hectare (one-fifth acre) lot with 30 percent plant cover, transpiration would absorb 1.2 million kilogram-calories of energy that would otherwise warm the environment. That is equivalent to the amount of cooling necessary to air condition two moderately sized houses 12 hours a day in the summer.

This evaporative cooling, however, is quickly diffused so that even on a hot summer day, the air temperature of an area of plants is not much lower than that of an area with few or no plants. Herrington, Bertolin, and Leonard (1972) found that shaded sites in a 30-hectare (75-acre) park in Syracuse, New York, averaged only 1.3°C (2.5°F) cooler than urban sites outside the park, even though solar radiation under the trees was only one-fourth of that in the urban sites. Temperatures outside the park would certainly have been higher had the park and other vegetation not been there, since temperature differences are evened out fairly effectively by air movement and diffusion. Fortunately, thermal comfort results from the interaction of the human body with a number of factors in the environment, including air temperature, relative humidity, wind speed, solar radiation, and infrared radiation. The most important influence of trees on the microclimate is their control of radiation, particularly solar.

Trees minimize heat reflection and reradiation by shading pavement and buildings and by reducing direct rays of the sun (Fig. 1-1). Shade from trees can reduce room temperatures in poorly insulated houses by as much as 11°C (20°F) in summer (Deering 1955). Deering and Brooks (1953) found that bare-ground surface temperatures of 56° to 67°C (130°–150°F) were cooled an average of 20°C (35°F) in five minutes after being shaded. Any barefoot youngster knows the value of shade on a summer afternoon. In Memphis, Tennessee, Dunn (1975) found that property owners, in appraising the value of a tree, were able to use the cost of constructing a structure that would provide equivalent shade, and had not been successfully challenged by the U.S. Internal Revenue Service. At the time, a modest sturdy structure would have placed the value of a shade tree at $27.50 to $33.00 per square meter ($2.50–$3.00/ft^2) of shade. In the tropics, evergreen trees provide year-round protection from the sun. In temperate regions, deciduous trees provide summer shade and permit the winter sun to warm the ground, although the sun can be considerably impeded: Light intensity under dormant trees is reduced 20 to 74 percent (Geiger 1961). Winter sun is essential for effective solar houses.

Plants modify wind by obstructing, guiding, deflecting, and filtering air flow (Robinette 1972). Air movement influences real and perceived temperatures; for example, with a chill factor of 30 kmph (20 mph), a wintry blast of 0°C (32°F) air will cool as if it were −14°C (7°F). Windbreak and shelterbelt plantings shelter smaller plants, animals, and property by reducing wind speed on both the windward and leeward sides (Fig. 5-6) (Geiger 1961). Nearby shrubs help protect buildings from the cold by reducing radiation loss and providing an insulating zone of relatively still air. A dense planting across a slope can create a frost pocket on its uphill side, however, by slowing the downward flow of cold night air (Fig. 4-4).

Plants, by their form and placement, can guide or funnel wind through openings or over their tops and in so doing can increase wind speeds by as much as 20 percent (Robinette 1972). A row of trees with dense heads can create quite a wind tunnel between the ground and the lowest branches.

Trees in particular can reduce fog density by condensing moisture on leaf and twig surfaces. Geiger (1961) reported that woods along the coasts of Japan protect inland areas from fog; six to ten times as much moisture was deposited under trees as on nearby open grassland. Similarly, the windward edge of a woods had 20 times as much fog precipitation as did the lee side. The summer fogs common in many coastal sites greatly influence the moisture economy of these areas. Not only does summer fog reduce evapotranspiration, but fog precipitation can account for 20 percent of the annual precipitation in wooded areas (Geiger).

By slowing air, a windbreak with a density of 50 percent can quite effectively cause snow to accumulate in front of, within, and behind the barrier. Conversely, plantings that channel air flow will keep the areas of increased wind relatively free of snow and other airborne particles.

Air Purification

In many urban areas the concentrations of air pollutants are so great that plants are not able to grow at their best, much less reduce pollution to acceptable levels. Air currents and the diverse sources of contaminants make air pollution a regional concern. The major effort must be to reduce emissions from vehicular and industrial sources (Schmid 1975). When air pollutants are at reasonably low levels, plants will be healthier, more active, and more effective in further reducing air impurities. Schmid found that ozone concentrations were reduced several times faster when plants were transpiring rapidly (stomates open) than at night, when transpiration was low.

Gaseous pollutants are absorbed into active plant tissue, primarily within leaves, and are adsorbed on plant surfaces. Lanphear (1971) estimated that 50 million Douglas fir trees 300 mm (1 ft) in trunk diameter would be required to mitigate the 410,000 metric tons (452,000 tons) of sulfur dioxide released each year in St. Louis. That many trees would cover about 5 percent of the city's land surface. Bernatzky (1978), however, cites two German reports of deciduous forests having no perceptible sulfur dioxide filter effect compared with coniferous forests. In addition, many of the gaseous pollutants are exhausted or rise above even the tallest trees. Little sulfur dioxide released in a city remains there to be absorbed by trees (Bernatzky, see Chapter 18).

Even though plants absorb carbon dioxide from the air and release oxygen, plants in a city have little effect on carbon dioxide and oxygen levels there. Plants in the surrounding countryside and remote areas, as well as plant organisms in the oceans, are the primary maintainers of the CO_2/O_2 balance. Photosynthesis in the oceans supplies 70 (Cole 1968) to 90 percent (Bonner and Galston 1952) of the world's total oxygen. Winds and convection currents help maintain these gases at fairly uniform levels.

The oxygen supply is exceedingly large and well-buffered (Broecker 1970). If all known fossil fuel reserves were burned, the oxygen level would be depleted by only 3 (Broecker 1970) to 8 percent (Cole 1970); even at the higher rate, the effect would be the same as exposing every person to a 600-m (2000-ft) increase in altitude. Cole (1970) also estimates that if all living things on earth died and decayed, atmospheric oxygen would be reduced by no more than 1 or 2 percent.

In the same article, Cole estimates that within the 48 contiguous United States, 40 percent more oxygen is being consumed than is being produced by green plants. He raises the concern that oxygen consumption is intensified in heavily industrialized cities, especially during times of reduced air circulation, and that this may cause transient oxygen crises. Broecker (1970) points out, however, that carbon monoxide

from automobile exhausts would reach serious physiological levels
before the oxygen content of the air dropped 2 percent. Other atmospheric conditions appear to be much more serious and threatening
than changes in the CO_2/O_2 balance.

Vegetation ameliorates air pollution most effectively through its
ability to reduce airborne particulates. This is primarily a windbreak
effect, reducing wind speed so that heavier particles settle out. In
addition, particles are adsorbed on plant surfaces, primarily the leaves.
Evergreens are recommended for this purpose since they are equally
effective the year around. Foliage can become so coated, however,
particularly along highways, that it must be spray-washed for both
health and appearance. Bernatzky (1978) reports that in Frankfurt/
Main streets with trees had 3000 particles per liter of air, compared
with 10,000 to 12,000 in streets without trees.

Plants also reduce the content of heavy metals polluting air. Sum-
marizing several studies, Schmid (1975) estimates that in Connecticut
one sugar maple tree of 300 mm (1 ft) diameter at breast height (DBH)
removes from the air in one growing season 60 mg of cadmium, 140 mg
of chromium, 820 mg of nickel, and 5200 mg of lead. He further states
that lead concentrations in plants taper off quickly with distance from
major highways. In spite of the several ways that plants reduce air pollution, their presence near polluting sources is not generally considered
to be of particular importance to urban air quality (Schmid).

Noise Reduction

Plants are not very effective in reducing noise. "Out of sight, out
of mind" applies somewhat and may be of importance in confined
spaces. However, in order to reduce noise levels appreciably, plantings
must be dense, tall, and wide (25–35 m; 80–115 ft) (Fig. 1-2). Plantings
close to the noise source are more effective than similar plantings
further from the source. Fleshly leaves and numerous branches increase
excess attenuation; that is, they reduce noise more than simple distance
will. Species do not differ greatly in their ability to reduce noise levels,

FIGURE 1-2
Thirty meters (100 ft) of trees and shrubs reduce truck noise about as effectively
as would a similar area of bare cultivated ground. (Cook and Van Haverbeke 1971)

although evergreen plants are best for year-round effectiveness (Cook and Van Haverbeke, 1971).

Reports conflict in their assessment of sounds of differing frequencies. Robinette (1972) indicates that excess attenuation by plants increases with frequency, which would be desirable, since the higher frequencies are most annoying to humans. Herrington (1974), however, finds no consistent correlation between excess attenuation and frequency. No ready explanation of these contradictory findings is available.

Plants can not only hide a noise source but can also diffuse noise so that it is more steady. Leaf and branch movement and wind mask other sounds by raising the background level of noise, much as music in stores and offices masks external sound.

Erosion Control

Bare soil can be seriously eroded if exposed to rain, flowing water, and wind. Water accounts for most topsoil erosion (Fig. 1-3). Plants intercept rain and thereby soften its impact, which can otherwise loosen soil particles and get them into suspension to be carried with the water. Plant litter on the soil further reduces raindrop impact. A mulch increases the rate of water infiltration and slows movement on the surface, so that water tends to enter the soil near where it falls, either as rain or as sprinkler irrigation. Fibrous roots hold soil and further reduce erosion.

FIGURE 1-3
Sloping bare soil is easily eroded when the application of water exceeds infiltration
(left); ground cover plants or a mulch could have prevented this erosion.
Plants, however, cannot prevent soil slumps and slides
(right); an engineering solution is necessary.
(Photos courtesy U.S. Soil Conservation Service)

The faster the speed of wind or water, the larger the particles that can be carried away. Wind erosion can be minimized by plantings that slow the movement of air.

Plants also conserve water by reducing runoff and by accumulating snow near windbreaks for later melting and infiltration into the soil. This latter phenomenon is important in areas of winter snow when precipitation has been low.

Other Uses

Plants can be used to direct pedestrian and vehicular traffic as well as to improve the appearance of roadways. Shrubs can screen headlight glare from oncoming traffic and can serve as barriers to slow vehicles that are out of control. Plants can reduce bothersome glare and reflection from the sun or artificial lights, just as they can reduce solar radiation, reradiation, and reflection.

VISUAL BENEFITS

Landscape Aesthetics

In our increasingly man-made world, the landscape aesthetics of plants are becoming more highly valued. Plants provide a basic contact with nature and heighten pleasure in our surroundings. Their aesthetic value is more difficult to quantify than the values already discussed and is difficult to describe, in fact, without seeming trite or overly sentimental. Without elaborating on the principles of landscape design, I list below some of the aesthetic advantages plants can provide (adapted from Bingham 1968):

Plants provide a variety of color, form, texture, and pattern in the landscape.

Plants soften architectural lines and accentuate structural details.

Plants can form vistas, frame views, provide focal points, and define spaces (Fig. 1-4).

Plants relieve the monotony of pavement and masonry.

Plants, particularly trees, make enticing play areas.

Plants offer cooling shade, pleasant fragrances, intriguing sounds, and serene settings.

Plants create the impression of a well-established place in new residential areas and minimize the raw unfinished look.

Plants unify, giving coherence to visually chaotic scenes.

Plants can emphasize the seasons.

FIGURE 1-4
Plants can form vistas or screen
unsightly views.

ECONOMIC BENEFITS

Economic Value

Landscape plantings in America have seldom been thought of as an economic resource from which revenue, other than taxes, can be realized. With increasing interest in conservation and alternative energy sources, public and private agencies are exploring the possibilities of utilizing what in the past have been regarded as waste wood products. Thorton (1971) estimates that the 1180 m^3 (0.5 million board-feet) of sawlogs and 10,900 m^3 (3000 cords) of pulpwood disposed of in Philadelphia in 1969 could have kept a small sawmill busy all year. That estimate does not include other plant wastes that could be used for fuel.

The International Society of Arboriculture (ISA) has developed a procedure for appraising the value of landscape trees on the basis of species, size, condition, and landscape location (Neely 1979b). In 1979 the Society established a basic value of $18 per square inch ($3.00/cm^2) of trunk cross-section at breast height (4.5 ft, 1.3 m above

11

the ground) (Chadwick 1980). For trees less than 300 mm (12 in) DBH, replacement value is to be used. Appraisals based on the ISA procedure have been used as the basis for settling insurance claims, civil suits, and property condemnation. The United States Internal Revenue Service has yet to recognize this procedure in determining value of casualty tree losses.

Landscape plants, particularly trees, add to the value of real estate, although there are few definitive examples of accurate assessment. Using photographs of wooded and open land in Amherst, Massachusetts, a U.S. Forest Service study asked real estate appraisers to estimate the value of trees (Payne 1973). Trees increased the appraised value of undeveloped land by 27 percent and that of 0.2 hectare ($\frac{1}{2}$ acre) lots with houses by 7 percent for an average property and up to 15 percent for some lots. Using assessors' records for 60 residential sales in Manchester, Connecticut, Morales (1975) calculated by multiple regression that a house and lot with at least 50 percent tree cover sold for 6 percent ($2686) more than a similar house and lot without trees. In 1965, Waller reported than an offer of $24,000 for a house and lot with a 300-year-old elm in its front yard was withdrawn after a hurricane destroyed the tree. The house eventually sold for $15,000.

PSYCHOLOGICAL BENEFITS

In a study by Talbott and others (1976), the researchers observed the effects of flowering plants upon the behavior of hospitalized psychiatric patients. When flowering plants were placed in the hospital dining room, there was a significant increase in talking among the patients, time spent in the dining room, and the amount of food consumed. A transient, though not significant, increase in the number of social gazes occurred during the first week the plants were present. Hospital rehabilitation programs increasingly include plant propagation and gardening.

Industry officials have found that attractive buildings and landscapes result in above-average labor productivity, lower absenteeism, and easier recruitment of workers with hard-to-find skills (Lovelady 1965). Equally important, handsome factories and offices build good community relations.

Many articles state that people have a basic desire for contact with plants, even that plants have a strong positive influence on human behavior. More studies are needed to verify such relationships and to find out how plants can be used more effectively for human well-being.

COSTS OF LANDSCAPE
PLANTINGS

Although landscape plantings in the United States perform some of the functions described in the preceding pages, Schmid (1975) states that "the plants which occupy Anglo-American front yards are present, neither because they enhance the mesoclimatic or acoustic environment, nor because they attract birds and small mammals, nor because they provide food or firewood, but rather because they are traditional ornaments necessary to maintain a public image of the appropriate setting for single family houses." To support that notion, he cites Gans (1967): "Homeowners are urged to support neighborhood norms and deviate neither toward shabbiness (which might lower neighborhood values) nor toward elaborate plantings (which could tend to push upward the general standard of maintenance and require more effort by others)." For all their values to our well-being and enjoyment, plants in the landscape have their price. If these plant problems are recognized, it is possible to eliminate or at least minimize them.

Plants not only have an initial cost but they must be maintained throughout their life—some plants grow easily from seed and seem to need little or no care; others require a king's ransom to plant and maintain. The California Department of Transportation estimates that in 1970 it cost about $10 to plant a 4-liter (1-gal) plant and an equal amount to care for it the first three years. A 1975 survey of 16 southern California cities found that the cost of maintaining a hectare of landscape park ranged from $4750 to $17,500 ($1900–$7000/acre).

Because of their large size, trees can be dangerous and costly. Trees and their limbs fall in fair weather or foul, injuring people and damaging property. A tree is a formidable object when struck by a vehicle or by lightning. Tree roots clog sewers (Fig. 1-5) and break and lift sidewalks, curbs, and pavement and damage house foundations (see Fig. 16-20) Electric power companies and districts in the United States spent more than $500 million in 1980 to keep trees and shrubs from interfering with power lines and rights of way (Robert Tackaberry, 1981, Electrical Power Research Institute, in a private communication).

Trees can seriously affect UHF (Ultra High Frequency) television reception (BBC 1979). Screening by trees can vary greatly with seasonal and weather conditions and is a special problem where television signals are weak. Trees can also reflect a signal so that a delayed image or ghost appears on the screen. Evergreen trees block television signals more completely than do deciduous trees, and conifers more so than broadleaved evergreens. When bare, deciduous trees have little effect

FIGURE 1-5
These sections of sewer pipe were replaced because roots clogged them.
Austin Carroll (with hat) of Sacramento, however, is touching two fruitless
mulberry roots that crushed the 100-mm (4-in) diameter orangeburg
(asphalt impregnated fiber) sewer line (to which the roots are wired
to show their positions). The tree was planted in the backfill
of a sewer trench that had been dug into hardpan.
The expanding roots crushed the sewer line.

except during wet weather; when covered with moisture, all trees can appreciably affect signals, particularly higher-frequency signals. UHF reception can be protected by mounting aerials above the tops of trees or so that no trees intervene between transmitter antennae and aerials. Cable television eliminates the problem altogether.

Many people are allergic to the pollen or leaf pubescence of many plants. Poison oak and ivy are quite dermally toxic when they are touched or their smoke is encountered. The leaves and fruits of many landscape plants are poisonous, although the number of fatalities is extremely low considering the number of poisonous plants.

Leaves, flowers, and fruit can be an expense and nuisance to clean up. Many plants seed profusely or root sucker and become weeds. People have requested that street trees be removed because of the litter they produce.

For whatever reasons, landscape plantings are important features of private and public properties and must be selected and cared for with skill. By understanding how plants grow and respond to cultural practices and their environment, we can increase our enjoyment of land-scapes. To that end, the following chapters are designed to provide the basis for a more complete understanding of the care of woody land-scape plants.

FURTHER READING

BERNATZKY, A. 1978. *Tree Ecology and Preservation.* New York: Elsevier Scientific Publishing Co.

GEIGER, R. 1961. *The Climate Near The Ground.* Cambridge: Harvard University Press.

GREY, G. W., and F. J. DENEKE. 1978. *Urban Forestry.* New York: John Wiley.

ROBINETTE, G. O. 1972. *Plants/People/and Environmental Quality.* Washington, D.C.: U.S. Government Printing Office.

SCHMID, J. A. 1975. *Urban Vegetation.* Univ. of Chicago, Dept. of Geography Res. Paper 161.

CHAPTER TWO

Plant Growth and Form

Arboriculture concerns the planting and care of woody perennial plants, primarily trees, but also shrubs, vines, and ground cover plants (Fig. 2-1). A tree is usually described as having one dominant vertical

FIGURE 2-1
Trees, shrubs, and vines offer a wide variety of landscape forms.

trunk and a height greater than 5 m (15 ft). In contrast a shrub has a number of vertical or semiupright branches originating at or near the ground and is smaller than a tree. A vine or wall shrub has a slender, flexible stem that requires support. Prostrate, spreading shrubs and vines are often considered to be ground cover plants. These are arbitrary definitions but afford some convenient characterizations of plant growth and form. The words *tree*, *shrub*, *vine*, and *ground cover* bring certain ready images to mind, although the differences among them are not always easy to specify. Multistemmed trees, when young, might be considered shrubs, and many plants we regard as shrubs eventually grow to be 10 m (30 ft) tall or more. The distinguishing feature is form rather than size.

Knowing the potential form and size of plants and the stages of their growth is essential to anyone who plans landscape plantings and maintenance practices. The genetic makeup of a plant influences bud formation and break, shoot elongation, and caliper growth in response to environmental conditions. The above-ground shoots and stems of woody perennial plants must survive the winter and have buds grow the following spring. Leaves of a deciduous plant last but one season, while those of evergreen plants persist longer, usually three to four years. An evergreen plant drops its older leaves each year, although this may occur over a long period and at any season.

LEAVES

Trees, shrubs, and vines compete more effectively with other plants and even with animals by getting their leaves up in the air for more exposure to the sun and protection from land-bound animals. Although almost every part of a plant is essential for its survival and well-being, leaves are the primary site for photosynthesis, which provide energy for nourishment, growth, and reproduction. In photosynthesis, carbon dioxide (from the air) and water (primarily from the soil) are converted into organic substances in the presence of chlorophyll and light. Chlorophyll (the green substance in plants) occurs primarily in leaves, some stems, and immature fruit.

Upright growth and branching of trees increase the amount of leaf surface exposed to sunlight. In order to grow upright, plants must resist the pull of gravity; this phenomenon is called *geotropism*. In addition, if plants are to grow tall without support they need a strong structure. In dicotyledonous plants (those with two-seed leaves), trunk, branch, and root are strengthened by secondary growth of the vascular cambium. Monocots (plants with one-seed leaves) such as palms and cycads may produce large stems by a special thickening meristem that is most active immediately below the top of the plant and may achieve root stability by increasing the number of roots but not their individual size.

SHOOT GROWTH: LENGTH

At the tip of each shoot in a tissue volume no larger than a pin-point, all of the aerial parts of a plant are determined, that is, the form of the mature plant is laid down. This region, called the shoot apical meristem (a meristem is a tissue in which cell division is relatively rapid and continuous), is a hemispherical mass of rapidly dividing and expanding small cells. Flower, leaf, and axillary bud primordia and the conductive tissues of the stem are initiated in this region. Usually the entire zone of stem elongation, in which cells divide and enlarge to their mature dimensions, is in the most apical 20 to 50 mm (1–2 in) part of the stem. This is true for branches as well as for the main stem. In some vines, stem elongation may be distributed over the terminal 100 mm (4 in), but this is uncommon.

Buds

A bud is an unexpanded shoot or flower. A shoot elongates by growth of its terminal bud and forms branches from lateral buds formed in leaf axils at nodes along the shoot (Fig. 2-2). When a terminal bud is removed by pruning or breakage, a bud or buds grow from below the break. In addition, *adventitious* buds may form on a root, a leaf, or occasionally an internodal region of a shoot or branch. Unlike terminal and lateral buds, adventitious buds have no pattern to their location.

FIGURE 2-2
Buds are located either terminally or laterally (axillary = in a leaf axil).
Lateral bud arrangement is alternate (left), opposite (center), or whorled (right).

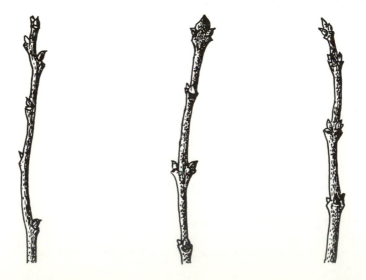

Buds may produce leaves (and a shoot) or flowers or both; they are respectively called leaf, flower, and mixed buds. More than one bud may be present in the same leaf axil.

The arrangement of leaves on a shoot is called *phyllotaxy*. Each species, and sometimes a genus or a family, has a specific pattern. Since buds form in the axils of leaves, buds and flowers will have the same phyllotaxy as the leaves. When two or more buds are opposed or spaced around the same node, their arrangement is said to be *opposite* (as in the ash or maple) or *whorled* (as in a catalpa) (Fig. 2-2). An imaginary line between two opposite buds at a node will usually not be parallel to similar lines at the nodes immediately above and below but rather will be nearly at right angles to such lines. Buds may be spirally arranged around the shoot when only one bud is at a node (as in an apple tree), or all the buds may be on one side at a node (as in a peach tree). The phyllotaxy of this *alternate* bud arrangement is usually expressed as a fraction, such as $\frac{1}{2}$, $\frac{1}{3}$, $\frac{2}{5}$, or $\frac{3}{8}$. The numerator is the number of revolutions around the shoot it will take to reach a bud directly above or below another; the denominator is the number of buds passed. The phyllotaxy of a plant is helpful in identifying it, particularly when the plant is not in leaf or flower.

Buds vary in their state of activity. The shoot apical meristem is an actively growing bud from which a shoot with leaves, buds, and possibly flowers develops. A *dormant* bud is an inactive bud on a one-year-old shoot between leaf drop or late fall and the start of new growth in the next season. Pruning, spraying, and transplanting may be scheduled during this dormant period, when the plant is most tolerant to many stresses. A *latent* or *suppressed* bud is one that is more than a year old but has grown enough each year so that its growing point is at or near the surface of the bark. When a terminal bud is removed by pruning or breakage, one or more buds near the tip of the remaining branch usually grows to take over the terminal role. Most shoots that grow from old branches, from the trunk, or near large pruning cuts are likely to be from latent buds, not from *adventitious* buds, as many people think. These are called *epicormic* shoots.

Types of Shoots

Annual shoot length is a good measure of plant vigor. Recent annual growth can be determined in plants that have one period of growth a year and set dormant terminal buds (Fig. 2-3). On deciduous plants the bud scale scars of the terminal buds and on conifers each whorl of branches indicate a year's growth when species have only one flush of growth a year or grow only from preformed initials. Current growth of deciduous plants can be determined from the leaves or buds that are present on the current season's shoot. On the basis of external appearances alone, however, current growth and the growth

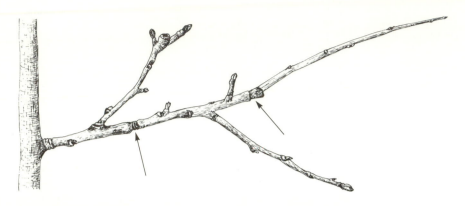

FIGURE 2-3
Three years' of shoot and spur growth from the vertical branch;
terminal bud scale scars (arrows) separate each year's growth.
Two short shoots and two short spurs have grown from the second-year spur.

of previous years may be almost impossible to distinguish on broad-leaved evergreens and plants that make multiple flushes of growth.

Many woody species form short as well as normal shoots (Fig. 2-3). Short shoots are usually less than 100 mm (4 in) long and are called *spurs* by fruit growers. Nodes are much closer together on spurs than on normal shoots and usually produce a higher proportion of flower buds. Most spurs have wide angles of attachment to the branch on which they arise. Annual spur growth may be terminated by a thorn (as on pyracantha and black locust), a normal vegetative bud, or a mixed bud that will produce both leaf and flower (as on apple). Spur-forming plants of a given size will usually have more flowers and fruit than species of a comparable size that lack spurs.

Watersprouts and *suckers* are vigorous, upright, epicormic shoots that grow from latent buds in older wood. They can be a problem in a number of fruit and landscape plants (Fig. 2-4) and can grow 3 to 5 m (10–15 ft) in one season. Often they outgrow the main leader of young trees. In some species they can be an annual problem for almost the entire life of a plant. If previous growths of this kind have not been pruned off close to the trunk or roots from which they arise, they can become particularly numerous.

Watersprouts and suckers differ primarily in their location or origin: Watersprouts arise above the ground or graft union, primarily from latent buds on the trunk or older branches; suckers arise below the graft union or the ground from the trunk or roots. Almost without exception, both should be removed as early as possible. In a few situations one might want to develop a watersprout into a main branch to fill a void in the plant, but on grafted plants, suckers can be quite different from the grafted top.

FIGURE 2-4
Vigorous shoots that grow from above ground or above the graft union
are called watersprouts (left); those from below the graft union (center),
or the ground (right) are called suckers. Note the black walnut leaves
and bark of the rootstock differ from those of the English walnut top (center).

Watersprouts are often forced into growth just below large pruning
cuts, particularly when branches have been cut to stubs. Watersprouts
are seldom firmly attached to the trunk or branch from which they
arise. Only the new layers of bark (phloem) and of wood (xylem)
formed since the sprout started growth unite it with the tree. Consequently, they can break out relatively easily for a number of years.

Watersprouts and suckers indicate that a plant is in vigorous condition, at least that part of the plant below the vigorous growth. They
may be forced into growth by injury or death of the top of a plant.
When an older tree is severely weakened by drought, deficiency, or
disease, many shoots grow along the main branches and trunk, but they
are of low vigor, usually less than 600 mm (2 ft) long (Fig. 17-4). It
almost seems as if they were the tree's last effort to stay alive, and
something more than pruning is needed. Similar shoots may grow if a
branch or trunk is exposed to sunlight because of defoliation, opening
up of the tree crown, severe pruning, or death of branches due to fire
or a similar catastrophe.

Control of Shoot Elongation

The rate of stem elongation depends on the rate and duration of
cell division and expansion. Daily and annual elongation rates differ
widely among different species. Faster-growing plants usually have
longer zones of cell division and expansion at their shoot tips than do
slower-growing plants.

Pines constitute an excellent example of genetic diversity in growth rates, often within a single species. Dwarf members may appear and grow at only a fraction of the rate for more vigorous plants in the species. Usually elongation rates among seedlings occur in a nearly continuous gradation. A major task for nurserymen working with seed-propagated plants is to *rogue out* or set aside the seedlings with extraordinarily high or low growth rates. When selecting young plants that have been seed-propagated, one should choose those of uniform size and shape to better assure mature plants with uniform characteristics.

Some plants elongate for a short period (3 to 4 weeks) in the spring, others in flushes during the growing season, while still others may elongate nearly continuously as long as conditions are favorable. The duration of shoot elongation may be controlled in several ways:

Initiation of Terminal Flowers   An inherent control of elongation occurs in species, such as crape myrtle and southern magnolia, that flower terminally on the current season's shoots (Fig. 2-14). Such control of elongation is similar to that of many herbaceous annuals whose growth habit is said to be *determinate*. This contrasts with plants of *indeterminate* habit (like peach and mulberry trees) that have lateral flower buds and a terminal that remains vegetative. The terms determinate and indeterminate are, however, usually confined to herbaceous growth habits.

Some woody species, such as apple and pear trees, form terminal mixed buds (with both leaf and flower primordia), usually on short shoots, which do not grow until the next spring. Although the terminal mixed bud does not control elongation in the way terminal flowering does, the effect on the next year's growth is similar. In both cases, growth begins from lateral buds, giving a somewhat zigzag (sympodial) pattern to the branches.

Preformation of Nodes in the Terminal Bud   In all deciduous and some evergreen plants, buds and nodes are formed in the terminal bud, but the internodes do not elongate until the following spring or when stimulated to do so. These are called *preformed shoot initials*. Some plants continue shoot growth after their shoot initials have expanded, while species like the fir, the spruce, most pine, and many oak stop shoot elongation after the preformed initials have expanded. No matter how favorable the growing conditions, these plants have only an early flush of growth. Although no additional buds or nodes will be formed in the same season, favorable growing conditions will result in longer internode length and in more buds and possibly more vigorous growth in the following season.

Species such as the juniper, the sequoia, and most angiosperms continue shoot growth as long as conditions are favorable. Within a single season, these plants respond to environmental conditions and to human practices designed to stimulate growth. Still other species, like some of the pines, are intermediate in that their shoot growth is controlled by the number of buds preformed in the terminal bud. Under normal or stress conditions, these plants have a single period of growth. Under more favorable conditions, shoot terminals will remain active and continue growing. Growth from preformed initials usually exhibits shorter internodes and smaller leaves than growth from currently formed initials.

Response to Environmental Signals Environmental conditions, such as temperature and daylength, must be satisfactory for growth to occur (see Chapter 5). Many plants slow or stop growth when daylength shortens. Species native to higher latitudes, where late spring and early summer days are especially long, may be dwarfed when grown closer to the equator because there are fewer long days.

Response to Internal Controls Even though buds form on currently growing shoots, many do not grow until the following spring. *Apical dominance* (see the latter part of this chapter) prevents newly formed buds from growing. Later in the growing season and into the winter, a condition called *rest* (see Chapter 4) keeps the buds of many temperate-climate plants from growing even if favorable growing conditions prevail. Bud growth will not occur until these internal controls are removed or satisfied. Finally, the plant may be ready to grow but may encounter late winter temperatures that are too cool.

SHOOT GROWTH: CALIPER

Radial growth of shoots in woody plants depends upon cell division and expansion in a thin cylinder of cells, the vascular *cambium*, which extends from the base to the tip of most plants. New cells are formed on both sides of the cambium (Fig. 2-5): the cells to the inside give rise to the woody water- and nutrient-conducting tissues of the stem, the *xylem;* the outside cells become part of another transport system, the *phloem*, which conducts organic substances (such as sugars, amino acids, vitamins, and hormones) from the mature leaves to other portions of the plant. The oldest xylem is nearest the center, the oldest phloem on the outside. In some plants, the phloem tissue alone forms the bark, but in many species, another layer of rapidly expanding and dividing cells, the cork cambium or *phelloderm*, forms outside the phloem, adding bulk to the bark and protection to the trunk.

Xylem not only conducts water and nutrients but also provides the main support to hold plants upright. In addition, xylem stores a wide range of organic materials for later use and can change chemically to retard or restrict wound decay. As xylem ages it loses its ability to

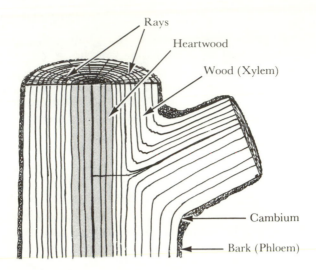

FIGURE 2-5
Cross- and longitudinal sections of a tree trunk and branch base. The cambium
is a thin layer of cells between the bark (phloem—the dark outer layer) and the
inner concentric rings of wood (xylem). Note that more growth (wider rings)
commonly occurs on the lower portion of the branch.

transport water and nutrients and to supply organic materials it has
stored, but retains its structural strength and infection-fighting ability.
The xylem of some trees, such as elm and walnut, discolors with age
and forms what is called *heartwood;* old xylem of other species does
not discolor with age but may after wounding.

Cambial activity is markedly seasonal. In temperate climates, it
begins shortly after bud break in the spring and proceeds down the
stem in an advancing wave that has been correlated with the movement
of auxin. Other hormones have also been connected with cambial
activity. As with shoot elongation, the duration of cambial activity
differs considerably among different species and different climatic conditions, especially variations in soil moisture. Wet and dry years are
clearly recorded in the xylem of many woody plants. The width of
annual rings can be used as a measure of vigor similar to shoot length.

Annual rings of xylem growth are evident because earlywood,
produced in the spring, usually has larger, lighter colored cells with
thinner walls than does latewood, formed later in the growing season.
In conifers, whose wood is composed primarily of tracheids (water-
conducting elements), the thick-walled small cells of the latewood of
one year are in marked contrast to the earlywood of the following
spring. In broadleaved plants, the boundaries between successive
growth rings may be quite distinct or hardly noticeable, depending on
the species. Annual growth rings in ash and oak are easily seen because
vessels are concentrated in the earlywood (Fig. 2-6). Trees with large

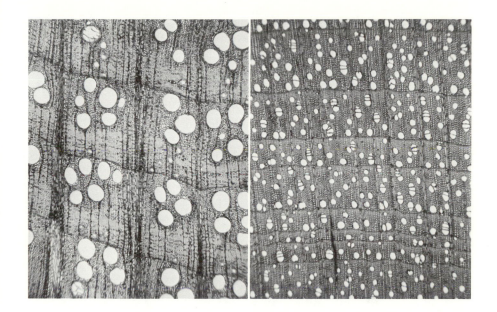

FIGURE 2-6
Variation in vessel sizes and distribution within annual growth rings. Trunk
cross-sections are taken from the oak, a ring-porous species (left), and the
maple, a diffuse-porous species (right). Annual growth rings of diffuse-porous species
are sometimes more difficult to discern than those of ring-porous species.
(Photos courtesy of Alex Shigo, U.S. Forest Service)

vessels in the rings are said to be *ring-porous*. In others, such as the
beech and the tulip tree, growth rings are not always clearly seen
without a hand lens because the large vessels are more evenly distributed throughout each ring (Fig. 2-6). These are referred to as
diffuse-porous trees. The last few rows of flattened cells may be the
only visible separation of the growth rings of two seasons. Late in the
season, many diffuse-porous trees, however, produce more fibers than
vessels; because the vessels are somewhat smaller, the annual rings in
these trees are more easily seen (Zimmermann and Brown 1974).

Lumber with relatively little earlywood is considered the most
desirable. Not only are there relatively more small cells, but the cell
walls are thicker, creating a harder, stronger wood. For a given thickness, large-celled wood in vigorous trees is not as strong, but overall
strength will usually be greater in a more vigorous plant because its
trunk will make more caliper growth. For a living plant, the relative
strength of these two types may be difficult to assess.

Perennial plants are highly compartmented, shedding plants (Shigo
1979). Xylem cells are grouped into compartments by the circumferential formation of late summer wood and the radial formation of xylem
rays. The circumferential walls are continuous around each growth

ring—except where sheets of ray cells pass through. The xylem rays are discontinuous walls because they vary in length, thickness, and height. These compartments can be quite tall. When xylem is wounded, the plant responds by plugging the vertical vascular system above and below the wound. The conducting elements—vessels in angiosperms and tracheids in gymnosperms—are plugged by tyloses, gum deposits, or pit asperations, and so forth. The plugged elements complete the transverse top and bottom walls of compartments.

Shigo numbered these compartment walls for identification purposes and difference in ability to resist discoloration and decay (Fig. 2-7). The plugged elements on the top and bottom walls he called *Wall 1*, the weakest wall in resisting decay. The closest annual circumferential ring in from a wound is *Wall 2* and it is the second weakest. The xylem rays form *Wall 3* and are considered the strongest walls in the plant at the time a wound occurs. Here chemical changes (primarily oxidation and polymerization of phenolic compounds in angiosperms and terpenes in gymnosperms) occur that increase the resistance of all of the compartment walls to discoloration and decay.

After a plant is wounded, the cambium forms xylem—*Wall 4*—that is both anatomically and chemically different from that formed at other times (Fig. 2-7B). Wall 4, also called the *barrier zone* (Sharon 1973), is pathologically the strongest of the walls and can usually confine injured and decay-infected wood within the diameter of the trunk at the time of wounding, or at the time a branch dies. Although Wall 4 is resistant to decay fungi, it is structurally weak and can lead to ring shakes (circumferential separation of xylem along an annual ring), which are undesirable in timber trees.

Perennial plants actively compartmentalize in response to wounding or wall off injured and infected wood. Injured plant tissue is not replaced or repaired to its previous healthy state (Shigo 1979), but instead is confined so that the rest of the plant does not become infected. Shigo termed this process CODIT, an acronym for *Compartmentalization Of Decay In Trees.*

Genetic makeup and vitality determine a plant's ability to compartmentalize. Clonal selections within several tree species are being made on their ability to compartmentalize wounds. Plant survival largely depends on ability to compartmentalize and to form new tissue.

When branches die, break, or are properly pruned off, discoloration and decay (should they occur) are usually confined to the branch stub zone by physical and chemical barriers. If a branch breaks off or is pruned close to but outside of its branch bark ridges, trunk tissue will envelope the branch stub and any decay will usually be confined to the stub base within the trunk. If a branch is cut off within the branch bark ridge (a flush cut), trunk tissue is opened to infection (Shigo and others 1979).

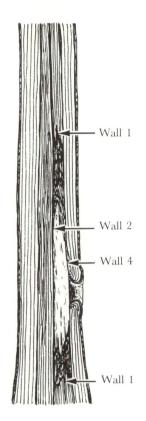

FIGURE 2-7
A longitudinal section through a tree trunk
with decay (light-colored area)
and discoloration (dark-colored areas)
originating from a wound six years before.
Decay and discoloration develop more
extensively vertically than in other direction
because Wall 1 is the weakest. Wall 2 has
not been very effective in resisting decay
in toward the center of the trunk. Wall 4
has kept infection from advancing into
xylem formed after the time of wounding.
(Adapted from Shigo and Marx, 1977)

Wall 1

Wall 2

Wall 4

Wall 1

Cross-section of a red maple trunk two
years after being injected (left). Wall 4
(darker ring) is visible around the entire
circumference connecting with the outer
edge of the injection wounds. Wall 2s were
not able to confine wood discoloration
from the two injections on the left, but
did on the right. Walls 3 and 4 were quite
effective in containing wood discoloration.
Wall 1 is not seen in cross-sections. Wall 4
is physically different from the xylem
present at the time of wounding or that
formed after the barrier zone is formed
(right). Note in the photomicrograph the
darker barrier zone formed late in the
springwood within an annual ring of
sugar maple. (Photos courtesy of Alex
Shigo, U.S. Forest Service)

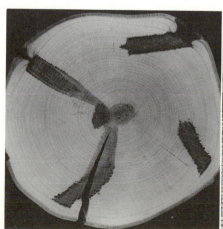

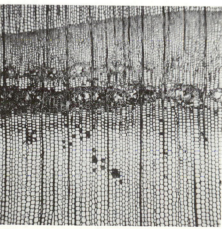

FACTORS THAT INFLUENCE
SHOOT GROWTH

Climatic influences on shoot growth will be discussed in a later section. Direct internal factors and responses to conditions other than climate, however, can affect shoot growth as well.

Dwarf Plants

Mature plants of small size have been long prized by plant growers (Fig. 2-8). Dwarf plants are more compact, more symmetrical, and usually more floriferous and fruitful than plants of normal size. Trees may be dwarfed in several ways:

FIGURE 2-8
Dwarf plants, such as the conifers in the left foreground,
appeal to many people. Normal-sized mature conifers are in the background
at Strybing Arboretum, San Francisco.

A small genetic change may dwarf a plant in relation to its normal counterpart. Dwarfs of perennial plants almost always have to be propagated vegetatively to maintain this characteristic. A few fruit species (such as peach) have dwarf varieties that will propagate from seed. In almost all of these plants, the internodes are much shorter than normal, accounting for their compact, symmetrical appearance.

A number of species can be dwarfed by grafting normal-sized varieties onto rootstocks that reduce the growth of the scion. Many citrus species and apples can be successfully dwarfed by proper rootstock selection; cherries and pears have less success. A number of landscape plants, such as lilac, can be similarly dwarfed.

Bonsai is based on top and root restriction achieved by pruning and the use of small containers. Some plants from the extreme northern or southern latitudes can be dwarfed under the short-day growing conditions of lower latitudes. With some species, gibberellins (growth-regulating hormones) will overcome dwarfism, and along with other substances probably regulate stem elongation and account for normal or dwarf habits of growth. Horticulturists continue to seek more genetic dwarfs and more techniques to manipulate plant size.

Intraplant Competition

Vegetative growth is reduced when plants flower and produce fruit. This is particularly noticeable if most of the blossoms are removed from a large limb of a fruit tree. Shoots will be much more vigorous on that branch than on branches that continue to nourish flowers and fruit.

Several branches arising near the same level on a trunk will often grow less than if one of them were dominant. This is probably due to competition for water and nutrients. More than one branch arising from the same level on the main leader of a broadleaved tree is undesirable, particularly if the tree is young. The main leader may be subdued and weak branch attachments produced.

Trunk Development

A strong, upright trunk is desirable in many trees for street, patio, and park use. By understanding the factors that influence trunk development, an arborist can train trees with such a structure, or with multiple stems or a more sculptured form.

Lateral branches along its length encourage rapid caliper growth of the trunk, but height growth will be inhibited if the laterals are not kept shortened (Chandler and Cornell 1952, Leiser and others 1972). Horizontally growing laterals usually grow slowly and are easily kept in bounds. As indicated earlier, vigorous shoots are more upright and usually form more laterals than do shoots of low vigor.

Reaction Wood

When an upright leader of a tree is forced from the vertical, the new xylem formed tends to return the leader to the upright. This is true not only for new growth at the shoot apex, but even back on the leader where secondary wood has already formed (Fig. 2-9).

When a main stem is grown at an angle away from the vertical, it produces so-called *reaction wood* to counteract the lean. In broadleaved trees, reaction wood is termed *tension wood* and forms on the upper surfaces of branches and leaning trunks. It exerts an internal contraction that tends to pull the trunk to the vertical or a branch to its original angle of growth. The corresponding type of wood in conifers, *compression wood*, develops on the underside of branches and leaning trunks. Compression wood exerts a compressive force due to the greater elongation of xylem cells on the lower side of the trunk that tends to force the trunk upright or a branch to its original angle of growth. This contraction (pull or tension) or expansion (push or compression) of the new xylem can be great, bending trunks 300 mm (1 ft) or more in diameter toward the vertical. For those who like memory aids, some distinguish between the tension wood formed in broadleaved plants and the **c**ompression wood of **c**onifers by the words that start with the letter **c** and remembering that **t**ension wood forms on the **t**op side of branches or leaning trunks.

Christmas-tree growers capitalize on this phenomenon by leaving one or more whorls of branches on the stump when they harvest trees.

FIGURE 2-9
The young growth of the trunks of many trees, such as this deodar cedar at Arundel castle, England, will become as vertically straight as the lower trunk due to reaction wood forcing it more upright.

The strongest growing branch of the remaining whorl will usually bend upright. A tree that grows from such an upright branch or a bud will develop more quickly than will a young seedling. Its trunk will not be as straight or its branches as symmetrical as growth that would come from a latent bud, but there are few if any such buds in most conifers.

When dry, reaction wood warps and twists and is weaker than straight-grained wood. Reaction wood is therefore undesirable for lumber, and forests are managed to minimize its formation. In the live state, however, reaction wood appears to be as strong or stronger than normal wood.

Each branch's particular angle of growth appears to be determined by its upward response to geotropism and the epinastic response of the terminal. Branches attempt to maintain their inherent angle by the formation of reaction wood. In most cases, more wood is formed on the lower than the upper side of a branch. If a branch is forced from its normal angle, whether up or down, reaction wood will form in an attempt to restore the original angle of growth. Arborists will do well to keep this response in mind when they cable or brace tree limbs so that they do not prevent normal reaction wood from forming to help support the branches. In fact, pulling a branch out of its inherent angle of growth may actually increase the load on the cables and inhibit the branch's ability to maintain its natural position.

In most cases the formation of reaction wood cannot completely overcome diversions of the trunk from the vertical or of branches from their normal angle of growth. If a trunk or a branch is held for a year or two so that it cannot correct the angle at which it is growing, the new wood laid down makes it more difficult for it to return to its former position when the external force is removed.

The differences in compression and tension wood can be observed under the microscope. Cells of compression wood are more elongated, while cells of tension wood are shorter than normal. Auxin concentrates on the undersides of branches or leaning trunks in some species and, along with other substances, may account for these differences in xylem elements. Differences also can be seen when a branch dies. In many conifers, a branch that was horizontal when alive will curve down as the wood dries because the lower side of the branch shrinks more than the upper side. In contrast, the branches of many angiosperms curve upward when they die; this is particularly noticeable in the branches that are more upright originally.

Trunk Movement

When a trunk has been kept from moving by guying, staking, or dense planting, the tree grows taller, exhibits less caliper growth near the ground, and develops with less taper (a decrease in trunk diameter with height). Each of these three growth responses lessens the ability of a staked tree to stand upright when its support is removed. The small trunk base is less able to support the top of the tree. The higher branches encounter more wind resistance and exert greater leverage on the trunk. The lack of taper tends to concentrate stress close to the ground instead of distributing it more uniformly along the trunk. These points are discussed further in Chapter 5.

The primary effect of staking and dense plantings seems to be the reduction of trunk movement. When the upper stems of greenhouse sweetgum and corn are moved as little as 30 seconds a day, height growth is reduced by 25 percent or more (Fig. 2-10) (Neel and Harris 1971). Bud dormancy is also hastened. Trunk movement seems to function somewhat similarly to diversion from the vertical, except that movement causes reaction wood to form more uniformly around the trunk.

Lack of movement may be one of the main causes of the tall, slender trunks on trees and shrubs grown close together in nurseries (see Fig. 4-11) (Harris and others 1972). While such trunks are undesirable for nursery trees, it is ideal in forests that trunks should be straight, have little taper along their main trunks, and have few or no side branches in the lower portion. This produces lumber of high quality, straight grain, relatively few knots. When a forested area is partially cleared for development, as for a road, subdivision, or recreation site, trees adjacent to the clearings may not be able to withstand the storms from which they had previously been protected. Such trees can be extremely hazardous.

FIGURE 2-10
The two liquidambar trees growing in a greenhouse were the same height (arrows) before the upper portion of the left tree was gently shaken for 30 seconds daily during the 21 days before the photograph was taken.

FIGURE 2-11
Young Japanese zelkova tied closely to opaque stakes
(left tree in left photo) bend away from the vertical (center),
even when turned upside down (right). The control trees
(on the right in each photo) were tied to transparent stakes.

Unidirectional Shading of the Trunk

Many times the trunk of a young tree is tied closely to a single stake. When the ties are removed, the tree almost invariably bends away from the stake. In extreme cases, the treetop may actually lie on the ground. Originally it was thought that this was due to a weak trunk, but when some of these trees, still in containers, were turned upside down, the trunk still bent away from what had been the vertical (Fig. 2-11) (Neel 1971). Examination of the wood showed that the cells of the xylem next to the stake were more elongated than the cells away from the stake. This asymmetrical development did not occur when transparent stakes were used. Studies have shown that the auxin content on the shaded side of a shoot tip is higher than that on the lighted side, causing the shoot tip to grow toward the light. The tendency of the trunk to bend away from an opaque stake is probably due to a redistribution of growth substances between the lighted and the shaded sides of the trunk.

ANGLE OF BRANCH
ATTACHMENTS

Vigorous leaders tend to grow more upright and produce laterals with wider angles of attachment than do leaders of lower vigor. Angle-of-branch attachment, however, is primarily influenced by the genetic makeup of the plant. Certain species or cultivars, like Modesto ash, are noted for their narrow angles of branch attachment.

Branch angle influences the form as well as the strength of the tree structure. Narrow-angled branches are more upright and therefore usually more vigorous than more horizontal branches; they may even outgrow or compete with the main leader. When a branch is nearly the same size as the leader, the strength of its attachment is less than if the branch were smaller (see Chapter 14). The imbedded bark at some narrow-angled branch attachments also weakens the union with the main trunk.

On young apple whips, the angles of attachment increase from the terminal toward the base (Verner 1955). On Modesto ash, however, almost the opposite may be true. The branch angle formed by a new shoot will be more acute if the main stem is broken or cut back, if buds are removed, or if the bark is scored (cut to the cambium) above the bud that gave rise to the new shoot. Fruit growers learned this when they disbudded newly planted apple and pear whips (with no side branches) and left well-spaced buds to develop into what they hoped would be well-distributed main branches. These practices apparently decrease the level of hormones (primarily of auxin) or at least the balance that favors a wide angle of branch attachment. Conversely, if the bark is scored below a bud, the branch angle will be wider.

Removing a lateral with a sharp angle of attachment from a young leader may force one or more accessory buds to grow from the same node (Harris and Balics 1963). The new shoots will be much smaller than the trunk, and their angles of attachment will usually be greater than that of the original shoot (see Fig. 14-19). By applying this information, one can train young trees to develop stronger lateral attachments.

PLANT FORM

The form or shape of woody plants, particularly of trees and shrubs, is determined by (1) the location of leaf and flower buds (terminal or lateral); (2) the pattern of bud break along the trunk and branches; (3) the angle at which branches grow; and (4) the differential elongation of buds and branches. For example, the absence of lateral buds in most of the arborescent monocots leads to a columnar growth habit in which an unbranched trunk ends in a tuft of leaves, as in palms. In most of the conifers and a few angiosperms, the main stem or leader outgrows and subdues the lateral branches beneath, giving rise to cone-shaped crowns with a central trunk. This branching habit is called *excurrent* (Fig. 2-12). In contrast, most angiosperm trees and shrubs are more round-headed and spreading, with no main leader to the top of the plant. This habit is called *decurrent*, deliquescent, or diffuse. The lateral branches of such plants grow almost as fast as the terminal shoot or may even outgrow it so that the central leader becomes lost among other branches (Fig. 2-13).

Lateral buds may be inhibited by the active growth of a terminal bud cluster on a currently elongating shoot; this is known as *apical dominance*. Trees with decurrent (spreading) branching habit have been thought to have weak apical dominance, while those of excurrent habit were thought to have strong apical dominance. Just the opposite is found to be true (Brown and others 1967). Decurrent species such as oak, elm, and maple exhibit strong apical dominance while the shoots are growing; in the year this growth occurs, few or no lateral buds grow on the developing shoots. The next year, however, one or more lateral buds are released and sometimes develop into branches that outgrow the original leader, thus producing a round-headed plant. The leader is not able to subdue and outgrow the laterals that develop after the first year. Such a tree is said to have weak *apical control* (Fig. 2-13).

On the other hand, excurrent species, such as sweetgum, tulip tree, and most of the conifers, have weak apical dominance. In such trees, varying numbers of lateral buds grow during the same season as the shoot on which they are formed. In most cases, however, the leader is able to keep ahead of the lateral branches, resulting in a typical "central leader" or conical form. In these cases, the leader has weak apical dominance but strong apical control (Fig. 2-12).

To understand the relations between apical dominance and apical control, it may help to remember that strong dominance is confined primarily to shoot growth of the current season. During the following season, lateral buds and the terminal bud formed the year before will start growth. The apex of each new shoot will have strong dominance over buds that form on current growth but little or none over shoots that are growing below. If the new lateral shoots outgrow the original terminal shoot year after year, a round-headed (decurrent) tree will result. Strong apical dominance of current growth leads to weak apical control of subsequent growth, and vice versa. Plants do not always fit neatly into one growth-habit category or the other but range between the easily distinguished extremes.

Plant vigor influences the expression of apical dominance and control. Within a single species, vigorous shoots exhibit less apical dominance than those of low vigor. Therefore, some of the lateral buds on more vigorous shoots may be released from shoot tip dominance and produce a more excurrent growth habit. In contrast, when plants mature or grow under conditions of stress or low fertility, they become less vigorous. Shoots on such trees increase in apical dominance, which then leads to a loss of apical control and gives rise to a more round-headed plant. A tree that has an excurrent growth habit while young may become round-headed as it reaches maturity.

FIGURE 2-12
A typical excurrent tree, such as liquidambar, after one (left),
three (center), and many years (right).
The terminal usually outgrows and subdues the branches below. Not to scale.

Young liquidambar seedlings, which normally have an upright central leader, completely change their habit when grown in dense shade (Brown and others 1967). They become almost prostrate with a decurrent growth habit because their vigor is so low that few or no lateral buds develop on the current year's growth.

Removal of the terminal bud cluster will not immediately change the excurrent or decurrent habit of an individual tree. Only by repeated pruning can excurrent forms be made to approximate decurrent forms, and vice versa. It is true, however, that removal of the terminal bud cluster on a shoot with strong apical dominance will release one or more buds immediately below the point of removal and keep buds

FIGURE 2-13
A typical decurrent tree, such as ash, after one (left),
three (center), and many years (right);
the terminal bud usually prevents lower buds from growing during the first season,
but the terminal shoot usually is outgrown in succeeding years.
Not to scale.

further below the cut from growing. Particularly if a dormant shoot has been cut back before growth begins, the new shoots are clustered just below the cut. On the other hand, if a shoot with no lateral branches is not cut back, more buds will develop the following season and will usually be distributed more evenly along the shoot.

Inhibition of lateral buds is apparently determined by a balance of growth factors. These relations are complex and vary with species. No single explanation is yet available, but dominance seems to reflect auxin levels produced by the growing point and by young leaves.

Other Factors

Species with terminal flowers or flower inflorescences are normally round-headed, since growth after blossoming comes from lateral buds below the flowering terminal (Fig. 2-14). More than one bud will usually break, thus speeding the process that leads to round-headedness. Competition between two or more shoots also reduces the growth of each, leading to a more compact plant.

Plant form can also be influenced by other plants growing nearby. Plants that are close together grow taller, with less spread and fewer or no branches along their lower trunks. In contrast, a similar plant in the open will be shorter and develop more branches with greater spread. The branches of a plant in the open will be more uniformly distributed from the ground upward.

In a mixed planting of trees, certain species dominate and are less affected by their neighbors. The more sensitive trees may be quite misshapen or devoid of lower branches (Fig. 2-15). Some large trees may form a canopy over slower-growing species without greatly affecting the symmetry or normal growth of the lower plants (Fig. 2-15). In fact, some species when young require such a protective canopy. Even in close plantings of a single species, certain trees will dominate and outcompete others. Under severe stress, the smaller trees may not survive.

FIGURE 2-14
Species that flower terminally, such as *Koelreuteria bipinnata*, often fork below the previous year's terminal flowers. This hastens a decurrent form and can lead to weak structure.

FIGURE 2-15
A young Chinese pistache misshapen by growing in the shade
on the west side of a large incense cedar (left); on the other hand,
Pinus patula grows upright even though it receives
most of its light from the west (right).

Juvenility and Maturity   A number of morphological and physiological changes occur in the development from seedling stage to maturity.   These changes have been of particular interest to propagators, breeders, and those who grow plants for their leaves or flowers and fruit; some plants may not flower until they are more than ten years old.   During its early life, a plant is considered *juvenile* if it will not flower even though all environmental conditions are favorable. A plant is considered sexually *mature* when it will respond to flower-inducing conditions.

Morphological traits associated with juvenility may include vigorous upright growth, vine form, leaf shape and phyllotaxy that differ from mature forms, excurrent growth habit, smooth bark, spines or thorns, and retention of leaves in deciduous plants (Fig. 2-16) (Zimmermann and Brown 1971).   The most common juvenile physiological traits are easy rooting of cuttings and lack of flowering response.   Differences in leaf form and arrangement have led to the misclassification of juvenile and mature forms as two different species.   Even though they are actually older, the lower and inner portions of woody plants may retain juvenile characteristics while the upper and outer growing points produce mature shoots and flowers.   These upper meristems tend to pro-

FIGURE 2-16
The juvenile form of English ivy is a vine with palmately lobed leaves
(right, inset); the mature form is shrub-like and upright with cordate
leaves (left, inset) and fruit.

duce mature shoots as they age or when levels of moisture, nutrients, and other substances that come from other parts of the plant are reduced (Zimmermann and Brown 1971).

Practices that will increase the size of a juvenile plant most rapidly will hasten flowering and fruiting; on the other hand, young plants that have been propagated from tissue of a mature plant, such as most fruit trees and many woody ornamentals, will be delayed in flowering if they are extremely vigorous (Hackett 1976). Young juvenile plants are much slower to mature and flower than are young plants propagated from mature tissue. To retain the juvenile form and foliage of a plant,

40

low to moderate plant vigor should be maintained and any mature growth pruned out as soon as it appears. If one realizes that some plants have distinctively different juvenile and mature characteristics, one may avoid some incorrect assumptions about their well-being. In vegetatively propagating plants, the proper selection of cutting or grafting wood can be quite important to rooting success and structural development. Not only can juvenile and mature tissues differ, but propagating wood taken from flattened horizontal branches may impart the flattened characteristic to the plant produced; this is true for some of the conifers, particularly araucaria.

Allelopathy is the term given to the phenomenon in which one plant directly harms another, particularly when young, through the production of chemical compounds released into the environment. Roots and leaves are the most common source of such inhibitors. Roots of one plant can be inhibited or repulsed by either genetically identical or different plants (Del Moral and Muller 1970). Allelopathy can keep root systems from intermixing or seedlings from growing near or under established plants. Competition from other plants is thereby reduced, enhancing its chances for survival in arid conditions.

Fescue sod has been shown to inhibit the growth and survival of young liquidambar (Walters and Gilmore 1976) and black walnut (Todhunter and Beineke 1979) (Fig. 2-17). Leachates from live fescue roots, from dead roots, and from dead leaves have been filtered through

FIGURE 2-17
The allelopathetic effect of fescue sod on black walnut seedling growth (right) is compared to that of a ground cover composed of mixed herbaceous species (left). (Photos courtesy of M. N. Todhunter and W. F. Beineke, Purdue University)

quartz sand in which month-old liquidambar trees were growing; they reduced the trees' dry weights by 19, 48, and 60 percent, respectively. This could raise some question about using fescue lawn clippings for mulching young plants (see Chapter 13).

An increasing number of herbaceous and woody plants are found to have allelopathic action on some other species. Not all species are equally inhibited by leachates from other plants. Some allelopathic toxins may remain in soil or dead roots for many years after a large plant has been removed, thus arguing against the use of certain replant species. Conversely, the allelopathic action of certain low-growing plants might reduce the growth of troublesome, tall species in utility rights of way. This technique is being tried in the northeastern United States.

ROOT SYSTEM

The above-ground parts of a plant depend upon roots for anchorage, absorption of water and mineral nutrients, storage of food reserves, and synthesis of certain organic materials, including those that may regulate activities in the top of the plant. The development, size, form, and function of root systems are controlled by the genetic makeup of a plant and by environmental and management conditions that modify the expression of these characteristics. Roots differ from shoots in a number of ways. The former have no regular branching pattern like that of stems whose branches develop from buds in the axils of leaves. Lateral roots develop from tissue several layers inside the epidermis but outside the phloem (Fig. 2-18); lateral branches grow from buds whose tips are usually exposed beyond the epidermis though they have vascular connections within the main stem (see Fig. 2-5). A root cap protects each root tip as it is forced through the soil by the elongating tissue behind; a growing shoot tip is composed of tender, expanding shoot, leaves, and buds with little or no protection from the elements. At its tip, the oldest cells of the root cap slough off and lubricate the movement through the soil.

Behind the root cap, the root apical meristem replenishes the root cap through cell division and forms undifferentiated cells away from the tip for later development into specialized root tissue. Behind the apical meristem of active cell division is the region of most intense cell elongation, which forces the terminal 1 to 5 mm (0.05–0.2 in) through the soil (Fig. 2-18). In and behind the zone of cell elongation, cells differentiate into xylem, cambium, phloem, pericycle, endodermis, and cortex. The epidermal cells in this region of primary tissue formation form root hairs up to 8 mm (0.3 in) long. Water is absorbed from the soil primarily through root hairs and the region of cell elongation, although the absorbing surface of the latter is much smaller than that of

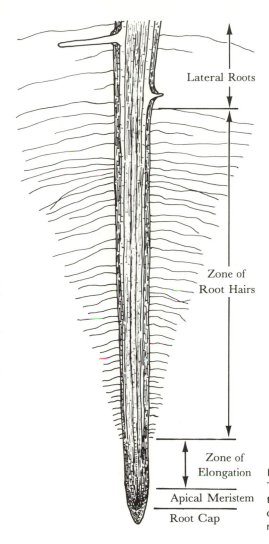

Lateral Roots

Zone of
Root Hairs

Zone of
Elongation

Apical Meristem

Root Cap

FIGURE 2-18
This longitudinal section of a root tip shows
the root cap, a zone of elongation, a region
of root hairs, and the beginning of lateral
root development.

the root hairs. The root hairs of a single mature rye plant have been estimated to number more than 14 billion, with a surface area greater than 370 sq m (4000 sq ft), and a rate of increase that exceeds 100 million per day (Robbins and others 1950). Root hairs do not become roots; in fact, those of most species live only a few weeks. The root hairs of plants like the honey locust and redbud, however, may persist for more than a year, while some of the conifers have few or no root hairs (Weier, Stocking, and Barbour 1974).

Lateral roots, which arise in the pericycle, and secondary growth of xylem and phloem from the root cambium occur in the region behind the root hairs. Adventitious roots, on the other hand, can arise from stems, rhizomes (underground stems), stolons, and other organs.

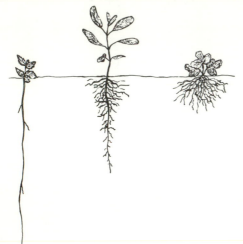

FIGURE 2-19
Roots of young plants vary from a single
tap root with few or no lateral roots (left)
to a fibrous root system with most roots
originating near the crown (right).

Root Form

Seedlings tend to have either one main *tap root* or a number of primary roots that form a *fibrous root system* (Fig. 2-19). In some species and soil conditions, primary roots may be almost devoid of secondary roots until they have grown a considerable distance or encountered an obstacle. In other plants, primary roots may readily form laterals as they grow. The tap roots of young seedlings can penetrate the soil to a considerable depth, while growth of the top is considerably less. Six months after a blue-oak acorn was planted in central California, the tap root of one seedling was excavated to a depth of one meter (40 in) before it broke; the top of the plant was only 75 mm (3 in) above the ground. Under arid conditions such root growth significantly aids the establishment of young plants. Tap-rooted plants, however, are considered difficult to transplant, either bare-root or balled and burlapped, because they lack a fibrous root system to hold the soil ball together and have few small roots to absorb moisture after transplanting.

The primary seedling roots, particularly tap roots, are strongly geotropic and will grow vertically if soil conditions are favorable. Lateral roots usually grow more horizontally. If the primary root tip is pinched or damaged, lateral roots will grow more vertically. If the tip of an unbranched tap root is pinched, laterals will develop above the pinch and grow in almost the same direction as the parent root (Fig. 2-20). As root systems enlarge, the primary roots have less influence over the growth of the other roots.

Roots other than the tap root may enlarge as they grow downward, giving stability to the top of the tree. These are called *heart roots.* Some trees have primarily horizontal roots near the surface and *sinker*

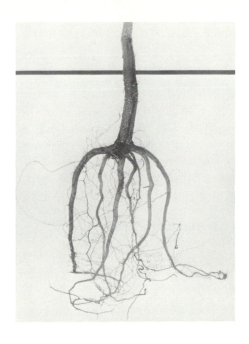

FIGURE 2-20
Branching of an oak tap root. The root
tip was pinched when the seedling was
transplanted from a seed flat into a small
container.

roots that grow vertically from the horizontal ones. Sinker roots usually
occur near the tree trunk, adding stability to the tree and increasing the
volume of deeper soil exploited by the roots (Fig. 2-21). In most mature trees, the tap root is outgrown by heart and lateral roots and is

FIGURE 2-21
Types of roots that may occur on a mature tree.
The vertical tap root is usually choked out by roots above.
The heart root grows at an angle from the buttress of the trunk.
Horizontal lateral roots are usually near the surface.
Sinker roots grow downward from lateral roots.
(Adapted from Fayle 1968)

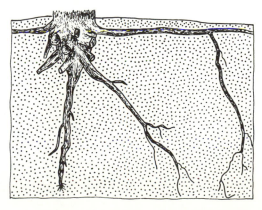

difficult to find. Root systems are much closer to the surface than is commonly thought.

Although root growth is greatly influenced by soil conditions, individual roots seem to have an inherent guidance mechanism. Large roots with vigorous tips usually grow horizontally (Wilson 1970). Smaller roots lateral to the large roots grow at many angles to the vertical, and some grow up into the surface soil. On the other hand, few roots in a root system actually grow down (Wilson). The growing tips of large, horizontal roots are often *plagiotropic;* that is, if the tip is displaced up, it grows down, and if it is displaced down, it grows up. In addition, these roots tend to grow at about the same soil depth regardless of the slope of the soil surface. More study of this phenomenon may help to minimize root-heaving of pavement and sidewalks.

The roots of some plants are able to maintain compass direction of growth. If a root grows into an obstacle and is deflected, it will tend to resume its original direction when the obstacle has been passed (Wilson 1970). This is called *exotropy*.

Size of Root Systems

Roots of most plants, including large trees, grow primarily in the top meter (3 ft) of soil. Most plants concentrate the majority of their small absorbing roots in the upper 150 mm (6 in) of soil if the surface is protected by a mulch or forest litter (Zimmermann and Brown 1971).

Of the woody plants, fruit trees have been the most intensively studied with regard to their roots. Even though roots have been found at depths of more than 10 m (30 ft), most roots of several different fruit species growing in sandy soil were found in the upper 0.6 to 1.5 m (2-5 ft). In soils of finer texture, a higher proportion of roots are closer to the surface. Most soils in orchards are disced, which eliminates roots in the upper 150 to 200 mm (6-8 in) and leaves the soil surface exposed to the sun. Exposed bare soil can become so hot near the surface that roots do not grow in the upper 200 to 250 mm (8-10 in).

Under forest and many landscape situations, however, the soil surface is not directly exposed to the sun and soil near the surface is most favorable for root growth. Management practices, such as mulching, that favor exploration by feeder roots near the soil surface foster good plant growth, particularly in shallow or fine-textured soils. The surface rooting of trees in lawn areas or near buildings and pavement, however, can be a serious problem, more especially with vigorous young trees than with mature trees that have almost reached their ultimate size (Morling 1963). Surface rooting will be increased either if poor irrigation practices keep the soil too wet on the surface and too dry at lower depths or if inadequate drainage keeps the soil too wet throughout.

Roots may extend laterally for considerable distances, depending on the plant and the soil conditions. Roots growing in moist, fertile, well-aerated soil branch more, and each root grows less than if there were fewer roots. Roots will be more numerous, more fibrous, and closer to the trunk than those that grow in less desirable soil. In infertile soil inadequately supplied with moisture, roots are fewer in number but larger and able to grow greater distances from the plant. This is certainly a good survival tactic.

Roots of trees grown in the open often extend two to three times the radius of the crown (Zimmermann and Brown 1971), while those growing in a close planting will be more confined. If there are no internal inhibitions, roots grow wherever aeration, moisture, temperature, nutrition, and soil tilth (structure and openness) are favorable. These factors are discussed in more detail in Chapters 6 and 13.

Top-to-Root Relationships

It has often been said that there is as much of the tree below ground as there is above. That is not true for most species, although their root systems are indeed substantial. The relative size and weight of top and root system can vary considerably with the conditions under which the plant is grown. In soils with low fertility and periods of inadequate moisture, the root system will be larger in relation to the top than it would be in more fertile soils with adequate water supplies.

In fact, the size of the top in relation to the root system is an important factor in the failure of many container-grown plants when they are set out in the landscape. In many nurseries, plants are irrigated and fertilized almost daily during the growing season and develop proportionately large tops, particularly when their roots are confined in containers. Even though no roots may be removed during transplanting, the moisture-holding capacity of the root balls is relatively small in relation to the transpiring surface of the plant tops. In addition, root balls in the soil will retain less water following irrigation than they did in the containers. Plants of moderate size in relation to their containers and those whose tops have been thinned out may better survive in the landscape. After any transplant, frequent watering is usually necessary at first.

FURTHER READING

KRAMER, P. J., and T. T. KOZLOWSKI. 1979. *Physiology of Woody Plants*. New York: Academic Press.

ROBBINS, W. W., T. E. WEIER, and C. R. STOCKING. *Botany: An Introduction to Plant Science*. New York: John Wiley.

WILSON, B. F. 1970. *The Growing Tree*. Amherst: University of Massachusetts Press.

ZIMMERMANN, M. H., and C. L. BROWN. 1974. *Trees: Structure and Function*. New York: Springer-Verlag.

CHAPTER THREE

Plant Selection

Selecting plants that will adapt well to their intended site and will fulfill their intended landscape function is extremely important to the success of a planting and the ease with which it can be maintained. The quality of young plants can also be a crucial factor.

SELECTING THE RIGHT PLANT

From earliest times, people have selected, propagated, and grown plants they thought superior to others. These included food plants, plants that produced fiber and other useful materials, and plants that enhanced their surroundings. Migrating people took seeds and cuttings of desirable plants to new areas. Early explorers and travelers brought back plants for ornamental purposes as well as for food and fiber. Estates of the nobility and early botanical gardens became repositories of a vast array of interesting plants, many of which have landscape value. Ornamental plants were selected for particular qualities of size, form, texture, leaf color, flowers, fruit, and many other characteristics.

Flowering shrubs and small trees like the rose, camellia, and lilac were bred by amateur and professional horticulturists. Almost all new trees and shrubs (without showy flowers), however, have come from native and locally developed plantings. The new cultivars were selected primarily because of aesthetic characteristics rather than for adaptability to unfavorable environments and resistance to pests. Even so, many of them were as vigorous as the species from which they were selected, or more so.

Plant explorers from arboreta and governmental agencies have looked primarily for food and fiber plants but have also collected seed from many plants with landscape potential. The U.S. Department of Agriculture (USDA), through its plant introduction stations and contacts in foreign embassies, has acquired many promising ornamental plants. Research workers, plant pathologists in particular, began screening landscape plants for their resistance to specific serious diseases, pests, and disorders. Nurserymen, botanical garden horticulturists, arborists, landscape architects, and amateur gardeners found plants that grew better than others of their kind. These they introduced or brought to the attention of research workers so the plants could be tested or observed.

The first efforts to breed superior landscape trees began in the late 1920s in an attempt to develop trees that would be resistant to Dutch elm disease (in Holland) and chestnut blight (at the Brooklyn Botanical Garden). Tree-breeding is a slow process; it took the Dutch 48 years to develop elms through the F_3 generation (Shurtleff 1980).

At the National Arboretum, Santamour (1972) began the first large-scale effort in the United States to develop genetically superior shade trees for use in urban areas. The goal is to produce superior trees in a wide range of genera with different growth characteristics to fit urban environments. Resistance to major diseases and insect pests and tolerance to environmental stress factors are basic requirements for these trees. Townsend (1977), Santamour (1979b), and others have used provenance and progeny testing, long used by foresters, to test variation within species that are native to different regions and to select superior individuals for further evaluation and possible breeding. For example, Bey (1980) found that black walnut seedlings grown in the U.S. midwest from seeds collected 320 km (200 mi) closer to the equator grew better than seedlings from local sources. Seedlings from seed collected yet closer to the equator were more likely to be injured by spring frosts and cold winters.

As these efforts continue and more people become involved in research and observation, we will be in a much better position to select trees and other plants that can perform well. Even then, the best way to estimate how a plant will perform and look when mature is to observe mature specimens in nearby parks, botanical gardens, arboreta, street plantings, and private gardens. People who care for the plants and live near them should be consulted.

Inherent Characteristics

Besides the many visual and functional characteristics of woody plants, other attributes must be considered if they are to satisfy their owner and/or caretaker. Plants should be selected on the basis of their functional uses, their adaptation to the site, and the amount of care they will require. The relative importance of a particular characteristic will vary. Table 3-1 rates plant characteristics in relation to function, adaptation, and care requirements.

Plants have several landscape functions. Their *architectural features* can afford privacy, progressively reveal a view, or articulate space. In an *engineering* sense, plants can be used to reduce glare, direct traffic, filter air, reduce soil erosion, and attenuate noise. Plants influence the *meso-* and *microclimates* by transpirational cooling; interception of solar radiation, reflection, and reradiation; and modification of rain, fog, and snow deposition. Plants can be placed to increase, decrease, or direct wind. On the other hand, certain plants irritate some people with their allergenic pollen and leaf pubescence, toxic sap or exuda-

TABLE 3-1

Characteristics of woody plants and suggested relative importance of their influence on landscape function, site adaptation, and plant care

Plant Characteristics	Function			Adaptation to Site	Plant Care
	Architectural and Engineering	Climate and Human Comfort	Aesthetic		
Growth Habit					
Tree, shrub, vine	***	***	***	**	**
Size	***	***	**	**	***
Form	**	**	***	*	
Growth rate	*			*	**
Branching	**	*	*	*	**
Wood strength	**			*	*
Rooting	**			**	***
Plant Features					
Leaves	**	***	***	**	***
Flowers		*	***		*
Fruit			**		**
Bark			**		
Environmental Tolerances					
Temperature				****	**
Drought				**	**
Wind				**	**
Light				**	*
Soil				***	**
Air				**	*
Pests				***	***
Fire	*				**

*** = major influence; no * = little or no influence.

tions, and odorous flowers and fruit. In our surroundings, however, the primary use of landscape plants has been aesthetic. Their diversity of form, color, and texture lend charm, grace, and interest to our landscapes.

Many growth habits and features of functional value may also influence a plant's *adaptation* to a particular site, its care requirements, and the ease of giving this care. Certain stress *tolerances* of plants directly influence their ability to adapt to a site and the amount of care they will require.

Plants will affect and be affected by a specific landscape site. Seldom will a plant have all the qualities desired; as with most decisions, the selection of a plant will usually be a compromise. Certain site-adaptation or maintenance problems may be accepted for the sake of a prized aesthetic effect or a special functional need. Some people in warm temperate regions put up with the pollen and messy fruit of the olive in order to have a gray-foliaged small tree that will grow with little maintenance under difficult conditions.

The important plant characteristics will be discussed with regard to their influence on landscape function, site adaptation, and plant care.

Growth Habit and Size

The *growth habit* of a plant, whether tree, shrub, vine, or ground cover, and its ultimate *size* should be considered in terms of intended landscape use and the nature of the site. Is the plant to shade, screen, enclose, soften, accentuate, direct, or protect?

Mature size is extremely important, because a plant should not outgrow the allotted space. Overgrown shrubs are a universal problem: Designs are spoiled, windows and views blocked, walks and drives crowded, and lower branches bereft of foliage (Fig. 3-1). Trees that become too large for their niche can break curbs, gutters, sidewalks,

FIGURE 3-1
All too often trees and shrubs
become too large for their intended use.

and paving; can unduly shade plants beneath them; are difficult to care for when mature; can be hazardous in storms; and are costly to remove. A large tree may be a poor selection for a shallow soil or a windy site. Vines can require special support and can damage structures if they grow beyond a certain point.

Considerable interest in smaller trees began in the 1940s as Dutch elm disease began to take its toll in the eastern United States, as home landscapes became more modest, as trees planted along downtown streets blocked store signs, and as utility companies experienced increasing costs in keeping trees out of their overhead lines. Some arborists, however, think that small trees have been too heavily planted in public and private landscapes.

More attention is now being given to landscape function. Size, shade intensity, leaf persistence, and plant placement greatly affect a tree's ability to shade paved areas and reduce air-conditioning costs in the summer, to allow proper functioning of solar heating and power units, and to allow sunlight to warm buildings directly in winter. Species or even clonal selection are the most satisfactory approaches; using pruning or chemicals to control plant size is a bother and a recurring expense.

Plant Form   Tree and shrub species and their clones may have distinctive forms that provide the landscape architect and arborist with an almost unlimited range of architectural and functional elements (see Fig. 2-1). The bark of many plants develops interesting patterns as it stretches and cracks with increases in trunk caliper. Bark color, though subtle, can accent an intimate landscape scene. Branching structure not only provides aesthetic potential but can be particularly important in determining the structural strength of trees and their daily and seasonal shade patterns. A tree will be inherently stronger if its branches are well-spaced along the trunk, if the trunk is larger in girth than its branches where each arises, and if the branches fork or have laterals. These characteristics are inherent in certain species, such as most conifers, pin oak, and plane trees, to a lesser extent, but not in others, such as Modesto ash and black locust. Undesirable traits can be overcome by proper training when the tree is young, but such early training is often overlooked or done improperly.

Rate of Growth   People usually want a fast-growing tree (growing more than 1.5 m or 5 ft per year) that will quickly provide shade, screen unpleasant views, protect from the wind, or just do its job in the landscape. Fast-growing trees are usually tolerant of difficult sites and neglect and will also grow rapidly in the nursery. As a result, nurserymen, landscape architects, developers, homeowners, and landscape maintenance people very much like these trees. How could one go wrong with such attributes?

In a number of ways! Most fast-growing trees become large and incur the related disadvantages; fast-growing trees are usually weak-wooded and subject to limb breakage, especially in storms; fast-growing trees usually live a shorter time than trees with more modest growth rates.

Species with moderate growth rates (300 to 600 mm or 1 to 2 ft per yr) can usually be encouraged to grow more rapidly than normal if plants are selected for good quality, are planted properly, and are given adequate water and generous nitrogen fertilization. Even so, in especially difficult landscape situations, fast-growing trees may be the best solution or at least part of the overall plan. A mix of trees for rapid effect and long-term results may be a reasonable compromise. Care in evaluating the landscape site and developing a design solution is essential in deciding the plant characteristics that will best do the job.

Wood Strength   Branch structure and wood strength are closely allied in determining the inherent ability of a tree to withstand wind, ice storms, and limited vandalism. Again, the faster-growing species tend to have weaker wood. The wood of some conifers is not as strong as that of hardwoods. In their ability to withstand the elements, however, the form and the branch structure of conifers often more than compensate for any lack of wood strength.

The inside limbs of some trees die and may be hazards unless removed. Dead branches up to 50 mm (2 in) in diameter are common in London plane, Japanese zelkova, and some elms and can cause injury when they fall, although they are not large. The ground under these trees is usually littered with twigs and small branches after a strong wind.

Rooting   Roots provide anchorage, nutrients, and water for plants. Deep or spreading root systems are essential to hold large plants against strong winds, particularly where soil is wet. Wet sites require species that are tolerant of low oxygen or high carbon dioxide levels in the soil and are resistant to root rots.

Shallow soils, soils with differing textural strata, and rainfall or irrigation practices can contribute to shallow rooting, but certain species (some ash, mulberry, and elm, for example) exhibit surface rooting more frequently than do other species (such as oak, pistache, and pear). Large, shallow-rooted trees are often subject to windfall. The surface roots of some species (elm and poplar) sucker heavily, particularly when injured, and can be an unsightly maintenance problem. Species with shallow roots that frequently raise the soil around

FIGURE 3-2
The soil is often raised near the trunks of large trees that have shallow roots.

the tree trunk (elm, Japanese pagoda, and mulberry) are usually poor choices for narrow planting strips and tree wells in pavement (Fig. 3-2).

Willow, poplar, and silver maple are renowned for invasive roots. Rapidly expanding roots can crack intact sewer lines and provide entry for finer roots, which can quickly form a fibrous mass inside the lines and plug them (see Fig. 1-5). These roots can be a problem near septic leach lines. On the other hand, such plants may be of value for transpiring water that is being discharged from water treatment facilities. The structure and growth rate of tree roots can pose special problems where utility lines are close together in easements in or near the tree planting strip along streets.

Plant Features

Leaves   Leaves are our primary source of food and fiber; in addition, they add interest and beauty to plants. Leaves also influence a plant's functional effectiveness and, to a certain extent, the maintenance it requires.

The many sizes, shapes, and colors of leaves provide *aesthetic* enrichment for landscapes. *Leaf size* offers a range of textures, from light and airy to bold and dark, and can influence the scale of a landscape. *Leaf shape* can add interest and character to a landscape. *Leaf color* can supplement the architect's palette of flower and fruit colors by providing ranges over a growing season, over the entire year, during the first burst of growth, or during the final days of autumn only. Leaf color influences the amount of radiation reflected by the leaf canopy. Gray and light-colored leaves reflect more than dark green and red leaves. Color, however, may have more effect on the perception of temperature than on the temperature itself. Red foliage produces warm or hot feelings and gray, cooler ones.

Thick, pubescent, and waxy leaves enhance the drought tolerance of plants. *Large, numerous, or dense leaves* enhance the depth of shading under a tree. Trees with large, overlapping leaves, such as mulberry, magnolia, and catalpa, cast a shade so dense that it is difficult to grow lawn or flowering plants under them. Other species, such as honey locust, silk tree, and acacia, have more open branching and small leaves or leaflets that produce light, filtered shade. *Fine leaves* may not need raking when they grow in open areas or over large shrubs or coarse grass but can make cleanup difficult on rough surfaces, around patio furniture, and in pools. Large leaves may be more voluminous, noticeable, and easily blown but are usually easy to rake and pick up.

Leaf persistence determines whether a plant will be deciduous or evergreen. Evergreen trees and shrubs provide year-round foliage and windbreak protection; their leaves normally stay attached for three years or more, depending on the species and the vigor of the plant. In contrast to deciduous plants, broadleaved evergreens may be more subject to winter wind burning and snow breakage. Leafless trees let only 26 to 80 percent of winter sunlight through (Geiger 1961), and species vary greatly as to when they produce and lose their leaves (see Appendix 4). Honey locust leafs out late but defoliates early, while Bradford pear does the opposite. Near Portland, Oregon, the Bradford pear is in leaf about four months longer than are some of the honeylocust cultivars (Ticknor 1981). Such information can be helpful in selecting plants where summer shade and winter sun are wanted to moderate landscape and building temperatures (see Chapter 5).

The *season and duration of leaf fall*, in both deciduous and evergreen trees, affect landscape appearance and maintenance. Many evergreens drop their leaves over a longer period than deciduous trees, and not necessarily in the fall. Cork oak, an evergreen, sheds leaves in early spring before new growth begins. London plane begins dropping leaves in midsummer and may continue until after the first hard frost. The leaf fall of southern magnolia varies greatly within the species: Some trees may shed leaves for only two weeks while trees close by lose leaves for six weeks or longer. Under dry summer conditions, trees such as honey locust and Japanese pagoda drop most of their leaves before the early fall when shade is often desired. The fronds of some palms do not fall when they die, but hang appressed to the trunk. Some think these palms are attractive, but dead fronds on the trunk can be a fire hazard. The base of the live fronds can be a haven for rats and birds. On occasion, palm trees thus infested have been removed as hazardous to public health.

Flowers and Fruit   A landscape is brought to life with the bright hues and varied forms of flowers and fruit. The flowers of many plants add fragrance to the garden and some cause allergies. In some locations, it is possible to have flowers or fruit year round. Witch hazel and forsythia flower in winter and early spring. Prunus species bloom late winter to late spring. Buddleia and crape myrtle flower in the summer. In general, flowering may be slim during the late fall and early winter.

Wildlife may be attracted by shrub berries. This is often a source of enjoyment but may have its negative aspects as well. Mistletoe can afflict susceptible trees more seriously near extensive plantings of fruiting shrubs. If other mistletoe is in the vicinity, birds attracted to the berries carry mistletoe seeds and leave some on nearby trees when they perch between feedings. The areas below these trees can also become quite messy with bird droppings.

The fruit of some plants, such as ginkgo and tree of heaven, is foul-smelling when crushed. The fruit of large shrubs and trees can be a maintenance problem if it falls on paved or turf areas. Such plants should be underplanted with shrubs or thick ground cover, so that the fruit is concealed when it drops. To eliminate the fruit problem, fruitless clones have been selected and propagated for a number of species. Fruitless trees may still produce heavy bloom and abundant pollen and therefore present problems for hayfever sufferers.

Environmental Tolerances

Climatic Adaptation Semitemperate regions sustain a wide range of plants and plant characteristics. Most broadleaved evergreens and palms are limited to the semitemperate and warmer regions, while the colder areas are populated primarily by conifers and deciduous broadleaved plants.

Before planting trees, check local weather records to learn the lowest *temperatures* on record, observe plants in nearby areas similar to your planting site, and talk to knowledgeable growers about which plants in the area may reach or exceed their hardiness range. It would be a serious loss to have a fifteen-year-old southern magnolia frozen to the ground during an unusually severe winter just when the tree was beginning to fulfill its landscape function. Rehder (1940), the USDA (1960), and Dunmire (1979) present plant hardiness zones that identify the climatic adaptation of many plants. Plants indigenous to areas having cold winters are usually more hardy than plants of the same species indigenous to warmer areas. Yet a few fine plants survive in areas much colder than their native locations.

Occasionally, low temperatures above the lowest recorded can be devastating if they occur in the fall or after a winter warm spell when plants are not at their maximum hardiness. In 1972, thousands of eucalyptus in the San Francisco Bay Area were killed or lost branches back to the trunk because of temperatures of −5°C (23°F) during December. More than $10 million of public funds were spent to clear dead branches and trees to reduce the serious fire hazard. Spring or early fall frosts can damage plants that begin growth early or do not harden soon enough in the fall.

Within an area or even within a single landscape site, microclimates may exert particularly favorable or unfavorable influences on certain plants. If tender vegetation in riskier portions of generally protected sites is smaller and less expensive, injuries or deaths will be less serious. It may, however, be worth growing certain plants for their unique beauty and taking a calculated risk that they might be killed or damaged once in 20 or 30 years. Without this kind of risk-taking many landscapes would be poor indeed.

Just as the temperature can become too cold for certain plants, areas with mild winters may not provide enough chilling to allow some plants to resume proper growth the following spring (see Chapter 4). Some varieties of apple and peach have been bred for mild-winter areas and require less chilling than many of the older varieties. The source (provenance) of seed and propagating stock can be quite important in the performance of species that naturally have a wide geographical range.

Moisture is another crucial factor. Landscape sites must be analyzed as to water reservoir potential, evapotranspiration, and the amount of supplemental water available. Drought tolerance, although not of great concern in many areas, can be crucial in unirrigated landscapes that have longer than normal periods with little or no rain. Woody plants can survive drought periods only through adaptations that enable them to obtain or conserve water. These plants can be classed into three groups:

> *Water spenders* use water freely but in deep soils have extensive root systems that absorb water from a large volume of soil. As long as some of their roots are in moist soil, they can survive. Many common landscape plants are of this type: black walnut, plane tree, and mulberry.
>
> *Drought evaders* avoid water stress by drying up or dropping their leaves, in some cases by shedding twigs and branches, or by becoming virtually dormant during the dry period. Examples are California buckeye and palo verde. While they have leaves, however, these plants usually transpire as rapidly as the water spenders. Most of them have limited value in our landscapes.
>
> *Water conservers* have ways of reducing water loss. Their leaves may be small, gray-colored, leathery, or arranged to reduce the amount of sunlight that strikes them. Their stomates may be structured to conserve moisture. Many plants from desert and Mediterranean climates are of this type.

Conversely, some planting sites may be waterlogged or flooded for varying periods during the year. Many reservoir shorelines and river levees that used to be kept free of vegetation are being planted for erosion control, fish habitat, and aesthetic reasons. Observation and experiments have identified many woody plants with inundation tolerance (Whitlow and Harris 1979).

Prevailing *wind* may sculpture or deform some trees and shrubs more than others. If the planting site is in an exposed location, plants that can withstand such conditions should be sought. For information, survey trees in the area and talk with local plantsmen. Planting techniques may be able to maximize protection so that a wider range of plants can be used (see Chapter 9).

Light (radiation) is essential for green plants, although full sunlight may cause many of the so-called shade plants to develop brown leaves or grow poorly. Full sunlight warms both the exposed leaves and the air. If transpiration becomes so great that the plant loses water faster than it is absorbed, the tips, margins, or entire leaves may be killed. Leaves that have grown in low-light conditions may become bleached under full sun through a loss of chlorophyll. It is high transpiration, not high irradiation or high temperature, that causes many shade plants to do poorly in full sun.

Certain plants perform well under low-light situations and are well-suited for plantings indoors, on the shady side of structures, or under large trees (see Fig. 10-6). Many plants lose their leaves when moved from high to low levels of radiation. Others become deformed when grown near a building or a larger plant. Their branches grow toward the light and mold the plants into asymmetrical forms. Plants should be observed in similar sites before selected for such use.

Unfavorable Soil   Poor soil can spell disaster for many plants. Soil can often be modified or amended to improve plant growth, but that may not be feasible in large landscape developments, so it is well in such situations to use plants, like fast-growing trees, that are fairly tolerant of poor soil conditions.

Shallow, poorly drained soil dooms many trees and shrubs to poor growth or to an uncertain future unless they are tolerant to flooding. Small to medium trees will be less subject to uprooting by wind in wet, shallow soil.

Alkaline soils can reduce the availability of iron and manganese so much that some landscape plants develop small, pale, chlorotic foliage that browns along the margins; growth may also be stunted. Particularly susceptible are liquidambar, camphor, and pin oak. Certain plants, on the other hand, are more tolerant of soils that are saline or high in certain minerals and some laboratories in the southwestern United States are studying this phenomenon.

Air Pollution   Many species will weaken, reduce their growth, and suffer leaf injury because of air pollution; other species are more tolerant. Air pollution may involve many chemicals from a number of sources. Fluorine or sulfur dioxide, especially toxic to plants, may

come from metal, ceramic, and chemical manufacturers or from petroleum refineries. Levels of these compounds should be checked at
locations downwind from such sources. Pollution from autos and
other industries is more widespread in some areas. Many pines are quite
sensitive to air pollutants, whereas ginkgo, London plane, and pin oak
are fairly tolerant.

Pest Resistance Genetic resistance to pests is the ultimate weapon.
A few woody plants, like the ginkgo and the goldenrain tree, are practically pest-free. Most plants, however, can fall victim to one or more
maladies. Even limited tolerance to serious pests will help plants to
look and perform their best and will reduce necessary maintenance.
Some insects and diseases cause little damage, whereas others can kill
within a season. Mildew may be only unsightly on some plants, but
Dutch elm disease is usually fatal. A pest can be serious in one region
but not in another: Mimosa webworm infests honey locust in Pennsylvania and further south, but is not a problem in northern regions where
winters are cold.

A few plant breeders are beginning to breed for pest and disease
resistance. An apple breeding program in the midwest, for example,
has produced a number of fine ornamental crabs that are resistant to
most common apple diseases. The National Arboretum has made
selections of crape myrtle that are more resistant to mildew than those
more commonly planted.

Inquiry among local cooperative extension agents, nurserymen,
or arborists, along with careful observation, can afford valuable information about the local pest problems of plants you would like to grow.

Longevity Long life is more important in trees than in shrubs
and vines, which can be replaced more easily and regrown to landscape
effectiveness more quickly than trees. Generally, fast-growing trees
and shrubs are not as long-lived as slower-growing ones. As a consequence, trees with different growth rates are sometimes planted in a
mixture: A rapid landscape effect is obtained with the fast-growing
trees, which can be successively pruned back and finally removed as
the slower-growing trees become large. All too often, however, an
owner or supervisor waits too long to prune or does not have the heart
to cut down good-sized trees in their prime.

Certain trees and shrubs are known to be fairly short-lived, and
others are noted for their longevity. Even though site and care are
important to a plant's longevity, its survival potential ought to influence selection for landscape use. Most trees, however, will outlive
their owners.

Fire   Fire is a threat in chaparral areas, particularly during hot, dry, windy weather. To reduce the jeopardy to homes in such areas, people have recommended landscaping around homes with so-called fire-resistant plants. Of course, there is no such thing as a plant that will not burn if there is enough heat.

Landscape design and management are the keys to minimizing fire hazard in high-risk areas. The main objective is to minimize the accumulation of dry fuel (Maire 1976) and to concentrate on low, slow-growing plants with high water content. Tall-growing conifers that retain dry needles and high volumes of resins should be avoided close to buildings and to one another. Certain other plants may also be high in resins or may accumulate large amounts of dead wood or litter and are to be avoided.

SOURCES OF PLANTING STOCK

A few individuals and agencies propagate and grow plants for their own use, but almost all plants for public and private landscapes are obtained from wholesale and retail nurseries. Wholesale nurseries usually grow plants in the field or in containers. Some nurseries special-ize in a few kinds of plants, whereas others grow a wide variety. Some propagate and grow plants for retail sale, while others buy small plants and grow them for resale. Most retail nurseries and garden centers sell a complete selection of landscape plants as well as landscape supplies and equipment. Many retail nurseries grow some portion of the plants they sell.

It is wise to get acquainted with one or more nurserymen in your area: Not only are they a source of plants, but the better ones can offer a wealth of information about plant adaptation and maintenance. They also can benefit from your experience and knowledge. Because many trees take several years to produce, you should contact a nursery several months or even a year before you plan to plant in order to be assured of the species, sizes, and quality you want. For recently introduced plants, in the quantity and size wanted, you may have to contract with a nursery to grow them. If the number of plants is large enough, most wholesale nurseries will grow by contract for future delivery.

Woody plants are usually sold or transplanted as *bare-root*, *balled and burlapped* (B & B), or *container* plants (Fig. 3-3). Most deciduous plants are undercut, dug, and handled with little or no soil on their roots (bare root) during late fall and winter. The plants are stored so the roots are kept moist and the tops dormant, and they should be planted in the landscape before growth begins.

FIGURE 3-3
Young trees are available for planting either as bare-root (left),
balled and burlapped (B & B, center), or container (right).
(B & B photo courtesy William Flemer, Princeton Nurseries, New Jersey)

Many trees and shrubs are field grown. Field-grown evergreens are dug with a ball of soil around the roots, which is then wrapped in burlap or some other material. Plants with balls that are larger than 500 mm (20 in) in diameter are usually packed into boxes that are 600 mm (24 in) or larger on a side; such plants are handled somewhat like container-grown plants until sold. B & B plants have a longer planting season than bare-root plants but should be planted before spring.

A number of deciduous plants cannot be handled bare-root but transplant well from containers. Container-plant production makes year-round planting possible in many areas and makes greater uniformity and mechanization possible in nursery operations. Since the 1950s, an increasing number of plants have been grown in containers. A 20-liter (5-gal) plant will start as a seed in a seedflat, where it germinates and grows from 40 to 100 mm (1.5–4 in) tall. It is transferred (hopefully, with root pruning and care) to a liner pot 55 to 75 mm (2.25–3 in) in diameter. After several weeks, the plant is transferred to a 4-liter (1-gal) can and moved from the greenhouse to the nursery yard. Later it is moved into a 20-liter can.

Because of increasing interest in planting large specimens, particularly trees, more plants are now grown to large size in nursery fields, are dug mechanically, and are placed directly in the landscape or into large containers for transport and later sale.

SELECTING QUALITY PLANTS

The quality of the trees and shrubs selected for planting can be as important in determining their success in the landscape as proper selection of species, planting, and maintenance. Certain characteristics have considerable influence on plant survival and growth, while other characteristics, although desirable, may not affect performance greatly. Branching pattern and size determine the immediate landscape effect of a plant but may have little influence on subsequent growth. Variations in vigor, laterals on the trunk, and height-to-caliper ratio will not jeopardize survival but can materially affect the effort necessary to obtain satisfactory growth. Root and shoot quality can determine not only performance but even survival.

Root Characteristics

A well-developed, healthy root system is essential to a vigorous plant, particularly to a tree that lives many years and becomes large. In a well-formed root system, branching is symmetrical and main roots grow down and out to provide trunk support. Container-grown and balled-and-burlapped plants should have fibrous roots sufficiently developed that the root mass will retain its shape and hold together when removed from the container or when handled during planting. The main roots should be free of kinks and circles (Fig. 3-4). The main roots of bare-root plants should be sound and free from breaks, torn or bruised bark, crown gall, and nematodes.

Root quality at planting cannot be overstressed, because certain defects can doom a plant to poor growth, breaking of the trunk at the ground, or death (Barrows 1970, Long 1961). The top of a young plant may not be a good indicator of the quality of the root system (Harris and others 1971). Although root defects may be primarily associated with container-grown plants, field-grown plants propagated in seedflats and liners or seeded directly in the field can also develop root problems.

Types of Root Defects   Two types of defects may occur: (1) *kinked roots*, in which the taproot, major branched roots, or both are sharply bent, and (2) *circling* or *girdling roots*, which form circles, generally horizontal, around the trunk or other roots (Fig. 3-4) (Harris, Long, and Davis 1967, Harris and others 1971). Both types of defects often occur within the same root system.

Location of Root Defects   Roots must be inspected to evaluate their condition.   Root deformities that are serious enough to affect

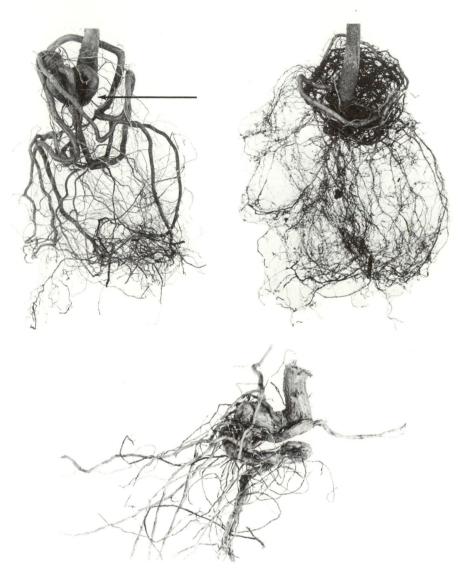

FIGURE 3-4
Kinked and circling roots are exposed when soil mix is washed from the roots
of trees in 4-l (1-gal) cans. The oak root (left) was kinked (arrow)
when transplanted from a seed flat to a 50-mm (2-in) peat pot;
the shape of the peat pot is revealed by the roots that were not pinched back
when the peat pot was put in a 4-l can. Pine roots (right)
still outline the shape of a 100-mm (4-in) square pot
because they were not pruned and straightened when moved up
to the 4-l can. The root of a field-grown Monterey pine (bottom) is severely
kinked. Kinking occurred at the first transplanting in the greenhouse.
None of these root systems could now be pruned
to make planting in the landscape worthwhile. (Harris and others 1971)

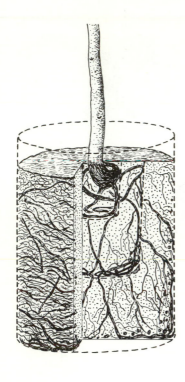

FIGURE 3-5
Kinked and circling roots can occur
in different zones of a root ball
of 14-l (3.5-gal) or larger;
each zone would represent one
of the earlier root ball sizes.

growth and survival may occur in any or all of three zones within the
root ball of a container-grown plant (Fig. 3-5):

> *The trunk-surface root zone* is within 75 mm (3 in) of the soil surface and
> 50 mm (2 in) from the trunk of the plant. This zone is generally that of
> the liner pot.
>
> *The center root zone* is within the root ball outside the trunk-surface root
> zone, but not at the periphery of the root ball. The location depends on
> the size of the intermediate container, if one has been used.
>
> *The peripheral root zone* occurs at the edge and bottom of the root ball.

Trunk-surface roots are often visible on or above the soil surface;
if not, you may examine them by washing the soil from the crown to
a depth of 50 to 75 mm (2–3 in) below the crown and 25 to 50 mm
(1–2 in) from the trunk. If no branch roots from the main root are
found, at least one plant should be sacrificed by washing deeper until
branch roots are found. If plants are planted too deep originally, crown
rot or other problems can result when they are planted out (Pirone
1978a). Plants with taproots that have not been pruned during transplantings might also have main roots without laterals near the surface.
Unpruned taproots may become twisted during transplantings, so that
they later become girdling.

You may see center roots that are extremely circled when you examine trunk-surface roots; otherwise, you may inspect them by washing out more soil from around the crown to a depth of 100 to 125 mm (4–5 in) and 50 to 75 mm (2–3 in) from the trunk. To examine the center root zone completely, however, you must remove the container and wash the soil from the roots. If such destructive inspection is necessary, it is done on a sample of the plants.

Peripheral roots can be seen by removing the container from the root ball. This is particularly easy when containers have tapered sides. Since this inspection is primarily designed to compare general root condition to the vigor of the tops, only a few plants usually need be inspected.

Except for those inspected destructively, plants whose roots have been partially exposed will usually not be affected adversely if the roots do not dry out and a soil mix equal to that removed is replaced, firmed, and moistened after inspection. When plants are accepted after such a root inspection, however, their performance in the landscape should be the responsibility of the purchaser, not of the supplier. Handling, planting (including judicious pruning), and irrigation should proceed with care.

Evaluating Root Defects   Observations of landscape plants indicate that many are restricted in growth by kinked and circling roots (Barrows 1970, Long 1961). The following root conditions may not only reduce growth but may kill plants, particularly trees:

Kinked roots with a sharp bend in the main root(s) of more than 90 degrees and less than 20 percent of the root system originating above the kink. A plant with such roots would probably bend at or below the soil line if untied from its supporting stake.

Circling roots at the *trunk-surface* or the *center* that circle 80 percent or more of the root system by 360 degrees or more. A plant with such roots would have less than 20 percent of its roots free for support should girdling occur.

Even though there is considerable evidence that kinked and circling roots cause girdling or constricting of the main root(s) and trunk, no conclusive studies have indicated how severe the kinking or circling has to be before a plant is jeopardized. A small portion (20 percent or more) of free roots may be adequate to maintain the plant, but this estimate is speculative. Agencies, landscape architects, and arborists are more frequently specifying that trees be inspected for root defects as a condition of acceptance.

Circling peripheral roots are usually not a problem, since they can be corrected at planting; container-grown plants with adequate root development will inevitably exhibit some circling of this sort. Peripheral roots, however, should be checked for indications that the plant has been in the container too long. Masses of large, entwined roots at the bottom and around the root ball may indicate an abnormally potbound condition (Bancroft-Whitney 1967). If the top of the plant is vigorous, however, the root system is not yet abnormally potbound.

The most difficult root defects to correct occur in the trunk-surface and center root zones. Since main roots are usually involved, correction will probably seriously weaken the plant. Even if the top is still vigorous, correcting entwined and matted roots at transplanting may be too time-consuming to be worthwhile. On the other hand, circling and matted peripheral roots can and should be thinned and straightened when the plant is transferred to a larger container or into the landscape. If the peripheral roots are large, entwined, and matted, cutting and straightening them may further retard growth that has already been slowed by prolonged confinement in the container. In some cases, however, if the top is cut back to balance the reduced root system, the plant may be invigorated. If the circling is not corrected, the plant will probably continue its poor growth until the circling roots finally girdle it.

Certain trees, such as eucalyptus and redwood, are known to form new roots from portions of the trunks that become covered with soil or rock. Whether this capacity is important in overcoming kinked and girdling roots is not known. That many plants, including eucalyptus, are lost from trunk-root breakage and root girdling would indicate that these new roots are of minor importance, but further investigation is needed.

Preliminary studies and field observations indicate that plants will grow rapidly when planted out if their growth has not been slowed during production. Even some plants that appear healthy have been left too long in containers that have begun to check growth. At any transplanting, the ideal root system is developed just enough to hold the root ball together. Any less development would jeopardize tree survival by loss of soil from around roots and root breakage; much further development at that particular stage might reduce future vigor.

Top and Trunk Characteristics

Most trees grown with adequate space and without staking or severe pruning are capable of supporting themselves even in high winds (Harris and Hamilton 1969, Leiser and others 1972). Most trees grow with certain height-to-caliper relationships, fairly uniform trunk taper, and crown configurations that distribute wind loading along the trunk (Leiser and Kemper 1973). Growing practices can modify these characteristics greatly.

Height-to-Caliper Ratio   Field-grown trees, whether they are sold bare-root or balled and burlapped, have long been specified according to height and trunk caliper. The American National Standards Institute (ANSI Z60.1, 1980) specifies height-to-caliper values for bareroot trees (Fig. 3-6). Heights that are greater than those listed for a given caliper indicate a spindly trunk that may not be capable of stand-

FIGURE 3-6
Height and caliper standards for different types of field-grown trees (ANSI Z60.1, 1980). The shaded area indicates range in heights and calipers for standard trees (Type 1). Note that the height-to-caliper ratio for small trees (Type 3) is almost a continuation of the lower portion of the curve for standard trees. The height of a slower-growing tree is two-thirds the height of a standard tree for the same caliper. The two parallelograms represent the height-to-caliper relationships for unstaked container-grown trees that are given adequate space. (Harris, Leiser, and Davis 1976)

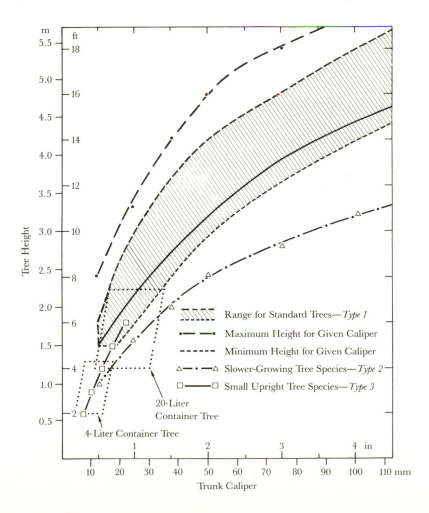

ing upright when in full leaf. Trees with heights considerably less than those listed for a particular caliper should be checked for adequate plant vigor.

Height-to-caliper values have not been established for container-grown trees. Experiments with trees growing in 20-l (5-gal) containers in three nursery areas in California, however, showed that certain staking and pruning practices will produce trees that meet or exceed the standards for field-grown trees (Leiser and others 1972). Caliper measurements ranged from 18 to 32 mm (0.62 to 1.17 in) for trees that were 2 m (6 ft) tall, provided these trees were not staked, were given adequate space, and received light pruning of lower trunk laterals. The calipers of these trees averaged 5 mm (0.2 in) or 30 percent greater than those of trees that were staked and had their low trunk laterals removed. Most of the staked trees were not able to stand upright when untied from their stakes (see Fig. 8-10).

The ANSI Z60.1 standards for height-to-caliper values (Fig. 3-6) have an upper caliper limit for a given tree height as well, though most of the emphasis is on lower caliper limits. Certain species normally develop large caliper for their height. If a tree is too short for given caliper, however, its growth may be so poor that the tree will stagnate. Such a tree probably would not do well when planted in the landscape.

Taper Taper is the decrease in trunk caliper with increasing height. Under wind stress, the tapered trunk of an unstaked tree bends uniformly along the stem, but a trunk with little or no taper bends essentially along one section near the ground (see Fig. 8-10) (Leiser and Kemper 1973). In the first case, stress is evenly distributed along the lower two-thirds of the trunk and decreases toward the tip, thus reducing the possibility of trunk deformation or breakage. In the latter case, most of the stress is concentrated, causing the trunk to break or remain bent even after the wind has stopped. Leiser and others (1972) discovered that staking and severe pruning produce a tree trunk with little or no taper or even with reverse taper; that is, trunk caliper may be greater at 2 m (6 ft) than at a lower level.

In the same study, trees grown in 20-l (5-gal) cans without stakes exhibited from 5 to 10 mm reduction in caliper for each meter ($\frac{1}{16}$ to $\frac{1}{8}$ in/ft) of trunk height below the permanent branches. A trunk with this degree of taper would have a caliper at half the tree height equal to 50 to 80 percent of the trunk caliper at 50 mm (2 in) from the ground. Trunks with branches along their length normally increase more in caliper at their base than at higher levels, so that as a young tree grows its taper increases (Harris and Hamilton 1969). Therefore, for the same species, a large young tree with low branches should have a greater taper than a smaller tree.

Crown Configuration   In a wind, a young tree with branching along its trunk will have a fairly uniform distribution of stress (Leiser and Kemper 1973). Removal of lower laterals or shading due to crowding, however, may produce a tree with a crown (and hence a center of wind loading) concentrated near the top. Such a tree may be unable to support itself without staking.

An ideal branch distribution would center the wind load acting on the tree at about two-thirds of the total height (Leiser and Kemper 1973). Thus, one-half or more of the foliage should be on branches originating on the lower two-thirds of the trunk, and one-half or less should originate on the upper one-third (Fig. 3-7).

In addition to equalizing stress distribution, lower limbs nourish the trunk and shade it, resulting in a tree with greater caliper and taper.

FIGURE 3-7
The wind load on the main trunk of the plane tree (left) will be more evenly distributed than on the magnolia because more than one-half of the foliage is on branches that originate along the lower two-thirds of the plane tree trunk (the stake was never needed). Most of the stress on the magnolia will be concentrated just below the lowest branch.

Low foliage may help obscure the trunk and protect the tree somewhat from vandalism. The lower laterals should be pinched (headed) to keep them small in relation to the higher permanent branches and the trunk. These laterals can be removed when the top more completely shades the trunk, when the trunk is larger and the bark thicker. Pruning wounds on young trees usually heal quickly.

If a tree can stand without support and can return to an upright position after being deflected by wind or hand, it will usually meet the aforementioned guidelines as to height-to-caliper ratio, taper, and top configuration. Such trees usually result when they are grown without staking, with lightly headed laterals left along their lower trunks, and with adequate space in the nursery.

Since such trees may not always be available, however, landscape staking and pruning can be used when trees are not able to stand upright by themselves. These practices are discussed in Chapter 8. The guidelines set forth here should be followed whenever possible, and suppliers should be encouraged to produce plants that meet these criteria. When such trees are available, support staking can be virtually eliminated, though protective or anchor staking may be needed.

Branching Pattern   Plants should have a branching pattern or shape that is in keeping with their intended use. Low main branches may be used as a windbreak, visual screen, or specimen in a shrub bed but may be unacceptable in situations where traffic or visual clearances are desired. Early removal of lower laterals will be at the expense of trunk development and wind stress distribution (Chandler and Cornell 1952, Leiser and Kemper 1973). When trees have been headed (their main leader cut back) (see Fig. 14-4) branches with close vertical spacing will usually develop immediately below the cut; if these are not corrected, they can lead to structural weakness as they become large (Harris and others 1969). Also, many trees that have been headed become misshapen when lower branches outgrow the branches above. For most landscape purposes, the main leader should not be headed, but if it is, a new leader should be selected to fill its place.

The main framework branches of a tree should form wide angles of attachment with the trunk. Since little or no connective wood forms in a sharp-angled crotch, it is inherently weak. If main branches with narrow angles of attachment are removed from young trees, they will usually be replaced by new shoots growing at wider angles of attachment (Harris and Balics 1963).

A young tree may not be tall enough to develop branches that will be the permanent ones. Once formed, a branch does not increase in height above the ground. As the tree develops, laterals below the lowest permanent branch should be handled as described in the section on Crown Configuration and main branches should be selected.

Top-to-Root Relationships

More public and private landscaped areas, including city streets, are being planted with 60-l (15-gal) and boxed trees in an effort to maximize desirable landscape effects and reduce vandalism and theft. These larger trees are not only more expensive and difficult to plant, but they usually require more intensive initial maintenance.

The size of a tree should not necessarily affect its performance in the landscape but, practically, it usually does. The larger the tree, particularly if it has been grown in a series of containers, the more critical is the vigor of its top. Skill and careful scheduling must foster tree growth during production. At the other extreme, the frequent watering and high fertility that are common in most nurseries may result in a larger top than the root ball can support without continued high maintenance in the more severe landscape environment. Such a condition is not necessarily undesirable, but it must be recognized and handled accordingly: High initial maintenance can be continued or the top can be thinned at planting to reduce the need for frequent water.

A tree whose top is of moderate size in proportion to the root system will usually grow more vigorously than a tree with a large top (Hendrickson 1918). The size of the root system or the volume of soil available influences the interval between irrigations. Reducing the size of the top by thinning (Harris and others 1969) will increase the proportion of roots to top and thereby usually improve moisture and nutritional relationships. Even so, careful attention to irrigation requirements is essential following planting. It should be remembered that too much water can be as detrimental as too little.

The American National Standards Institute (ANSI Z60.1, 1980) recommends that the root spread of bare-root nursery-grown trees should be about one-sixth the height of trees up to 3 m (10 ft) tall. The ANSI Z60.1 also specifies minimum ball sizes for balled-and-burlapped trees, depending on tree caliper or height (Table 3-2, Fig. 3-8). Until other information is available, these can also be used as guidelines for trees that are moved mechanically. The specifications in Table 3-2 may be modified in that plants whose roots are coarse or widespreading because of natural habits of growth, soil condition, or infrequent transplanting and plants that are moved out of season would require balls in excess of the recommended sizes. Also, when stock has been grown in pots or other containers, when field plants have recently been planted out from containers or with smaller balls, and when material has been frequently transplanted or root pruned, the sizes recommended may be excessive (ANSI Z60.1, 1980).

Plants grown in containers are usually sold by the size of the container: One might buy a 20-l (5-gal) liquidambar. Container size does

TABLE 3-2

Recommended minimum root-ball sizes for different types of field-grown trees (ANSI Z60.1 1980)

Shade Trees Types 1 and 2[a]				Smaller Trees Types 3 and 4			
Metric		English		Metric		English	
Caliper	Minimum[b] Diameter Ball	Caliper	Minimum Diameter Ball	Height / Caliper	Minimum Diameter Ball	Height / Caliper	Minimum Diameter Ball
mm	cm	in	in	m/mm	cm	ft/in	in
15[c]	30	0.5	12	0.5	20	2	10
20	35	0.75	14	1.0	30	3	12
25	40	1.0	16	1.25	35	4	14
30	45	1.25	18	1.5 [d]	40	5	16
35	50	1.5	20	20	40	0.75	16
40	55	1.75	22	25	45	1.0	18
50	60	2.0	24	35	50	1.5	20
65	70	2.5	28	40	55	1.75	22
80	80	3.0	32	50	60	2.0	24
95	95	3.5	38	65	70	2.5	28
105	105	4.0	42	80	80	3.0	32
115	120	4.5	48	95	95	3.5	38
125	135	5.0	54	105	105	4.0	42
				115	120	4.5	48
				125	135	5.0	54

[a]Types 1 and 2 are standard and slower-growing shade trees. Type 3 is a small upright tree. Type 4 is a small spreading tree.

[b]Trees having a coarse or wide-spreading root system or that are moved out of season would require a root ball in excess of the above recommended sizes.

[c]The metric measurements have been rounded for convenience.

[d]Figures above the broken line are height (m or ft); those below the line are caliper (mm or in).

FIGURE 3-8
The standards of the American National Standards Institute (ANSI Z60.1, 1980)
for balled and burlapped trees base the size of the root ball on trunk caliper for
single-stemmed standard and slower-growing trees (left) and on height for small
trees (right).

not necessarily indicate the size of the plant. In regions where year-round production is common, plants transplanted to larger containers in the fall may by spring have relatively small tops but extensive root systems, which have been able to develop during cool weather. On the other hand, plants transplanted in the spring may by fall have relatively large tops but few roots. Seasonal variation should be taken into account when one evaluates plant size.

Health

The health of a tree is characterized by its vigor and freedom from injury and pests.

Vigor Vigor is a good measure of a plant's ability to perform when planted out. A knowledge of the species is of value in making this rather subjective evaluation. Leaf color should be green to dark green, depending on the species and time of year. Large leaves and dense foliage denote vigor. Shoot growth is a more quantitative indicator, but also varies greatly with species. Almost all trees to be used in the landscape should grow at least 300 mm (1 ft) a year; many kinds will grow 1 to 2 m (4–6 ft). This is not undesirable if the trunk is large and well-tapered enough to hold the top upright. Certain dwarf plants and species that grow only from preformed shoot initials may grow

less vigorously. Shoot growth that has been pruned will be more difficult to evaluate, but the age and size of the pruned shoots can be used
in estimating vigor.

Smooth, bright bark on the trunk and main branches of young
plants usually signifies good vigor. Rough, cracked, dull, and dark
bark, although characteristic of some species, could be indicative of
poor vigor, so other indicators should be checked.

Young roots of most plants will be light in color. Their diameters
vary greatly with species and are not good indicators of vigor. Large,
dark-colored roots, however, may be an indication of low vigor; a plant
with such roots may be old for its size.

Injury   Select plants free from injured bark, but be sure to distinguish normal pruning wounds from actual injury. Bark injury can be
a problem when trees are staked (Harris and others 1969, Leiser and
others 1972); tree ties and labels sometimes girdle the trunk or main
branches so that growth is reduced and slow to recover even when the
restriction is removed.

Sunburn is common when plants, particularly trees of low vigor,
are exposed to the afternoon sun or have trunks with few or no leafy
shoots. Sunburned trunks are extremely slow to heal and are subject
to borer infestation.

Pests   Pests should seldom be a problem because most nursery
stock is inspected or comes from nurseries that are periodically inspected. Delivery of plants should be in compliance with nursery
inspection and state nursery stock grades, standards, laws, and regulations. If there is reason to suspect a problem, the purchaser should
call the local county agricultural commissioner's office or another
responsible nursery regulatory official.

It should be remembered that some organisms or indications of
their presence are not necessarily harmful. Many organisms have little
or no effect on plants; some are beneficial.

Galls or swellings on roots and trunks are normal on a number of
plants (Fig. 3-9). Sometimes these swellings are confused with those
caused by nematodes or crown gall bacteria. Mycorrhizal fungi can
cause enlarged feeder roots with areas that are lobed, beaded, or brainlike in shape. The roots of legumes may have nodules with bacteria
that fix nitrogen from the air into nutrient forms. Swellings may occur
at the trunk base of redwood, manzanita, eucalyptus (Kelly 1969), and
other species; these are referred to as *burls* or *lignotubers*. These swollen masses of woody tissue form reservoirs of latent buds. A plant with
well-developed lignotubers is able to sprout quickly if its top is injured.
In contrast, crown galls and knots caused by nematodes do not, to our
knowledge, produce sprouts.

FIGURE 3-9
Lignotubers on some eucalyptus and
swellings on other species are normal
and should not cause concern. Some trees
may have lignotubers while others of the
same species do not.

Inspecting Plants

All plants, or a representative sample, should be inspected before they are accepted to be sure the plants meet criteria for size and quality. When a large number of plants, particularly trees, are involved, arborists should inspect them at the nursery before delivery. It is much easier to hold to the specifications or make substitutions at the nursery than at a planting site. Many times it is possible to select the plants you want at the growing grounds, just as you would in a retail nursery.

The value of clearly stated specifications is most evident at the time of acceptance. Realistic specifications should ensure good growth, ease of maintenance, and landscape fulfillment but should minimize unnecessary restrictions. Some characteristics of young plants are essential for future vigorous growth or even for survival. Others may encourage better performance or easier maintenance, but their absence will not necessarily jeopardize the life of the plants. Depending on the use of the plants, other characteristics may be desirable but not critical for survival or well-being. These characteristics are categorized by their probable importance:

Characteristics That May Jeopardize Plant Life and Growth	Characteristics That May Improve Growth or Ease Maintenance	Characteristics Desirable but not Critical and Characteristics That Can be Corrected
Severely kinked roots	Moderate to good vigor	Correct species
Severely circling roots	Moderate top-to-root ratio	Branching pattern
Noxious pests	Caliper and taper (ability to stand without stake)	
Desiccation of shoots or roots	Laterals on trunk	

The portions of trees and shrubs that are easiest and most obvious to inspect are the tops: Is the plant the size you want? Is it vigorous? Are shoots and leaves turgid and healthy? Will the plant stand without support? Is it free of pests and injuries? Does it have the desired shape and structure? These characteristics are fairly easy to determine.

If the top is acceptable, then check the roots. For bare-root plants this is easy; for container and B & B plants it is more difficult. First check the trunk-surface zone for kinking and circling roots, no matter how the plants have been grown. That is about as far as you can go with B & B plants: If the trunk-surface roots are acceptable, the rest of the root structure is probably free of problems because the plants have been dug from field plantings. Root development in container plants may be tested in two simple ways. If the plant is staked, untie it from its support: If the plant bends over sharply at the soil line or if the trunk is loose in the soil, the main roots are probably circling or kinked below the soil surface (Fig. 3-10). Check further. With the

FIGURE 3-10
Simple tests to check the root system of a container-grown tree. Circling or kinked roots on the surface indicate serious root problems (extreme left).
If the tree is staked, untie it from the stake; if the trunk bends sharply from the top of the container (left center), suspect kinked or circling roots;

plant untied from any support, lift the plant slowly by grasping its trunk.   If the top moves up 20 or 30 mm (1 in) before the container and soil do, the roots are either underdeveloped or circling (Fig. 3-10).

If the plant has passed these tests, the center roots may be checked by washing out some of the soil at and just below the trunk-surface-root zone (Fig. 3-5).   It may be necessary to wash the soil from the roots to be sure they are free from kinking and circling.   Some agencies destructively inspect the root systems of at least 2 percent of the trees and shrubs they buy.   The peripheral roots can be inspected by cutting the container or sliding the root ball from its tapered container.   The size and condition of the visible roots will indicate to a fair degree how well the plant will grow when planted.   Small- to medium-diameter roots that are supple and not densely entwined around the sides are ideal.   Numerous large roots that are rigid and matted around the sides and at the bottom of the root ball will need to be pruned severely at planting.   A plant with such roots will have to be watered carefully and, even then, may have little vitality.   If there are only a few small roots at the periphery of the root ball and the root ball does not hold together, you will be paying for a larger plant than you receive.

Additional Thoughts

Species   At least one plant of each clone, cultivar, or species should be labeled with the correct botanical name when delivered to the purchaser.   If more than one clone of a species is being ordered, every plant should be labeled to ensure against mix-ups in planting.

if it arches above the container (right center), the roots may be all right. Alternately, lift gently on the trunk (extreme right); the container should rise with it. If the trunk can be raised 25 to 50 mm (1–2 in) before the container moves, the roots may be circling or poorly developed.

If the particular kind of plant is important for a specific land-scape purpose and is to be used in quantity, its correct identification should be verified by the purchaser or his qualified representative. A number of species have strains with differing flowering and growth characteristics. Replacement guarantees usually cover only the size originally supplied. Depending on the expected losses, a certain percentage of extra plants should be purchased and moved to larger containers or planted out to be used as replacements should plants suffer vandalism, do poorly, die, or turn out to be of the wrong kind.

Acceptance Delivery   Plants are usually ordered for delivery on a certain date. If the purchaser wants to delay delivery, the plants should be inspected on the original delivery date or soon thereafter to determine their acceptability. If acceptable at that time, they should not be turned down later for being potbound, overgrown, or of low vigor. The retailer or wholesaler can reasonably charge additional fees for the extended care of the plants.

Plant Protection   During handling, plants must be protected from injury and exposure to desiccation and temperature extremes. The roots should be kept moist and cool, particularly if plants are bare-root or in dark containers.

FURTHER READING

BAILEY, L. H. 1949. *Manual of Cultivated Plants*. New York: Macmillan.

————, E. Z. BAILEY, and STAFF OF THE L. H. BAILEY HORTORIUM. 1976. *Hortus Third*. New York: Macmillan.

BEAN, W. J. 1950. *Trees and Shrubs Hardy in the British Isles*. 3 vols. London: Butler and Tanner.

BENSON, L., and R. A. DARROW. 1954. *Trees and Shrubs of the Southwestern Deserts*. Tucson: University of Arizona Press.

CHITTENDON, F. J., ed. 1951. *Dictionary of Gardening* (Royal Horticultural Society). Oxford: Clarendon Press.

CLARK, D. E., ed. 1979. *Sunset Western Garden Book*. Menlo Park, Calif.: Lane.

CLOUSTON, B., ed. 1977. *Landscape Design with Plants*. London: Heinemann.

CREASY, R. 1982. *Complete Book of Edible Landscaping*. San Francisco: Sierra Club.

FISHER, M. E., E. SATCHELL, and J. M. WATKINS. 1975. *Gardening with New Zealand Plants, Shrubs and Trees*. Auckland and London: William Collins N.Z.

GRANT, J. A., and C. L. GRANT. 1955. *Trees and Shrubs for Pacific Northwest Gardens*. San Carlos, Calif.: Brown and Nourse.

HILLIER, H. C. 1973. *Hilliers' Manual of Trees and Shrubs*. Newton Abbot, England: David and Charles.

Hoyt, R. S. 1978. *Ornamental Plants for Subtropical Regions*. Anaheim, Calif.: Livingston Press.

Kelly, S. 1969. *Eucalypts*. Melbourne: Thomas Nelson.

Lord, E. E. 1948. *Shrubs and Trees for Australian Gardens*. Melbourne: Lothian.

Menneinger, E. A. 1964. *Seaside Plants of the World*. New York: Hearthside Press.

McMinn, H. E. 1939. *An Illustrated Manual of California Shrubs*. San Francisco: J. W. Stacy.

McMinn, H. E., and E. Maino. 1963. *An Illustrated Manual of Pacific Coast Trees*. Berkeley: University of California Press.

Metcalf, L. J. 1975. *The Cultivation of New Zealand Trees and Shrubs*. Wellington, N.Z.: A. H. and A. W. Reed.

Mullins, M. G., ed. 1979. *Reader's Digest Illustrated Guide to Gardening*. Sydney: Reader's Digest Services.

Osborne, R. 1975. *Garden Trees*. Menlo Park, Calif.: Lane.

Ray, R. M. 1977. *The World of Trees*. San Francisco: Ortho.

Rehder, A. 1940. *Manual of Cultivated Trees and Shrubs Hardy in North America*. 2nd ed. New York: Macmillan.

Watkins, J. V. 1961. *Your Guide to Florida Landscape Plants*. Gainesville: University of Florida Press.

Wigginton, B. E. 1963. *Trees and Shrubs for the Southeast*. Athens: University of Georgia Press.

Wyman, D. 1965. *Trees for American Gardens*. New York: Macmillan.

———. 1969. *Shrubs and Vines for American Gardens*. New York: Macmillan.

———. 1971. *Wyman's Gardening Encyclopedia*. New York: Macmillan.

CHAPTER FOUR

Planting Site: Climate

Less than 8 percent of the earth's land surface is cultivated. Most of the rest is either too cold, too hot, too dry, too salty, too rocky, or too steep to support cultivation. Climate is probably more important than soil in determining the growth and well-being of plants. In fact, climate is an important factor in the development of soil. To the old cliche "Everybody talks about the weather, but nobody does anything about it" could be added "Very few people understand the weather and its effects on plants." A horticulturist should be one of those few who understand and use it to advantage.

Climatic factors include temperature, radiation, light, precipitation, fog, humidity, wind, and air pressure. Given a particular site, climate is a key factor in deciding what plants are best adapted to it and how they can best be nurtured by spacing, irrigation, timing of fertilizer application, transplanting, pruning, and pest control. In fact, some cultural practices can modify the microclimate to benefit the plants to be grown or the people who will frequent the area. Woody perennials must be able to withstand the many vicissitudes of the weather not only for an entire year but for a succession of years.

An individual plant or group of plants undergoes climatic influences on three different scales. The *macroclimate* extends over a relatively large area defined by fairly uniform climatic conditions. These conditions are largely determined by the air masses moving over the earth's surface and modified by latitude, elevation, large bodies of water, mountain ranges, and the season of the year.

On a much smaller scale, the *mesoclimate* is the local weather of a neighborhood or a city, a large park, one or more farms in a similar setting, or a woods. The mesoclimate is the macroclimate as modified by local influences of terrain, bodies of water, cloud cover, wind, and land cover.

On an even smaller scale, the *microclimate* includes the conditions around or within an individual plant or planting. It is a further refinement of the mesoclimate by influences in the immediate area. The term microclimate is often misapplied to the mesoclimate, but you can avoid that error if you know the scale with which you are dealing.

Precipitation and temperature are most commonly the limiting climatic factors in plant distribution and performance. In many landscape situations the hazards of drought or too much water can be minimized by irrigation, drainage, or other cultural practices. Even when favorable sites are chosen, adaptable plant species selected, and the hazards of cold injury reduced by proper cultural practices, these efforts toward providing a favorable microclimate remain at the mercy of the macroclimate. Temperature remains the least controllable of environmental factors in the landscape.

TEMPERATURE

The effects of temperature vary with plant species, stage of growth, age, condition, and even with particular plant parts or tissues. Although plants may grow poorly or be killed by high temperatures, low temperatures are much more commonly a problem.

Cold Temperatures

Plants can be injured or killed by low temperatures at almost any season of the year. The most critical periods, however, are (1) spring or autumn, (2) the coldest portion of winter, and (3) occasions when minimum temperatures occur after a warm winter period, even though plants had been at maximum hardiness before the warm period.

The more comon types of cold injury include (Weiser 1970):

Black heart in stems of trees and shrubs

Winter kill of dormant flower buds

Sun scald and frost splitting of tree trunks

Soil-heaving damage of young plants

Winter burn of conifer foliage

Dieback of overwintering broadleaved plants

Spring and fall frost damage of tender shoots, flowers, and fruit

These types of injury are especially prominent in the higher latitudes and altitudes. Several of them, however, limit plant distribution or impair performance in all but the warmest regions. Even in regions with mild winters, dieback and frost injury constitute real dangers.

Some species are more hardy than others or develop sufficient hardiness early in the autumn and retain it long enough to withstand critical temperatures. When healthy plants of moderate vigor produce a moderate fruit crop, they are usually hardier than excessively vigorous or heavily fruiting plants. Not only do species vary considerably in their inherent capacity to resist freezing injury, but the resistance of individual plants changes during the year. A plant that would be killed at temperatures slightly below freezing in summer may be able to survive the coldest of winter temperatures. This change from a susceptible (tender) to a resistant (hardy) condition is referred to as *acclimation* or *hardening*.

A further complication is the difference in hardiness between adjacent tissues or parts of a plant. In stems, for example, the living xylem is hardier in early winter than phloem but is often several degrees less resistant in midwinter than are adjacent cambium or phloem tissues. The xylem of the smaller branches on trees and shrubs may be damaged by low temperatures, creating a condition called *black heart* because of the darkening of the xylem. The problem is not evident until injured shoots are cut for examination, but will be signalled by poor growth, reduced flowering, and possibly death of some shoots in the following spring.

Plant organs also differ in hardiness. Leaves of deciduous plants acclimate little and abscise when subjected to cold. Flower buds are usually less hardy than leaf buds and thus may suffer winter-kill before vegetative buds do. Roots are consistently less resistant to freezing than are overwintering stems but are usually adequately insulated by the soil.

On sunny days in winter, a tree trunk may be warmed as much as 10°C (18°F) above air temperature. If a trunk or branch so warmed becomes shaded by a dense cloud or opaque object, the bark temperature may drop quickly to a critical level, causing injury or death to the bark and cambium; this is generally called *sun scald* even though it is actually a freezing injury and may develop into a *frost canker*.

Wide fluctuations above and below the freezing temperatures of wood may also cause *cup shakes*, or separations of the wood along one or more annual rings. These occur when a frozen trunk warms quickly upon exposure to the sun; the warmed outer wood expands and separates from the inner wood, which has expanded less rapidly. Cup shakes are not evident until the trunk is cut or breaks, but they cause serious defects in lumber and may weaken a tree while it is living.

Longitudinal *frost cracks* usually occur in the bark and wood parallel to the grain and extend to the center of the trunk (Fig. 4-1).

FIGURE 4-1
Frost cracks on London plane tree may occur when the temperature drops to
−13°C (8°F) or below (left). (Photo courtesy of Eugene Himelick, Illinois
Natural History Survey) Radial shake or frost crack is associated with an old
wound. (Butin and Shigo 1981)

Frost cracks (radial shakes) in oak are associated with a variety of
wounds and stubs of branches and basal sprouts (Butin and Shigo 1981).
Himelick (1970) found that frost cracks opened on London plane trees
when the air temperature dropped to −13°C (8°F). The colder the
temperature, the wider these cracks opened. Once these frost cracks
occurred, at least on London plane, they continued to open each winter
and close in the spring. It did not seem to matter whether the temperature went down suddenly or gradually, but when it dropped to −13°C,
all the old cracks reopened within half an hour of one another. After
repeated splitting, considerable callus forms along the edges of the
cracks, and wood decay may also begin. Trees 150 to 450 mm (6–18 in)
in diameter are more likely to be affected than trees that are either
smaller or larger (Pirone 1978a).

   Even in cold-winter areas, frost cracking can be minimized by
careful species selection, planting, and maintenance. Evergreens are
less subject to cracking than are deciduous trees. The most commonly
affected species are London plane, elm, horse chestnut, linden, oak,
willow (Himelick 1970), apple, beech, crabapple, goldenrain tree, and
walnut (Pirone 1978a). Trees protected by thick planting or other
structures are less subject to cracking than are lone trees.

Soil moisture levels in the fall and winter do not seem to influence the occurrence of frost cracking (Himelick), but shading a trunk or painting it with white latex exterior paint will tend to moderate daily temperature extremes and may reduce frost cracks as well as sun scald and cup shakes.

Frost cracks can usually be kept from reopening if one inserts screw rods through the trunk. Himelick found that 12 mm ($\frac{1}{2}$ in) rods screwed into 10 mm ($\frac{7}{16}$ in) holes drilled through the crack every 300 mm (12 in) stopped most cracks from reopening the following winter. He also found it necessary to use an oval washer recessed against the wood on each end of the rod.

Surface soil that has little snow or mulch cover may alternately freeze and thaw in winter, particularly at the beginning and near the end of the usual cold period. Recently planted small trees, shrubs, and other plants can be lifted by freezing soil, returned part way to their original position during a thaw, and then lifted further by the next freeze. This movement can break roots and partially unearth the plant so that it can no longer stand by itself. Damage from *soil heaving* can be avoided by spring planting or by use of a mulch cover to moderate temperature extremes in the soil. Plant parts immediately above the mulch, however, may be injured by greater temperature extremes.

Even though roots are not as hardy as the tops of trees and shrubs, they are less often injured by low temperatures. Dry soil, organic mulch, or snow can insulate roots from winter cold. Trees are more likely to be injured by frost in poorly drained soils than in well-drained soils. The damage to roots will not be apparent until the first warm days of spring, when new shoots and leaves will wilt and die.

Cold weather that freezes the soil may reduce water absorption to such a low level that it cannot keep up with the transpiration of conifers, whose more exposed foliage will then die. Wind can aggravate the situation by increasing heat loss from the soil and water loss from the leaves. Because the dead foliage appears as if it had been burned, such injury is called *winter burn* although it is caused by desiccation. The needles begin to turn brown at the tip and brown further backward depending on severity.

During cold winters, trees in raised individual planters or planting containers in elevated malls may be subject to temperatures that injure roots directly or reduce water absorption to injurious levels. Roots themselves do not harden more than a few degrees Celsius (Kramer and Kozlowski 1979). Injury is more likely in small and poorly insulated containers. Where winter cold is severe and prolonged, heating cables are sometimes put in raised planters to keep roots from freezing. The soil should be well-drained (see Chapter 10) and should be moist but not wet. The soil surface should be mulched to minimize temperature fluctuations (see Chapter 13). Anti-transpirant spray on foliage and bark can reduce desiccation (see Chapter 12), and sensitive plants or those in small planters can be moved to protected storage.

Attempts to reduce freezing injury to marginally hardy plants have generally involved breeding for adapted varieties, modifying the weather, providing mechanical protection, attempting to lengthen the dormant period, and hoping for the best. Although these attempts have been only partially successful, it is tempting to think that we may eventually find ways to manipulate plants so they can resist cold injury.

Lowest Temperature Expectable

In selecting and planting woody plants for a given area, one should know that they can withstand the coldest temperature that might be expected. Local horticulturists or weather bureau records will be able to supply information on expectable temperatures. General information for the United States can be obtained from the hardiness-zone map prepared by the U.S. Department of Agriculture (Fig. 4-2) (USDA 1960). State-by-state statistics are compiled by the National Climatic Center in Asheville, North Carolina, and are available in most large libraries. Information for most foreign countries is available from a number of sources (World Meteorological Organization 1971). Although a geographic region may have relatively consistent temperature regimes overall, meso- and microenvironments can vary considerably. Weather stations may be in the center of a city, where temperatures are often higher than in the surrounding countryside (Peterson 1969). Hardiness maps must be used with discretion, particularly with regard to boundary zones.

Temperatures several degrees above the winter minimums can be devastating if they occur in late fall before plants have reached their maximum hardiness; even deciduous plants that have already dropped their leaves. Every few years, certain plants may be killed or seriously injured by early cold weather. A warmer than normal period in late fall or winter will cause a plant to lose some of its ability to withstand low temperatures. Somewhat similarly, plants are more susceptible to injury if the temperature drops quickly rather than slowly and continuously, particularly if it continues to decrease for several days. Plant tissues adjust to decreasing temperatures by water moving out of the cells into the intercellular spaces and thus concentrating cell protoplast contents.

Conditions that favor early cessation of growth, such as low levels of nitrogen and water in late summer, favor early development of hardiness. A plant that is vigorous early in the season will usually have a higher food supply at the onset of the cold weather and will be better able to withstand cold. Older plants are usually more hardy than young plants of the same cultivar.

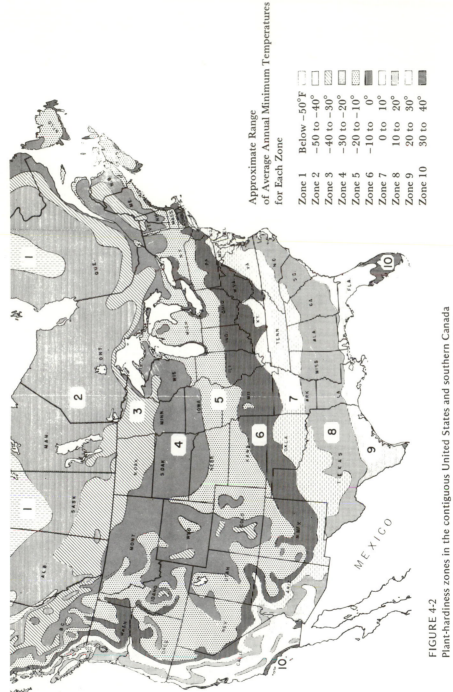

Approximate Range
of Average Annual Minimum Temperatures
for Each Zone

Zone 1   Below −50°F
Zone 2   −50 to −40°
Zone 3   −40 to −30°
Zone 4   −30 to −20°
Zone 5   −20 to −10°
Zone 6   −10 to   0°
Zone 7   0 to   10°
Zone 8   10 to   20°
Zone 9   20 to   30°
Zone 10   30 to   40°

FIGURE 4-2
Plant-hardiness zones in the contiguous United States and southern Canada
show the range of average annual minimum temperatures. (USDA 1960)

Hardiness cannot always be determined by noting which species reach maturity and survive. Temperatures in the area may not have reached their coldest possible levels since the plants were introduced, or plants may have regrown after being seriously damaged. Research horticulturists are testing the relative hardiness of the more important landscape plants so that more intelligent selections can be made by growers. This information will be of particular importance for large, long-lived trees.

Spring and Fall Cold

Temperatures at or below freezing can cause wilting or flagging, darkening, and the death of new shoots and leaves. Opening buds, blossoms, and young fruit can be killed by spring frost, and late-growing shoots and maturing fruit can be damaged in autumn. Needles of conifers may redden and drop if injured by cold. If the cold occurs when there is little air movement, the portions first and most seriously injured will be the lower parts of plants and parts exposed to the sky. In the spring and fall, even the older tissues are less hardy than they are in midwinter. Intelligent selection and care of plants depends on understanding the factors that contribute to low temperatures.

First, it is well to understand how heat is transferred. *Conduction* is the transfer of heat through contact of one molecule with another, as heat moves up the handle of a spoon in a hot cup of coffee. *Convection* is the transfer of heat by currents in gases and liquids; this may be caused by differences in density, as when heated air rises from a hot surface, or by an outside force, such as wind. *Radiation* is the transfer of radiant energy, such as heat from the sun or an open fire. Radiant energy passes through the atmosphere with a negligible direct effect on the air temperature. Conduction and convection influence air temperature more directly. Radiant energy heats (or its loss cools) objects, which in turn influence air temperature by conduction and convection. Although heat is usually transferred by all three pathways at the same time, one may predominate.

Local weather conditions (largely determined by the macroclimate), are modified by interaction with the meso- and microclimates. This interaction is particularly important to plant survival and well-being when the macroclimate approaches critical temperatures at either end of the growing season when shoots, flowers, and fruit are sensitive to cold. If a particularly intense cold front moves into an area, local conditions may not provide an adequate buffer to protect plants. This type of cold is called an *advective freeze*. Midcontinental areas are more susceptible to polar air masses than are western coastal areas. Freezes are usually associated with wind, low humidity, and no visible frost deposition. If temperatures drop rapidly, they can quickly cool exposed plants. In such situations, sites exposed to the afternoon sun and protected from the wind will be the warmest. In coastal areas where freezes are uncommon, any freezes that do occur can be particularly devastating: In December 1972, a cold front killed the branches and small trunks of hundreds of thousands of eucalyptus trees in the Oakland and Berkeley hills along the central California coast. The damage created a potential fire hazard of catastrophic proportions, and the removal of dead branches and trees cost millions of dollars.

In the spring, when plants begin to grow, flower, and set fruit, the radiant loss of heat accompanied by little or no air movement can depress near-critical temperatures to injurious levels close to the ground. Depending on the temperature and humidity, moisture condenses on exposed surfaces and may freeze, resulting in a visible *frost*. Since the exposed surfaces radiate more energy than they receive, they become colder than the air, and frost forms on them before air temperature reaches the freezing point of water. Plant parts exposed to the sky can, through radiation loss, become as much as $10°C$ ($18°F$) colder than the surrounding air. On the other hand, if the humidity is low, freezing temperatures due to radiation cooling can occur without a visible frost; this is called a *black frost*.

Some plants are susceptible to temperatures just above freezing; all but the youngest shoots of most temperate and subtropical plants can withstand temperatures of -1 to $-3°C$ ($26-30°F$). The amount of injury a plant sustains will depend on the extremity, duration, and speed of a temperature drop. The longer the exposure to low temperatures, the greater the damage and, usually, the more slowly the temperature drops, the lower the temperature a plant can withstand before injury occurs. A calm liquid, including cellular contents, can be *undercooled* (supercooled) 1 to $4°C$ ($2-8°F$) below the temperature at which freezing would be expected. If the temperature rises above the freezing point before freezing occurs, the tissue will not be injured. If undercooled tissue is jarred, however, or an ice crystal is introduced, the tissue freezes immediately.

Radiation Frost Conditions A number of factors determine whether a radiation frost will occur. The temperature at the start of the evening must be in the critical range (near $5°C$ or $40°F$) so that the air can be cooled to or below the freezing point. Almost without exception, the sky must be clear and the air calm during the night. Under clear skies, surfaces that were warmed by the sun during the day radiate more energy to the cold surroundings than they receive. Air in contact with the cold surfaces is cooled primarily by conduction from air molecule to air molecule. Since air is a poor conductor of heat, the process is slow and air near the ground can become several degrees colder than the air a short distance above. Clouds at night, however, absorb and reradiate heat back to the earth and thus reduce net heat loss (Fig. 4-3).

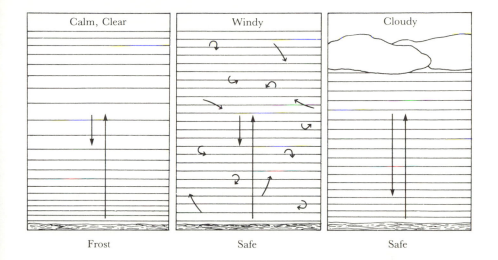

Calm, Clear	Windy	Cloudy
Frost	Safe	Safe

FIGURE 4-3

Radiation conditions before sunrise can result in frost if temperatures are in the critical range and conditions are clear and calm (left). Conductive cooling from the soil or other surfaces may create a temperature inversion, with colder air (horizontal lines close together) near the ground and the warmer air (lines further apart) higher (left). A breeze of 10 km/hr (6 mph) or more mixes warmer air above with the colder air below to produce an average temperature that is warmer than if there were no wind (center). Clouds reradiate most of the heat received and thereby slow the cooling process (right). The straight vertical arrows represent the radiation received and lost by the soil and other surfaces.

Above level ground during the night, the air will be cooled to ever higher elevations and, as it cools, become heavier and tend to settle. The air will be coldest near the ground and become progressively warmer with height until, at a certain elevation, the air has not been cooled by the air below. The air cools in place and does not involve warm air rising from near the ground or cooler air settling. Air temperatures can be affected to heights of 60 to 180 m (200-600 ft) (Chang 1968). This increase in temperature with increasing elevation contrasts to the normal daytime situation, in which the air temperature decreases 0.6°C with every 100 m (1°F/300 ft) of altitude, and is therefore called a *temperature inversion* (Fig. 4-3).

Under radiation conditions, the air will be coldest and the height to which it is cooled the greatest just before sunrise, since radiation loss and conductive cooling continue all night. A temperature inversion is considered "large" or "strong" if there is a marked increase in temperature with height, if the air, for example, is 6°C (10°F) warmer at 12 to 15 m (40-50 ft) than at ground level. If the change is less than 3°C (5°F), the inversion is considered "weak" (Gardner, Bradford, and

Hooker 1939). A strong inversion is most likely to occur when the air is humid, surfaces are cold at the beginning of the night, and the air is still and clear. Wind machines to mix the warmer air above with the cold air near the ground and heaters are more effective under strong than weak inversion conditions. Calm conditions allow the coldest air to remain near the surface and reach the critical temperature more quickly (Fig. 4-3).

Just as clouds or wind at night reduce the likelihood of a radiation frost, cloud cover or wind during the day reduces radiation input that would warm the air, plants, the soil, and other heat reservoirs. After such a day, the night begins at a lower temperature, with less heat available to slow cooling should the skies clear and the wind die during the night. A frost is more likely the next morning than if it had been clear and calm during the day.

Other factors also influence air temperature under radiation cooling conditions. Dry air favors more rapid temperature changes than does moist air, since water vapor is the largest absorber of long-wave radiation. Even with no visible fog or clouds, high moisture content in the air will somewhat reduce radiation loss from the earth's surface. More important, as the moisture is cooled, it must change from a vapor to a liquid and then to a solid. Relatively large amounts of heat are liberated when vapor turns to liquid, and even more when water freezes. The amount of moisture in the air is commonly expressed in terms of a *dew point*, the temperature at which the air is saturated with water vapor (100% relative humidity). Vapor pressure, or humidity, can be calculated from the dew point and the temperature. If the air temperature drops below the dew point, moisture in the air condenses as water droplets (fog or rain). The higher the dew point, the more moisture there is in the air and the more slowly the temperature will drop, other things being equal.

Dew will condense on many surfaces because they cool to the dew point sooner than does the air. If the dew point is above freezing, the moisture that has condensed on surfaces will change to ice crystals as these surfaces reach the freezing point and will form typical frost. The higher the dew point, the more moisture will condense and the thicker the frost will be at a given temperature.

The heat capacity and conductivity of the material exposed to radiation cooling greatly influence surface temperature and, in turn, the degree to which the air is cooled. Soil is commonly exposed to sky, and its surface condition can have an important influence on air temperatures. A soil that is firm and moist will absorb more energy from the sun during the day and reradiate more during the night than will a soil that is drier or less firm. While there is cooling due to evaporation from moist soil surfaces (similar to the action of an evaporative cooler), it is more than offset by the heat capacity and conductivity of moist soil. Compact soil is a more efficient conductor of heat than loose soil. Therefore, the surface of a moist, firm soil will cool less rapidly than will one that is drier and looser: As heat is radiated from the surface of a moist soil, it is replaced with heat that has been absorbed in lower depths during the day.

A soil cover such as an organic mulch, a wooden deck, turf, weeds, or ground cover has poor heat conductivity and low heat capacity. A mulch surface absorbs energy but because of poor conduction reemits much of it and stores less heat during the day than does bare soil.

Weeds are often plowed under in the late winter so that the soil will be exposed and can firm during later rains. Sparse grassy weeds of 25 mm (1 in) or less in height, however, may minimize air movement and yet allow the soil to absorb and radiate heat. This can more effectively reduce the likelihood of a frost than will a more complete weed cover or recent cultivation, both of which have an insulating influence on the soil. On the other hand, lower temperatures will occur over freshly cultivated soil and may kill new shoots while warmer temperatures over uncultivated soil result in no injury. This combination of circumstances has been recorded in a number of instances.

Asphalt and concrete have higher heat capacities than do organic mulches or turf and will more readily radiate accumulated heat and conduct heat from the soil. Frost damage to plants is therefore less likely over soil with a concrete or asphalt covering than over soil covered with turf or organic mulch or over a concrete or asphalt structure without soil in contact with it, such as a bridge.

Frost Pocket   Cold air is denser than warm air; it flows down a slope and accumulates in the lowest areas, displacing the warmer air upward. When some of the warmer air replaces the cold air that has moved down (Fig. 4-4), it creates a warm zone up the slope. In many areas where spring frosts and winter killing are problems, susceptible fruits and vegetables are grown on hillsides and tops, even though the soil there may be less desirable than in the valleys.

FIGURE 4-4
A frost pocket can result when cold (heavier) air flows down slopes and forces warmer air to rise. An obstruction on the slope can form a localized cold pocket.

Warmer

Colder

A row of closely planted trees or shrubs, a solid fence, or a long building at right angles to the slope can act as a barrier to the flow of cool air and can create localized cold pockets (Fig. 4-4). If these will constitute hazards at a particular site, try not to design them into the landscape. If they are already present, plants should be chosen with this phenomenon in mind.

Protecting against Frosts

A number of measures can reduce the likelihood of frost injury to plants. Under radiation frost conditions, only 1 to 2°C (2-4°F) can make a substantial difference to the health and survival of plants.

Select a Site to Avoid the Cold The modifying influences of large bodies of water are well-known and can be used to advantage in reducing cold hazard. The closer the body of water, the greater will be the protection it affords. It is best to plant on the lee side, so that prevailing winds come from over the water (Fig. 4-5). In winter, air that is colder than the water or ice will be warmed as it passes over. In spring, warm air passing over the water is cooled, thereby delaying the time when the plants will lose their hardiness and begin to grow. Should a late cold spell develop, the water will again warm the passing air and reduce the likelihood of critical temperatures. Seldom do horticulturists have the opportunity to choose such major landscape features, but these conditions should be kept in mind. They also explain why most of the important northern fruit-growing areas of the United States are adjacent to the Great Lakes.

As has been discussed, planting sites can be chosen to avoid frost pockets. Within a given landscape, it may even be possible to select specific sites that provide more protection and warmth than the area usually affords. For example, a concrete or rock wall that faces the afternoon sun will reduce radiation loss at night and will reradiate heat stored during the day. To *espalier* is to train plants against a wall, and the circumstances just described constitute one of the advantages of this practice (see Chapter 14).

FIGURE 4-5
A large body of water moderates air temperatures, particularly on the leeward (downwind) side of the water.

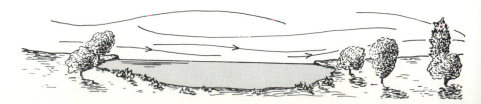

Injury from low temperatures can be further minimized if one selects varieties or species that are hardy in winter cold, that begin growth or bloom late in spring, and that cease growth and mature their fruit before fall cold. When fruit matures during the time that frost may be a danger, the top and outside fruit, which is exposed to the sky and is usually riper, should be picked first. The inside fruit, protected by leaves, can be harvested later with greater safety. Plants sensitive to fall cold should be watered and fertilized sparingly so that growth will end early, enhancing the maturation of shoots and fruit. When the cold period begins, however, the soil should be moist to ensure adequate water to the plant and to improve heat absorption, transmission, and release.

Reduce Heat Loss   Plants can be protected if they are partially or completely covered to reduce the loss of long-wave radiation (heat). The larger the soil area that can be encompassed, the larger will be the heat reservoir for the plant. Canvas, plastic, wood, or evergreen boughs can provide effective covers (Fig. 4-6), as can the sunny side of a roof overhang, a wall, a solid fence, or a large evergreen tree. In extreme situations, low-growing plants and the bases of others can be protected by covering them with mulch or soil.

By reflecting and reradiating heat back toward the earth, clouds almost invariably prevent a frost. Some people think that a smoke or haze cover will work the same way. It is true that smoke can reduce incoming (short-wave) radiation from the sun, but it is fairly ineffective in preventing (long-wave) reradiation. Ten meters (30 ft) of smoke will

FIGURE 4-6
Covering a plant for frost protection.
Extending the polar side of the cover to the ground and leaving the equator side
partially open to the sun allows for the warming of soil and lower tree
during the day (left) and conservation of heat at night (center).
If the cover is tied to the trunk, night temperatures within the cover
will be colder than without the cover (right).

reduce transmission of infrared radiation less than 20 percent, but the same depth of fog will reduce transmission more than 90 percent (Mee and Bartholic 1979). Orchard heaters (smudge pots) are effective primarily because they radiate heat to the area not because their smoke significantly reduces radiation loss. In fact, to increase efficiency and reduce pollution, present-day heaters must emit only minimal amounts of smoke.

Utilize Heat from the Immediate Surroundings A cover can reduce radiation loss from plants as well as from soil and other surfaces (Fig. 4-6). In order to build up as much heat as possible during the day, plants, soil, and their surroundings must be exposed to the sun. Covers are most effective if they can easily be removed from plants or at least from the soil. If a cover must be left on during the day, an opening near the top will minimize warming of the plant and consequent loss of cold hardiness. A plant cover that is gathered at the trunk often results in more cold damage than no cover at all because it excludes heat from the soil (Fig. 4-6).

As pointed out earlier, a firm, bare, moist soil absorbs more heat and conducts it more rapidly than soil that is loose, dry, or covered with mulch or vegetation. Walls and benches exposed to the sun during the day are good sources of heat at night.

Air near the ground can be warmed under radiation-frost conditions by mixing the warmer air from above with the cold surface air. The mixing is done naturally by a breeze—5 km (3 mi) per hour or more—or mechanically by wind machines. In some commercial fruit and vegetable areas, helicopters are used to mix the air. The effectiveness of wind machines depends upon the strength of the inversion. The stronger an inversion, the greater the protection, if the mixed air is above freezing.

Add Heat Shrubs and small trees close to an electric source can be warmed by a 100-watt bulb hung inside the plant canopy. In some landscape situations, special orchard heaters may reduce or prevent frost injury. These heaters heat primarily by radiation from their hot stacks, not by warming the air, and a number of small heat sources are more effective than a few large ones. A large fire may be counterproductive, since it can create considerable updraft, taking the heat up above the plants and drawing in cold air from surrounding areas (Kepner 1950).

Flood or sprinkler irrigation can help to protect plants because of the heat capacity of water. The real warming potential of water, however, is realized when ice is formed: For example, only 1 calorie of heat is given off by 1 gm of water cooled 1°C, but 80 calories of heat are released when the same amount of water is changed from liquid to ice.

Since the freezing temperature of most plant tissues is several degrees below 0°C (32°F), plants can be encased in ice without being frozen (Schultz and Lider 1968). If cold persists for several days, additional water must be added, which will cause a problem in all but the best-drained soils. For plants that can withstand the weight of ice, sprinkler irrigation may provide greater protection because of the more rapid cooling and freezing of the water so distributed. Sprinkling must be continued, however, until the air temperature is above the critical range.

These frost-protecting techniques are more effective when there is a strong inversion, that is, when temperature increases markedly with elevation above the ground. Under this condition, also referred to as a low ceiling, both heating and wind machines are most effective. In a weak inversion (high ceiling), temperature increases little with elevation and heated air will rise higher before reaching equilibrium, so more heat is dissipated above the plants (Fig. 4-3).

Inhibit Ice Formation   When certain species of bacteria (*Pseudomonas syringae*, *P. fluoroscens*, and *Erwinia herbicolia*) commonly on plant surfaces are reduced in number or inactivated, the plant can supercool, that is, cell contents remain in a liquid state at temperatures as cold as -9°C (15°F), to avoid injury by temperatures that would normally freeze the tissue (Lindow 1980). The bacteria, up to 100 million per gram of leaf, are nuclei for ice crystal formation. Ice nucleation-active bacteria (called INA bacteria) can be greatly reduced by applying copper-containing compounds, Streptomycin, or certain other bactericides. Significant frost control has been achieved in experimental applications of bactericides on several vegetable and fruit crops. Antagonistic bacteria are another way to reduce the INA bacteria population but their action may not be rapid enough to provide protection. A third possible control is to inactivate the bacteria with chemical inhibitors that cause various physical and chemical stresses. These chemicals almost immediately inactivate the nucleus produced by the INA bacteria without necessarily killing the bacteria. Although inhibiting ice nucleation is still experimental, it holds considerable promise in protecting plants from frost.

Winter Chilling and Rest

Vegetative and flower buds may be prevented from growing for a number of reasons, depending on the season of the year. While a shoot is actively growing, lateral buds will be inhibited by the strong apical dominance of the terminal. As the days begin to shorten, the buds of many temperate-zone plants begin to enter a condition of rest and will not grow even though all other conditions are favorable. Even after rest is no longer a factor, buds may still not begin growth if temperatures are below 10°C (50°F). To distinguish between these last two dormant conditions, we use the terms *rest* and *quiescence*. Dormant-quiescent plants will begin to grow under favorable environmental conditions, whereas dormant-resting plants will not.

A condition of rest develops within each bud or seed of certain temperate-zone plants. In most woody plants, dormant buds not only inhibit shoot elongation but slow the activity of other growth centers, roots, and cambium. Rest develops gradually during the summer as the days become shorter, intensifies in the fall, and reaches a maximum in early winter. By late winter, rest is overcome in most temperate regions.

To overcome rest, buds must be subjected for four to eight weeks to low temperatures, -4 to 10°C (24–50°F). As might be expected, species and subspecies differ not only in the temperature but also in the length of time this is necessary to overcome rest. Extremely low temperatures are little or no more effective than the threshold temperature in overcoming rest. The amount of necessary chilling is expressed as the number of hours at or below the critical temperature for a given species. The air-temperature threshold for most fruit tree species is 7°C (45°F). It is the bud temperature, however, that is important; a bud temperature of 10°C (50°F) would probably be effective in overcoming rest. Within fruit species, apples tend to require a longer chilling than do peaches, and peaches a longer chilling than almonds. Even so, peach varieties may require anywhere from 350 to 1200 hours at or below 7°C (45°F). Varieties with shorter chilling requirements have been bred to grow satisfactorily in the southern United States and other mild-winter areas.

After a certain amount of chilling, the overcoming of rest can be hastened by pruning close to a bud, massaging individual buds, pricking buds with a pin, or treating the plant with ethylene. These treatments, however, cause injury to the plant. The only practical treatment for overcoming rest is used by some pear growers, who add a dinitrophenol compound to their dormant oil sprays.

In years when the chilling period is brief, fog, cloudy weather, and wind are important in keeping buds at or near air temperature. On sunny days, buds may be warmed several degrees above air temperature. The central valley of California would not be the important fruit-growing area it is if it were not for the winter fog.

Following winters with inadequate chilling, trees are slow to leaf out and bloom is prolonged. Leaf buds are affected more seriously than flower buds. In extreme cases, so few buds grow that branches are sunburned and the tree weakened from lack of food production. Since flower buds are less affected, a weak tree may bloom and set fruit, which further weakens the tree. Trees so affected are said to be suffering from *delayed foliation*.

The setting of fruit may be enhanced by a longer bloom period. Prolonged bloom, however, increases the danger of disease, particularly on plants like the pear, pyracantha, and hawthorn. If a certain amount of chilling does not occur in the fall, the flower buds of certain species, such as apple, apricot, and pear, may die before opening the following spring. The chilling requirements of deciduous fruit trees have been extensively studied (Chandler and Brown 1951). A number of fruit-growing areas experience winters that are not cold enough for the profitable production of certain species.

The dormant-resting condition is the key to winter survival for many plants in the temperate region. Growth is prevented if periods of warm weather begin to induce growth that will be killed by later cold weather. It has been thought that, since the rest period prevents buds from starting to grow during a warm period, rest would also prevent the loss of hardiness during unseasonably warm weather in midwinter. Unfortunately, that may not be the case. Raspberry canes in a state of rest have been found to lose their hardiness under high temperatures more rapidly than equally hardy canes that were quiescent (Weiser 1970). Whether this applies to other plants is not known.

Root dormancy or rest presents a confusing picture with regard to different species of woody plants. The roots of some pines in Northern California have been observed to grow throughout the year. The roots of other species seem to grow whenever food reserves and soil conditions are favorable. The roots of still other species exhibit considerable periodicity even during the growing season. Japanese yew roots were recently observed to grow four times as much after nine weeks of root chilling as roots not exposed to such cold (Lathrop and Mecklenberg 1971).

Root activity is difficult to study since it depends on the health and vigor of the top as well as the temperature, aeration, moisture, compaction, and toxicity of soil. Evidence to date suggests that the roots of most woody plants do not have a dormant-resting condition independent of the tops. A number of species must have developing buds before roots will begin growth and must have active leaves if root elongation is to continue (Richardson 1958). In addition to these hormonal controls, root growth is dependent on carbohydrates, either stored or manufactured in the top.

Other conditions being favorable, the roots of most plants will grow at lower temperatures than will the shoots. The roots of silver maple will start growth at 4°C (40°F), but leaf buds do not expand until the temperature is at least 10°C (50°F) for 20 days. This can be an important factor in determining the time to transplant trees and shrubs.

It is not one particular temperature at one particular time that makes an ideal climate for a species. Many plants have different optimal temperature requirements at certain stages of growth. Almost all plants grow, flower, and fruit more abundantly when day temperatures are higher than night temperatures, yet night temperature is more commonly the controlling influence during the growing season, if water is not a limiting factor (Kimball and Brooks 1959). This is particularly evident during bloom, when fruit set may be adversely affected by either low or high night temperatures.

Citrus and some deciduous fruits develop richer color and better flavor in areas where night temperatures are considerably cooler than day temperatures. The intensity of leaf color on many deciduous trees is similarly affected. Some perennials require periods of different night temperatures for different purposes. For example, camellia flower bud formation requires a period of 15 to 18°C (60–65°F) at night followed by 10°C (50°F) for best bloom. Peach tree growth and fruit bud formation occur during the summer at night temperatures of 18°C (60°F) and above, but rest is broken in the winter by periods of 7°C (45°F) or lower. In this latter case, only the night and early morning periods may be cold enough to break the rest.

The different optimal temperatures for different plant processes support the use of landscape plants in climates that differ from those to which the plants are native. Experiments under controlled conditions and observations in the field support the following statements: Plants will make satisfactory vegetative growth over a wider range of temperatures than is satisfactory for flower bud formation and fruit setting (Fig. 4-7); the temperature range for fruit set is usually wider than that for producing viable seed. This would imply that many plants can be used for landscape purposes where vegetative growth is the prime consideration in areas where they may not be able to reproduce themselves.

For example, a number of so-called "fruitless" olive trees have been discovered over the years and have been propagated vegetatively. The lack of fruit actually enhanced their value for landscape purposes. Until recently, when these trees were planted in areas where olives were grown commercially, the trees set fruit, although less abundantly than commercial varieties. In localities where the original fruitless trees were found, conditions during flower bud formation or bloom were such that no fruit formed, even though the tree was able to make satisfactory vegetative growth. In 1961, Hartmann (1967) introduced the *Swan Hill* olive tree. This is apparently a true fruitless variety, though it does flower profusely, much to the discomfort of hayfever sufferers.

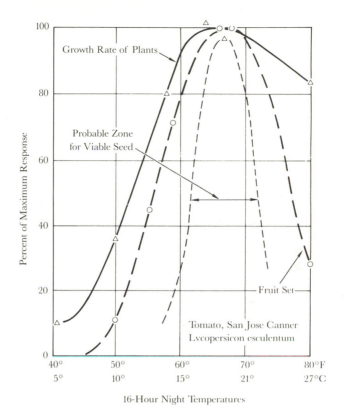

FIGURE 4-7
Response to tomato growing under controlled conditions demonstrates that for viable seed to be produced, the night temperature range is more limited than for satisfactory plant growth. (Went 1944, as adapted by Kimball and Brooks 1959) Most other plants probably respond similarly to temperature and other environmental constraints on growth and flowering.

Increasing knowledge about diurnal and seasonal temperature influences on plants should be an invaluable tool in determining favorable environments for landscape plants. Even now, much worthwhile information is summarized on a county or regional basis and is available from local weather bureau offices or agricultural agencies.

High Temperatures

All plants will be killed or seriously injured above a certain maximum temperature, which varies with species, stage of development, and previous environmental history. What may be the maximum temperature for one plant may be the optimum for another.

High-temperature injury is primarily a matter of desiccation, when transpiration exceeds moisture absorption. Water content in the leaves is lowered, leaves wilt, transpiration is reduced, leaf temperatures are increased by reduced evaporative cooling, and the cycle can intensify to kill a portion or all of the leaf. Under cloudless skies, the leaves of well-watered plants in low vapor pressure (low humidity) and high net radiation (high temperatures) will be cooler than the air, while leaves in low net radiation and higher vapor pressure will be warmer than air. Leaf temperature will usually reach a temperature equilibrium below the lethal level, but it may rise to the lethal level if the leaf comes under water stress.

Small, thick leaves with few stomates are usually more tolerant of high temperatures than are large, thin leaves. Within the same species or clone, a plant grown under cool conditions with low light intensity (irradiance) will develop larger, thinner leaves that can be quite susceptible to heat. A cool, cloudy spring may cause newly planted plants to develop large, thin leaves. Should the weather change abruptly to hot, dry, and slightly breezy conditions, transpiration may exceed the ability of a plant to absorb and transport water from the soil to the leaves. Wilting and possible injury to the exposed leaves will result. Continued high temperatures, particularly at night, can increase the rate of respiration to nearly that of photosynthesis. High night temperatures have been found to slow the translocation of carbohydrates from the leaves of many plants. If such conditions persist for several weeks, a plant may do quite poorly.

Roots are more sensitive to temperature extremes than are the tops of plants. Wong, Harris, and Fissell (1971) found that young roots in a number of plants were killed by a single 4-hour exposure to temperatures between 40 and 45°C (104–113°F). Temperatures in the surface 100 to 150 mm (4–6 in) of soil exposed to the sun can become lethal to roots. The soil 25 mm (1 in) in from the side of a 4-liter (1-gal) container exposed to the afternoon sun will reach 43 to 46°C (110–115°F). This can occur any sunny day in spring, summer, or fall. Plants so exposed may have no roots in the west third of the container (Harris 1967) (Fig. 4-8).

Summer Branch Drop

Tree limbs up to 0.6 m (24 in) and trunks up to 1.2 m (48 in) in diameter have occasionally broken and fallen during hot, calm summer afternoons and subsequent evenings (Kellogg 1882, Harris 1972). This phenomenon, known as *summer branch drop* in England (Rushforth 1979), has been reported in Australia, England, South Africa, and California. I know of only one such instance in the central and eastern United States. Kellogg reported in California in 1882:

FIGURE 4-8
Roots of European white birch were killed
or prevented from growing on the side of
the rootball whose container was in the
outside row of a nursery block exposed
to the afternoon sun.

> Sometimes, most unaccountably, it [valley oak] is said to burst with a loud
> explosion, and strong limbs that had hitherto withstood centuries of
> storms, in the calm airs of late summer and early autumn crash unexpect-
> edly down, the fracture disclosing not the least cause of weakness. I be-
> lieve this happens in the hotter valleys and exposures, and is never known
> of any evergreen oak at any season.

As described, most branches fall with little or no warning. Branches
that drop are at least 100 mm (4 in) in diameter and usually extend to
or beyond the edge of the crown of the tree. Such large limbs can cause
serious injury and property damage. The break most often occurs some
distance from the branch attachment, where the wood is frequently
sound. Even limbs with internal decay or brash (weak) wood exhibit
no outward appearance of a hazardous condition.

In England, summer branch drop has been reported in oak, sweet
chestnut, beech, ash, poplar, willow, and horse chestnut (Rushforth
1979). Harris (1972) has reported similar limb breakage in several
species of elm, eucalyptus, deciduous oak, and pine, and in plane tree,
deodar cedar, Japanese pagoda, olive, and Indian laurel fig. Young and
vigorous mature trees of susceptible species appear to be less prone
to summer branch drop than are over-mature and senescent trees, which
may shed branches repeatedly (Rushforth).

The phenomenon appears to be caused by some type of water stress. In England it seems to be associated with calm weather conditions that follow heavy rain after a period of increasing soil dryness. Similar conditions are known to cause cracks in conifer stems (Rushforth). In California, where summers are dry on the other hand, limbs have fallen from native and planted trees in both irrigated and nonirrigated landscapes (Harris).

No explanation is readily apparent. Treetrunks and branches decrease in diameter during the afternoons (Kozlowski and Winget 1964). Horizontal limbs have been observed to rise during the afternoons and return to their original positions during the night. Both observations indicate that the limbs weigh less in the afternoons because more water is being transpired than is received. In some cases of limb failure in oak, moisture has been reported to exude from both sides of the break. This appears to contradict the above observations but is probably due to wetwood (see Chapter 19). Until the phenomenon is better understood, it may be wise to shorten or reduce the weight of long, horizontal branches.

LIGHT (RADIATION)

Plants are affected by quality (wavelength or color), intensity (irradiance), and duration of light, whether the light comes from the sun or from the artificial sources that are commonly used for interior landscaping (see Chapter 10) and night lighting.

Visible light, that portion of the electromagnetic spectrum to which the human eye is sensitive, extends from roughly 380 to 760 nanometers (nm); the wavelengths that stimulate visual sensation most effectively are between about 500 and 600 nm (green to orange). The amount of illumination incident on a surface (illuminance) is expressed as lux (lumens per m^2) or foot-candles (ft-c), its English counterpart; 10.8 lux equals 1 ft-c. Plants, by contrast, respond to radiant energy instead of luminous energy. Effective wavelengths vary for the different photobiological processes in plants:

For photosynthesis—blue (400–450 nm) and red (625–700 nm)

For photoperiodic responses—red (625–700 nm) and far red (700–850)

For phototropic movements—blue (400–450 nm)

Illuminance and irradiance measure the amount of energy incident on a given surface and not the intensity of the energy source. Irradiance that is photosynthetically active, called PAR (radiation between 400 and 700 nm), is most accurately measured and expressed in microeinsteins per m^2 per second ($\mu E/m^2/sec$). (Irradiance, when expressed as $erg/sec/cm^2$, $watt/m^2$, and $calorie/min/cm^2$, typically includes more of the electromagnetic spectrum than PAR.) Irradiance, however, is commonly expressed in lux or ft-c, which are really measures of illuminance. Measurements of these two energies could be converted from one to the other easily and accurately if we were concerned with only one or two very similar energy sources, such as sunlight or incandescent lamplight. Depending on the light source, however, the relative amounts of illuminance and irradiance can differ. Different lights have different emission spectra: One lamp gives off more light of certain wavelengths (color) than another lamp does; fluorescent light is high in blue and some red, and incandescent light is high in red and far red. Also, visual sensation is at a maximum at 550 nm of illuminance energy, whereas PAR per unit of radiant energy varies inversely with wavelength. For example, for a single wave cycle (1 photon), irradiance at 400 nm (blue) has 75 percent greater radiant energy than at 700 nm (red). Therefore, light meters are not particularly suitable for comparing irradiance from different light sources. Radiometers can measure irradiance, but they are expensive.

Even though illuminance does not accurately characterize irradiance, it has been and will continue to be used for this purpose. Therefore, measurements of illuminance or irradiance and recommendations for minimum light intensity (given in lux or ft-c) for growing plants under artificial light should include full descriptions of lamps and their irradiance characteristics.

Photosynthesis

Irradiance (intensity) during the sunny days of the growing season may be five to 10 times that needed for maximum photosynthesis of an individual leaf (Table 4-1). Within the canopy of the outer foliage of a plant, however, irradiance falls off rather quickly. Depending on branching habit, leaf density, and leaf size, irradiance levels 250 to 500 mm (10–20 in) within the canopy surface may be so low that leaves are not able to persist; this is particularly true with sheared hedges (see Fig. 14-47). Many plants have a dense shell of foliage and a center devoid of leaves. Such plants often present just dense masses of green with little character or interest. It may be difficult to grow plants under trees that have a dense canopy of foliage. Training and pruning practices can either intensify or help to overcome this situation. Irradiances below the optimum will cause many plants to produce branches with long, spindly internodes, few laterals, and large, thin, pale-colored leaves.

TABLE 4-1

A. Illuminance (irradiance) for various types of plants, individual leaves, and plant responses

	Irradiance	Illuminance	
	Microeinsteins[a] per Meter^2 per Second ($\mu E/m^2/sec$)	Lux	Footcandles (ft-c)
Adequate for most plants	1000	54,000	5000
Individual leaf	240	13,000	1200
Shade plants	200	10,800	1000
Indoor plants	10–100	540–5400	50–500
Photoperiodic induction	0.06–3	3.2–160	0.3–15
Indoors (sufficient for reading)	—	200–300	20–30

B. Illuminance (irradiance) estimates for various situations

		Irradiance	Illuminance	
		$\mu E/m^2/sec$	Lux	ft-c
Full sun[b]	Summer	2400	130,000	12,000
	Winter	800	43,000	4000
Cloudy[c]	Summer	1000	54,000	5000
	Winter	200	10,800	1000
Shady side of two-story building (sunny)	Summer	400	21,600	2000
	Winter	100	5400	500
Moonlight		—	0.22	0.02
Incandescent light[d] (100 watt at 1.5 m or 5 ft)		5	240	22
Fluorescent-Deluxe cool white (150 watt at 1.5 m or 5 ft)		17	1050	100

[a]For sunlight 1 $\mu E/m^2/sec$ = 54 lux = 5 ft-c

1 lux = 0.093 ft-c

1 ft-c = 10.8 lux

[b]Sunlight for noon on a typical summer and winter day in Davis, California (38.5° N).

[c]Due to cloud density, irradiance could vary 50 to 75 percent.

[d]Bulb with 250-mm (10-in) diameter reflector.

Photoperiod

Although an individual leaf may require at least 20,000 lux (1850 ft-c) for maximum photosynthesis, photoperiod responses may be induced with less than 160 lux (15 ft-c) (Table 4-1). In plants sensitive to photoperiod, day length usually controls vegetative and reproductive activity. In some plants, photoperiod can also influence leaf shape and abscission, pigment formation, pubescence, root development, and onset of dormancy. As mentioned previously, the irradiance required for photoperiod response is only a fraction of that required for photosynthesis. In almost all sensitive woody species, long days and short nights promote vegetative growth. The shoot elongation of many woody plants native to temperate regions will stop when days are short. Shoot elongation of plants in the higher latitudes may be inhibited when the photoperiod is less than 18 hours, while others are not affected until there are less than 10 hours of light per day. As mentioned in Chapter 2, when plants are grown at lower latitudes than those to which they are native, they may experience a shortened growing season and will therefore be smaller than normal. In general, day length can be an important factor influencing plant size and form.

Experiments have shown that the length of the dark period is more important in inducing photoperiod responses than the length of the light period. When plants that are vegetative during long days are grown during short days, many can be kept growing if they are exposed to half an hour of light in the middle of the night. Nurserymen in warm areas have found that they can encourage vigorous growth in many woody plants by extending day length with incandescent lights. If plants are to respond to increased day length, however, all growing conditions must be favorable, including temperatures above 10°C (50°F). For many plants, growth will be most vigorous when the temperatures prevailing during the extra light period are 21 to 27°C (70-80°F).

Night Lighting

Many people have wondered about the influence of street lights and other night lighting on the growth of nearby plants. As should be apparent from the earlier discussion, such low-intensity lighting will have little or no influence on photosynthesis but might affect plants that are sensitive to day length. Light quality, wavelength, is an important factor in plant growth. (A nanometer is a measure of the wavelength of light, as are angstroms and microns. 1 nanometer = 10 angstroms = 0.001 micron = 0.0000001 millimeter.) Phytochrome, a photo-reversible blue pigment, by means of its red (625-700 nm) and far red (700-850 nm) absorbing forms, is involved in the regulation of plant responses. A delay in dormancy induced by long days may result in continued elongation of vegetative shoots. Far red light (700-850 nm), which promotes shoot elongation, may delay dormancy if present in supplementary light. Red light (625-700 nm) promotes flowering.

Cathey and Campbell (1975a, 1975b) found that five light sources used for 16 hours at a level of 11 lux (1 ft-c) at a night temperature of 20°C (68°F) promoted vegetative growth (delayed dormancy) of woody plants. These researchers used two species each of elm and maple and one each of *Koelreuteria*, *Rhododendron*, *Rhus*, and Japanese zelkova. From most to least effective, the light sources used were: incandescent (INC), > high pressure sodium (HPS), >> metal halide, = cool white fluorescent, and >> clear mercury. The degree of delay in dormancy was dependent on the quality of light emitted by the light source. Incandescent sources, emitting the greatest amount of far red light, promoted elongation the most. At intensities higher than 11 lux (1 ft-c) or with different species, the relative effectiveness of light sources can change.

Poinsettia, *Betula*, *Catalpa*, *Platanus*, and *Tilia* were the most sensitive to light and continued to grow vegetatively in response to all sources. Holly and two species of pine did not respond to any of the light sources at the 11-lux level, but did respond if the intensity of the light was increased to 200 lux; these species are said to have low sensitivity to extended photoperiod. HPS lighting had to be increased at least 4- to 8-fold (on a lux basis) to be as effective as incandescent lamps in regulating vegetative growth. In most landscape and street situations, HPS, incandescent, and (to a lesser extent) metal halide lamps may have red and far red intensities high enough to prolong the vegetative growth of sensitive plants.

Plants sensitive to day length may thus have their growth prolonged in the fall of the year and their first stage of acclimation for the cold winter ahead delayed. Such plants, particularly their young growth, will be more sensitive than usual to early freezing temperatures. In Maryland, European white birch whose tops are exposed to incandescent floodlights have been observed to be killed back in the top almost every year. Young plants, because of greater vigor, will be more subject than older plants to cold injury as a result of prolonged growth from illumination.

Continuous lighting depresses the formation and maintenance of chlorophyll in leaves and promotes lengthening of shoot internodes and leaf expansion. The foliage of plants grown in continuous lighting may be more susceptible to air pollution and water stress during the growing season. Cathey and Campbell (1975a, 1975b) suggest using HPS where high visibility is desired and only light-tolerant plants in the vicinity. Although less efficient than HPS lamps, metal halide lamps may be preferred in malls, parks, and residential areas with fairly dense plantings when color rendering of plants, people, and buildings is desired.

In contrast, Andresen (1977) surveyed 200 trees along streets in Chicago, Cleveland, and Milwaukee and their response to high-pressure sodium vapor (HPS) outdoor lighting. He also contacted officials in 16 other cities where HPS lamps have illuminated trees for several years and states, "All indicators, to date, suggest that HPS has no detrimental influence on trees grown in cities of the eastern half of the United States in general and our mid-west in particular." Andresen concludes that it is safe to install HPS lamps up to 1000 watts in the presence of transplanted or mature trees. In view of the temperature requirement (at least 10°C) for photoperiod response, it would be interesting to examine the effects of HPS lighting at lower latitudes.

If necessary, outdoor lights can be shielded to direct the light away from plants. To avoid cold injury resulting from prolonged growth, light-sensitive container plants that are still growing in late summer should not be planted until growth stops.

Transpiration

Day length can also influence the amount of water transpired by plants, particularly evergreens. At Davis, California, the average evapotranspiration for the month of June is 212 mm (8.3 in), which is almost 30 percent more than the August rate (164 mm). Even though May in the northern hemisphere can be relatively cool, day length may cause evapotranspiration to be almost as high then as it is in August. Where needed, irrigation should be more frequent than is generally appreciated for turf and plants in leaf in late spring.

MOISTURE

Rain

Rainfall amount and seasonal distribution are important even in irrigated landscapes. The more the weather is dominated by continental air masses, the more likely precipitation will be at any season of the year. In contrast, the Mediterranean-type climates of many western continental areas in the temperate zone have rainfall limited almost exclusively to late fall, winter, and early spring. Under such conditions the moisture reservoirs of the soil become quite important to plants during the growing season.

Weather records can be central to determining the monthly rainfall and evapotranspiration (moisture lost by evaporation and transpiration) for an area. By estimating the amount of moisture in the soil, seasonal rainfall, the average date of last rainfall, and evapotranspiration, you can estimate the minimum irrigation needed to grow plants in a given area. Provisions for irrigation of valuable plantings are becoming more common in areas that normally receive summer rain, since periods of drought may have devastating effects in such areas.

Fog

Fog can be an important source of moisture for plant growth, particularly if it occurs during the growing season. Not only does it add to the actual soil moisture supply, but fog also adds moisture to leaf surfaces and delays increases in temperature, which increase evapotranspiration. During the growing season, fog is most common along the coast, where warm, moist incoming air is chilled when it passes over cold ocean surfaces. Large trees in foliage can intercept considerable moisture as fog condenses on their leaf and branch surfaces. During a four-day period of heavy fog, Geiger (1961) observed about 20 mm (0.75 in) of water to fall in open country, while over 150 mm (6 in) fell under fir trees on the edge of a forest. At Berkeley, California, 0.7 to 1 m (30-40 in) of fog precipitation was measured under individual eucalyptus trees during a single growing season.

Therefore, a grove of trees can effectively protect land areas from incoming fog. Geiger reports 20 times as much fog precipitation on the windward as on the leeward side of a wooded area. The Japanese have studied the species and grove dimensions most effective in reducing the moisture of fog from the sea.

In interior areas, ground fog may occur in winter under conditions that favor radiation loss of heat and high humidity. Winter fog can keep the buds of trees close to air temperature and make winter chilling more effective. Warm weather fog, however, may increase the incidence of certain fungal diseases.

Dew

Dew is another source of moisture that can either reduce evapotranspiration or, in some cases, add moisture to the soil reservoir. Water condenses on a surface when its temperature drops below the dew point of the surrounding air. This happens under conditions favorable for radiation loss of heat: Dew condenses on surfaces that radiate more energy than they absorb.

The maximum amount of dew that might form in a single night has been calculated to be equivalent to water about 1 mm (0.04 in) deep (Geiger 1961). Grass will produce only 0.1 mm (0.004 in). The maximum amount of moisture is possible only in arid and semiarid tropical climates, where rainfall is infrequent and dew formation is favored by extreme radiation cooling at night. This moisture would not be significant to the soil moisture supply, but dew on the leaf surfaces can be absorbed by a number of species and can delay leaf warming on the following day. More dew will form on an isolated plant than on an identical plant that is close to others.

Dew can partly replenish the soil reservoir in certain cases. In the deserts of North Africa, gravel mounds about 1.5 m (5 ft) in diameter and 0.6 m (2 ft) high that were built centuries ago are fairly common. They are thought to represent a method by which dew is made to condense on the radiation-cooled surfaces, allowing excess moisture to flow down to the soil. A variation of this is the desert survival technique that involves digging a hole in the soil, covering it with a plastic sheet, and placing a stone in the center to provide a sloping surface so that water may flow to the lowest point and drip into a container placed in the bottom of the hole. This has been adapted experimentally in Montana to provide moisture for new plants (see Fig. 10-11) (Jensen and Holder 1979). Both the rock mulch and the plastic-covered hole can be effective only if good radiation cooling occurs at night. When plants grow big enough, they may reduce radiation losses from the surface of the gravel or plastic and consequently the amount of moisture condensed.

Drought

In areas that have summer rains, a period of three weeks without rain will begin to create problems for sensitive plants and those with small reservoirs of soil moisture. By contrast, regions that have Mediterranean-type climates normally receive little or no rain for four to five months of the growing season. Unless such regions have supplemental sources of water or deep, well-drained soil, their vegetation will be quite different from that in areas that receive rain in summer. Plants will be smaller but will have relatively large root systems, and they will be farther apart. Some may have special or modified structures that reduce transpiration, and others may drop most of their leaves when water is limited.

Many landscaped areas have restricted supplies of water. Paving and buildings can reduce the amount of moisture that reaches root zones and can also increase transpiration through increased radiation. When the water supply to plants is insufficient, leaves wilt and young shoots droop; if the situation persists, the tips and margins of leaves begin to brown and the condition spreads to the veins. The oldest leaves on weak branches begin to fall. This is discussed in more detail in Chapter 18.

Excess Moisture

Intense rainfall can cause serious erosion and flooding. Soil and plants can be washed away, slopes can erode or slump as plants are lost, and plants can be buried by slides and sediment. If soils become waterlogged because of long periods of rain or runoff from melted snow, all but the most hardy plants will fail or grow poorly. Landscape plantings should be designed to handle the heaviest rains that might be expected (see Chapter 8).

Hail

Hail can defoliate plants, tatter leaves, injure bark, damage or knock off fruit, and break twigs and small branches, depending on hailstone size and firmness and wind speed. Damage will be greatest at the tops and on the windward side of plants. Hail damage can be quite unsightly, but plant appearance usually returns to normal when new growth begins.

Snow and Glaze Ice

Wet snow, freezing rain, and glaze ice can build up on branches of trees and shrubs until they break under the increased weight. Wet snow sticks to branches, particularly to those of conifers, and is much heavier than drier powder snow. Young trees can be severely deformed by the weight of the first wet snows of winter, which often adhere to them until spring. Even deciduous trees and shrubs can become encased in ice by rain that freezes when it contacts cold branches (Fig. 4-9). Entire trees have been lost because of the weight of ice. Semonin (1978) reported an estimated 45 metric tons (50 tons) of ice on a 15-m (50-ft) evergreen tree and a load of ice on a 6-mm (0.25 in) deciduous

FIGURE 4-9
Rain that freezes to branches can cause serious limb breakage.
(Photos courtesy Kim Morse, Dallas Park Department)

twig that was 12 to 17 times the weight of the twig. Pirone (1978a) reported an ice covered twig that weighed about 40 times as much as the twig alone.

WIND

Tornados, hurricanes, and strong winds can devastate landscape plantings by dismembering and uprooting trees and shrubs. Winds of lesser intensity can damage leaves and branches and deform plants. Wind affects plant growth by increasing transpiration and carbon dioxide uptake and decreasing shoot elongation.

Plant Damage

High prevailing winds can not only deform plants but seriously damage them, particularly if the plants are in foliage and if soil is moist (Fig. 4-10). When winds from one direction strongly prevail, a moderately strong wind from the opposite direction can blow down trees that had withstood much stronger winds. Winds can cause more damage when trees have a full canopy of leaves than when they are without leaves. Winds in Wellington, New Zealand, are so strong in the spring that few deciduous trees are grown there, except in protected areas, because the tender new leaves are seriously damaged. In areas where the ground may freeze in winter, wind can increase the transpiration of conifers and cause winter burn of the needles.

FIGURE 4-10
Wind during or following heavy rains can be devastating.
About half of the 100- to 125-mm (4–5 in) diameter plane trees were blown down
during an early fall California rain (left); it is interesting that the only trees
that were blown down were those that were staked. (Harris and others 1976)
Many mature elms were uprooted in a severe Midwest storm (right).
(Right photo courtesy Eugene Himelick, Illinois Natural History Survey)

The gale forces of hurricanes and the fierce twisting winds of tornados lay waste to thousands of trees and other plants each year. A healthy, well-structured tree with a sound root system is best able to withstand strong winds. Trees in a grove protect one another from strong winds, but should some trees be lost or removed the survivors are more seriously endangered when their protection is reduced. Methods of controlling wind by plantings are discussed in Chapter 5.

Transpiration

The direct influence of wind on transpiration is difficult to quantify, not only because transpiration varies with plant species and depth and density of leaf canopy but because changes in wind speed are usually accompanied by changes in air temperature and humidity. Reviewing several controlled experiments, Chang (1968) reported that transpiration increases with wind speed up to a certain point, beyond which it stabilizes or decreases slightly. Isolated tall plants will be most seriously affected, since wind increases with height above the ground and the foliage of isolated plants is more exposed to air movement. According to Chang, Copeland (1906) found that an 8-km/hr (5-mph) breeze approximately doubled the transpiration rate of coconut palms in full sunlight. A number of papers conclude that if the foliage canopy is complete and even, the effect of wind on transpiration is usually small. As wind speeds increase, however, drier and warmer air is usually brought deeper into the leaf canopy, thereby increasing transpiration. Wind influence will be greatest when humidity is low, temperatures high, and soil moisture adequate. Pruitt (1971) calculated that a 16-km/hr (10-mph) wind would increase evapotranspiration about 25 percent at noon on a clear, fairly calm midsummer day in Davis, California (a typical temperature of 30°C [87°F] and a relative humidity of 30 percent were assumed). At typical temperatures and humidities, a 16-km/hr wind would increase evapotranspiration about 50 percent in early morning and about 100 percent in late afternoon. A 32-km/hr wind would approximately double the amount of evapotranspiration.

When relative humidities are near saturation, wind has less effect on transpiration. Wind will also affect transpiration only slightly if soil moisture is limiting or if stomatal openings restrict vapor movement.

Photosynthesis

Under high light, the rate of photosynthesis increases as levels of carbon dioxide increase up to about four times the normal concentration of 0.03 percent in the atmosphere. Under calm daylight conditions, the level of CO_2 in the plant canopy drops below that in the atmosphere, but air movement will keep the CO_2 levels more nearly equal. If the CO_2 level in the leaf canopy is to be maintained, wind speed may need to be at least 20 km/hr (12 mph). This may increase photosynthesis by 10 to 20 percent (Chang 1968).

Shoot Elongation

Movement of the growing point of a plant can reduce height growth as much as 25 percent, although it will increase the growth of basal stem diameter (Leiser and others 1972) (see Chapter 8). Harris and others (1972) discovered that shoot growth of liquidambar was increased 50 percent when container trees were grown close together so that wind movement of the tops was reduced (Fig. 4-11); basal trunk growth and taper were decreased. Forest trees grow taller with less basal trunk growth and less taper than their isolated counterparts. Studies of herbaceous plants show that those grown in high-wind areas have a greater proportion of roots to top than do more protected plants (Whitehead 1963).

FIGURE 4-11
Effects of spacing liquidambar (in 14-l or 3.5-gal containers) during the first growing season in a nursery: 600 mm (24 in) apart (left), 400 mm (17 in) apart (center), and 250 mm (10 in) apart (right). (Harris and others 1972) Close spacing restricted movement of the trees and shaded lower foliage.

Wind is one of several factors that affect human comfort through the individual microclimate. In heat, air movement cools a person by causing moisture to evaporate from the skin and by removing heat directly. In cold weather, removal of body heat by wind is increased. For example, a temperature of -15°C (5°F) accompanied by wind at 25 km/h (15 mph) has the same cooling effect as calm air at -31°C (-24°F) (Foster 1978). This is true only if the object is warmer than the air. Once the object is at air temperature, wind will tend to maintain this equilibrium by removing heat if it is added by radiation or respiration and by adding heat if it is lost by evaporation or radiation.

In cold weather, the wind-chill factor is detrimental to exposed animals, buildings, and plants whose comfort or well-being depends on being warmer than the air. Windbreaks can reduce wind-chill to increase animal comfort and to reduce the cost of heating buildings (see Chapter 5). Wind will not, however, increase cold injury to plants if they can withstand the air temperature. In fact, wind will keep exposed plant parts at air temperature when they might have been further cooled by radiation loss under clear-sky conditions. On the other hand, wind can increase plant injury under freezing conditions by increasing moisture loss and causing desiccation as already discussed. Wind may also prevent undercooling.

URBAN CLIMATE

Urban development certainly modifies the mesoclimate of a city: Vegetation is reduced and pavement and buildings take up much of the denuded spaced. As a result, temperature (termed *urban heat island*), precipitation, and cloud cover increase relative to the surrounding rural areas; wind, humidity, and radiation decrease (Table 4-2). The extent of these influences depends on the size of the city and the amount of vegetation. Most cities have considerable vegetative cover. The surface covered by trees and grass in three Kansas cities has been found to be 53 percent (in Kansas City, pop. 168,000), 63 percent (in Lawrence, pop. 46,000), and 85 percent (in Baldwin, pop. 2500) respectively (Marotz and Coiner 1973). Areas within a city, however, can vary greatly—Outcalt (1972) has identified what he calls an "urban cold doughnut." In fact, there may be concentric areas with differing climates. For example, going from the countryside into the city center, the outer area will normally consist of farmland in crops. New residential areas follow, with little vegetation and expanses of exposed pavement, buildings, and bare ground, which will be warmer. Next, older residential areas with considerable tree and grass cover will be

TABLE 4-2

Average changes in mesoclimate elements caused by urbanization
(Adapted from Landsberg, H. E., 1970. Climates and Urban Planning.
In *Urban Climates*. World Meteorological Organization Tech. Note 108:
364–74.)

Element	Comparison with Rural Environment
Wind Speed	
Annual Mean	20 to 30% less
Extreme Gusts	10 to 20% less
Calms	5 to 20% more
Temperature	
Annual Mean	0.5 to 1.0°C (0.9–1.8°F) higher
Winter Minima (average)	1.0 to 2.0°C (1.8–3.6°F) higher
Heating Degree Days	10% lower
Precipitation	
Total	5 to 10% more
Days with less than 5 mm	10% more
Snowfall	5% less
Relative Humidity	
Winter	2% lower
Summer	8% lower
Cloudiness	
Cover	5 to 10% more
Fog, Winter	100% more
Fog, Summer	30% more
Radiation	
Total	15 to 20% less
Ultraviolet, Winter	30% less
Ultraviolet, Summer	5% less
Sunshine Duration	5 to 15% less
Contaminants	
Condensation Nuclei and Particulates	10 times more
Gaseous Admixtures	5 to 25 times more

similar in temperature to farmland. The city center, composed almost entirely of pavement and buildings, will be the warmest area. Wind will modify the location and configuration of these mesoclimatic areas. Within the city center, areas with tall buildings usually will be cooler in summer, warmer in winter, and more protected from wind than are the areas with open pavement and low buildings of fairly uniform height (Federer 1971). On the other hand, isolated tall buildings can deflect strong winds downward and concentrate their force at the base and corners of buildings (see Chapter 5). Tall buildings also funnel wind into the streets between them, greatly increasing wind velocity. Many cities plant trees more resistant to wind deformation on streets most subject to increased wind velocity. Buildings also create permanent shade, reflect light, and radiate heat to form their own micro-climates. The different sides of a single building have quite different microclimates that affect plant growth and human comfort.

Near large areas of pavement and building surfaces, plants must endure a harsh environment. Summer temperatures will be higher, the relative humidity lower, soil moisture and aeration limited, and air quality poorer. Trees and shrubs more tolerant to such conditions should be planted in such areas, and special care should be given to get them established (see Appendix 7).

FURTHER READING

BARFIELD, B. J., and J. F. GERBER, eds. 1979. *Modification of the Aerial Environment of Plants*. St. Joseph, Mich.: Amer. Soc. of Agr. Engineers.

CHANG, J. 1968. *Climate and Agriculture*. Chicago: Aldine.

GEIGER, R. 1961. *The Climate Near the Ground*. Cambridge: Harvard University Press.

WEISER, C. J. 1970. Cold Resistance and Injury in Woody Plants. *Science* 169: 1269-78.

CHAPTER FIVE

Modifying Climatic Influences

Climate or matters related to climate have seldom been adequately considered when new landscapes are being planned and developed. Considerations that do not fit neatly under the traditional discussions of climate in Chapter 4 are presented here.

PLANTING FOR CLIMATE CONTROL AND ENERGY CONSERVATION

Increasing energy costs have generated interest in conservation through improved subdivision plans, building design and siting, and landscape construction and planting. These methods can involve little or no sacrifice of human comfort. Much of the following discussion is based on a manual prepared by Robert Thayer (1981).

The use of solar energy for heating, cooling, and power requires that solar collectors and areas to be passively heated have access to the sun. Federal, state, and local governments have enacted legislation to protect such access for solar facilities, particularly between 10 A.M. and 2 P.M. (sun time).

Plants can reduce building energy consumption by

Intercepting summer sunlight

Absorbing and reflecting summer energy

Reducing ambient summer temperature through transpiration

Shielding buildings from cold winds (see windbreak section)

Channeling cool breezes to ventilate warm buildings (see windbreak section)

Plants should be kept from shading solar collectors and areas to be heated passively by the sun.

Intercepting Sunlight

During summer, buildings become hot when early morning and late afternoon sun strikes walls that face east or west, particularly when those walls are poorly insulated or made of glass. Although a house may become warmer if its west wall is exposed to the afternoon sun, it will be heated earlier in the day and will stay hot if the east wall is equally exposed to morning sun. Deciduous trees, hedges, and vines on trellises can shade these walls from summer sun while allowing the winter sun to penetrate. At high latitudes ($>30°$), evergreen plants are also suitable since the sun angles in winter are such that east and west walls will not effectively collect heat.

Walls and roofs that face the equator should be protected from the high-angle summer sun. Tall, spreading deciduous trees are most effective in screening roofs. Trees can be placed close to foundations on the east and west sides and even on the polar sides of buildings, but they should shade neither equator-facing glass in winter nor rooftop solar collectors. Even the bare branches of deciduous trees block 20 to 74 percent of the sunlight (Geiger 1961, Zanetto 1978).

Screening Reflecting and
Radiating Surfaces

Reflecting and radiating surfaces, such as swimming pools, concrete patios, and pavement, can heat east, west, and equator-facing glass at undesirable times, even though this glass is shielded from direct sunlight. To prevent excessive heat gain in a building, these reflecting surfaces should be screened with a hedge or a trellised vine, shaded by trees or a trellis, or replaced by a shrub or ground-cover planting.

Paved surfaces (particularly asphalt) act as heat sinks if exposed to the direct sun for long; they absorb heat and reradiate it to heat the surrounding environment. This is fine in the winter but unacceptable during hot summers. Pavement location, size, and shape and strategic placement of shade trees can be used to minimize summer heat and take advantage of winter sun. Tall, spreading deciduous trees on the equator-facing, east, and west sides of the pavement will provide the most effective summer shade but will still allow some penetration of winter sun (Fig. 5-1).

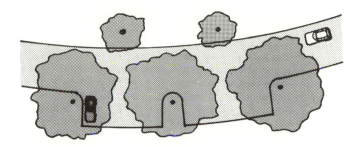

FIGURE 5-1
Parking bays along the equator side of narrow streets can be effectively shaded by large trees. Smaller trees on the polar side provide solar access to houses there. (Thayer 1981)

Reducing Ambient Summer Temperatures

Plants cool the air by transpiration, though breezes disperse the effect so that the temperature in a given landscape is usually lowered by only a few degrees.

Developing Sites for Solar Access

Much can be done in site development to help ensure solar access and to conserve heating and cooling energy. Proper planning of a subdivision is particularly important, because few changes are possible once the subdivision is established. Passive or natural heating and cooling can be facilitated in the following ways (Thayer 1981):

Residential streets should be oriented so that the maximum number of lots face streets that run east and west. This will maximize the number of houses with solar access to the equator end of the lots. Equator-facing wall glass is crucial for maximum passive solar heating.

Setback requirements should allow houses to be placed to the extreme polar end of a lot to allow for maximum solar access.

Usually houses should be sited so that equator-facing glass is maximized, east- and west-facing glass is minimized, and interference with the skyspace of equator-facing glass and rooftop collectors is negligible. Solar access can be provided on less-than-desirable lots if there is flexibility in siting of buildings. Attached and multistory housing units require special consideration to ensure adequate solar access.

Streets should be narrow, with little or no parallel parking; these can be shaded more easily and quickly than wide streets. Older residential areas with mature trees are several degrees cooler in summer than new subdivisions with much exposed asphalt and concrete. Perpendicular parking

119

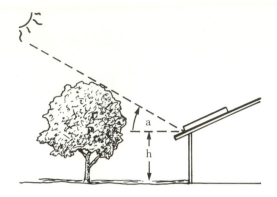

FIGURE 5-2
The *bulk plane height* (h) provides that no structure or plant shall protrude
beyond a plane that extends toward the equator at an angle (a) equal to
that of the sun at noon on the first day of winter. (Thayer 1981)

bays along the equator side of narrow streets can be planned so that
both the parking areas and the street are shaded by trees of moderate
size (Fig. 5-1).

Setback regulations should allow fences to be placed so that they provide
privacy but do not interfere with equator-facing glass collectors.

Height restrictions on buildings and collectors may be necessary to ensure
solar access to one-story houses. Taller buildings must be situated far
enough away from solar collectors so that they do not shade them. A
reasonable restriction is the so-called *bulk plane height*, which provides
that no building or plants shall protrude above a plane in space that extends toward the equator at an angle equal to that of the sun at 12:00
noon on the first day of winter (Fig. 5-2). The angle of the plane depends
on latitude.

Because the slope orientation of building sites affects solar access, solar access
requirements should allow for taller trees and closer spacing between
buildings on equator-facing slopes. Just the opposite is true on polar-facing
slopes.

No matter how good solar orientation and access are, solar houses should not
be located in frost pockets, in cold-air drainages, or on ridges exposed to
cold winds.

Planting to Accommodate
Solar Energy Use

In all but the warmest climates, solar collectors for heating water
should have unobstructed access to the sun between 9 A.M. and 3 P.M.,
particularly in winter. Climates differ in the amount of energy needed
to maintain a building at a comfortable temperature. Certain coastal,

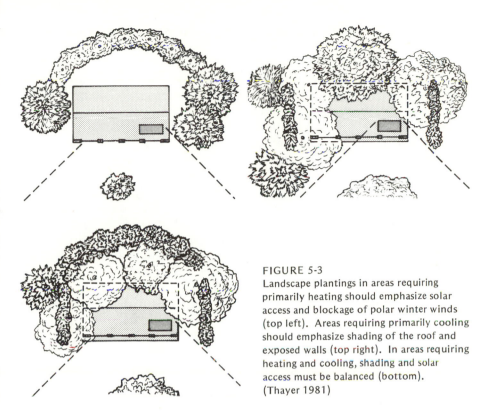

FIGURE 5-3
Landscape plantings in areas requiring primarily heating should emphasize solar access and blockage of polar winter winds (top left). Areas requiring primarily cooling should emphasize shading of the roof and exposed walls (top right). In areas requiring heating and cooling, shading and solar access must be balanced (bottom). (Thayer 1981)

high-elevation, and high-latitude areas require considerable heating in the winter but little or no cooling in summer. Low-elevation and low-latitude areas may need cooling in the summer but little or no heating in the winter. Much of the temperate zone requires considerable energy for both heating in winter and cooling in summer. In cold-winter areas, a windbreak can screen buildings and solar collectors from cold polar winds. In hot climates, walls and roofs can be screened from the summer sun. Where both heating and cooling are needed, plantings for shade must be balanced with provisions for solar access (Fig. 5-3).

The amount of heating or cooling necessary can be expressed by the number of *heating degree days* or *cooling degree days*. If the mean daily temperature is below 18°C (65°F), each degree of difference between the two is called a heating degree day unit (Berdahl and others 1978). A monthly or annual value is the sum of the degree day units for the days in the month or year. Heating degree days are a fairly good measure of the heating requirements for a conventionally constructed building. Similarly, a cooling degree day unit indicates that the mean daily temperature has been one degree above 18°C for one day. Cooling degree days are not a particularly good measure of a building's cooling requirements, because they do not account for humidity, which greatly influences human comfort, particularly at high temperatures.

Whether more effort should go into reducing the cost of heating or the cost of cooling depends on the relative amounts of energy needed for each, the form of energy used (gas, oil, coal, or electricity), and the relative cost of energy per unit of heating or cooling. Cooling primarily requires electricity, which is usually the most expensive energy source. The amount of daytime cloud, fog, and smog cover influences the effectiveness of solar collectors and shade trees. Even though the heating requirement may be considerably greater than the cooling requirement, the wise use of shade trees for summer protection may save considerable energy in areas that are subject to long periods of winter fog and clouds and bright summer sun. On the other hand, for greatest economy in a coastal valley with relatively clear winter weather and foggy summer mornings, the property owner should maximize solar access for collectors and (in winter) the equator-facing walls. Minimizing east glass and planting a sun-screening hedge along the east wall of a building would not conserve energy under such conditions.

Even though it may be more economical for an individual property owner to strive for maximum solar access in the winter at the expense of summer shade protection, such a plan may not be in the best interests of an electric utility. The latter must accommodate peak demand, which occurs in many areas on hot summer afternoons because of the use of air conditioners. In such situations, deciduous trees should be used to block sunlight from wall and roof surfaces, even if the trees shade part of the equator-facing glass and solar collectors in the winter.

To provide maximum summer shade with minimum shading of solar collectors, structures and plants must be placed according to vertical (altitude) and horizontal (azimuth) angles of the sun at different times of the day and days of the year. Although the *bulk plane height* method is simple to use in determining locations for plants of expected mature heights (Fig. 5-2), it may not provide maximum summer shade and solar access or may unduly limit the size and placement of trees and trellised vines.

The solar access zone for a collector or equator-facing glass is most accurately described by a complex surface (Fig. 5-4) formed primarily by the lowest (winter) sunrays striking the lower and east and west edges of the collector between 9 A.M. and 3 P.M. On the east and west edges of the zone surface, however, this surface is modified so that tall structures or trees do not shade the collector at other seasons (Fig. 5-4). As long as no object projects above this imaginary surface, the collectors should have full solar access during the critical midday hours throughout the year. Solar access on the east and west edges of the zone is more important for protecting solar water heaters than for space heating. Tall obstructions along the zone edges will shade the collectors at the beginning and end of the critical period during summer but not

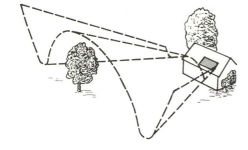

FIGURE 5-4
A complex surface rising toward the equator as outlined by the broken lines should not be penetrated by buildings or plants if solar access is to be provided. (Thayer 1981)

during the cold part of the year, because of the winter azimuth of the sun.

The solar access zone can be estimated by using a *sky chart* of the sun for a particular latitude. Charts should be available from the local weather bureau or energy agency. The angles (both vertical and horizontal) of the sun are obtained from the chart for 9 A.M., noon, and 3 P.M. on June 21 and December 21; by using the collector dimensions and these points, a surface can be depicted that describes the solar access zone. Computer graphic analyses of solar skyspace are feasible for large housing projects and land developments and will help in analyzing spatial solar access planting and configurations. Analyses for a desired latitude should be available soon at solar equipment distributors, weather bureau offices, or planning agencies, if they are not currently available. Such analyses may become the only accurate way to establish solar easements over adjacent properties in order to protect solar access. Some state and local governments have enacted legislation in this realm; it will be most applicable to new subdivision developments.

Tall, upright trees with long, bare trunks can shade equator-facing glass in the summer and allow winter sun to reach the glass (Fig. 5-5). However, high crowns will shade the roof and any solar collectors there in winter. Trellises and overhangs may be more satisfactory than trees for protecting an equator-facing glass wall, particularly if a rooftop solar collector is nearby.

East-west streets planted with tall trees along the equator side of the street and shorter trees along the polar side will shade the street in the summertime but will leave solar collectors exposed on the polar side of the street (Fig. 5-1). Existing and newly planted trees can be pruned severely in the early fall to ensure that winter sun reaches collectors or areas to be heated. Appropriate species can be pollarded to minimize winter shading, provide firewood, and encourage vigorous growth during the summer (see Chapter 14).

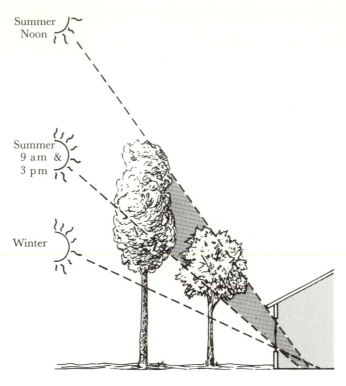

Summer
Noon

Summer
9 am &
3 pm

Winter

FIGURE 5-5
Trees can be planted and pruned to provide solar access in the winter and shade in the summer. (Adapted from Foster, R. S. 1978. *Homeowner's Guide to Landscaping that Saves Energy Dollars*. New York: David McKay. Ruth Foster is a landscape consultant in Belmont, Mass.)

CONTROLLING WIND

Air movement can be pleasant, uncomfortable, or destructive, depending on its velocity, the temperature, and its involvement with rain, snow, hail, or dust. Plants can be used alone or in combination with landforms and manmade structures to influence airflow in the landscape and around or through buildings. This phenomenon has been studied most extensively for use in windbreaks or shelterbelts that will control strong winds, primarily in the open countryside. Much less is known about wind effects in urban areas, where there are canyon-like streets in the central city, tree-covered streets in the suburbs, and plants and structures of various sizes, shapes, and densities. More concern and study has been directed to reducing the destructive effects of wind than to enhancing the pleasant aspects of air movement, but both aspects are receiving more attention as interest in energy conservation grows. White (1945) has studied the use of landscape plantings to enhance the natural ventilation of buildings in warm regions.

Windbreaks and Shelterbelts

Windbreaks and shelterbelts can influence a wide range of activities (Caborn 1965). They can

Reduce the destructiveness of wind

Improve site conditions for plants, people, and other animals

Increase growth and yields of farm and range plants

Reduce heat loss from houses and other structures

Mitigate salty ocean winds

Minimize wind erosion

Control snow accumulation to keep roads open or to enhance soil moisture supply

Provide wildlife habitat

Increase the ecological diversity of an area

Although the terms *windbreak* and *shelterbelt* are used somewhat interchangeably, the former is more appropriate for a structure (a fence or a screen) or a single row or narrow width of plants, while *shelterbelt* refers to several rows of trees or shrubs. The term *windbreak* as used herein includes *shelterbelt*; the term *shelterbelt* refers specifically to multiple rows of plants.

Plants have been grown for wind protection for centuries. Their effectiveness depends on the width of planting only to the extent that it is necessary to achieve height and density. In regions of strong prevailing winds, trees and shrubs adapted to other environmental constraints of the area may not grow tall enough without some protection. Several rows of different plants may be needed to provide that protection. On the other hand, a single row of trees can be satisfactory if they are narrow and upright with dense foliage from the ground to the top and can grow upright under the winds that are characteristic of the area.

Although windbreaks are most effective when placed at right angles to the wind, they are almost equally effective against winds that are within 45° of the perpendicular (Caborn 1965). The *distance* over which wind speed is reduced is determined primarily by the height of the windbreak and secondarily by its density. The *amount* of wind reduction is controlled by the density of the windbreak. A windbreak porosity of 40 to 50 percent will give the most acceptable combination of wind reduction and distance over which that reduction occurs. An open windbreak is less likely to trap cold air, which can create a frost pocket (see Fig. 4-4). Protection will extend downwind for a distance

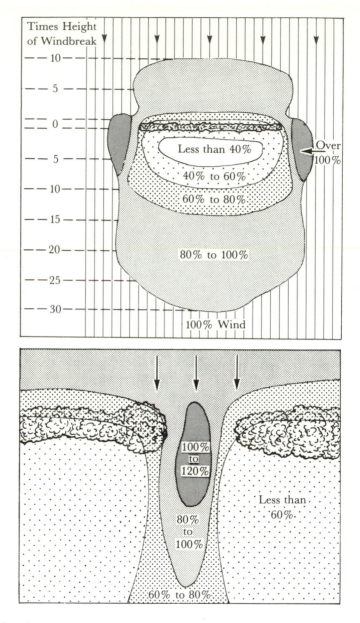

FIGURE 5-6
A windbreak of medium density will reduce wind speed 50 percent or more
in an area that extends from just behind the windbreak to eight times its height
(top). The speed of wind passing over and around windbreaks can be 20 percent
greater than the wind in the open (bottom). (Adapted from Caborn, J. M. 1965.
Shelterbelts and Windbreaks. London: Faber & Faber)

up to 30 times the windbreak height, and upwind up to five times the windbreak height (Fig. 5-6) (Caborn 1965). For a windbreak of medium density, wind speed 1 to 1.5 m (3-5 ft) above the ground is reduced about 20 percent for an upwind distance about two times the windbreak height and a downwind distance 15 times the height. A reduction of 50 percent extends from just behind the windbreak to eight times its height downwind. A denser windbreak will reduce wind near the windbreak more greatly but less farther away. Wind that comes through the windbreak reduces the amount of air that must go over and around it and also maintains a higher pressure behind the windbreak so that wind moving over the top does not form a suction eddy but stays aloft for a greater distance downwind.

For maximum distance protection against oblique winds, the windbreak should be at least 12 times as long as it is high (Caborn). Openings through windbreaks not only provide for pedestrian or vehicular access but also reduce any tendency for the wind to flow parallel to the windbreak. Wind passing over the top and around the ends of windbreaks may be 20 percent faster than the wind in the open. Even though wind passing through an opening in a windbreak increases in velocity, if the windbreak is of medium density, that extra speed only affects windbreak effectiveness near the opening.

A windbreak designed to protect a building, a garden, or other facility also provides shading that may not be desirable. Soil and roads shaded by a windbreak may remain icy or wet much longer than soil in the open. In the winter at 50° north or south latitude, the midday (shortest) shadow of an east-west windbreak will be about 3 times its height.

To protect a large area from winds that come from one direction or two opposing directions, parallel windbreaks are usually spaced 20 to 25 heights apart. However, if objectionable winds come from various directions, a system of parallel and perpendicular windbelts provides better protection. The absolute orientation of the windbreaks is less important when some are at right angles to others; thus border plantings of small fields or properties are particularly effective in reducing wind. Although windbreaks used to protect farms and pastures occupy only 1 to 5 percent of the land, shading or root competition may render a much greater area unsuitable for growing.

A single row of trees or shrubs or a mixture of both can provide adequate protection in some instances and may be necessary because of restricted space. The choice of plants is limited for single-row plantings if height but not much width is wanted. In severe locations, a single row can be difficult to establish and maintain. Each plant is essential for the integrity of a single-row windbreak, and gaps occur if plants die or lose lower branches. Single rows cannot be renewed without taking up more land or sacrificing the existing windbreak. Even so, single-row windbreaks are common throughout the world and will continue to be used in spite of some shortcomings. Caborn (1965) reports a number of species used for single-row plantings around the world, including *Caragana arborescens*, cypresses, eucalyptus, poplars, and Sitka and white spruce.

Multiple-row shelterbelts have fewer restrictions than single rows and will accommodate replacement as needed. Three to five rows of belt plantings can be kept within 6 to 9 m (20–30 ft). Species that tend to spread are planted in the center. Shrubs and smaller trees planted windward will help maintain density at lower levels; this is important because low openings can increase wind speed in their vicinity. Detailed planting arrangements, specifying species and distances, are given in a number of windbreak publications, many of which are listed by van Eimern (1964).

Windbreaks are more effective and take up less land if their windward and leeward (downwind) faces are vertical instead of sloped. In areas of strong prevailing winds, however, a sloping windward face may be the only way to obtain enough height. Windbreaks of irregular height normally reduce wind speed more than those of uniform height but will be less effective if the ground surface upwind is rough.

Where snows occur, plants, landforms, buildings, and other obstacles greatly influence snowfall, accumulation, and melt. Plants can intercept snow and delay its accumulation on the ground. They can influence wind velocities and patterns and thus the drifting of snow. The shade of plants will delay snow melt. Plants, structures, and temporary barriers or snow fences can be placed so as to keep an area relatively free of drifting snow, to accumulate snow in a fairly restricted area, or to distribute it more evenly over a large area. Snow control becomes more difficult in smaller sites that have obstacles and uneven terrain.

Wind is accelerated through and downwind of openings as well as around the ends of windbreaks. Increased wind speed will keep a road parallel to the wind free of snow. Roads at right angles to the wind and larger areas can be kept relatively free of snow by placing a windbreak upwind to catch most of the snow before it arrives. Snowdrift patterns follow wind speeds closely; areas that are calmest or have the greatest drop in wind speeds tend to accumulate the most snow, and those with the highest speeds will have the least snow. A more porous windbreak will distribute snow more uniformly for a greater distance downwind from the windbreak. As wind speed increases, a windbreak becomes relatively more effective, so that more snow accumulates near it. Dense barriers accumulate some snow upwind, while more open barriers accumulate most of the snow to leeward. Caborn (1965) developed an equation for estimating snowdrift length as a function of windbreak height.

A number of site factors influence wind patterns and, in turn, snow accumulation. A rough surface upwind of the windbreak will cause more snow to accumulate on that side, with shallower drifts over a larger area downwind (Caborn). Landforms and structures can greatly modify wind patterns and complicate the prediction of windbreak functioning.

Protecting Landscape Sites

Most landscape sites are restricted in size, and many are surrounded by plantings and structures that affect air flow greatly. Buildings, which are solid windbreaks, create the greatest differentials in wind speed. Isolated tall buildings can create wind speeds on the ground that are twice the normal. Wind speed normally increases with elevation; when strong winds encounter the tops of tall buildings, they are deflected parallel to the building face, and some travel down the building to create strong winds. The taller the isolated building, the stronger the winds become at the bottom. On a calm, hot summer day, this phenomenon can create refreshing breezes in open areas at the base of tall buildings, but under strong wind conditions people and property can be damaged. People are frequently knocked off their feet, and in Cleveland once at the base of a 39-story building a trailer truck was blown over (MacGregor 1971). Plants can be damaged or at least misshapen by these strong side- and down-drafts, particularly near building corners (Fig. 5-7). In a group of tall buildings, only the windward buildings are affected. The bases of isolated tall buildings

FIGURE 5-7
Trees near corners on the windward side
of tall buildings can be misshapen.
The tulip trees near the corner of a
five-story building have been deformed
by polar winds (left);
a tulip tree planted the same time
near the center of the building
on the polar side has developed normally
(right).

and their landscaping must be redesigned to contend with this problem. The solution to date has been to use wind-resistant plants and handrails.

Buildings and large trees can divert wind into areas that were intended to be calm. Streets and side yards between buildings can become wind tunnels. In residential areas with moderate-to-large trees, the wind coming under the tree crowns may be the most troublesome. Plants should be placed to block winds around the side of a house and to form screens on the windward side of the house and garden to reduce wind velocities (Fig. 5-3). Depending on orientation, it may be necessary to divide protection unequally between the house and garden.

On the polar (north or south) side of the house, Caborn (1965) recommends that windbreak distance from the house be at least $2\frac{1}{2}$ and no more than 5 times the ultimate height of the barrier. If wind protection is wanted in the other quadrants but not at the sacrifice of light, Caborn suggests windbreak distances from the house of 4 to 5 times the height of the plants. Because the strongest winds usually occur in winter, evergreens will be needed for maximum protection. A deciduous windbreak is only about 60 percent as effective in winter as when it is in leaf. Summer shade, however, may be more important than full winter sun, so another compromise must be evaluated. A residential area may be so heavily populated with large trees that only low-growing windbreaks are needed. Additional protection can be provided if one wraps the ends of the main windbreak around the windward corners of the house.

As a windbreak grows in height, the area of maximum protection increases in size but moves further downwind. As a windbreak thins out, the degree of protection decreases, while the protected area increases in size and moves further downwind. Pruning, revitalization, and replacement are needed to maintain a windbreak in proper condition.

COASTAL PLANTINGS

Plantings on or near the coast, particularly along western continental shores, endure prevailing onshore winds that carry salty mists and, on occasion, salt spray for considerable distances inland. Although coastal winds may not be any stronger than continental winds, they are more continuous. Plant establishment, formation, and survival are problems because salty wind desiccates and deforms exposed plants, coats them with salt, and keeps sand dunes constantly shifting.

Wind and salt markedly curtail the number of species, particularly tree species, that can withstand direct exposure close to the shore. Small plants adapt to a shore site more quickly than large ones because they are protected by being close to the surface. If a front line defense is to deflect the wind in an upward direction, plants known to tolerate local seacoast conditions (see Appendix 7) should be chosen for

planting.   Tenacity is more important than appearance.   Once the
wind deflection has begun, less rugged plants can be introduced farther
back to increase the height of the windbreak.   Plants that spread by
means of suckers do well in coastal sites (Caborn 1965).

Sand must be stabilized and, for most plants, enriched with
nutrients and organic matter. Small areas of sand are usually protected
to some degree, but large expanses often have shifting windblown dunes
that can quickly bury obstacles in their path. Beach grass, ice plant,
Rosa rugosa, and broom will usually grow well in almost sterile sand,
begin to slow blowing sand, and start the development of a more
permanent dune.   As sand accumulates, these plants can grow up
through it and stay on top.   Organic matter added to sand behind
the protective dune, where some protection from the wind is found,
will improve conditions for other plants. John McLaren created Golden
Gate Park on sand dunes by developing coastal windbreaks and enriching the sand with manure from San Francisco stables.

Sand can be slowed and most exposed plants established if temporary windbreaks are formed of staked-down brush or a low snow
fence.   Such protection should have a porosity of at least 40 to 50
percent. Because some upwind protection is provided, plants should
be set out in front of the windbreak as well as behind it. Upwind plants
will not only protect the windbreak but will replace it when it is removed or deteriorates.

Individual young trees or shrubs are often protected by almost
solid screen structures set close to windward (Fig. 5-8). Plants protected in this fashion are not conditioned to the accelerated winds that
overtop the screens or to the winds that will reach them when the

FIGURE 5-8
An impermeable, closely-placed windbreak does not protect a plant once it grows
above the windbreak. In fact, wind speed is increased at this height. (See Fig. 5-6)

screen is removed or becomes ineffective. Screens for single plants or groups of plants should provide only enough protection to keep shoot terminals from being killed or severely deformed, because the plants will eventually have to endure unabated winds. Screens can be placed at a distance two or three times the screen height upwind from the plants unless the wind varies considerably from a single prevailing direction. If the screen forms a shallow V with its point into the wind, it will tend to deflect more of the wind to the sides and less over the top. Establishing individual or isolated groups of plants in the brunt of coastal winds may be an exercise in frustration and futility unless the plants are very hardy and the grower very determined.

Strong coastal onshore winds, through desiccation and salt toxicity, kill exposed shoot terminals. Several laterals will appear from behind each killed twig tip, only to be killed back as they themselves grow. Soon a thick mat of leafless twigs, with leaves below, will form a barrier that forces the winds upward. Each successive plant grows taller than the one in front, following the line of wind deflection. Few hedges are sheared more evenly than plants on a wind-swept coast. The compact windward surface of exposed plants provides little of the filtering action expected of a good windbreak. The sheltered area is short and close to the plants, with little wind protection extending to leeward beyond 12 times the height of the tallest plants (Caborn 1965). The effectiveness of a wind-swept planting can be increased if clumps of vigorous trees can be grown taller than the general plant surface. Temporary protection may be necessary to allow for the taller growth.

Onshore winds of 25 km (15 mi) per hour or greater can carry considerable ocean salt spray (Boyce 1954). When rain does not accompany strong onshore winds, the winds can be particularly toxic to plants. Ben-Jaacov, Natanson, and Hagiladi (1980) were able to prevent leaf scorching of oleander by activating an overhead sprinkling system when onshore wind speed exceeded 22 km (13 mi) per hour. Even when plants were well-watered, foliage that was not sprinkled was severely damaged by only one dry wind storm. The sprinklers applied 50 mm (2 in) of water per hour during strong winds and for five minutes after the wind slowed below the critical speed. A lower application rate may be just as effective and will reduce water use and the possibility of waterlogged soil. Sprinklers should emit large droplets so that less water will become windborne.

LIGHTNING

Lightning is an extremely powerful electrical discharge from one cloud to another or from a cloud to the ground. Thunderstorms account for the most serious lightning strikes from clouds to the ground.   Raindrops and ice crystals are broken up and churned about by the turbulence in a thunderhead; electrical charges develop such that the upper portion of a thunderhead has a net positive charge and the cloud base has a net negative charge.   The electrical charge at the base of a thunderhead increases until it can overcome the resistance of the air between the cloud and the earth.   A typical lightning discharge has 10 million to 100 million volts at 1000 to 300,000 amperes and may travel 1.6 km (1 mi) or more (Frydenlund 1977).

Thunderstorms are counted by the number of days per year when thunder is heard.   In North America, the annual number ranges from five thunderstorms on the Pacific coast and in northern Canada to more than 100 in south central Florida.   Even though the New England and midwestern states and the Cascade, Rocky, and Sierra Nevada mountain ranges have only 20 to 40 thunderstorm days, the storms are unusually long and severe.

The trend toward increased liability litigation may mean that more deaths and injuries from lightning in areas open to the public may be blamed on management for failing to provide adequate protection.   In the United States, more than 40 percent of the human deaths caused by lightning occur at recreational sites; about 20 percent occur in homes and 20 percent on farms (Frydenlund).   About one-third of human deaths by lightning occur under or near unprotected trees.

Lightning can cause property damage and death either by direct strike or by a major side flash, in which the lightning charge jumps from a stricken object (pole, tree, or structure) before being partly grounded.   Step voltage (a lightning current through the soil that goes up one leg and down the other) can injure victims but is usually not fatal.

The nature and extent of lightning damage to a tree or structure can vary enormously.   Branches and trunk may be blown completely asunder (Brown 1972).   The crowns of trees may be killed or large limbs broken out (Fig. 5-9).   Trunks can be split open.   In hardwood trees, such as oak, a continuous groove of bark or bark and wood may be stripped out along the entire length of a main branch and trunk (Brown).   In other cases, the xylem may be burned or injured without any external evidence, or part or all of the root system may be killed (Pirone 1978a).   These variations appear to be related to the lightning intensity, the species involved, the amount of water on and in the bark, and the character of the branch and trunk tissues.

Lightning is most likely to strike a lone tree, the tallest tree in a group, a tall tree at the end of a row or on the edge of a grove, a tree growing in moist soil or adjacent to bodies of water, and a tree closest to a building (Cripe 1978).   Pirone (1978a) notes that deep-rooted or decaying trees appear more likely to be struck than shallow-rooted or healthy trees.   Plants under or near lightning-struck trees are often injured or killed.

FIGURE 5-9
The trunk and branches of this maple tree were split asunder by lightning (left).
Lightning stripped a groove of bark from the top of this oak down the trunk to the
ground (right). (Photos courtesy of Independent Protection Co., Goshen, Indiana)

Ten species of trees have been reported to be hit by lightning more
often than others (Cripe):

Maple	Sycamore
Ash	Poplar
Tulip tree	Oak
Pine	Hemlock
Spruce	Elm

On the other hand, horse chestnut, birch, and beech are struck less
often than average. Pirone (1978a) suggests that the difference among
species may be due to chemical constituents in the trees. Trees high in
starch (ash, maple, and oak) may be better conductors than trees high
in oils (beech and birch). Bernatzky (1978), however, reasons that
trees with smooth bark (beech) lead rainwater from the branches to
the trunk, which becomes moist; the bark then increases in conductivity
so that lightning is led off down the trunk without damaging the tree.
On the other hand, trees with deeply fissured bark have a poorly
moistened trunk surface; lightning can penetrate into the water-
saturated, conductive cambium and cleave the trunk. This would
account for less injury to a smooth-bark tree, but a moist bark surface
should increase the likelihood of a tree being struck.

Protection against Lightning

People, structures, and trees can be protected against lightning if
means are provided by which a lightning discharge can enter or leave

the ground without causing damage or loss. Because of the danger of side flash, trees that have trunks within 3 m (10 ft) of a building and branches that extend above the structure should be equipped with a lightning protection system. Specimen trees, historic trees, and those under which people might seek shelter from a storm should also be so equipped.

The National Fire Protection Association in Boston (NFPA 1980) and the National Arborist Association (N.A.A. 1979) have published standards for protecting trees from lightning. The salient feature of these standards is a single 32-strand copper (17-gauge) or aluminum (14-gauge) conductor attached to an air terminal (tree point) installed in the highest part of a tree and then fastened along the trunk down to the ground connections (Fig. 5-10). If the tree is round-headed, several

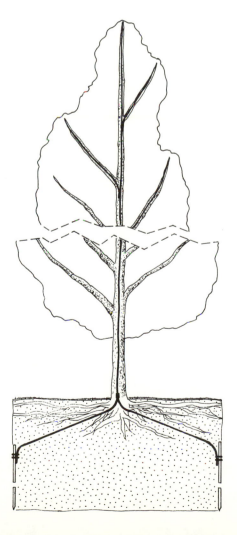

FIGURE 5-10
A tree can be protected from lightning if heavy copper or aluminum wire is installed from the highest parts of the tree to 3-m (10-ft) ground rods at the dripline of the tree or beyond it.

smaller air terminals and branch conductors should be placed in the highest parts of the main limbs. If the trunk is 0.9 m (3 ft) in diameter or larger, two down conductors should be installed on opposite sides of the trunk and interconnected. The conductors should be attached securely to the tree to allow for wind sway and growth without danger of breakage. Each conductor should extend from the trunk 300 mm (1 ft) below the ground out to the dripline of the tree or 7.5 m (25 ft) from the trunk, whichever is farther. Each conductor should be attached to a copperweld ground rod 12.5 mm (0.5 in) in diameter and 3 m (10 ft) long driven into the soil at least 3 m. In sandy soils, three con- ductors should be attached to each down conductor a short distance from the trunk; each of the three lateral conductors should extend at least 3 m (10 ft) from the down conductor. Special forked ground con- nectors should be used to ensure that the connection is secure. The end of each lateral conductor should be attached to a 3-m ground rod. In rocky areas where ground rods cannot be driven 3 m deep, two or more ground rods should be driven as deep as possible; these rods should be at least 2 m (6 ft) apart. If a grounding system is within 7.5 m (25 ft) of a water pipe or a similar grounding system for a building or another tree, the systems should be connected to provide common grounding and increased grounding power (N.A.A. 1979).

In a group of trees, only tall trees need be protected; they will themselves protect smaller adjacent trees. In a row of tall trees, down conductors can be connected as long as a 3-m ground rod is placed at least every 25 m (80 ft).

Lightning protection systems should be inspected each dormant season to make sure that they are intact and that the air terminals are high enough. The air terminals may need to be raised every two or three years. None of the conductors should be bent more sharply than 90° or have a radius of less than 200 mm (8 in).

A tree protected with a lightning rod system will actually be more likely to be struck by lightning, but a properly installed system should protect the tree and people underneath from harm.

Caring for Lightning-Struck Trees

Many trees that are struck by lightning must be removed; others may be worth attempts to save them. Since the extent of internal damage cannot be assessed immediately, repairs should be limited to safety pruning and cleanup until tree survival seems assured. If the cambium is still moist, loose bark should be tacked back into place, covered with moist cloth or burlap and polyethylene film to slow dry- ing, and shaded to keep them cool. Nitrogen should be applied under the tree as recommended on page 285. Depending on the season and expected rainfall, fertilizer should be watered into the root zone. Root growth should be stimulated so as to improve the supply of water and nutrients to the top of the tree. If, after a full growing season, the tree appears to be in reasonable health, more careful pruning and wound repair may be worthwhile. Struck trees, however, may continue to fail over a period of years, eventually requiring removal.

FURTHER READING

CABORN, J. M. 1965. *Shelterbelts and Windbreaks.* London: Faber and Faber.

DEWALLE, D. R., and E. P. FARRAND. 1978. *Windbreaks and Shade Trees: Their Use in Home Energy Conservation.* Penn. State Univ. Agr. Special Circular 245.

GEIGER, R. 1961. *The Climate Near the Ground.* Cambridge: Harvard University Press.

GEORGE, E. J. 1956. *Cultural Practices for Growing Shelterbelt Trees on the Northern Great Plains.* U.S. Dept. of Agr. Tech. Bull. 1138.

———. 1966. *Shelterbelts for the Northern Great Plains.* U.S. Dept. of Agr. Farmers Bull. 2109.

NATIONAL FIRE PROTECTION ASSOCIATION. 1980. *Lightning Protection Code 1980.* Boston: Natl. Fire Protection Assn.

THAYER, R. L., JR. 1981. *Solar Access: It's the Law.* Univ. of Calif., Davis, Inst. of Govt. Affairs and Ecology Environmental Quality Series 34.

CHAPTER SIX

Planting Site: Soil

Soil provides plants with water, nutrients, and root anchorage. Soils are responsible for the poor performance of landscape plants more often than any other single factor but are often given very little consideration when a site is being selected for landscape plantings. Further, the development of a site and the construction of facilities compact the soil so that the surface can be practically impervious to air and water. Even when proper soils can be selected and imported for special planting situations, poor plant growth often reveals that landscape developers lack the necessary understanding of soil properties and their importance to plants.

Soil is a complex physical, chemical, and biological system. About half of its volume is composed of solid matter, primarily mineral particles with some organic matter; the other half consists of pore spaces filled in varying proportions with air and water. Soils are also alive with a wide range of bacteria, fungi, and other organisms.

SOIL FORMATION

Soils develop from the partial weathering of rocks, the interaction of rock with water, climate, and organisms. Soils are dynamic and evolving: A number of changes occur over time, giving rise to more or less distinctly visible layers below and parallel to the soil surface. These layers are called *horizons*, and their sum makes up the *profile* of a soil (Fig. 6-1). Soil horizons are designated by the letters *A*, *B*, and *C*, moving downward from the soil surface. A young soil usually has a thin *A* horizon, a *C* horizon of loose or weathered rock material, and little or no *B* horizon.

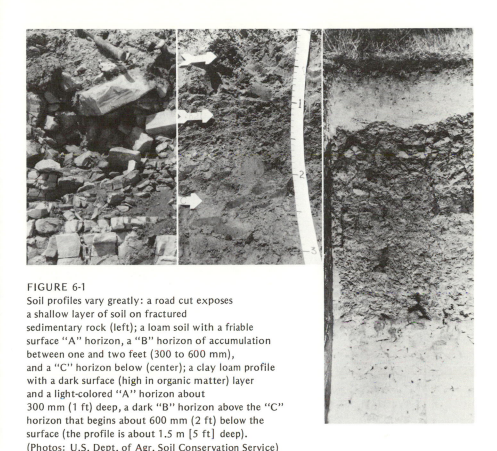

FIGURE 6-1
Soil profiles vary greatly: a road cut exposes
a shallow layer of soil on fractured
sedimentary rock (left); a loam soil with a friable
surface "A" horizon, a "B" horizon of accumulation
between one and two feet (300 to 600 mm),
and a "C" horizon below (center); a clay loam profile
with a dark surface (high in organic matter) layer
and a light-colored "A" horizon about
300 mm (1 ft) deep, a dark "B" horizon above the "C"
horizon that begins about 600 mm (2 ft) below the
surface (the profile is about 1.5 m [5 ft] deep).
(Photos: U.S. Dept. of Agr. Soil Conservation Service)

The first distinct layer to emerge is the dark-colored layer of decomposing organic matter on the surface. Its influence on the developing soil below will depend in large measure on the climate. In colder temperate regions, where rainfall exceeds evapotranspiration and the organic matter is usually acid, the surface soil is leached of soluble material and colloids. A light-colored *A* horizon develops under the organic layer. In this type of soil formation, called *podsolization*, calcium carbonates are leached out of well-drained soils, leaving them somewhat acidic. Poorly drained soils are alkaline and usually high in calcium.

In areas with high rainfall and high temperatures, soils form by *laterization*. Decomposing organic matter in warm climates is usually alkaline. The alkalinity combines with high temperatures to leach silica from the surface soil. Iron and aluminum compounds remain, giving the soil a reddish-brown color.

In areas where evapotranspiration exceeds rainfall for part of the year, materials from the surface soil may be leached only a short distance, depending on the amount of rainfall. Calcium carbonate is predominant among the deposited salts, and the process is called *calcification*. If sodium is the primary cation, *alkali* soils are formed. The zone of deposition can become almost impervious to water, air, and roots. This *hardpan*, which can be a few centimeters or more than ten meters (30 ft) thick, is the extreme example of a *B* horizon, a zone of accumulation. Soil depth above a hardpan varies from less than 100 mm (4 in) to several meters (10–15 ft), depending on rainfall and the age of the soil. Organic muck soils and peat form in marshy areas from decomposed plants.

It is useful to know whether a soil has developed in place or has been transported some distance from the original site. *Residual* soil usually occurs on mountain and foothill slopes; depositional or *alluvial* soil occurs in valley and outwash fans. Most alluvial soils are deposited by streams that flow out of mountains, carrying sand, silt, and clay from erosion of residual soils and the breakup of rocks in stream beds. These soils are usually deeper, more level, and more permeable to water than are residual soils. In some places, soil—called *aeolian* soil—is deposited by wind.

Soil color sometimes suggests past history or soil properties that may influence plants (Wildman 1976):

Dark soil near the surface usually indicates a higher organic matter content there. This soil often has better tilth (ease of cultivation) and higher nutrient content than the subsoil.

A uniformly colored profile is typical of unweathered young soils and usually uniformity of other soil properties and thus an excellent root environment.

Increasing yellow and red colors in the subsoil are typical of older, weathered soils that contain free iron oxides. These subsoils often have a high clay content, a blocky structure, and, hence, decreased permeability for roots, water, and air.

Whitish or light-gray zones or white networks following soil cracks usually indicate a high content of calcium carbonate (lime), which may induce iron and manganese deficiencies, particularly in certain woody plants. If fizzing or foaming results when a drop or two of 1 percent hydrochloric acid is applied to the soil, lime content is confirmed. Small white crystals that glitter in the sun but do not foam with acid are usually gypsum (calcium sulfate).

Blue or gray zones or gray-and-rust-colored mottling of the soil indicate poor drainage and a lack of good aeration. Methane gas from natural gas or a landfill may cause a similar coloration of the soil (see Chapter 18).

Soils are grouped into *series* to aid in their classification and description. A series includes soils that have developed from similar materials by similar processes and have similar horizons. The soils within a series vary primarily in the texture (particle size) of the *A* (surface) horizon. Soil series are usually designated by the geographical names of the locations in which they were first found and described. The series name is modified by the texture of the surface soil, as in *Lansing silt*, *Hanford fine sand*, and *Dublin clay loam*.

Soil surveys have been conducted in the agricultural and forested areas of many countries. In the United States they are usually available on a county basis.

PHYSICAL SOIL CHARACTERISTICS

A number of soil characteristics affect plant growth and well-being. These include soil texture (size distribution of particles), structure (arrangement of soil particles), and soil depth. These factors, in turn, influence water infiltration and movement in the soil, water and nutrient retention capacities, and aeration. Topography can influence soil and plants through its effect on water accumulation or runoff, erosion, and temperature.

Texture

The larger mineral particles, rock and gravel, add little to a soil chemically or physically but do take up space and affect its permeability. Sand, silt, and clay particles compose most of the solid material of a soil. The smallest sand grain is 25 times the diameter of the largest clay particle (Table 6-1). A silt particle is between the two in size. Sand and silt are spherical or cubical in shape, while clay particles are more wafer-like. This difference in shape and the differences in particle

TABLE 6-1

Classification of soil texture according to particle size (Hartmann, H. T., W. J. Flocker, and A. M. Kofranek. *Plant Science: Growth, Development, and Utilization of Cultivated Plants,* © 1981, p. 171. Reprinted by permission of Prentice-Hall, Inc., Englewood Cliffs, N.J.)

	Diameter Size (mm)[a]			
Particle	*USDA System*		*International System*	
Gravel	>2.0	(>0.08)	—	—
Very coarse sand	2.0–1.0	(0.08–0.04)	—	—
Coarse sand	1.0–0.5	(0.04–0.02)	2.0–0.2	(0.08–0.008)
Medium sand	0.5–0.25	(0.02–0.01)	—	—
Fine sand	0.25–0.10	(0.01–0.004)	0.2–0.02	(0.008–0.0008)
Very fine sand	0.10–0.05	(0.004–0.002)	—	—
Silt	0.05–0.002	(0.002–0.00008)	0.02–0.002	(0.0008–0.00008)
Clay	<0.002	(<0.00008)	<0.002	(<0.00008)

[a]Numbers in parentheses are in inches.

size account for the much greater surface area of clay soil than of silt or sand. The greater particle size and smaller surface-to-volume ratio account for the greater permeability and workability of sandy soils and also for their lower water- and nutrient-holding capacities.

A given soil will usually comprise particles of varying sizes, and its characteristics are determined largely by the proportion of the different sizes of particles. A soil is described by its dominant particle sizes (Fig. 6-2). Note that a soil is not considered "sandy" until it contains more than 45 percent of sand. Similarly, a "silt" soil must be 40 percent silt, but only 20 percent clay makes a "clay" soil because it takes less clay to dominate the characteristics of a soil. Thus, a soil that is 60 percent sand would nevertheless be a "clay" soil if it were 35 percent clay.

Certain soils exhibit intermediate characteristics. These *loam* soils are ideal for growing a wide variety of plants because they combine the

FIGURE 6-2
A soil textural triangle to determine the type (texture)
of soil depending on the percentages of sand, silt, and clay. (U.S. Dept. of Agr.)
The lines for a soil composed of 20 percent clay (left),
40 percent silt (right), and 40 percent sand (bottom) intersect
to place the soil in the loam category. Note that it takes less clay
than silt or sand to impart its characteristics to a soil.

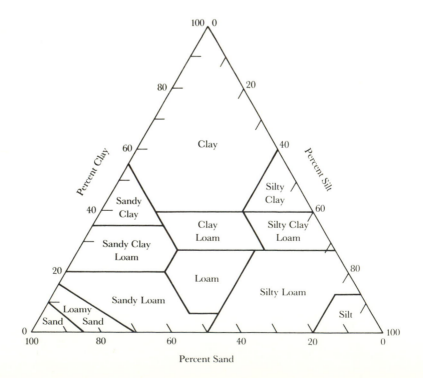

Percent Sand

desirable attributes of each of the particle sizes. Note again that it takes less than half as much clay as either silt or sand to impart its characteristics to a loam soil.

Structure

Water and air movement and ease of root growth through soil are determined by pore size, not necessarily by the size of soil particles. Most particles in the soil are not discrete units but are grouped into aggregates or granules, each of which acts as a large particle as far as water and air movement are concerned. The water- and nutrient-holding capacities of the smaller particles, however, are still retained within each aggregate. The particles are held together by a combination of electrical and chemical bonds, gelatinous materials produced by plant roots and microorganisms, and iron and aluminum hydroxides.

Soil structure is a subjective term used to indicate the degree to which soil particles are aggregated. The more aggregated a soil, the more desirable is its structure for plants. Decomposing organic matter aids in the development and stability of soil aggregates. Soils of hot, arid regions usually have less organic matter than those of milder, humid regions. Soils that are primarily silt or sand aggregate poorly, so that their properties depend essentially on texture. Soil aggregates are fragile and easily compressed or destroyed by physical compaction, shearing, or chemical dispersion (as in high-sodium soils). Soil aggregation is a slow process, but aggregates can be destroyed quickly.

Soil Depth

The depth of a given soil determines its moisture- and nutrient-holding capacities and influences the depth of rooting. Usually, the deeper the soil, the greater will be the water and nutrient supplies for plant use. Depending on texture and structure, depth can affect the moisture content of shallow soils even when drainage is provided.

The term *effective root depth* has been used to describe the portion of the soil that is favorable for roots. In an alluvial soil with no noticeable stratification, effective root depth may be more than 1.5 m (5 ft); in a clay pan soil it may be as little as 300 mm (12 in) or the depth of soil above the clay layer. To determine soil depth, it is necessary to determine which layers in the soil restrict root and water penetration. Table 6-2 identifies the terms used to describe soil depth in relation to average root zones.

Soil depth in landscape situations can be increased by the use of raised planters or mounds. Mounds provide surface drainage during periods of high rainfall as well as more complete internal drainage for better aeration. If the soil volume available in low-maintenance landscapes will not provide sufficient physical, nutrient, and moisture

TABLE 6-2

Terms used to describe soil depths (Wildman and Gowans 1975)

| | Depth Range According to: | | | |
| | Storie (1932) | | USDA | |
Terms	mm	in	mm	in
Very shallow	<300	<12	<250	<10
Shallow	300–600	12–24	250–500	10–20
Moderately deep	600–900	24–36	500–900	20–36
Deep	900–1200	36–48	>900	>36
Very deep	>1200	>48	>1500	>60

support, plants can be spaced farther apart to increase soil volume per plant.

Root and water penetration through a soil are altered by a layer that has a distinctly different texture from the soil above or below. If a subsoil layer has a noticeable increase in clay, water may accumulate for a time above this layer, forming a "perched" water table, and roots may be injured because of poor aeration. This condition is often called *waterlogging*. Very sandy or gravelly layers can also interrupt the normal downward penetration of roots and percolation of water. For example, water does not drain freely from a silt layer into sand until the silt layer becomes saturated for some depth above the coarser layer. This saturated layer can damage roots. The saturated zone will remain even after drainage has occurred because particle-to-particle flow of water is extremely slow and capillarity holds the water between the silt particles above the sand.

Topography

Hills and swales add variety and interest to landscapes and are often created on sites that would otherwise be fairly level. Slopes, however, offer special problems to both the landscape architect and the maintenance person and should be artificially introduced only for good reasons. Earth mounds are more effective sound barriers than plants alone on a level site. Mounds provide surface drainage during periods of high rainfall and can thus protect roots from waterlogging. Adequate drainage must be provided, however, to carry away the surface runoff so that it does not accumulate at the base of the slope. Mounds are often placed in the landscape primarily for aesthetic reasons. If these are to be covered with plants rather than turf alone, an easier solution would be to establish taller plants on level ground.

The most obvious problem with slopes is soil erosion from rain or irrigation. During irrigation, water is usually applied faster than it can enter the soil; this results not only in erosion but also in uneven water distribution, with underwatering at the top and overwatering at the base.

Mounds should be of fairly uniform soil texture so that their full depth can be utilized by the roots of trees and large shrubs. If the surface soil is a shallow layer on top of compacted soil or soil of different texture, water will be drawn down through the top layer of soil toward the bottom of the slope, leaving little or no moisture to enter the mound for the benefit of growing plants (Fig. 6-3).

Both the soil and the air close to the surface will vary in temperature with the direction of the slope. At a highway interchange in central California, soil temperatures in mid-autumn at 25 mm (1 in) below the surface were 34°C (93°F) on the slope facing the sun and 17°C (62°F) on the slope facing away from the sun. This range of temperatures can greatly affect the distribution of plants and their transpiration rates during the growing season. Near the top of a mound, these temperature extremes can exist within three meters (10 ft) of each other. Two different exposures should be irrigated by different lines if they are sprinkler- or drip-irrigated.

FIGURE 6-3
When water is applied to a slope more quickly than it can infiltrate
the surface or penetrate deeper soil layers, it will flow to the bottom
of the slope resulting in underwatering at the top, overwatering at the bottom,
or both.

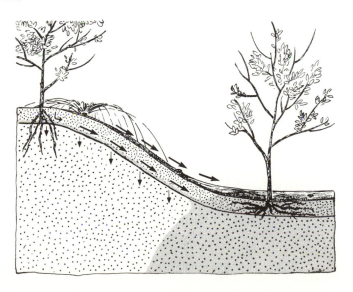

CHEMICAL PROPERTIES OF SOILS

Soil Fertility

Inorganic minerals make up about half the soil volume, organic matter makes up 1 to 6 percent, and pore space makes up the balance. Mineral particles, particularly the clays, and organic matter are the principal reservoirs of 13 of the 16 elements essential to plant growth (Table 6-3).

Nutrients occur in soil in various states:

Dissolved ions in the liquid phase of the soil (soil solution)

Crystalline precipitates (calcium phosphates, lime, gypsum) and rock minerals (feldspars and micas)

Soil organic matter

Exchangeable ions adsorbed on the surface of soil particles (especially on clays and organic matter)

TABLE 6-3

Elements essential for plants

Source and Element	Symbol	Form Available to Plants	
Air and Water			
Carbon	C		CO_2
Oxygen	O		O_2, H_2O
Hydrogen	H		H_2O
Soil			
Macronutrients			
Nitrogen	N	Nitrate	NO_3^-
		Ammonium	NH_4^+
		Urea	$CO(NH_2)_2$
Phosphorus	P	Phosphate	$H_2PO_4^-$
Potassium	K	Potassium	K^+
Calcium	Ca	Calcium	Ca^{++}
Sulfur	S	Sulfate	$SO_4^=$
Magnesium	Mg	Magnesium	Mg^{++}
Micronutrients			
Manganese	Mn	Manganese	Mn^{++}
Zinc	Zn	Zinc	Zn^{++}
Boron	B	Borate	$H_2BO_3^-$
Copper	Cu	Copper	Cu^{++}
Iron	Fe	Iron	Fe^{+++}
Molybdenum	Mo	Molybdate	$MoO_4^=$
Chlorine	Cl	Chloride	Cl^-

Nutrients in solution are available to plant roots and are also easily leached. As nutrients are absorbed by plants or are leached, more nutrients usually come into solution. Nutrients become available as soil particles weather (a slow process) and as organic matter decomposes. The capacity of a soil to adsorb nutrients on its particle surfaces, however, determines in large measure its fertility and productivity.

Cation Exchange Capacity   Nutrients most readily available to roots are those in solution and those adsorbed on the surface of clays. Soil particles are negatively charged and therefore attract positively charged ions, or cations. The capacity of a soil to adsorb cations is called its *cation exchange capacity*. The finer the texture of a soil, the more clay it contains and the greater its cation exchange capacity. Fine-textured soils are usually more fertile than coarse-textured ones.

Cations held on soil are exchangeable: They go into solution and are replaced by others in solution. Even though adsorbed ions are exchangeable, they are held fairly securely against leaching unless they are replaced by ions in solution. In contrast with ions in solution, exchangeable cations are not readily leached.

Anions   The soil solution contains equivalent anions and cations. Nitrate, sulfate, and phosphate are the principal nutrient anions in a soil solution. Nitrate is almost completely soluble; it is readily available to plant roots and is also leachable. Sulfate is moderately soluble and is less subject to leaching than the nitrate ion is. Phosphates are held in most soils as precipitates, largely as insoluble iron and aluminum phosphates in acid soils and as calcium phosphates in neutral and alkaline soils (Pritchett 1979). In many soils, phosphates are not readily available to roots, nor are they readily leached.

Soil Reaction

Soil reaction, expressed as pH, refers to the acidity or alkalinity of a soil (Fig. 6-4), or the relative proportion of hydrogen (acid) and hydroxide (alkaline) ions. Equal concentrations of the two produce a neutral reaction—a pH of 7.0. As a soil becomes more acid, its pH decreases; as it becomes more alkaline, its pH increases. A one-unit change in pH indicates a tenfold change in hydrogen and hydroxide ion concentrations.

Soil reaction primarily influences plant growth indirectly through its effects on the solubility of ions and the activity of microorganisms. The availability of a number of nutrients, particularly phosphorus and the micronutrients, is influenced greatly by soil pH (Fig. 6-4). Except for molybdenum and chloride, micronutrients become less available as soil alkalinity increases. In soils with high pH, usually only iron and manganese are seriously deficient. On the other hand, manganese and

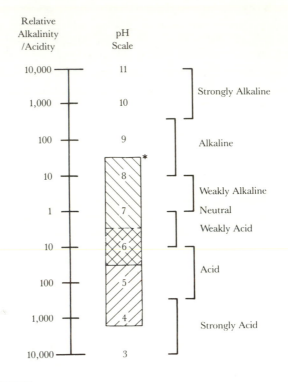

FIGURE 6-4
Relative alkalinity and acidity (pH scale).
The scale on the left shows alkalinity or acidity relative to the neutral (pH 7).
For example, pH 4 is 100 times more acid than pH 6. (Harris, Paul, and
Leiser 1977)

aluminum can reach toxic levels under moderately to highly acidic conditions. Most microbial activity in the soil, especially that of nitrogen-fixing bacteria, is reduced as soil acidity increases (below pH 5.5). Most leguminous plants grow better in neutral or alkaline soils, where nitrogen-fixing bacteria can thrive. Conversely, soils of coniferous nurseries are often kept below pH 5.5 to reduce "damping-off" (a root-killing disease) and other fungal diseases (Pritchett 1979).

Many plants that grow well in soil between pH 4.0 and 6.5 also grow best in well-aerated (well-drained) soil. Most "acid-loving" plants have developed in well-drained soil under high humidity or heavy rainfall (above 500 mm or 20 in annually), which leach out the basic ions, primarily calcium and magnesium, and replace them with hydrogen ions. Poorly drained soils retain most basic ions and are alkaline. Even if drainage is good, most soils in arid regions (those that receive less than 500 mm of rain annually) are neutral or alkaline because rainfall is not heavy enough to leach the basic ions. In fact, ions accumulate under most arid conditions.

Soil reaction is influenced by vegetation as well as by rainfall and soil drainage. Grasses contain higher concentrations of calcium and magnesium than does conifer litter; hence, more hydrogen ions are displaced and leached from the surface soil under grass by the recycling "base" elements. Grasses tend to keep soils more alkaline than forest plants, particularly coniferous ones (Halfacre and Barden 1979). Using conifer needles as mulch is one way to help maintain acid soil.

Most plants grow well in soils with a wide range of pH—from about 5.5 to 8.3—particularly if the soil is well drained. Except for some acid-loving plants that grow well in soils as low as pH 4.0, a soil pH between 6.0 and 6.5 is thought to be best.

Adjusting Soil Reaction   If a soil is too acid or alkaline, its pH can be increased by adding lime or decreased by adding sulfur. The amount of lime or sulfur needed depends on soil texture, organic matter, material used, and how much the pH is to be changed. The finer the soil and the higher the organic matter content, the more amendment must be added to achieve the same pH change.

Ground limestone, consisting of calcium carbonate or calcium magnesium carbonate (dolomite), is commonly used to increase soil pH (Table 6-4). Quicklime (calcium and magnesium oxides) and slaked

TABLE 6-4

Approximate amounts of finely ground limestone[a] needed to increase pH of the surface 200 mm (8 in) of an acid soil to pH 6.5 (Adapted from USDA Handbooks 18 and 19 and from Dickey 1977)

	Kilograms of Limestone per 100 Square Meters[b, c]					
Desired Change in pH	*Sand*	*Sandy Loam*	*Loam*	*Silt Loam*	*Clay Loam*	*Muck*
From 4.0 to 6.5	130	250	350	420	500	950
From 4.5 to 6.5	110	210	290	350	420	810
From 5.0 to 6.5	90	170	230	280	330	630
From 5.5 to 6.5	60	130	170	200	230	430
From 6.0 to 6.5	30	70	90	110	120	220

[a]Or dolomite lime

[b]Double the kg/100 m^2 to approximate the lbs/1000 ft^2.

[c]For soils with considerable organic matter, increase the amount of limestone applied about 15 percent for clay soils and up to 50 percent for sandy soils.

TABLE 6-5

The approximate amounts of soil sulfur required to increase the acidity of the surface 200 mm (8 in) of a carbonate-free soil (Adapted from Shaw, E. J., ed. 1980. *Western Fertilizer Handbook*. 6th ed. Sacramento, Calif.: Calif. Fertilizer Assoc.)

	Kilograms of Sulfur per 100 Square Meters[a]		
Desired Change in pH	*Sand*	*Loam*	*Clay*
From 8.5 to 6.5	100	125	150
From 8.0 to 6.5	60	75	100
From 7.5 to 6.5	25	40	50
From 7.0 to 6.5	5	8	15

[a]Double the kg/100 m^2 to approximate the lbs/1000 ft^2.

TABLE 6-6

The relative acidifying effects of certain amendments and fertilizers (Adapted from Shaw, E. J., ed. 1980. *Western Fertilizer Handbook*. 6th ed. Sacramento, Calif.: Calif. Fertilizer Assoc.)

Material	*Grams Equivalent to 100 Grams of Sulfur*	*Grams of Calcium Carbonate Neutralized by 100 Grams of the Material*	*Grams of Calcium Carbonate Required to Neutralize Material per 100 Grams of Nitrogen*
Sulfur	100	312	—
Sulfuric acid	306	104	—
Gypsum	538	58	—
Aluminum sulfate	694	45	—
Iron sulfate	869	36	—
Anhydrous ammonia	212	148	180
Ammonium sulfate	283	110	530
Urea	439	71	160
Ammonium nitrate	503	62	180
Aqua ammonia	867	36	180
Single & treble superphosphate	—	0	—
Most potassium salts	—	0	—
Calcium nitrate	(1560)[a]	(20)[b]	(130)
Potassium nitrate	(1357)	(23)	(180)

[a]Alkaline reaction; grams of material necessary to neutralize 100 grams of sulfur.

[b]Alkaline reaction; equivalent to the same number of grams of calcium carbonate.

lime (calcium and magnesium hydroxides) act more quickly than lime-stone but are more expensive and disagreeable to handle. Small particles of limestone will increase pH more quickly than large ones. Lime should be spread evenly on the soil and worked into the surface for 50 to 100 mm (2-4 in) or deeper if there are no plants. When it is needed, lime should be incorporated into the soil during site prepara-tion before planting. The increased pH will improve soil structure and will increase phosphorus and molybdenum, reduce harmful levels of soluble manganese and aluminum, and enhance microbial activity.

Wood ashes are 30 to 50 percent as effective as calcium carbonate in neutralizing acid soil, since they are composed of approximately 45 percent calcium carbonate (American Horticultural Society 1980). Hardwood ash contains about one-third more calcium than does soft-wood ash. It also contains up to 30 percent potassium and can be used as a source of K in acid soils (see Chapter 11). Wood ash should not be applied directly to alkaline soils but can be composted along with leaves and other organic matter where organic acids will partially neutralize soil alkalinity.

Alkaline soils are usually acidified by the application of agricul-tural sulfur (Table 6-5) or sulfur-containing materials such as iron or aluminum sulfate (Table 6-6). The amount of sulfur required to lower pH is influenced greatly by the carbonate content of a soil. If a soil is high in carbonate, it should be analyzed to determine the precise amount of sulfur needed to produce the desired pH; how to take a soil sample is described in Chapter 11. This is a fairly simple test for a chemical laboratory. Finely ground dusting and wettable sulfurs act faster than coarser agricultural sulfur but cost more. Like lime, sulphur should be applied evenly on the soil and worked into the surface 50 to 100 mm (2-4 in) or deeper if possible.

Alkali Soils   In arid and semiarid regions where rainfall is not ade-quate for leaching, soil is usually alkaline and high in salts. Corrective measures involve leaching salts from the soil (see Chapter 12). If sodium represents more than 15 percent of the exchange capacity, a medium- or fine-textured soil will lose its granular structure due to deflocculation of clay particles. Such soils are called *alkali soils* and are exceedingly impervious to air and water. Gypsum or sulfur should be added to alkali soils before they are leached.

BIOLOGICAL PROPERTIES OF SOILS

A teaspoonful (5 ml) of soil may contain billions of living organ-isms (Clark 1957). Among the soil inhabitants are those that decom-pose organic matter, transform nitrogen, improve soil tilth, produce antibiotics, and otherwise affect plant welfare.

Soil Animals   The usefulness of animals to the soil is essentially inversely proportional to their size (Pritchett 1979). The more important animals are insects, worms, protozoa, and nematodes. A wide range of insects and spiders chew and move organic matter into the soil.

Earthworms are probably the best known of the soil fauna and occupy the greatest volume. They contribute to moist, friable, and fertile soil by aerating, mixing, and enriching it in tremendous quantities. This does not mean, however, that earthworms can be added to a dry, compacted, or infertile soil to transform it into a decent growing medium. The worms most likely will crawl away in the night to seek a more favorable site.

Protozoa—one-celled organisms—are the most numerous of soil fauna. They decompose organic matter and bacteria, primarily in the surface soil. Nematodes, microscopic roundworms, include both saprophytic and parasitic species. The saprophytes are free-living and are generally beneficial in that they decompose organic matter (Pritchett 1979). Some of the parasitic nematodes attack other microorganisms in the soil and are considered desirable. Others, however, invade the roots of plants and can inflict considerable damage (see Chapter 20).

Soil Microflora   Microflora in soil include algae, bacteria, actinomycetes, and fungi. Algae and a few species of bacteria contain chlorophyll and are able to form carbohydrates. Algae are most abundant in fertile soil that is neutral or slightly alkaline. They assist in dissolving soil minerals and in forming soil.

Bacteria, even in anaerobic soils, are involved in many biological and chemical processes that contribute to decomposition of the organic matter. Certain bacteria, alone and in symbiosis with leguminous plants, fix atmospheric nitrogen; others denitrify nitrates. Still other soil bacteria may cause serious plant diseases, such as crown gall (see Chapter 19).

Actinomycetes are similar to bacteria in their environmental requirements, and they are similar to fungi in that they form mycelia, masses of hyphae (hair-like filaments), and in the method by which they decompose cellulose. Some species, in symbiosis with the roots of certain plants, such as alder and ceanothus, can fix nitrogen.

Fungi are probably the most important decomposing microflora, particularly in well-aerated, acid soils. A number of soil fungi cause serious plant diseases: damping off of seedlings, phytophthora root rot, armillaria root rot, and verticillium wilt (see Chapter 19). Fungi also form symbiotic relationships with roots, which can be vital to the well-being of a plant. These associations are called *mycorrhizae.*

Mycorrhizae

Many trees, including those on previously unforested land, fail because certain fungi are not present to form mycorrhizae with the roots of the trees. Mycorrhizae (*myco* means fungus and *rhiza* means root) are root structures created when young lateral roots are invaded by specific fungi that form symbiotic associations to the advantage of each. Mycorrhizal plants are reported to grow more vigorously and remain healthier than do noninfected plants under stressful conditions, as in infertile soils, at arid sites, in the presence of root-disease organisms, and in other harsh environments (Merrill and Solomonson 1977).

Mycorrhizae are numerous and common in most higher plants. Pritchett (1979) has estimated that more than 2000 species of mycorrhizal fungi exist on trees in North America. Merrill and Solomonson report that mycorrhizae occur in more than 80 percent of vascular plant taxa investigated. Although almost all researchers report that each tree species tends to produce mycorrhizae only with characteristic groups of fungi (Pritchett), Iyer, Cavey, and Wilde (1980) state otherwise. They report that all currently or previously forested soil contains fungi that can form mycorrhizae with all tree species and that prairie and other grassland soils that have never been forested do not have such fungi.

Mycorrhizae generally divide into two major types, depending on the relationship of the fungal hyphae and the root cells. The fungal hyphae of *ectomycorrhizae* grow between the cortical cells of short lateral roots and form a sheath or mantle around the root invaded. The fungal hyphae of *endomycorrhizae*, on the other hand, primarily invade individual cells within the cortex and do not form a mantle around the roots. Even though fungi invade the cortex vigorously, they do not invade the endodermis or the more interior tissues of the root. Mycorrhizal fungal mycelial strands and their hyphae will extend from the roots they have invaded and function very much like the root hairs of normal roots, except that they are better able to extract nutrients that are otherwise not readily available (Merrill and Solomonson). Ectomycorrhizae are found almost exclusively on trees, whereas endomycorrhizae are more numerous and widespread, occurring on most families of angiosperms and gymnosperms (Pritchett). Ectomycorrhizae infect lateral roots, which are greatly shortened by their presence, are swollen, frequently dichotomously branched, and usually devoid of root hairs (Fig. 6-5). In contrast, endomycorrhizal infection does not grossly affect root appearance. Even though it involves the invasion of roots and even of individual cells, it does not kill cortical cells but often prolongs their effectiveness.

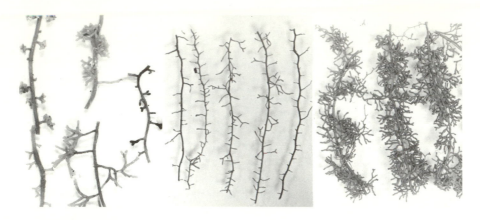

FIGURE 6-5
Mycorrhizae increase the ability of roots to absorb water and nutrients.
Examples of five ectomycorrhizal types on loblolly pine roots (each root
segment is about 25 mm [1 in] long) from a conventional nursery (left).
Nonmycorrhizal roots of loblolly pine (center);
loblolly pine ectomycorrhizae formed by *Pisolithus tinetorius* (right)
(center and right root segments about 50 mm [2 in] long).
(Photos courtesy Donald H. Marx, U.S. Forest Service)

Plants can benefit from mycorrhizae in several ways. Nutrient
uptake, particularly that of phosphorus, is enhanced in infertile soils
because mycelial strands and their projecting hyphae explore the soil
more extensively and are apparently able to dissolve and absorb nutrients that are unavailable to normal roots (Bowen 1973). Even though
a given section of mycorrhizae may not absorb water as quickly as a
similarly situated section of normal roots, the extensive fungal exploration of the soil nevertheless increases the total amount of water available to the plant. The fungal hyphae may also exude toxins, and the
fungus mantle may serve as a physical barrier that reduces infection by
soil-borne pathogens (Marx 1973).

Fortunately, tree mycorrhizal fungi are legion; their spores spread
extensively by wind but do not form mycorrhizae unless they find host
trees (Iyer, Cavey, and Wilde 1980). Attempts to improve plant performance by enhancing the variety and concentration of fungi have
rarely been successful (Mikola 1973). Appropriate fungi are usually
already present in the plant, the soil, or both if the site accommodates
similar plants. On the other hand, mycorrhizal enrichment would be
beneficial in some situations, as when exotic tree species are introduced,
particularly by seed, into grasslands; when noninfected plants are
planted in subsoil or sterilized soil in landscape planters or mine spoils
isolated from sources of inoculum; or when noninfected plants are
grown in fumigated nursery soil.

Seemingly noninfected plants will grow satisfactorily if they are wisely fertilized and watered and if diseases are controlled. Iyer, Cavey, and Wilde, however, report that seed of nearly all tree species planted in grassland soils, including those of high fertility, usually produce trees that die within two years. Mycorrhizae do not develop as extensively on plant roots that are growing in fertile, moist soil as on those in infertile and inadequately watered soil. Attempts to grow noninfected plants in an infertile soil devoid of appropriate fungi would be devastating. In some infertile soils, even abundant mycorrhizal development may not supply enough nutrients for vigorous growth, particularly if plants are young. Generous fertilization will decrease mycorrhizal development. If soils low in phosphorus are fertilized heavily with nitrogen, plant growth may actually be suppressed. This is thought to occur because fewer mycorrhizae develop under high-nitrogen conditions and thus absorb less phosphorus for the plant.

Mycorrhizal fungi can be introduced by means of soil or surface litter from woody plantings. The soil or litter should be raked into the surface under the plants to be inoculated. If plants ready for planting are not already infected with the appropriate fungi, they may have their roots dipped in a slurry of inoculated soil. Seeds can be inoculated by dipping them in a slurry of inoculum. Alternately, pure cultures of mycorrhizal fungi may be sprinkled on the soil and raked in around the plants. The most advantageous method will depend on the number of trees or seeds involved and how far the inoculum must be transported.

Even though mycorrhizae are necessary to the survival of certain plants under infertile conditions, appropriate mycorrhizal fungi are present at almost all sites. Mycorrhizal inoculation may be extremely valuable in low-maintenance landscapes known to be devoid of appropriate fungi.

SOIL WATER

Landscape plants probably suffer more from moisture-related problems than from any other cause. For them it is either feast or famine, flood or drought, air or suffocation, acceptable water or saline water.

No organic process occurs in the absence of water. Water is a primary constituent in the photosynthetic production of organic matter. Water is the solvent for nutrient and food transport within plants. Transpiration cools plants. Roots can extend into soil and shoot tips grow only when water is absorbed and the consequent turgor produced. Nevertheless, though water is vital to the well-being of a plant, excess water is often responsible for decline and death.

Soil is the reservoir that supplies water to terrestrial plants. Air and water occupy the pore space between soil particles. Pore space occupies about half the total soil volume. Porosity is determined by soil texture and structure. Fine-textured soils, particularly if they are well aggregated, have more total pore space than do coarse-textured soils. Total porosity, however, is less important than the size of individual pores. Large pores allow for faster water movement and better aeration. In a well-aggregated and fine-textured soil, the soil will not only be well drained but will also have a high moisture-holding capacity. Sandy soil has a high percentage of large particles and pores, is well drained and well aerated, but has a limited water-holding capacity.

The size and number of pores determine the amount of water that a soil can hold. Ideally, 30 to 50 percent of the pores will be large enough to allow water to drain from them after rain or irrigation. Such pores are called *noncapillary pores*, and the water that drains from them is *gravitational water*. The amount of water remaining in the soil after drainage has taken place is called the *field capacity*. The moisture remaining in the soil is held against the pull of gravity in the smaller pores by *capillarity*, on the particle surfaces by *adsorption*, and even between the layers of single clay particles. About half of this *capillary water* is available to plant roots. As water is absorbed by plants, smaller and smaller pores are drained and the films of soil water become thinner until the *tension* or force with which water is held becomes so great that the roots are no longer able to absorb water fast enough to keep the plant from wilting. The soil is said to be at the *permanent wilting point* if the plant will not recover overnight (or in 100 percent relative humidity) unless water is added to the soil (Fig. 6-6).

The water between field capacity (FC) and the permanent wilting point or percentage (PWP) is *available water*. The water that remains in a soil at PWP is essentially unavailable to plants and is termed *hygro-*

FIGURE 6-6
Soil moisture conditions. At saturation, pores in the soil are filled
with water (left); at field capacity, water adheres to soil particles after drainage
and fills smaller pores (center); at the permanent wilting point,
a thin film of water remains on soil particles and roots cannot absorb water
fast enough to prevent wilting (right).

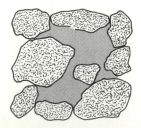 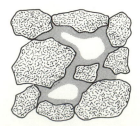 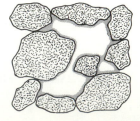

scopic water. Hygroscopic water is adsorbed on the surface of soil particles and is present in the tiniest capillary pores. It can be removed by oven-drying the soil or exposing thin layers to the sun, although some water will still remain chemically combined with clay particles. At the other extreme, when the soil pores are completely filled with water the soil is said to be *saturated*.

Another term, *container capacity*, is useful for considering water relationships in shallow container soils. As discussed in the following section, water does not drain freely from a fine-textured soil to one that is coarser or to air. Therefore, the amount of water left in a container soil after it has stopped draining will be somewhere between field capacity and saturation, depending on soil depth (Fig. 6-7).

The amounts of water held by a soil at saturation, at container capacity, at field capacity, and at permanent wilting, and the water available to plants are characteristics of the soil, not of the plants growing in it. These amounts of water can be expressed in different ways. Most often, they are determined by weighing a representative sample of soil, including its water, and oven-drying it until the weight is constant for an hour or more. The water content or weight is then expressed as a percentage of the dry weight of the soil. The amount of water can also be expressed on a percent-by-volume basis, but this is more difficult to determine, since the volume of the soil sample must be measured along with the volume of water, which is calculated from the weight of the soil before and after drying. Water content is also

FIGURE 6-7
The water content of soil in a planter varies with depth.
In soils of similar texture, the water content at a given distance
from the bottom will be about the same.

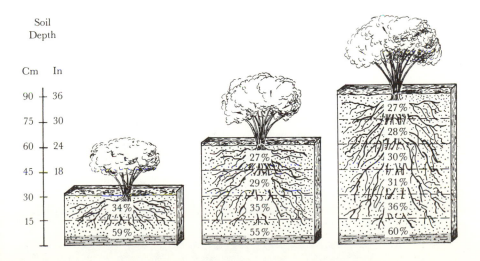

expressed as the depth of water per unit depth of soil, for example, cm of water per meter of soil or inches of water per foot of soil. Water use or loss (evapotranspiration) is usually expressed in depth of water lost per day, week, or year. Rainfall and water application, particularly from sprinklers, are also expressed as depth of water. The water available to a plant is directly related to the depth of soil and plant roots.

The force with which a soil holds moisture is called the *soil moisture tension* and is most commonly expressed in *bars* or *centibars* (cb). One bar is equal to a standard atmosphere (1030 g/cm^2 or 14.7 lbs/in^2). At the various soil moisture levels, moisture tensions are

	Centibars
Saturation	0
Field capacity	30
Reduced water availability	70
Permanent wilting point	1500

Moisture was originally thought to be equally available to plants at all levels between field capacity and the permanent wilting point (Veihmeyer and Hendrickson 1955). For most practical purposes, that is indeed the case, even though moisture tension increases from 30 cb at field capacity to 1500 cb at permanent wilting. Moisture tension can essentially be ignored in most field soils because 60 to 90 percent of the available water is held between 30 and 100 cb and only 10 to 40 percent above 100 cb tension (Fig. 6-8). The amount of available water may vary in different soil horizons and therefore influence rooting density and extent. As a result, a plant may be able to absorb enough water even though much of the root zone is at PWP. The water available to trees and large shrubs may be reduced at moisture tensions of 70 cb.

Certain types of clay particles, particularly montmorillonite, swell and shrink as soil moisture fluctuates between field capacity and PWP. This causes an internal loosening or cultivation of the soil that helps to retain or improve soil structure. It can, however, play havoc with all but the strongest small building foundations and pavements, particularly if many plant roots also invade the underlying soil (see Chapter 16).

Movement of Water

Water movement in soil depends primarily on the height of the water column (head or potential), the texture and structure of the soil, and its initial water content. If water does not infiltrate a soil as quickly as it is applied by rain or irrigation, the run-off may cause soil erosion, uneven water distribution, and on a slope, less water than applied will accumulate in the root zone. Water will enter grass- or

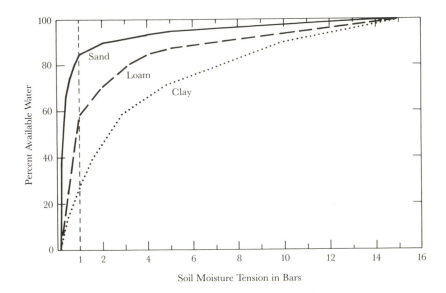

FIGURE 6-8
Available moisture at different soil moisture tensions for three soil types.
(Richards and Marsh, 1961) At 1 bar, 85 percent of the available water
in a loamy sand is available to the plant; 58 percent of available water
in a loam soil is available to plants; and 30 percent in a clay soil.

mulch-covered soil more rapidly than it will a bare soil (Table 6-7).
The surfaces of bare soils often crust from *puddling* (dispersion of soil
aggregates by drops of water) during rain or sprinkler irrigation (see
Fig. 13-5); these crusts slow water intake.

Water penetrates a soil more rapidly before the surface 20 to 30 mm
(1 in) are wetted. Water moistens the surface soil to field capacity or
above before it moves farther down in the soil. Thirty-five mm (1.4 in)
of water applied to a sandy loam soil (see Table 12-2) at the wilting

TABLE 6-7

Typical water infiltration rates into three soils, each with a surface area
of 148 square meters (1600 square feet) (Harris 1962)

	Soil with Cover				Soil without Cover			
Soil	mm/ hour	in/ hour	l/ hour	gal/ hour	mm/ hour	in/ hour	l/ hour	gal/ hour
Sandy loam	13	0.50	2000	530	8	0.33	1350	355
Silt loam	8	0.33	1350	355	6	0.25	1000	265
Clay loam	6	0.25	1000	265	5	0.20	800	210

point will wet the top 600 mm (2 ft) of soil to field capacity. The soil below this level, however, will remain at the wilting point.

Above field capacity, water moves down through the soil in response to gravity. The greater the head (potential) and the larger the pores, the faster the water moves. Below field capacity, water moves by capillarity from zones of higher moisture content (low tension or high potential) to those of lower content (high tension or low potential). Capillary movement can be in any direction. Water can move farther (though it will move more slowly) in soils with small pores than in soils with large pores. Unless it is near a water table, capillary flow is extremely slow and restricted in distance covered. Capillary movement decreases rapidly as soils dry below field capacity. A water table will supply the soil above it by capillarity for some distance, depending on soil pore size. In laboratory experiments, the upward movement of water was significant in a fine, sandy loam from water tables as deep as 8 to 9 m (27–30 ft) (Gardner and Fireman 1958). Rate of movement was fairly uniform until the water table reached to within 2 m (6 ft) of the surface; it increased by a factor of eight at 1 m (3 ft). If a water table remains at about the same level (within about 9 m [30 ft] of the surface) throughout the year, it can be a continuous source of water for plants.

As the permanent wilting point is approached in a soil, capillary water movement all but ceases and water moves only in the form of vapor. Although vapor movement in soil is extremely slow, it is more rapid than was once thought. Vapor can condense on one surface of water held between soil particles while water is evaporating from the other surface (Philip and DeVries 1957). Vapor movement is greatly increased by temperature gradients within the soil. Other things being equal, water vapor moves toward colder zones, where vapor pressure is lower than at higher temperatures. Surface soil is warmer in summer and colder in winter than soil at greater depths. A similar cycle can happen daily: Moisture vapor moves to the surface and condenses during the winter and during each night, particularly if strong radiation cooling occurs. The movement will be downward during the summer and during each day. Only in special circumstances, however, will the amount of water condensed in the root zone be sufficient to sustain plants. The more extensive the foliage canopy, the less radiation cooling of the soil surface will occur at night, and the less water vapor will rise. The water supply of small plants, however, may be augmented if a clear plastic sheet is used to condense water vapor from a 3 to 4 m^2 (30–40 ft^2) area of soil and concentrate it at the base of the plant (see Fig. 10-11).

Water movement in soils can be inhibited not only by compaction but also by differently textured strata. If a coarse stratum is on top of a finer-textured soil, water may accumulate above the lower layer until it can be more slowly absorbed. Should the interface slope, water will flow to lower levels above the interface, resulting in uneven watering and a possible low wet spot. When drainage is complete, the finer-textured soil will be at field capacity, but the coarser soil above will be somewhat below its field capacity because of the greater tension with which water is held by the finer-textured soil.

A more serious situation occurs when a coarse-textured stratum is located below soil of finer texture. Water will not move from the fine-textured soil into the coarse soil below until it saturates the top layer and develops enough of a head (potential) to overcome the cohesive and adhesive forces holding it there. Even after drainage has been completed, the soil immediately above the interface will be saturated. This essentially constitutes a *perched water table*. Roots penetrate this interface with difficulty (Fig. 6-9). This phenomenon cannot be stressed too strongly because it comes into play in a number of land-scape situations: Drainage pipes are sometimes lined with gravel to ensure lateral water movement to the drain; the bottoms of planters are

FIGURE 6-9
An extreme example of stratification with abrupt boundaries between the top 300 mm (1 ft) of sandy loam, the second 300 mm (1 ft) of sand, and the underlying silty clay loam. Although soil density was not a problem, the roots of young pear trees would not grow across the boundary into the coarse white sand (left). Compare this root growth with that two years later when trees had been replanted in a soil mixed uniformly with the use of a trenching machine (right). (Wildman, Meyer, and Neja 1975)

usually covered with gravel; a container rootball may experience decreased moisture when planted in landscape soil; and gardeners will sometimes place gravel in the bottom of a planting hole with the fallacious notion that it will improve drainage.

LANDSCAPE SOILS

Many soils may adversely affect plant performance unless they are modified or precautions are taken. The original soil may be difficult to manage, and many landscapes are manipulated during site development by the removal of topsoil, addition of soil, compaction of soil, burial of construction materials, or construction of pavement. It is a wonder that plants grow at all in some soils. In addition to the following discussion, many soil problems are considered in the first parts of Chapters 10 and 18.

Soil Compaction

Soil aggregates are delicate and easily damaged by cultivation and other types of compaction. Few soils can withstand traffic without becoming severely compacted. Aggregates are compressed and sheared into smaller aggregates or individual particles, which then retard the movement of water, gas, and roots.

Aggregates are particularly vulnerable when wet. Water not only lubricates the soil particles but also causes the aggregate-cementing agents to become more fluid, thus allowing the particles to slip easily into positions that take up less space. The amount of compaction a given force or weight will cause depends on the original soil structure, texture, and moisture content. A loose, well-aggregated soil will compact more than a soil that is less well aggregated. Soils that have a broad range of particle sizes can be more severely compacted than soils of uniform texture, regardless of particle size, because small particles fill the pores between larger particles. The higher the moisture content, the more easily a soil will compact, until the pores are so compressed that little air remains. Maximum compaction is most likely at field capacity, when soil particles move fairly easily with pressure, and in soils where there are air-filled pores that can be compressed or filled with smaller particles. A saturated soil cannot be further compacted, because the water in the pores does not compress. Because most pressures on soil are localized, as are those caused by feet and tires, muddy soils move out from under the area of compression and in so doing shear aggregates apart and rearrange particles. Soils that undergo this abuse are said to be *puddled*, and upon drying they shrink and become compact and practically impermeable.

The pressure of a compacting force spreads with depth, and its damaging effects are dissipated within 100 to 150 mm (4–6 in) of the surface. Even a shoe heel can exert a force per unit area equal to that of heavy equipment. A vehicle with tires inflated to a given pressure exerts a similar pressure on the soil. Turning, starting, and stopping increase the force. Heavily treaded tires exert more pressure near the soil surface than do smooth tires. Compaction depends not only on the pressure exerted but also on the duration and frequency of the exertion.

The compacting effects of pedestrian and vehicular traffic are usually confined to the surface 100 mm (4 in) of soil (Chancellor 1976). The greatest compaction occurs about 20 mm (0.75 in) below the surface. In a similar manner, cultivating to the same depth year after year, as is common in many farming operations, compacts the soil just below the depth of plowing. This can create a very dense layer 25 to 50 mm (1–2 in) thick, known as a *plowpan*. Because plowpans interfere with water and air movement, they can be a problem when farmland is converted to landscape plantings, particularly if landscapers add fill soil without first breaking up the plowpan. Heavy equipment, trucks, and tractors can affect the soil at greater depths, but the most serious compaction will usually be confined to the surface 150 to 200 mm (6–8 in), unless the soil is so wet that it flows from under the tires.

Aggregates on the surface of bare soil can also be broken down by the impact of rain or sprinkler irrigation. Finer particles may seal the surface, thus reducing water infiltration and air exchange.

Preventing Soil Compaction   Although some breakdown of structure within the surface 300 mm (1 ft) of soil may be inevitable when construction is taking place, an understanding of soil texture and structure can be applied to minimize structural breakdown. Compaction is easier to prevent than to remedy. The following practices should be observed:

> Cultivate soil when it is dry or moderately moist; avoid wet soils.
>
> Avoid recompaction of freshly plowed or loosened soil; the less tillage after loosening, the better.
>
> Schedule landscape maintenance work when the soil is as dry as possible.
>
> Keep travel over the site to a minimum; confine it to a few paths and keep it away from trees.
>
> Use lightweight vehicles with large, smooth, low-pressure tires.
>
> Spread thick, coarse mulch on the soil surface to disperse the load.

Rejuvenating Soil of Poor Structure   Before a compacted soil is brought to final grade, it should be broken up by ripping or deep plowing when it is relatively dry. The area should then be rough-graded.

Incorporating organic matter into the soil is a good general practice, but most people add too little organic material to be very effective in rejuvenating soil structure. In hot regions, organic materials decompose rapidly, so that large amounts must be added more than once if they are to have much effect on soil structure.

Should additional soil be necessary to bring the surface up to grade, its texture should be similar to or coarser than the soil already there. After rough-grading or the addition of fill soil, the area should be thoroughly irrigated to settle the soil before final grading.

In the effort to reduce the cost and time of construction, fill soil is often put on top of compacted soil without any prior preparation. An uneven and compacted surface overlaid with a shallow fill can lead to moisture and aeration problems. High areas of compacted soil under the fill will cause water to flow toward low areas under the fill, resulting in wet spots. Plantings in such an area will be watered unevenly and will consequently grow unevenly or poorly. If the fill is fairly deep, there will be less of a problem.

Amending Soil

Amendments are sometimes added to improve the aeration and tilth of fine-textured or compacted soils, to increase somewhat the moisture and nutrient capacities of sandy soils, or to decrease bulk density (weight per unit volume) when excessive weight may be a problem. Mineral amendments (pumice, perlite, vermiculite) are fairly permanent in the soil, if they are not compacted. Organic matter (peat moss, bark, sawdust, manure, compost) is more commonly used, but it decomposes with time. To have much effect, an amendment must constitute 25 to 50 percent of soil volume.

Amending an entire soil area for tree and shrub planting is seldom justified. It would probably be just as effective and less expensive to break up the compacted surface, bring it to rough grade, irrigate to settle the soil, and then, when the soil is dry enough, bring it to final grade, plant, and mulch. The first two steps should be performed when the soil is fairly dry. In most cases, amending the backfill with organic matter will not be of any particular benefit (see Chapter 8).

If a soil is to be amended, use a slowly decomposing organic material. It may include sphagnum peat, redwood sawdust, untreated fir bark, rice hulls, or certain wood products treated and fired so they are high in lignin and slow to break down in soil. Fresh redwood sawdust should be leached with water to remove a phytotoxic constituent before it is mixed with the soil. Rice hulls can be a source of weed seeds unless they are treated. Even though these materials decompose slowly, they can reduce the nitrogen level in the soil soon after they are added. Since peat and treated wood products usually have a low cellulose content, they should not deplete the nitrogen supply. With other amendments, however, it might be well to spread nitrogen on the surface (5–10 gm/m^2 or 1–2 lbs/1000 ft^2) after the organic matter has been added. The organic matter should be spread 75 to 100 mm (3–4 in) deep and rototilled into the surface 100 mm (4 in) of the soil.

Amending a soil with organic matter or other materials essentially dilutes the soil so that some of the particles are spread farther apart. Sand is occasionally suggested as an amendment. As sand is added, however, clay particles of the soil fill the spaces between the sand grains, producing a denser, less porous mixture. Not until sand constitutes 45 percent of the volume will the mixture begin to have some of the properties attributed to sandy soils (Fig. 6-2). It is thus an unsatisfactory amendment for clay soils.

FURTHER READING

HAUSENBUILLER, R. L. 1978. *Soil Science: Principles and Practices*. 2nd ed. Dubuque, Iowa: Wm. C. Brown Co.

PRITCHETT, W. L. 1979. *Properties and Management of Forest Soils*. New York: John Wiley.

UNITED STATES DEPARTMENT OF AGRICULTURE. 1957. *Soil: The 1957 Yearbook of Agriculture*. Washington, D.C.: U.S. Superintendent of Documents.

CHAPTER SEVEN

Planting Site: Preparation

Landscapes range from intensive plantings in indoor and rooftop containers to natural areas that receive little or no maintenance; from deep, fertile agricultural soils to shallow, landfill soils of varied origin, subject to toxic gases and high temperatures; from compacted soil in planting wells surrounded by asphalt and buildings to soil on an exposed seacoast. You name it; someone will want to grow plants there.

The performance of plants in the landscape depends on how well the species are adapted to the specific environment in which they are to grow, the quality of the planting stock, preparation of the site, planting methods, and later care. A site is usually determined in advance, so site preparation and species selection are quite important. A site that is less than ideal can often be modified to improve plant performance and ease of maintenance. Many of these aspects have already been covered; some others must be dealt with now.

PROTECTING EXISTING TREES

When land is developed for residential, industrial, business, park, roadway, or other specific uses, existing trees may deserve to be saved. In many cases, the placement and design of buildings and pavement can capitalize on some of the finer specimens and provide construction details that will improve the future well-being of remaining trees. With early planning and some simple precautions, many trees can be saved with little effort or expense. Certain trees in strategic locations may justify considerable effort on their behalf.

Choosing Trees to Save

Trees must be evaluated on the basis of their desirability in the new landscape and the effort that will be needed to save them. Tree species, location, health, and present and future sizes should be considered. Characteristics to be evaluated will be the same as those one considers when a new tree is to be planted (see Table 3-1). Certain species can adapt better than others to changes in their environment. The USDA (1975) identifies elm, poplar, willow, plane tree, and locust as being tolerant to changes in their surroundings. Less adaptable trees include beech, birch, hickory, tulip tree, some oaks, most maples, and most conifers.

Decisions about preservation will be based on desirable locations and both present and ultimate sizes of trees. Young, small trees adapt to change more easily than do older, larger trees. It may be easier and cheaper, however, to remove trees that are below a certain size (100–150 mm, for example, or 4–6 in) and to replace them later.

If buildings and landscape have been developed to feature certain trees, those trees should be in good health with a life expectancy that will justify the effort of selecting and preserving them. Other trees may be featured in small groves, where the overall effect is more important than the condition and appearance of any one tree.

Trees to be saved should be marked prominently so there will be no possibility that they will be bulldozed down or seriously damaged. A handsome 20-m (65-ft) black walnut, 1.5 m (5 ft) DBH (diameter at breast height), that was to be the key feature of a new apartment complex in Davis was mistakenly bulldozed down in the last month of construction. Such blunders will continue to happen unless steps are taken to minimize error.

Trees and shrubs that are not to be saved should be removed before construction begins if the trees to be saved can thus be better protected. In other cases, trees scheduled for removal can be left during construction to protect those that are to remain. Trees to be removed should be cut near the ground and their stumps ground rather than pulled, so that roots of plants to be kept are not injured. Remaining trees may need pruning and even guying to reduce the likelihood of windthrow when the removal of adjacent trees leaves them more exposed. Previously shaded treetrunks exposed to the afternoon sun should be shaded or painted with white latex paint. The bark of large trees not previously exposed to the sun can be killed if there is no transition period from shaded protection to full sun.

Grading and Construction Considerations

Almost every building and landscape development will involve some grading and excavating. The consequences of these activities and of wind patterns, exposure to sun and reflected heat, space limitations, and changes in the soil must be considered. Underground construction may block water flow in a soil stratum so that water accumulates on the upper side and is reduced on the lower; such changes may affect trees adversely. Unless construction or later maintenance is designed to modify these effects, serious injury could result before the cause is determined.

Raising the Soil Level   Soil placed on the surface around existing trees can create problems involving gaseous exchange in the root zone and soil moisture. The latter problem can be eliminated or greatly minimized if the soil added is of the same or a coarser texture and structure and is properly prepared beforehand. Soil aeration is the critical factor. The roots must receive adequate oxygen and not be subjected to buildups of carbon dioxide or toxic gases. In most cases, measures that will assure this aeration need be employed only until new tree roots become established in the fill soil. Another concern is to protect trees from crown rot, which can occur if the trunks of certain species are kept moist for long periods. Fills of 150 mm (6 in) or less with fair to good drainage will not harm most species of trees (USDA 1975) and will need no aeration channels.

There seems to be some uncertainty among arborists as to whether tree roots will grow upward into fill soil. No thorough study has examined that question, although numerous reports do indicate rooting into fill soil by many species. Duling (1969) cites a 1964 National Arborist Association Newsletter observation:

> Most everyone who has dug out stumps of trees around which there had been earth fills has occasionally found two root systems: the original roots, and another set that had developed in the filled soil. In such instances the original roots are usually dead and the new upper story roots are keeping the tree alive and growing. These new roots developed from the original root collar at the base of the tree or from the trunk itself.

The phenomenon may be widespread. Wilson (1970) states, "The small, short roots that grow lateral to the large roots grow at many angles to the vertical and a good many of them actually grow up into the forest floor." Zimmermann and Brown (1974) observe, "Each successive generation of lateral roots becomes less and less responsive to gravity, and other factors including available soil moisture and temperature often influence and control the direction of root growth. Moisture from above causes the roots to grow upward into the moist substrate instead of downward."

For the best evidence that roots can and do grow upward, one can lift a mulch that has been on the soil under a plant for several years.

Where a bare soil exposed to the sun was previously devoid of surface roots, it will be alive with active roots in and under the mulch. If fill soil conditions are favorable, the roots of most plants will grow up into this new soil. More information is needed to determine whether there might be species exceptions.

Tree roots have a difficult time rooting into fill soil, however, if the fills are installed as commonly recommended: "Install a layer of gravel and a system of drain tiles over the roots of the tree" (USDA 1975). Variations of these recommendations include the use of stones under gravel and straw, burlap, or fiberglass matting to hold the 300 mm (1 ft) or so of fill soil (Bernatzky 1978, Duling 1969, Pirone 1978a). Until some of the fill soil works down through the stone and gravel to the original soil or unless there are roots beyond the gravel layer, only roots of the most aggressive trees are likely to grow into fill soil installed this way. Although water can reach the original soil directly through the drain tile from the surface, the fill soil over the gravel layer must become saturated before water will move into the gravel and down to the original soil. After draining, the fill soil or at least the lower portion of it will remain saturated (see Chapter 6). Water will flow along the bottom of the fill soil (if it slopes away from the trunk) to the edge of the gravel layer and then down through the original soil. In periods of excessive rain, the soil below the periphery of the gravel layer will become waterlogged.

A modification of the recommendations above will save money and give better results in the long run. The stone and gravel layer can be eliminated over the soil between aeration spokes radiating from the trunk dry well (Fig. 7-1). Fill soil between the aeration spokes will provide contact with the original soil, so that water will drain from the fill soil into the soil below and roots can grow more easily into the fill soil. The aeration spokes will allow oxygen to diffuse into the adjacent soil, and carbon dioxide and other gaseous products to diffuse out. The favorable root environment thus established will allow a root system to become actively stabilized in the two soils and balanced with the top. The aeration channels will progressively lose their effectiveness as they fill with soil, debris, and roots, and the essential continuing maintenance task will be to keep the dry well around the trunk functioning so that water does not accumulate there.

Certain steps will foster the successful adaptation of a tree's root system to a soil fill. The ground within the drip line of the tree and a little beyond it should be cleared of all plants, leaves, twigs, and debris. The top 250 mm (10 in) of soil should be checked for signs of compaction, slow water penetration, or layers of different soil textures. Any problems should be corrected as described later in this chapter, but procedures should be modified to minimize damage to roots. If the ground has been greatly disturbed, wet it to settle the loose soil and

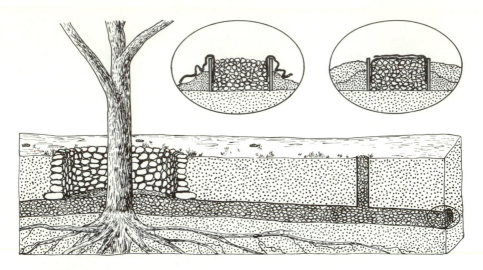

FIGURE 7-1

Aeration spokes of gravel radiating from a dry well 1 to 1.5 m (3–5 ft) in diameter
provide for gaseous exchange until new roots can grow into the fill soil.
Boards hold the gravel in place (left inset) until the gravel channel is covered
with fiberglass matting and some fill soil holds the gravel and fiberglass (right
inset); the boards are removed before the rest of the fill soil is applied.
Each gravel channel is vented to the surface near the drip line of the tree.

leave it undisturbed for several days. Without moving much soil or
placing any around the trunk, shape the soil surface so that no area
drains to the trunk. The surface soil should be loosened, but do not
cultivate or rototill deeply enough to injure roots. Mark on the surface
where the gravel aeration channels will radiate from or bypass the dry
well at the trunk. If the surface grade is such that water will be trapped
near the base of the tree, a drain to remove accumulated surface water
may be required. If necessary, remove soil from the aeration channel
paths (about 300 mm or 12 in wide) to provide a slope of 2 to 3
percent.

The gravel for aeration channels should be of uniform size (about
10 mm or 0.5 in) and covered with fiberglass filter matting (6–12 mm
or $\frac{1}{4}$–$\frac{1}{2}$ in thick). The channels should be 150 to 200 mm (6–8 in) in
depth (Fig. 7-1 insets). Boards can be used as forms for holding the
gravel. Line them with fiberglass matting and use some soil or gravel
on the outside to hold the boards in place as gravel is dumped into the
forms. Overlap the fiberglass matting on top of the gravel. Place fill
soil in the outside of the forms to hold the gravel in place when the
forms are lifted out.

The gravel channels should radiate from a dry well around the
trunk or be connected to channels that do. The inner wall of the dry

well should be 500 to 750 mm (20–40 in) from the treetrunk; the deeper the fill, the farther the dry-well wall should be from the trunk. The wall should extend the complete depth of the well. It may be necessary to get into the well to clean or service it. An open-jointed dry-well wall can be made of stone, brick, block, or broken concrete. The gravel channels penetrate the dry wall. At each place where the radiating channels intersect with the circular ring, a perforated PVC pipe 100 mm (4 in) in diameter should be placed vertically so as to protrude 20 to 40 mm (1–2 in) above the final surface. The other end of the vertical pipe will sit on the gravel below. Air will flow through the aeration channels by convection, and gases will diffuse from areas of high concentration to those of low. The dry well can be left open, covered with a grate, or filled with gravel. If it is filled with gravel, two vertical pipes with caps should be placed on opposite sides of the well so that they can be inspected periodically for water.

Because aeration is the primary concern, a sandy loam fill with good structure would be ideal. At the least, soil of a texture and structure identical to or coarser than the original soil should be used to bring the fill up to final grade. Before the fill soil is added, 5 g of nitrogen per m^2 (1 lb/1000 ft^2) should be broadcast on the loosened soil surface. When the fill is in place, the entire area should be well watered and the soil allowed to settle. If needed, additional soil can be applied in a week or two to bring the fill up to final grade. The new surface should drain away from the treetrunk and the vent pipes. During the soil filling, vent pipes should be covered and later filled with gravel or covered with perforated caps.

When grades are raised, a simple alternative to a dry well and aeration system may sometimes work. A number of reports indicate that some species form roots when fill soil is placed against or near the trunk. Duling (1969) reports that Charles Schmaltz of Rochester, New York, induced trees (including Norway spruce, black oak, and native varieties of maple and poplar) to root into fill soil. In May or June of 1936, he cut the bark of each tree to the cambium in several places about 300 mm (1 ft) below the final fill level. Even though the fill was as deep as 4 m (14 ft) on some trees, all were still alive 27 years later, when the report was made.

A number of species are known to develop new roots at places where the trunk is newly buried or just kept moist and dark. Wilson (1970) reports that poplar, willow, and black spruce root easily after the trunks have been buried during a flood. Coast redwood is known to root readily when the trunk is engulfed by river silt. Walter Barrows (private communication, 1978) observed that several large eucalyptus growing in a ravine behind the civic center at Whittier, California, were doing well after the soil level had been raised around them by 6 to 9 m (20–30 ft) more than ten years before (Fig. 7-2). Pieces of broken side-

FIGURE 7-2
More than 10 years before this photograph was taken in Whittier, California, the soil level around these eucalyptus trees had been raised 6 to 9 m (20–30 ft).

walk had been stacked around each trunk as the soil level was raised. A utility trench dug several years later revealed that roots had grown from the trunks through the concrete pieces into the new surface soil.

Vigorous young trees that root easily and are tolerant of or resistant to crown rot will best survive raised soil levels. Preparation corresponds to that for a dry well and gravel aeration system. Water infiltration and penetration of the existing soil should be checked and corrected. The surface soil should be shaped so that water will not accumulate near the trunk and loosened so the interface with the fill soil will not interfere with water movement and root growth. Five g of nitrogen per m^2 (1 lb/1000 ft^2) should be broadcast on the surface of the soil.

In place of the dry well, put moist, uniform, medium sand (0.25–0.5 mm diameter) around the trunk so that it extends about 300 mm (1 ft) out from the trunk a little above the height of final grade. Firm, but do not compact, the moist sand so that its outer surface slopes away from the trunk but not more than 45° from the vertical (Fig. 7-3). Add fill soil of good structure and coarser texture than the original soil. To avoid a hard-surfaced interface, loosen the surface of the moist sand slightly as the soil is filled against it. As water moves down near the trunk, it will follow the interface between sand and fill soil, thus keeping the soil near the trunk drier than it would be without the sand. The toe of the sand cone should be within 1 to 1.3 m (3–4 ft) of the trunk; otherwise, an excessively large volume of soil around the

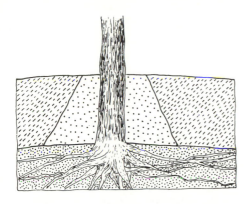

FIGURE 7-3
Medium sand around a trunk can often protect it from excessive moisture. The sand should extend about 300 mm (1 ft) from the trunk at the top and should slope away from the trunk. The sand should be higher at the trunk than the fill soil.

trunk may be inadequately watered. After the final grade is established to drain away from the trunk, be sure the sand collar is not covered with fill soil. Water will not enter the sand if its surface is covered with a finer-textured material, and aeration will be impeded.

Soil levels can probably be raised around trees with equal success at any season of the year. Less soil compaction will result when the surface soil is dry or frozen. On the other hand, trunk rooting will be best encouraged if the bark is cut about 300 mm (1 ft) below the height of final grade in early spring.

To encourage rooting into the fill soil and to ease the transition after raising a grade, thin out 20 to 25 percent of the leaf area and apply about 5 g of nitrogen per m^2 (1 lb/1000 ft^2) on the new soil surface. Rain or subsequent irrigation will move the nitrogen into the soil. This fertilizer recommendation differs from that of others. Bernatzky (1978) and Pirone (1978a) recommend that a complete fertilizer, containing nitrogen (N), phosphorus (P), and potassium (K), be applied to the original soil before the fill is installed. Tattar (1978) advises that "Phosphate fertilizer should be applied to encourage root growth. Addition of nitrogen, which primarily stimulates foliage production, should be avoided." Bernatzky states that "Trees require an abundance of K_2O and P_2O_5 to further root growth and drought resistance." Although phosphorus fertilization is beneficial to most shallow-rooted herbaceous crops and for trees that grow on old soils in certain parts of the world, almost no experiments have shown that fruit or landscape trees respond to phosphorus fertilization (see Chapter 11). Phosphorus and potassium are essential nutrients, but I am aware of no experimental evidence that they are any more essential for root growth than are other nutrients or that high levels of either of them promote rooting. On the other hand, reasonable amounts of nitrogen will invigorate trees and increase leaf surface, which will supply roots with food and hormones for increased growth. Whether the increased leaf area requires additional irrigation is open to question (see Chapter 12).

Once the fill soil has been well watered to settle it, the usual irrigation practice (if any) should be sufficient to sustain the tree. It is neither necessary nor beneficial to apply water through the dry well–gravel channel system in order to irrigate the roots in the original soil. Water applied to the new surface soil will move down into the root zone where needed.

Stabilizing Steep Slopes   Steep slopes often occur along roads, at construction sites, and in subdivisions on hilly terrain and may jeopardize established trees, roads, or buildings. *Breast walls* and vegetation can be used to stabilize these slopes and river banks (Leiser and others 1974). Breast walls differ from retaining walls in that they are built of loose, dry-laid rock placed against more or less undisturbed earth and receive little force from the earth behind them (Fig. 7-4). In addition to stabilizing a slope, they decrease the area covered by the slope or reduce its steepness.

Rock for breast walls should be as large as possible given the height of the wall and should be laid in a stable manner on undisturbed soil or rock sloping into the bank. The width of the wall should be at least one-half the height at its base. The bedding or base for each rock should slope into the bank, and the batter or outer face of the wall should incline toward the bank by one horizontal unit for each eight units of wall height.

As the rock is set in place, branches of woody species that root readily, such as willow and poplar, can be laid in the crevices of the wall, extended back into the slope, and covered with soil so that few

FIGURE 7-4
A breast wall of loose, dry-laid rock is used to retain the original grade near a tree trunk when the soil level is lowered near the tree. (USDA 1975)

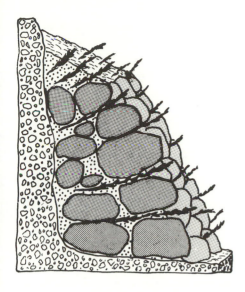

FIGURE 7-5
If small branches of easy-to-root trees and shrubs are placed between the rock and covered with soil as a breast wall is constructed, they will help to stabilize the wall and soften its appearance.

voids are left (Fig. 7-5). Cut branches no more than two days before placement and keep them from drying. Tamp the backfill firmly in place as the wall is built. Smaller rocks can be used in the larger interstices to reduce the amount of exposed soil. They must, of course, be held firmly in the wall. The branches, particularly if they root, increase stability, prevent or retard the sloughing of soil or rock from the slope, and soften the appearance of the wall. Plants that grow in the wall are particularly valuable for protection along stream banks.

Brush layering (or *buschlagenbau* in Schiechtl 1978) of woody branches can help stabilize fill slopes (see also Leiser and others 1974), both small and large, with grades of 3:1 and steeper. The technique involves placing alternate layers of fill and brush on the contour (Fig. 7-6). Depending on the size and steepness of the slope, the fill layer

FIGURE 7-6
If branches of easy-to-root trees are laid on the contour with their tips protruding as successive layers of fill are placed, they will stabilize steep slopes (left). Fill slope in New Zealand four years after willows planted by the brush layer method (right). (Photo courtesy R. L. Hathaway, Ministry of Works and Development, Palmerston North, N.Z.)

(lift) can be 0.6 to 1.5 m (2–5 ft) deep between two adjacent brush layers. The fill layer on which the brush is placed should slope into the hill at least 10 degrees from the horizontal. Branches of about 1 meter (3 ft) long (for shallow fills) to 2 or 3 meters long (for deeper fills) are placed more or less randomly, with some criss-crossing of stems, on top of each layer of fill soil. About one-fourth of the brush length should protrude beyond the fill when covered. Branches with butts 100 mm (4 in) thick or thicker can be used on the deeper fills. Each layer of fill is compacted as it is placed. Species that have flexible stems and root easily are preferred because the resulting root and top growth gives more or less permanent protection to the slopes. The branches of species that do not root easily can provide interim protection until other vegetation is established.

Brush layering can be used for original construction or as a remedial action on seriously eroding areas. For remedial work, one may have to cut into the slope in order to place the brush deep enough. Shorter brush can be used or the brush laid diagonally. In both remedial and original construction, one should start at the base of the slope and work up. Properly installed, brush layering can be an effective, relatively economical solution to a variety of problems concerning slope stability and erosion. Wattling is a variation of brush layering more commonly used for planting (see Chapter 10).

Severely compacted fill slopes are often required by law so that slopes will be stable. Developers favor steep slopes because they require less land area. It is difficult to establish vegetation on a steep, compacted slope, and the toe of such a slope may encroach on an existing tree and cause problems. A compacted slope is usually more impervious than a more level fill and can seriously interfere with aeration and rooting. Water from the slope may accumulate near its toe, further decreasing aeration and increasing the danger of crown rot for any trees located there. Most serious of all, a 450 to 600 mm (18–24 in) excavation is usually made to anchor the toe of a steep slope; this can sever a large proportion of active roots and jeopardize those below due to compaction.

If a tree near the slope is worth saving, several steps can be taken to protect it. Instead of excavating to anchor the slope toe close to the trunk, build a retaining wall up to 1.5 m (5 ft) high and support it with metal posts (2–2.4 m or 7–8 ft on center) set 3 m (10 ft) in the ground (Fig. 7-7). Such a retaining wall and the slope anchor excavation can be placed at least 2 m (7 ft) farther from the treetrunk than an excavation would be placed without a retaining wall. A retaining wall may mean that the slope above can be less steep, so that contour planting can be used in slope construction. A vertical perforated pipe or drain tile (100 mm or 4 in) can be used at or beyond the drip line of the tree and connected to a gravel channel leading through the retaining wall to provide some aeration for the covered roots and encourage root

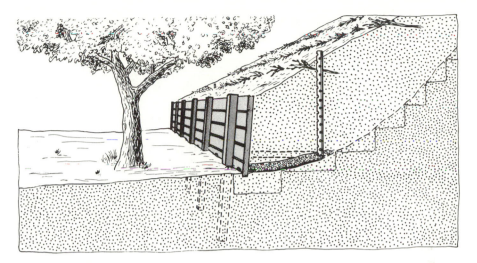

FIGURE 7-7

A well-anchored retaining wall can shorten the toe of a slope to protect a tree
from encroachment. The fill is further stabilized with an anchor excavation
and step terraces on the slope. A gravel channel and vertical vent pipe provide
aeration. Brush layers have been placed to stabilize the slope.

growth into the fill. The gravel channels should be covered with fiberglass filter matting and should emerge about 2 m (5–6 ft) apart along
the retaining wall. More than one gravel channel can be served by each
vertical vent pipe, but not more than 25 percent of the original ground
surface should be covered with gravel. The gravel channels need only
be 100 to 150 mm (4–6 in) deep and 200 to 300 mm (8–12 in) wide.
See Figure 7-7 for construction details.

Lowering the Soil Level   Removing soil from under a tree canopy
can seriously damage roots and may even impair the stability of the tree.
The root system of most trees in medium-textured soils exists primarily
within a meter (3 ft) of the surface (Brown 1971). Most of the feeder
roots occur in the upper 150 mm (6 in) of most forest and many landscape soils. Each tree must be analyzed to determine the location of
its major roots and the bulk of its absorbing roots. Isolated large trees
may extend shallow roots two to three times the radius of the canopy.
Horizontal roots tend to be close to the surface, particularly if the soil
has been shaded or covered with a mulch of leaves. Even the long
horizontal roots of oak, which are considered to be relatively deep,
are within 300 mm (1 ft) of the surface (Wilson 1970). Most long
horizontal roots of maples and birches are within 150 mm (6 in) of the
surface. On the other hand, many species have "sinkers," or large roots
that grow down from large horizontal roots, most generally close to the
trunk of the tree (Wilson). There are seldom more than three to five

sinkers per tree. Although their main function may be moisture supply in times of drought, they also provide anchorage.

If a tree is stable, a reasonable amount of root loss can usually be compensated for by top pruning, nitrogen fertilization, and more frequent watering. At a fairly uniform site, the rooting zone is considered to extend beyond the drip line a distance equal to at least one-half the radius. Lowering the grade on one side of a tree to a line tangent to the drip line would cut off less than 15 percent of the roots near the surface (Fig. 7-8). Excavating back to within half the distance between the drip line and the trunk would remove about 30 percent of the shallow roots. Most healthy trees should be able to withstand removal of 30 percent of their shallow roots without serious effect, particularly if the trees are thinned proportionately and watered more frequently. After mechanical excavation, fork the soil from the roots along the embankment back toward the trunk about 450 mm (18 in) and cut the root stubs back to within 25 to 50 mm (1–2 in) of the soil. Bernatzky (1978) suggests that pruning cuts be made obliquely so that the cut surfaces face down. There is even less reason to paint root pruning wounds than to treat above-ground wounds (see Chapter 14). Moisten the embankment soil and the exposed root ends and then cover to keep the bank and roots moist and cool. Construct a retaining wall parallel to and about 500 mm (20 in) from the soil embankment. Be sure that water can pass through the wall; weep holes at the base of solid walls are necessary. Fill between the embankment and the wall with a sandy loam soil, and water well to settle the soil. The sandy fill soil should enhance the formation of new roots.

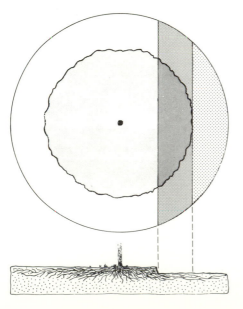

FIGURE 7-8
This plan sketch of a tree outlines the drip line, and the circle indicates the extent of most of the roots. Lowering the soil grade back to the drip line on one side would cut off less than 15 percent of the roots; lowering the grade halfway from the drip line to the trunk on one side would eliminate about 30 percent of the shallow roots.

If you want to remove soil closer to the trunk than is shown in Figure 7-8, carefully fork the soil away from the roots, working toward the trunk. In many cases, you will encounter no large horizontal or sinker roots until within 2 to 3 m (6–10 ft) of the trunk. For tree stability and continued well-being, do not cut sinker roots unless they are a considerable distance from the trunk. Horizontal roots can usually be safely cut nearly to the point where their caliper begins to increase markedly toward the trunk. A sandy soil rooting zone should be established behind the retaining wall. When soil is removed this close to the trunk, top-thinning, fertilization, and irrigation should be more extensive.

Seldom would one want to lower the entire surface beneath a tree. An excavation to accommodate a roadway, landscape feature, or building foundation can usually be accomplished by terracing to maintain the original level around the treetrunk and then building a retaining wall out from the trunk to lower the soil level the desired amount (Fig. 7-9). If the entire area around a tree is to be lowered much more than 150 mm (6 in), the tree may not survive unless soil is retained within a fair distance of the trunk. Just how much of the original soil must be left intact will depend on the species of tree, its particular rooting pattern, and soil and moisture conditions at the site. Obviously, the more soil that can be left undisturbed, the better. For maximum excavation, however, careful root exploration will be needed. Using equipment, you can lower the soil one-third of the distance in from the drip line toward the trunk. If possible, place the retaining wall there. Fork the soil from the cut roots in toward the trunk, although no further than halfway between the drip line and the trunk. Stop sooner if you encounter large roots. The shape of the tree island can vary to retain the big roots, particularly those growing downward. Handle the roots, the retaining wall, and the rooting soil as described previously. Keep the soil surface and roots moist and shade the soil with plants or mulch. Aeration will be needed if the excavation area is to be paved (see Fig. 7-12).

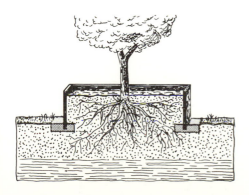

FIGURE 7-9
The largest possible area should be left at the original soil level when a grade is lowered on all sides of a tree. A retaining wall around the original soil will hold it in place and serve as a raised planter.

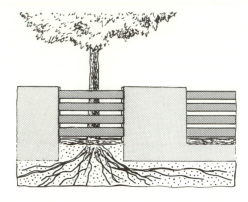

FIGURE 7-10
Interrupt the base of walls and fences
to minimize root injury and to allow
for root expansion.

Excavating for Buildings   Steps similar to those just discussed can be used when building excavations are dug near trees, although building excavations are usually deeper than excavations designed to lower a grade and will not be completely refilled until the building is completed, which may take weeks or even months. Excavate first to the planned location of the building foundation. Fork the soil from the cut roots toward the trunk for another meter (3 ft) or so. Instead of a retaining wall, install a temporary "root curtain" about 300 mm (1 ft) from the root embankment, and fill soil in behind it (Bernatzky 1978). A root curtain is made with posts and heavy wire fencing that support burlap or canvas to hold the soil fill. The soil fill will allow room for both root growth and building construction. The entire area can be filled when the building is completed and the root curtain removed.

Large buttress or surface roots that might be in the way of a wall or building foundation can often be protected by bridging (Fig. 7-10). A wall and its foundation can often be interrupted or modified to allow room for roots and their future expansion. A similar technique can be used for walks or paved areas near existing trees with large buttress roots.

Utility Trenching   For years, trenching for underground utility lines has caused serious root injury to trees that are restricted to the same easement as are utilities. Today, power-driven soil augers can bore tunnels through the lower portion of a root system with minimum root disturbance. Open trenches can be dug mechanically between trees, but near the drip line the trench should be continued by hand until the roots encountered are 50 mm (2 in) or larger in diameter. A similar trench should be dug from the opposite side of the tree and a tunnel augered under the tree between the two trenches (Fig. 7-11). If leeway permits, the line of the trench and tunnel should be to one side of the trunk by 300 to 600 mm (1-2 ft) so as to miss the tap root, should

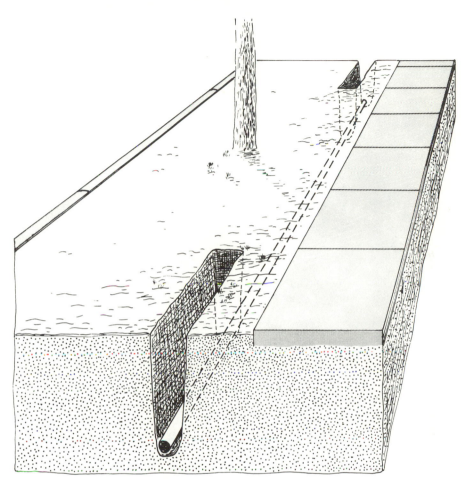

FIGURE 7-11
Trenching for utility pipes and cables can be stopped when major tree
roots are encountered; a hole is then tunneled below the major portion
of the roots to connect trenches on opposite sides of a tree.

there be one. Because most of a tree's roots are in the upper meter
(3 ft) of soil, the tunnel and the bottom of the trench may not need to
be below 1.2 m (4 ft).

When trenches are dug, roots over 100 mm (4 in) in diameter
should be carefully dug under. Again, if many roots must be cut in
the trenching and tunneling operation, the top of the tree should be
thinned proportionately. Any damaged roots should be cut back to
uninjured tissue and kept from drying out until the trench is refilled
with soil.

181

TABLE 7-1

Specifications for tunneling or trenching near trees for the installation
of utility lines (Jack Kimmel, Toronto, 1978, personal communication)

Tree Diameter		Distance of Tunnel from Each Side of Treetrunk		Distance of Open Bypass Trench from Treetrunk		Depth of Tunnel or Trench	
mm	in	m	ft	m	ft	m	ft
50	2	0.6	2	0.9	3	0.9	3.0
75	3	0.9	3	1.8	6	1.0	3.25
150	6	1.5	5	3.0	10	1.0	3.33
300	12	1.8	6	3.6	12	1.1	3.67
450	18	2.1	7	4.2	14	1.2	4.0
600	24	2.4	8	4.8	16	1.3	4.33
750	30	2.7	9	5.5	18	1.4	4.67
900	36	3.0	10	6.0	20	1.5	4.83
1050	42	3.6	12	6.6	22	1.5	5.0

Jack Kimmel of Toronto, Canada, has established specifications for the installation of utilities near city trees (Table 7-1). Experience will show whether these specifications protect existing trees.

Protecting Trees during Construction If the soil around trees will not be disturbed during construction, protection from compaction and mechanical injury should be all that is required. Fences should be constructed around each tree or group of trees at a distance from the treetrunk that depends on tree species and tree density, the depth and porosity of the soil, and the landscape or farming history of the site. Some species are relatively tolerant of abuse and compaction (USDA 1975). Roots of individual trees that grew in a dense planting will tend to be more closely confined than will roots of trees that grew in isolation, in infertile soil, or under arid conditions. Roots are usually more extensive in natural areas than in cultivated soils.

Recommendations vary considerably as to placement of protective fences. Tattar (1978) and Bernatzky (1978) recommend that a fence protect the soil within the drip line of the tree. A USDA publication (1975) advises that an enclosed area be "at least 10 feet (3 m) square with the tree in the middle." Bernatzky also suggests that the fenced area can be as small as four times the trunk diameter if the root area outside the enclosure is covered with 200 mm (8 in) of gravel with steel plates on top. To the detriment of many trees, protection at most construction sites falls short of even the minimal specifications. A safe compromise would be to place a sturdy protective fence at least 1.5 m (5 ft) from the treetrunk and to cover the remaining root area with a wood chip mulch 100 to 150 mm (4–6 in) deep. Use interlocking metal sections on top of the mulch for driveways. Mulch should be easier to handle than fine gravel after construction is over; any that cannot be separated from the soil will tend to improve soil structure near the surface, while any fine gravel left would slightly decrease soil porosity. If low branches are to be kept, the fence should be placed outside the drip line to prevent injury from equipment.

Materials, equipment, temporary buildings and work areas, fuels, soils, paints, and other construction items should not be placed within the drip line of a tree unless precautions are taken to protect both soil and tree. If branches must be pruned to make room for construction, this should be done by an arborist under the direction of the architect.

The construction site environment may differ significantly from the trees' previous surroundings. Trees should be examined at least once a month during the growing season to check on their condition. Water excess and deficiency commonly afflict trees in building areas. If site conditions allow water to accumulate around plants due to heavy rains or underground seepage, surface ditches may be needed to drain excess water away. In arid regions, during a dry spell, or when underground water flow has been interrupted, plants may need extra irrigation. The building contract should clearly outline responsibilities for safeguarding trees and should include penalty clauses for noncompliance.

Paving around Trees   Installation of pavement around trees is probably the most frequent of potentially damaging construction activities. The common question is, "How close can the pavement be to the treetrunk?" or "How large an area must be left in the pavement for the tree?" And of course the answer is, "It all depends." It all depends on the species of tree and its present health, the conditions under which it has been growing, the soil porosity and drainage, and the measures taken to provide aeration and water. Some trees survive when paving is within 50 mm (a few inches) of their trunks, while others die quickly when only a small area within their drip lines have been covered.

Aeration is the main concern when soil is filled around trees. Lack of roots to absorb water can be critical when the soil level is lowered around trees. Both problems are of concern when the soil within the drip line of a tree is covered with pavement or is compacted by traffic. Both aeration and water, along with nutrition, can be provided by a gravel diffusion layer installed under the pavement (Fig. 7-12).

If the area is fairly level, grade the soil surface to a 2 or 3 percent slope away from the treetrunk, without raising the soil around the trunk. Check and correct compacted layers that might interfere with water or air movement. Loosen the surface soil that will be under the pavement. Form a slight furrow (100 mm wide and 20–30 mm deep or 4 by 1 in) at each 45° radius. Apply gravel of uniform size (about

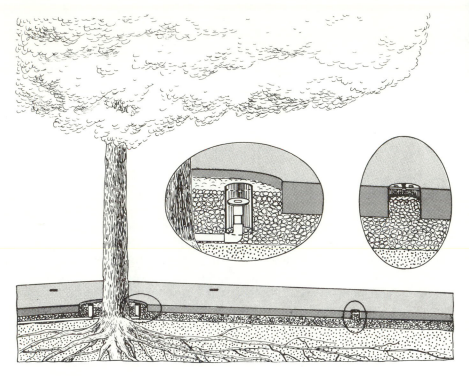

FIGURE 7-12

Most trees can be kept in good condition when paving for patios and courtyards (light traffic) is placed around them. The soil is graded to a slight slope away from the trunk, a layer of gravel is firmed in place, and the pavement installed. Irrigation bubblers at the tree trunk (left inset) and vents in the pavement at the drip line (right inset) should take care of irrigation and aeration. This would not be suitable for vehicular traffic.

20 mm or $\frac{3}{4}$ in. in diameter) over the soil in a layer 100 to 150 mm (4–6 in) deep. Firm it in place with a roller before putting down the pavement. Leave at least a 200-mm (8-in) gap between the pavement and the treetrunk. Around the treetrunk, gravel can be filled to within 10 to 20 mm of the pavement surface. Near the drip line of the tree, cut holes 100 mm (4 in) in diameter through the pavement approximately in line with the furrows at each 45° radius. Insert a pipe in the holes to hold them open and to seat a perforated cover. The pavement should not drain into the treetrunk basins or the holes through the pavement.

The trees can be irrigated by watering the gravel-filled basin by hose or by an underground bubbler system with two or more outlets at each treetrunk. Water should flow down the furrows to distribute the water more uniformly. As the water drains through and out of the gravel, air will be drawn in behind the water. The holes in the pavement at the drip line and trunk of the tree will provide some natural air circulation as well. Diffusion of the various gases in the gravel and soil will also actively keep oxygen up near a safe level and will prevent carbon dioxide and toxic gases from becoming too concentrated. When needed, nitrogen fertilizer (nitrate or urea) can be dissolved in the water used to irrigate the trees (see Chapter 11). The aeration-irrigation system must function for the life of the tree, but most trees will adjust within a few years to soil fills and cuts.

Pavement can be laid on cuts and fills under trees if the procedures followed are the same as outlined above. On slopes, the surface must be shaped so that water will not drain to treetrunks but will bypass them. The irrigation system should be adapted so that the water applied will flow fairly uniformly under the pavement. If the soil to be paved must be compacted to less than 10 percent pore space, a much larger area under the tree must be unpaved or special footings placed to support the pavement.

SOIL PREPARATION

Many soils become severely compacted during landscape construction, particularly if buildings are involved. As already discussed, soil compaction reduces water infiltration and movement, drainage, and aeration. These factors can greatly impair root growth and function and can increase the difficulty of later maintenance, particularly irrigation.

If surface soil has been compacted, this is readily apparent during irrigation or tillage. Former agricultural soils usually have a compacted layer (plowpan) 150 to 250 mm (6–10 in) below the surface, depending on the depth of earlier cultivations. Alluvial soils and soils graded during site preparation may have some layers that seriously interfere with root penetration and water movement. Fine-textured soils may be naturally tight, restricting root growth and function as well as drainage and aeration. In many soils in the southwestern United States and other arid regions, hardpan of varying thicknesses appears at various depths below the soil surface. Bedrock beneath shallow soils also restricts rooting and drainage.

The interdependent factors of water infiltration, movement, and drainage are all influenced by one or more of the above-mentioned factors and in turn affect soil moisture, aeration, and root growth and function. Moisture infiltration can be taken care of most effectively after the surface soil has been altered during planting, although efforts should be made to keep soil compaction to a minimum.

Soil Water Movement

Most water movement and drainage problems can be identified if you dig one or more holes or determine water percolation rates into different layers of soil. Note variations in the ease of soil penetration and the texture found with depth. Compacted layers or lenses of different textures indicate possible problems in rooting, irrigation, and drainage. Any hardpan and bedrock present can be readily recognized. Even though the soil profile appears to be uniform, portions of it may be tight and will restrict water movement.

Water movement can be timed in a particular soil layer. Wildman (1969) devised simple infiltrometers by using 75-mm (3-in) aluminum irrigation pipe, some glass jugs, rubber stoppers, and plexiglass tubing. He drove six sections of pipe, varying in length from about 0.3 to 1.1 m (12–44 in) down about 70 or 80 mm (3 in) into the bottom of auger holes spaced about 600 to 900 mm (2–3 ft) apart and drilled to various depths in the soil. The water reservoir for each infiltration pipe consisted of a glass jug fitted with a rubber stopper through which a plexiglass tube was inserted. When the jug was placed as shown in Figure 7-13, water dribbled out of the tube until its level in the aluminum pipe rose above the end of the tube. No further flow took place until water moved into the soil and allowed some air to enter the tube. A scale

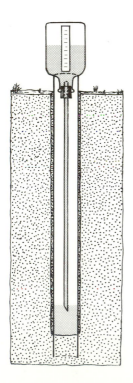

FIGURE 7-13
A set of inexpensive infiltrometers can be used to determine water movement at different depths in a soil and to locate any layers of slow penetration. An aluminum pipe 75 mm (3 in) in diameter is inserted into a 90 mm (3.5 in) augered hole in the soil and driven about 75 mm (3 in) into the soil at the bottom of the hole. A 4-liter (1-gal) jug with a rubber stopper and a length of plexiglass tubing about 200 mm (8 in) shorter than the length of the pipe is filled with water and inverted. The tubing is inserted into the pipe. The rate of water movement can then be measured by the rate of water loss from the jug. A series of infiltrometers at different depths will yield a profile of water penetration for a given soil.

taped to the side of the bottle was used to take readings at various intervals. The plexiglass tube arrangement provided a fairly constant head of water in the hole, and the jug held enough water to obtain sufficient readings without refilling.

The initial moisture content of the soil does not seem to greatly affect the usefulness of the readings. Water-level readings taken at 15, 30, 60, and 120 minutes will be sufficient to construct an informative infiltration curve. To obtain the depth of water infiltrated from the aluminum pipe, multiply the difference in water level readings by the radius of the jug squared divided by the radius of the pipe squared.

Strata with infiltration rates of less than 50 mm (2 in) an hour from a 75-mm pipe are considered slow and in need of improvement. Since several infiltrometers at the same depth at the same site are likely to give different readings, at least two or three replications are desirable. This technique gives only an approximate water infiltration rate, one that is higher than the water percolation rate in the soil at large. It does, however, give a fairly accurate measure of the relative differences in water movement at different soil depths. Thus layers that restrict water penetration can be identified.

Improving Soil Moisture Conditions

Reducing Water Accumulation   Although water may move through a soil at a faster rate than the minimum given above, the soil may still be too wet to allow most plants to grow well. In such a case, water is entering the soil faster than it can move through the soil and drain away. Water may be accumulating in a low area on the surface, moving into the soil from an outside water source at a higher elevation, or collecting from pavement and building runoff. Initial grading should anticipate surface water flow, so that low spots do not occur or are provided with drainage. Where grading alone will not remove surface water, a French drain may remedy an existing wet spot or intercept underground flow if the amount of water is not great. A French drain is a slit trench, 50 to 100 mm (2-4 in) wide, dug through the wet area toward an outlet, a lower area, or a more permeable soil. The bottom of the trench should slope (1%-3%) toward the low area. Fill the trench with medium sand (0.25-0.5 mm in diameter) or fine gravel. Crown the fill above the surrounding soil level to delay the sealing that will occur when fine soil is washed into the fill (Fig. 7-14). In a turf area, the grass will soon grow and hide the trench; in shrub beds, plants or mulch will obscure the fill. Even if the trench does not lead to a drainage way, water will be able to move laterally from the trench into more permeable soil than that near the surface. Aeration will also be improved along the trench. To intercept water flowing underground into the planting area, locate the French drain along the upper side and essentially at right angles to the slope, though allowing for some fall.

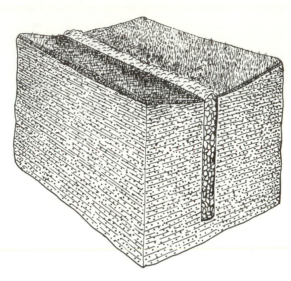

FIGURE 7-14
A French drain is used to remove surface water or to intercept underground flow.

Preventing or Reducing Soil Moisture Problems   Soil problems can often be avoided or at least reduced if certain precautions are taken before a site is graded or filled. If you plan much cut-and-fill grading, you must avoid locating impervious layers or infertile subsoil close to the surface (see Fig. 6-1) or creating new textural layers that could interfere with water movement and rooting. For these purposes, a knowledge of the soil profile is essential.

Compacted layers, particularly roadways, should be broken up and the area rough-graded and disced or rototilled before it is filled or graded. This will make grading easier, reduce the interface between the fill and original soil, and minimize impervious layers or pockets below fills that could impair water movement and drainage. Fill soil should be of a texture similar to or slightly coarser than the soil on which it is being placed. The site should be well watered to settle the soil and then graded when dry if the surface is too uneven.

Correcting Slow Water Movement   If the tests described above reveal compacted soil only near the surface, discing or rototilling to a depth of 150 to 200 mm (6–8 in) will loosen the soil for easier planting and improved aeration and water infiltration. A plowpan at 200 to 250 mm (8–10 in) can usually be broken up by deep plowing or ripping (subsoiling) in two directions, intersecting at tree planting locations if possible. Ripping at directions that are diagonal to one another is somewhat more effective than 90° cross-ripping (Trouse and Humbert

188

1959). Subsoiling is most effectively done when the soil is dry, so that
the compacted layers shatter. A good crop of weeds or other plants
can dry a soil fairly effectively if there is no rain. Subsoiling naturally
layered soils, however, has had only limited success (Aljibury, Meyer,
and Wildman 1979), since the lenses often reseal in a short time.

Subsoiling may not be possible in many confined landscape situations. Although it takes longer, a backhoe may be the best equipment
for preparing such a planting site. A backhoe can be used to dig a cubed
hole of 1.2 to 1.5 m (4–5 ft) for each tree; it can break up the compacted layers, mix strata of different textures, and replace the soil in
the hole (Fig. 7-15). A shrub bed can be similarly prepared. Chunks
of hardpan, rock, or blue-black strata resulting from anaerobic decomposition should be removed and discarded. If the excavated soil is wet,
it should be dried for several weeks in its original piles. When dry, the
individual piles of soil should be mixed and returned to the holes.
Some arborists add 25 to 30 percent by volume of a slowly decomposing organic matter. Studies, however, have questioned the value of
amending soil in shrub and tree planting (Pellet 1971, Whitcomb 1979b)
and this doubt may also apply to the preparation of somewhat larger
areas.   Backhoe preparation has been successful in fruit orchards,

FIGURE 7-15
A tractor-mounted backhoe prepares a planting site by breaking through
compacted layers and mixing strata of different textures.

even though organic matter has not been used. In addition, holes dug with a backhoe provide an opportunity to check the soil profile for compacted layers, changes in soil texture, structure, or color, and rooting from nearby plants. This information can be used to determine ongoing site preparation.

After deep cultivation or backhoe soil preparation, the disturbed soil should be wetted to settle it and remove air pockets. It may then take one to two weeks before the soil will be dry enough for planting. If the loose soil is not settled before planting, it will settle later, thereby creating a basin that will be slow to dry after rain or irrigation. This may lead to crown rot, poor growth, and even plant death. Soil preparation before settling should be done when the soil is not too wet; soil structure could otherwise be damaged.

Soil Drainage

If rain is frequent or irrigation available, general soil drainage can be estimated by a method similar to that used for checking water movement through soil. Dig or bore a hole in the soil 1.2 to 1.5 m (4–5 ft) deep or to hardpan or bedrock. Leave the hole open to see whether water accumulates after a heavy rain or irrigation. If water stands in the hole more than five or six days, a drainage system or other measures may be necessary. If possible, dig the hole a meter or two (3–6 ft) deeper to see if a permeable layer can be located to provide drainage. Observe the drainage of water from the deeper hole after rain or irrigation to determine whether a permeable layer has been found.

Providing Drainage

Penetrating Semi-Impervious Layers If the test holes provide drainage when deepened, as described above, bore a hole 100 to 150 mm (4–6 in) in diameter at each tree location, or space holes about 2 m (6 ft) apart for shrub beds. Actual experience and experimentation in this area have been sparse and have provided almost no recommendations for the spacing and fill material to be used in such drainage holes, which are somewhat akin to dry wells although their purpose is to drain wet subsurface strata. After several holes are installed, an observation hole 1.2 to 1.5 m deep can be placed among them for observing water in it after the soil has been wetted. The spacing of remaining holes can be determined by this observation.

The drain holes should be filled with medium sand (0.25–0.50 mm in diameter) or fine gravel. If coarser fill is used and the surrounding soil is silty, saturated soil may slake into the fill material and plug the drain. If the soil is fluid when saturated, the outside of a perforated plastic pipe 70 to 100 mm (3–4 in) in diameter should be covered with fiberglass matting about 10 mm ($\frac{1}{2}$ in) thick and inserted in each drain hole. The fiberglass will filter out the fine sand and larger particles so that they do not plug the pipe. The pipe should be capped at the surface to keep out soil and debris but does not need to be filled. Such a pipe drain can be back-flushed with water under pressure if the fiberglass filter seems to be plugged.

Penetrating Hardpan   Hardpan within 750 mm (30 in) of the surface should be broken up or penetrated if possible. If the site is large enough and the hardpan not more than 600 mm (2 ft) thick, large subsoilers or slipplows can rip to depths of 1.5 to 2 m (5–7 ft). The hardpan must be broken to its full depth in order to provide drainage. Ripping is most effective in dry, brittle soils and hardpans and least effective in moist loams and clays. Compact sandy layers will shatter even when they are moderately moist (Wildman, Meyer, and Neja 1975). For best results the parallel ripping passes should not be more than 2.5 m (8 ft) apart, and passes in a second direction should be diagonal to the first direction.

Unfortunately, deeply tilled soil should be allowed to settle until several rains or sprinkler irrigations have occurred. If the soil is worked before it has settled, the final surfaces will be uneven. If the site is allowed to settle over a winter before it is extensively worked and planted, the plants will probably grow better and be more evenly wetted during rains or irrigations.

If the hardpan is too thick to rip or if ripping is impractical, holes should be dug or drilled through the hardpan to provide drainage and possibly some rooting into the soil below (Fig. 7-16) (Gowans and Hall 1971). Power augers of the sort used to set utility poles can drill through most hardpan. If the soil is quite permeable above the hardpan, a clump of trees can be drained by less than one drain hole per tree. The spacing of holes in shrub beds should vary from 2 to 3 m (6–10 ft), depending on soil texture and depth. The finer the texture and the shallower the soil, the closer the holes must be. For such drain holes to be effective, strata under the hardpan must be permeable.

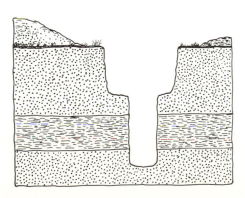

FIGURE 7-16
In shallow soils with underlying hardpan it may be necessary to drill through the hardpan to provide adequate drainage and rooting volume. Fill with soil of the same or coarser texture.

Unless the soil above the hardpan conducts water too slowly, the holes should be filled with soil of a texture similar to or coarser than that soil. This will provide one less interface to impede water movement and will create a greater head (potential) at the bottom of the hole, which will aid drainage. After the soil has been repeatedly wetted, water in the hole should lower at least 2 mm (0.1 in) per hour, or it will be of little value. If soil is well drained generally, roots may grow down the holes and into the pervious soil below the hardpan. This will increase the drought tolerance of the plant and lengthen the periods between irrigations.

Blasting hardpan or compacted layers with dynamite before planting trees and shrubs is not a good idea. It can form impervious basins where standing water can kill plant roots.

Providing an Internal Drainage System   If the downward movement of water is impeded by tight soil, rock, or hardpan, an internal drainage system may provide the only suitable rooting zone (Fig. 7-17). (Planters without natural drainage are discussed in Chapter 10 as a special case.) Landscape soils are often composed of fill material of various textures or are compacted during site preparation and may drain more slowly than similar soil nearby. The previously described methods can be used to determine the presence of soil layers that are permeated slowly.

The desirable depth and spacing of drain lines depend on soil texture and structure, depth of impervious soil or rock, depth of drainage necessary for the plants and the amount of water (from rain, irrigation, or seepage) that must be removed. To grow and stabilize adequately, most trees need deeper "drained" soil than do shrubs. The more slowly water moves in a soil, the deeper the drained soil must be; the more water that must be drained, the deeper and closer together the drain lines must be. The drain line should be placed above any impervious layer.

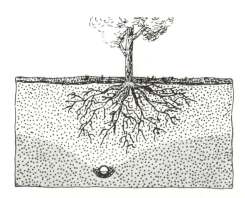

FIGURE 7-17
Drain tile may be necessary to provide adequate soil for root growth and function.

TABLE 7-2

Recommended drain spacing[a] for agricultural soils of different textures
and permeability (Beauchamp 1955)[b]

		Spacing	
Soil	*Permeability*	*meters*	*feet*
Clay and clay loam	Very slow	9–20	30–70
Silt and silty clay loam	Slow to moderately slow	18–30	60–100
Sandy loam	Moderately slow to rapid	30–90	100–300
Muck and peat	Variable	15–60	50–200

[a]These spacings are quite similar to those given by Hillel (1980).

[b]In poorly drained landscape soils, reduce spacing by half. Drain lines should be
1 to 1.5 m (3–5 ft) below the soil surface.

Drain lines should be at least 1 m (3 ft) below the soil surface,
and plants will grow even better if the lines are at a depth of 1.5 m
(5 ft) (Beauchamp 1955). If the lines cannot be placed this deep,
they should be placed closer together. Table 7-2 lists recommended
spacing between laterals for agricultural soils; in poorly drained landscape soils, the distances should be reduced by at least half.

To work properly, a drain system must have

An outlet from which the water can flow or be pumped

An even fall throughout its length, 50 to 300 mm per 10 m (6–16 in/100 ft)

A filter over the drain line to minimize clogging, particularly if soil is silty or
sandy

Drain lines can be made of clay, concrete, or plastic pipe. Clay
and concrete tiles 100 mm (4 in) in diameter and larger are commonly
used for draining agricultural soils. Because most landscape sites are
less than 0.5 hectare (1.35 acres), plastic pipe 50 to 100 mm (2–4 in) in
diameter is large enough to carry water; easier to install; more stable in
soils that vary in texture, compaction, and settling characteristics; and
less likely to be invaded by tree roots. The larger the pipe diameter, the
longer the drainage system will function without clogging. Plastic pipe
should be used in important landscape areas, and provisions should be
made so the drain lines can be reamed and cleaned out as well as back-
flushed. Tile (600–900 mm or 2–3 ft long) should be laid end to end
on a thin layer of coarse sand or fine gravel, and a small crack (about
3 mm or 0.125 in wide) left between each tile for entrance of water
(Fig. 7-18). Plastic pipe with three to four 3-mm holes per 100 mm
(10 0.125-in holes/ft) can be laid in a similar manner, but the pipe
sections should be cemented together.

FIGURE 7-18
Drain tiles are usually laid on a bed of sand or fine gravel and covered with
50 mm (2 in) of coarse material to allow water to move to the tile joints.
The joints are covered to reduce the amount of sediment entering the drain.

Tile joints should be covered with tar paper or plastic on top so that water must enter the tile from below; this allows some of the larger suspended particles to settle out of the water before it reaches the drain. Some plastic pipe has holes in only one-third of its circumference; this portion is placed on the bottom. Pipe with holes around the entire circumference should be covered on top with a strip of plastic if soil is silty; the holes on top should pose no problem in sandy or clay soils.

To allow water to flow to the drain openings and to keep soil particles from entering the drain, the drain line should be enveloped in 50 mm (2 in) of coarse sand or fine gravel or a synthetic filter material. A synthetic filter is preferable in sandy soils, and coarse sand or fine gravel in clay soils. In the eastern United States, drain lines in clay soil are commonly not encased in any coarse material; most clay aggregates are fairly stable, and clay particles that do enter drain lines seem to be carried along with the drain water. Silty soils, on the contrary, can cause problems in drains: Silt particles slough easily and clog fine-mesh filters. Drain lines in silty soils should probably be encased in 50 mm (2 in) of coarse sand or fine gravel and covered on the top and sides with this material and synthetic filter material, although this method has not been tested. An irrigation or drainage engineer should be consulted for all but the smallest drainage systems.

Drainage for an Existing Planting

Few landscapes have internal drainage systems, because most drainage problems are not recognized before a site is developed or are created during development. Also, the expense and delay of installing an adequate system may be such that many people hope to minimize problems with proper maintenance alone.

The procedures already discussed for site preparation will also apply to drainage problems in an existing landscape. Care should be taken to minimize damage to the root systems of important trees and large shrubs. Grading and French drains can improve surface drainage. Holes can be drilled near trees by power auger or water jet (see Chapter 12) to penetrate layers that restrict water movement, whether they be lenses of different texture, compacted layers, or hardpan. Internal drains can be installed if needed. Drainage trenches can be dug by machine until root density and size indicate that digging should be continued with hand tools so that roots larger than 5 cm (2 in) in diameter will not be cut.

IRRIGATION SYSTEMS

Landscape plantings can be irrigated in a number of ways: basin, furrow, sprinkler, soaker, and drip. The method used will depend on the type of plantings; amount, quality, and source of water; terrain; available funding; and sources of labor. In order to save water and labor, most intensive landscape plantings are being developed with automatically controlled systems. Adequate design of the system is imperative.

Basin

A basin is a level area surrounded by a berm to hold water. Basins are well adapted to level areas for shrub and flower beds and trees where foot traffic is infrequent. The soil should slope away from the trunks of plants so that water does not stand around their bases after irrigation or rain. Any pavement surrounding a basin should be at the same level or higher than the top of the berm to minimize the washing of mud onto the pavement (Fig. 7-19). In areas of high rainfall, however, or where the area of pavement is large compared to that of the basin, excess water may accumulate in the basin. In some arid situations, this may be an advantage. Runoff control should be considered in the design of paved areas and building downspouts.

Large basins in sandy soil may not be uniformly watered because of rapid infiltration near the water inlet. Smaller basins, more water

FIGURE 7-19
Berms around tree holes in pavement should be no higher than the pavement
or the concrete border so that irrigation may be simplified and mud prevented
from washing onto the pavement.

outlets, or faster water flow will overcome this. Water can be applied
to basins by hose, soaker, or built-in bubblers (small sprinkler-like
heads) (Fig. 7-20).

FIGURE 7-20
An irrigation bubbler (near corner)
at each tree greatly reduces the labor
needed to water plantings in paved areas.
Light and electrical outlets are also present.

Furrow

Furrows (shallow ditches) are useful for long, narrow plantings of shrubs, trees, and flowers. The furrow slope will depend on the amount of water available and the rate of infiltration into the soil. In existing furrows, water can be distributed evenly by adjusting the rate of application.

Furrows are most effective when little foot traffic will occur across them. On uneven ground, furrows should follow the contour. Wide, shallow ditches give better water penetration than narrow ones. Salt may accumulate on the ridge tops (see Chapter 12).

Sprinkler

Sprinklers are generally used on shrub, flower, and tree plantings. If engineered properly, a sprinkler system provides fairly uniform water distribution even on hilly terrains. The rate of application may need to be slow on uneven areas to allow for penetration; otherwise, high areas will be underwatered and low areas overwatered. Many plants do well with overhead watering, which keeps humidity somewhat higher and washes off dust and insects. Sprinkler irrigation moistens mulches but does not wash or float it away. Sprinklers can reduce maximum air temperatures by 6 to 9°C (10-15°F) (Hamilton 1978).

Sprinkler irrigation may pack the surface of bare soil and reduce infiltration. Wind may result in uneven water distribution. Flowers and tall plants may be damaged by the weight or force of the water. Frequent light sprinklings with saline water may cause an unsightly or even toxic buildup of salt. Less frequent, longer applications usually correct or reduce this. Sprinkler application does not appear to favor mildew and rust diseases, particularly in arid regions when water is applied early in the morning so that the foliage dries quickly.

Leaf burn may be caused by sodium and chloride absorption when the rate of evaporation is high. Conditions of low humidity, high temperature, and winds can increase the concentration of these ions in the water on the leaves between rotations of the sprinklers. Sometimes this can be corrected by increasing the rate of rotation. When the weather is hot and dry it may help to irrigate at night. Usually no problem arises if sodium and chloride are each 3 me/l (milliequivalents/liter) or less (Stromberg 1975).

Soaker

Consisting of canvas or porous plastic tubes, of hoses, or of plastic tubing (sometimes two-walled) with fine holes, soakers are used to apply water at slow rates. They are useful in difficult-to-irrigate areas: across slopes; near long, narrow plantings; or in soils with low infiltration rates. Water distribution along the soaker often varies from one end to the other, but will usually be more uniform at higher rates (pressure). Distribution will be greater along the lower portion of a soaker that is not lying level.

Drip Irrigation

Drip or trickle irrigation is the frequent, slow application of water through emitters located along a delivery line (Marsh 1977). Wetting of soil between emitters occurs primarily by capillarity. A drip system consists of emitters, lateral and main lines, a control station, and a water source. Emitters range from small-diameter tubing to complicated mechanical devices that reduce and stabilize water flow somewhere between 2 and 8 liters (0.5–2 gal) per hour. Two liters per hour is the most common rate.

Main lines are usually buried; laterals may be buried, covered with mulch, or left on the surface. The control station filters or screens the water and regulates its pressure and application time either manually or automatically (Fig. 7-21). A meter usually measures the water applied. Injectors can also be installed to apply fertilizer and other materials.

FIGURE 7-21
Underground installation of a control station for a drip irrigation system:
(from left to right) back-flow valve, manual and electric control valves,
filter (also inset), pressure regulator, and pressure gauge.
(See Fig. 12-7 showing distribution line and emitters.)

Drip irrigation has accounted for water savings as high as 90 percent over flood irrigation on sandy soil with widely spaced plants, or around 20 percent over sprinkling on clay soil with more closely spaced plants (Hamilton 1978a). The drip method, like soakers, permits irrigation on steep banks, shallow soils, soils with slow infiltration rates, and sandy soils. Saltier water can be used in drip irrigation than in other means of application, mainly because the salt level in the soil solution is kept near that of the water applied instead of becoming more concentrated as water is extracted by plants. Only a portion of the surface soil is wetted, so compaction will be less severe and the area can be used soon after irrigation. Water wastage through evaporation from soil and weeds is also reduced.

The most troublesome problem is clogging of the emitters by mineral or organic particles. Plants may suffer before a reduced flow of water is corrected, so an effective filter system, regularly inspected and serviced, is essential.

The soil must have an infiltration capacity of at least 13 mm per hour (0.5 in/hr) with an emitter discharge of 2 liters per hour ($\frac{1}{2}$ gph) to keep the pool of free water around the emitter from exceeding 600 mm (2 ft) in diameter. Slower discharge rates can be used or the water turned off and on to allow time for infiltration, although the latter method may not be feasible when drip irrigation must apply water over long periods of time.

Salts tend to accumulate at the soil surface and may be hazardous because light rains can move them into the root zone. Continuing irrigation on schedule until 50 mm (2 in) of rain have fallen will usually leach the salts below the roots. Covering the soil with a mulch will reduce surface evaporation and salt buildup.

If drip irrigation is interrupted, plant damage can occur rather quickly when reserve moisture in the soil is low, particularly in mid- to late summer. The soil volume normally wetted in drip irrigation is less than half of that wetted by flood or sprinkler methods. Even when only half of the rooting volume is irrigated, water uptake is uniformly transported in trees and shrubs (Hamilton 1978a). Experience indicates that roots not wetted by drip irrigation become inactive when they have dried the soil to near the permanent wilting point. Root hairs and fine roots may die, but regrowth is rapid under favorable moisture and temperature conditions. Individual plant installation of emitters and drip irrigation management are described in Chapter 12.

General Considerations

Whatever methods of water application are chosen, the main delivery system to individual planters or beds should be planned and installed, along with the other utilities at the site, before paving and planting take place.

CHAPTER EIGHT

Planting

Planting is the culmination of much preliminary work: analyzing the site, developing landscape plans, selecting appropriate and healthy plants, and preparing the site to preserve existing plants and accommodate new ones. Proper handling of stock, planting, and after-care can be fairly simple but are immensely important. Woody plants can be started in the landscape directly from seed, as young plants easily handled by one person, or as larger specimens that must be transplanted with special equipment. Each method has its proper place.

Most direct-seeded plants develop in balance with their surroundings and require little or no maintenance once established. Direct seeding may be an economical way to plant extensive areas that will receive minimum care and need not produce an immediate effect: highway and utility rights-of-way, reservoir areas, future recreation areas, urban forests, and open spaces.

Young plants are most commonly used in home, commercial, industrial, and public landscapes because their cost is moderate and they have an immediate modest effect on the landscape. The landscaping industry is based almost entirely on designing, producing, marketing, planting, and caring for these young plants.

Large shrubs and trees are key elements in new, intensively developed landscapes in buildings and shopping malls, on roof gardens, around office buildings, shopping centers, and public facilities. Large specimens are grown specifically for this purpose, become available when land use changes, or are harvested in woodlots, farms, and forests. Specialized equipment and techniques have been developed for handling large plants.

SEEDS, LINER PLANTS, AND CUTTINGS

In California, the seeds of more than 50 species of trees and shrubs have been found to germinate, survive, and grow satisfactorily with little care (Chan, Harris, and Leiser 1977). Several species were alive ten years after seeding in areas with less than 250 mm (10 in) of annual rainfall and little or no summer rain or irrigation (Fig. 8-1). Not all species did well in all locations, but at least five or six did well in 12 out of 13 test plantings. These ranged from ocean sites to elevations higher than 1.2 km (4000 ft) and to the interior valleys of northern and southern California. At higher elevations (2 km or 6000 ft), liner plants and cuttings were more successful than seeds, which got a slow start because of the short growing season. The following procedures make it possible to establish and maintain extensive landscape plantings as economically and naturally as possible.

Planting Seed

Since landscape trees and shrubs are usually wanted in specific locations and are not planted close together, it is best to seed them

FIGURE 8-1
Black locust (left) and flannel bush (right) 11 years after being direct-seeded in the San Joaquin Valley of California, where the rainfall averages less than 250 mm (10 in) annually. For the first six years the black locust was cut to the ground by mowing equipment.

in small planting holes rather than broadcasting or seeding in furrows. Broadcast seeding is useful primarily for ground cover and erosion control plantings of grass and herbaceous species. Even though seed germination is reduced by damage from spray equipment used in the hydroseeding and hydromulch seeding of grasses, spray application is widely used, particularly on difficult slopes. Mulch will protect the seed to some extent, both in the sprayer and on the ground, and will conserve moisture. Hydromulch seeding of woody plants, although fairly successful in sprinkler-irrigated landscapes, has been unsatisfactory in most unirrigated, low-rainfall locations. The larger seeds are damaged or cannot be handled by the spray equipment. The mulch may form a cover that is not complete enough to conserve moisture effectively but is sufficient to prevent or delay the covering of the seed by soil. Consequently, seedling establishment has been poor. In addition, fertilizer (usually added to the spray) stimulates weeds as well as the seeded plants. Broadcasting by hand or with a cyclone seeder on a loose, rough surface is usually quite satisfactory.

Selection of Woody Species for Direct Seeding   A direct-seeded plant must germinate, grow, and survive under the environmental conditions that will occur at the site. Vigorous species will usually compete with weeds more effectively. A species need not be native to the area; it may germinate and grow satisfactorily even in an environment where it cannot regenerate itself.

The creation of a favorable microenvironment for the duration of germination and seedling stages can increase the variety of viable plants in a given area. Many species have been successfully established without irrigation and with little or no summer rain (Chan, Harris, and Leiser 1977).

Seed Availability and Viability   The U.S. Forest Service publishes lists of seed dealers and of planting stock available in the United States (USDA 1979). Some dealers will collect, on contract, seed not normally carried if sufficient time is allowed to locate fruiting plants. Seed can be collected wherever desired species with a viable seed crop can be found and permission obtained. Plants grown from locally collected seed are more likely to be adapted to the area. Records of the plants from which seeds are collected will be of help in future collection.

Before undertaking an extensive seeding program, you should determine the viability of the seed. Sometimes the seed dealer can supply this information. If not, cutting tests, tetrazolium, or actual germination tests can be used (Chan, Harris, and Leiser 1977; Bonner 1974). Seeds of some species germinate readily whenever moisture, temperature, oxygen, and light are favorable. Germination of other species, however, can be delayed or even prevented by seed coat or embryo dormancies, chemical inhibition, or a combination of these conditions.

Time of Seeding Seeds should be planted when conditions will favor both germination and seedling growth after emergence. In areas with low to moderate rainfall, autumn seeding will be successful for many species. Where soils freeze in the winter, species that germinate readily but cannot withstand low winter temperatures after emergence and species that require warm temperatures for germination are best sown in the spring. Initial watering can extend the period during which seeding can be done.

Suggested Planting Procedures Dig a hole deep enough (100 mm or 4 in) to allow for the fertilizer, backfill (90 mm or 3.6 in) to separate seed and fertilizer, soil to cover the seed (Table 8-1), and a slight depression to collect water (Fig. 8-2). In most situations, a small seeding hole can be made without extensive soil preparation. If the soil is compacted, pulverize it with a shovel or soil auger so that roots can penetrate the surface soil more easily. Firm the loosened soil before planting.

On rocky surfaces, locate the seeding hole in a depression or crevice that offers a pocket of soil or fine rock fragments and one or more fissures for rooting. On slopes, construct pockets so that the seeding hole will not be covered by loose soil from above. A backslope will accumulate water so it will be available to the seedling.

TABLE 8-1

Suggested seeding guidelines (Chan, Harris, and Leiser 1977)

Seed Diameter (mm)	Seeds per Hole[a]	Depth of Seeding Depression below Surface (mm)	Depth of Soil over Seed (mm)	Depth of Depression for Water (mm)
1.5	20	6–9	3	3–6
1.5–3	10	9–12	3–6	6
3–6	5	12–18	6–9	6–9
6–12	3	21–24	9–12	9–12
(in)		*(in)*	*(in)*	*(in)*
1/16	20	1/4–3/8	1/8	1/8–1/4
1/16–1/8	10	3/8–1/2	1/8–1/4	1/4
1/8–1/4	5	1/2–3/4	1/4–3/8	1/4–3/8
1/4–1/2	3	7/8–1	1/2	3/8–1/2

[a]Based on a 50 percent rate of germination.

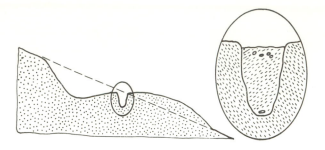

FIGURE 8-2

Direct-seeding site on a slope. The seed hole is placed near the original slope
surface so that it will not be covered by soil eroding into the planting basin or
kept too wet during rainy periods. The planting hole (inset) is similar for level
and sloping areas; fertilizer is placed in the bottom of a hole 100 mm (4 in)
deep and covered with soil. Seed is planted near the surface.

Newly germinated seedlings will grow more rapidly and compete
more effectively if they are fertilized. Place 1 g (0.03 oz) of nitrogen
as a slow-release fertilizer in the bottom of the hole. Some soils also
require potassium and sulfur. If sulfur is not a component of the
nitrogen fertilizer, add 2 g (0.07 oz) of potassium sulfate.

Replace and firm a portion of the soil in the hole. Leave a slight
depression, depending on the size and number of the seeds to be
planted (Table 8-1). The larger the seed, the deeper the depression.
Place the seeds in the planting depression and cover with pulverized
soil. Remove any excess soil, so that it will not wash or blow into the
seeding hole.

Frank Chan, horticulturist for Pacific Gas and Electric Company,

FIGURE 8-3

Spot-seeding technique using burlap-covered black plastic, a cylindrical carton
with soil mix, and protective wire screen; the individual items (left), items in
place with screen folded to keep out rodents, birds, and insects (center), and
screen opened to allow seedling to increase in height (right). Rocks from the
site help hold burlap in place and protect seed hole from large animals.

has refined spot seeding of woody species to ensure successful stocking of plants over a wide range of California sites (Fig. 8-3). Although more expensive and time consuming to install than the procedure just described, his seeding technique improves seedling emergence, reduces weed competition, improves moisture conditions, and minimizes animal damage. Chan recommends that the seeding site be cleared of rocks and weeds so that a 600 × 900-mm (2 × 3-ft) sheet of 9-mil black plastic enclosed in a burlap bag can be stapled through the corners of the burlap to hold the bag in place. The burlap will protect the plastic from sun and wind; the plastic will prevent weed growth and conserve moisture. A 100 mm (4 in) X is cut in the center of the plastic and burlap over a hole of similar diameter and depth, and a liter (quart) cottage cheese carton (without top or bottom) is inserted in the hole. Fertilizer is put at the bottom of the carton, and about 100 mm (4 in) of sandy soil mix is added and firmed. After planting the seed, a 450 × 600 mm (18 × 24 in) rectangle of aluminum window screen formed into a cylinder and secured with iron wire is placed around the carton. The top of the cylinder of wire is then flattened and folded over to provide a seal (Fig. 8-3). The wire cylinder can be opened for inspection and maintenance, and can be removed easily before the plant begins to be restricted.

Weed control is essential until the seedlings can successfully compete for moisture and sunlight. Herbicides or mulching can usually provide adequate control. If herbicides are used, contact are safer than preemergent herbicides around seed holes because young seedling roots are near the surface. Apply a contact herbicide after seeding but before seedlings emerge and when weeds are less than 75 mm (3 in) high. When applying contact herbicides, follow the manufacturer's recommendations and safety precautions carefully. Spray in a circle about a meter (3–4 ft) in diameter around each seed spot. If seedlings have emerged, invert an empty can over each of them before spraying. A second application may be needed if new weeds appear. Two applications will usually provide adequate control for the entire growing season on unirrigated plantings in arid regions. If necessary, hand weed around the seedlings after herbicide application.

Mulches not only control weeds but moderate soil temperatures and reduce moisture loss. The mulch must not interfere with seed germination and seedling growth, yet it should prevent weed emergence. A collar will prevent mulch from covering the seed spot. Tin cans, size $2\frac{1}{2}$, make sturdy collars; 2-liter ($\frac{1}{2}$-gal) milk cartons and asphalt or heavy kraft paper cylinders 100 mm (4 in) in diameter also work well.

Thin plantings to retain the strongest, most vigorous seedling in each hole. Remove the others carefully by cutting them off at ground level. Pulling them out can damage the root system of the remaining seedling. Oaks are an exception: Since they may resprout from the acorn, the acorns must be removed.

Liner plants make it possible to plant young seedlings or cuttings where the growing season may be too short to allow for germination and sufficient growth during the first summer. Liners also provide a faster start than direct seeding where the soil is slow to warm in the spring. The young plants are able to take advantage of the more favorable moisture conditions early in the growing season and to become established early, so they are better able to withstand the heat and drought of summer.

As with seeding, the plants established by the liner method must be able to survive the expected site conditions. The seedlings can be grown in many different kinds of small containers, including various tubes, blocks, and plugs. Tubes made of plastics, paper, or peat will vary in biodegradability. After tubes are filled with a growing medium, seeds or young seedlings are planted in them, and tubes and seedlings are planted together. Blocks are similar in size and shape to tubes and constitute the planting medium into which seeds are planted. The block and the seedling are then planted in the soil. The molds that hold plugs are similar to tubes in that they are filled with a growing medium, but the seedling and its rooting medium are slipped out of the mold or container when planted. The growing medium must be handled carefully since it is held together only by the seedling roots. Forest seedlings are handled by the million in each of these ways. Containers, methods for growing the seedlings, and techniques for planting vary considerably. Many annual herbaceous and ground cover plants are grown in small liner containers for direct planting into the landscape. Liner plants offer an economical landscaping alternative where traffic and animal activity will be light and the planting need not be conspicious in the first year.

Because cold winter and spring temperatures are the main reasons for planting liners instead of seed, planting will usually be more successful in spring than in the fall. At seven Sierra roadside sites (elevation 2 km or 6000 ft) involving more than 13,000 peat pot liner plants, Leiser and others (1974) confirmed the superiority of spring planting. Where winter temperatures may fall to $-10°C$ ($10°F$) and spring frosts are common, four species of native shrubs survived at a rate of 2 percent or less for fall plantings and 11 to 62 percent for spring plantings. Survival doubled in the spring for a fifth species, while another species exhibited over 50 percent survival in both fall and spring plantings. On the other hand, if winters and early springs are not too cold or wet, late fall and early winter plantings will help liner plants become established even earlier before the summer.

Liner plants should be "hardened" for a few weeks near the site or under conditions similar to those at the site so that heat, cold, or full sun does not injure or retard the liner plants seriously. Harden plants by exposing them to greater temperature extremes and more intense sunlight than they are used to, though not as severe as conditions after planting.

Liner plants should be planted in much the same way as seeds are planted, except the depth of the planting hole will depend on the length of the liner root ball. Current information suggests that fertilizer should not be used in the hole or on the surface at planting. Weed control, mulching, and protection from animals are to be carried out as with seeding.

Sticking Cuttings

A number of woody species that grow in thickets along streams or in meadows (willow, poplar, and aspen, for example) will root readily from cuttings placed in moist soil. Newly cut fence posts have occasionally been observed to grow. Leiser and others (1974) found about 30 percent survival of 320 *Salix lemmonii* tip cuttings after two years in a moist decomposed granite. Cuttings were 150 to 200 mm (6-8 in) long and 6 mm ($\frac{1}{4}$ in) in diameter and had been collected at an elevation of 2.1 km (7000 ft) in either the fall or early spring. When cutting size was increased to 250 mm (10 in) long and 10 mm ($\frac{3}{8}$ in) in diameter, the survival of 5500 cuttings was 47 percent at seven mountain sites. Fall and early spring plantings were equally successful. Hormone treatment with indolebutyric acid at 2000 ppm was of no benefit in these plantings. Cuttings placed in undisturbed, more level soil failed, primarily from lack of moisture in summer.

Willow cuttings have been successfully stuck in drawdown areas within California reservoirs to enhance recreation areas, improve fish habitat, and reduce shore erosion. Cuttings 300 to 450 mm (12-18 in) long and 25 to 40 mm (1-1$\frac{1}{2}$ in) in diameter should be driven into the muddy shore as the water level recedes in summer and fall. To indicate the basal ends and to ease cutting and driving, cut the base on the diagonal and the top at right angles. Many cuttings will root and grow tops before flooding occurs in the following spring. A high percentage of cuttings will survive more than four months of flooding in following years (Whitlow and Harris 1979).

Cuttings can be an economical way to establish plantings in certain locations, if the soil is kept fairly moist during the growing season. Easy-to-root species may pioneer an area until other plants invade the drier sites protected by the cutting plants.

YOUNG PLANTS

Handling Plants before Planting

Plants should be carefully inspected upon delivery to see that they meet specifications as to root quality and top conformation. If the planting site is colder or hotter or more exposed than the location where plants have been kept, they should be "hardened off" or "acclimatized." This may take a few days to a few weeks. The plants must be kept moist and initially protected from temperature extremes.

Many deciduous trees and shrubs are received bare root during the dormant season. Though it is best to plant them soon after delivery, bare root plants can be held up to a few weeks if they are kept cool so that neither roots nor buds begin to grow. The roots must remain moist. The plants can be "heeled in" in the shade with their roots in moist sawdust, peat moss, or sand. If they must be kept in the open, plants can be "heeled in" by placing the roots in a trench running at right angles to the early-afternoon sun. The tops of the plants should point toward the early-afternoon sun so bud warming is minimized (Fig. 8-4). Cover the roots with moist soil and work it in around the roots to avoid air pockets.

Plants can be kept in cold storage and bundled with moist packing material around their roots. The low temperature, by more completely satisfying the cold requirement that many plants have to overcome, will assure more uniform bud break when the plant is put into the landscape. Cold temperatures (0°–4°C or 32°–40°F), by inhibiting bud break, will allow bare root plants to be planted two to four weeks later than would otherwise be the case.

Balled-and-burlapped (B & B) trees and container-grown plants can be planted almost year-round, although B & B plants are usually planted soon after they are dug. Containers for most plants are dark in color. In the afternoon sun, soil on the exposed side can reach 49°C (120°F) and remain above 38°C (100°F) for six to eight hours. In one

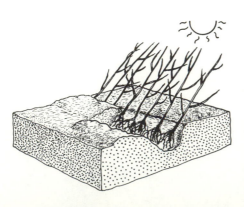

FIGURE 8-4
When bare root trees must be held in the open for longer than a few days, the roots should be placed in an east-west trench with the tops of the trees pointing toward the afternoon sun. Moist soil should be worked around the roots to cover them and minimize air pockets. Although it is not recommended, the plants can be "heeled-in" for several weeks before being planted if the weather is cool. Pointing the tree tops toward the afternoon sun exposes the least surface to the sun so the buds will be less likely to begin growth.

experiment, the growth of black locust roots was reduced 75 percent by soil temperatures of 35°C (95°F) for six hours on four consecutive days (Wong, Harris, and Fissell 1971). Temperatures of 40°C (104°F) for only four hours will kill the root tips of most plants. In another study, no roots were found in the western third of 4-liter (1-gal) containers exposed daily to the afternoon sun (see Fig. 4-8) (Harris 1967). The tops of these container plants may show little or no immediate effect because most are watered and fertilized daily. If the soil volume available to a plant, however, is only a portion of that in the container, it will have to be irrigated more frequently when transferred to the landscape than a plant whose roots had filled the entire container. Even these latter plants are seldom watered often enough.

While plants are being held for planting, place them close together, preferably in the shade. If they must be in the open, protect the outside cans from the sun. A 25 × 200 mm (1 × 8 in) board, a strip of cardboard, aluminum foil, or a dike of mulch or soil will do the job. Keep the plants from wilting. Placing them close together will reduce transpiration somewhat, will reduce injury from excessive movement of the tops, and will help to keep the containers upright. In sunny weather, minimize root damage by placing the plants in the landscape just a short time before they are to be planted.

Preparing the Planting Hole

The following discussion assumes that soil at the planting site is of good tilth or that it has been prepared as described in the preceding chapter.

In soils that have good structure, the planting hole need only be deep enough to hold the root ball of the plant. Plant "high" or "proud" in all but sandy soils. The hole can be dug or augered 50 mm (2 in) less than the depth of the soil in a 20-liter (5-gal) container plant (25 mm or 1 in. less for a 4-liter plant). If a deep hole is dug and loose soil returned, the plant usually settles after a few irrigations or rains. If the top of the root ball is below the level of the surrounding soil, water will collect around the trunk in all but very well-drained soils. Crown rot will frequently occur in such situations. In sandy soil, planting at the original depth or 20 to 50 mm (1–2 in) deeper will keep soil around the roots from drying out so quickly.

Each hole should be about twice the diameter of the container or root ball, so that the backfill soil can be worked in easily around the plant. For bare root plants, the hole need only be large enough to take the roots without crowding. To minimize "glazing" of the sides when holes are dug with a power auger, digging should be done when soil is at or below field capacity. The glazed sides of many planting holes are almost impenetrable to roots. The sides and bottom of the hole should be scarified or roughened with a shovel to intermingle the

FIGURE 8-5
The fallacy of placing a layer of coarse sand or gravel in the bottom
of a planting hole with no drain outlet is shown by the above laboratory soil profile.
The soil above the sand layer must become saturated
before water would move into the sand.
Water flowed around the sand before penetrating it.

backfill and field soil and to provide easier access for developing roots
(Smith 1977).

Unfortunately, a few publications still recommend placing about
100 mm (4 in) of gravel in the bottom of planting holes if the soil is
not well drained. The gravel layer is supposed to provide drainage
for the soil above. In fact, it will have just the opposite effect, causing
the soil above to become saturated when it otherwise would not (Fig.
8-5). As the amount of water exceeds field capacity in the upper soil, it
will begin to flow around the ends of the gravel layer, leaving that
layer essentially dry until the water rises from below or the head of
water above becomes greater than the water potential in the soil. If
drainage is needed, there must be an outlet for the flow or pumping
of excess water.

Pruning Roots

Remove the dead, diseased, broken, and twisted roots of bare root
plants by pruning to healthy tissue. Roots matted at the bottom or
circling around the root ball of container-grown plants should be cut
and removed or straightened. Some arborists cut the root ball vertically
on opposite sides for at least half the distance to the trunk to decrease
the chance that hidden circling roots will girdle later. When freeing the
roots at the periphery of the root ball, break away some of the soil to

provide better contact between the roughened root ball and the backfill soil. If the periphery roots are straightened so they extend 50 mm (2 in) or more into the backfill soil, the soil available volume will be double that of a 20-liter (5-gal) root system. In addition, the roots will extend into the fill soil and will grow more easily than when they have to grow out of an undisturbed root ball. Removing one-quarter to one-half of the roots in the outer 25 mm (1 in) of a root ball should set back none but the most sensitive plants. Most plants will in fact be stimulated.

Setting the Plant and Stakes

Just before setting the plant in the hole, loosen about 25 mm (1 in) of the soil at the bottom of the hole and create a slight mound in the center on which to set the root ball or spread bare roots. Firm the mound by stepping on it lightly with one foot.

You may also want to consider a number of less-than-critical factors when setting a plant in the hole (you will soon observe that some recommendations may conflict with others in certain circumstances and will then have to judge their relative importance and appropriateness): Orient the plant to best advantage in the direction from which it will be viewed most often. Place the scion of budded (grafted) plants, particularly trees, toward the afternoon sun to reduce the possibility of sunburn in the crook just above the bud union (Fig. 8-6). Low foliage may shade this area, but if the trunk is exposed, it can be painted with white exterior latex paint.

Orient the side with lowest branches toward areas of least activity and higher branches toward areas needing greater headroom or clearance. This will reduce the amount of subsequent pruning. Aim

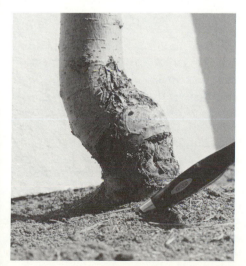

FIGURE 8-6
When planting a budded tree, orient it so the bud faces toward the afternoon sun. If exposed to the sun, the bark on the inside of the curve just above the bud can be killed or seriously injured. Shading this area or painting it with white latex paint will minimize the possibilities of later injury.

the side with the most branches into strong prevailing winds; for deciduous plants, winds during the growing season are more important than winds of winter. On the other hand, Stribling (1966) recommends orienting a young, budded fruit tree, particularly a whip (a trunk 1.2 to 2 m or 4 to 6 ft high with no laterals), so that the scion faces downwind and the tree remains upright. "We expect a greater concentration of strong branches facing into the prevailing winds," he explains. If this is a valid observation, it may be that an uneven development of phloem above the bud union will influence the hormone balance so that more buds are released or stimulated into vigorous growth on the windward side.

If wind, sunburn, and appearance are not factors in plant orientation, place the largest branch or the side with the most branches away from the afternoon sun. The less developed side of the plant will then be favored with more light. With bare root plants, place the largest root in the direction away from prevailing winds. This seems to be just the opposite of common sense. However, when the wind blows, the roots of young trees tend to be pulled out of the soil lengthwise on the windward side and forced downward on the leeward side. Since the latter encounters more resistance, particularly when the soil is wet, the largest roots should be on the downwind side to stabilize the tree. If the top of the tree is not vertical when the root ball or the trunk base is straight, tip the root ball in the hole to angle the trunk as you want it.

After the tree has been positioned in the planting hole, any stakes that are to be used should be placed close to the root ball or through the bare roots. Usually stakes are driven into undisturbed soil at the bottom of the planting hole to provide adequate support. Information presented later in this chapter will help you to determine whether staking is necessary, and what kind should be used.

Backfilling

In most cases, the soil dug from the planting hole is satisfactory for backfilling around the roots. This is particularly true for bare root plants, which would encounter no soil interfaces. If the soil has been previously loosened and mixed, it should be fairly uniform with depth and can be used directly for backfilling. If the soil has not been loosened and exhibits textural or compacted layers on the sides of the planting hole, adjacent topsoil or soil similar in texture to that of the coarsest layer should be used for backfill.

Although adding organic matter to backfill soil is often recommended, several studies in different parts of the United States have shown the practice to be of no benefit (Pellett 1971, Whitcomb 1979a). At five Oklahoma test sites with a range of soil types, Whitcomb found that fewer roots grew into the surrounding soil from amended than from unamended backfill soil. Whitcomb supplies no growth measurements for trees and shrubs planted in backfill soil amended with peat moss but states that the practice is detrimental. Wilson (1970), however, found that roots will often branch more profusely in friable, well-aerated, moist, fertile soil than in drier, infertile soil. The many small, short roots that form will not grow as far as roots in drier, less well-aerated soil. This probably accounts for Whitcomb's findings: He correctly observed that under drought conditions a plant whose roots were largely confined to the amended backfill soil would not fare as well as plants whose roots spread farther even though they were fewer in number. At any rate, in none of his trials did amended backfill soil produce a net benefit.

In England, Derek Patch (1980, personal communication) has observed poor results when backfill soil has been amended with organic matter. The main problem he identified was inadequate mixing: Large wet globs of organic matter interfered with root growth.

From the evidence to date, there seems to be little reason for amending backfill soil when planting trees and shrubs in the landscape. Commercial fruit tree growers, even those who plant subtropicals B & B, do not use organic matter at this stage. If organic matter is to be used, however, it should be added 20 to 40 percent by volume with the backfill soil and thoroughly mixed.

If fertilizer is placed in the planting hole or mixed with the backfill soil, the process is inconvenient and time-consuming and can injure the plants. Most plants will grow well for part or all of the first growing season without additional fertilizer. If nutrient deficiencies exist, nitrogen is usually the only element to be seriously lacking. If nitrate or urea is applied to the surface after planting, nitrogen will move into the soil with rain or irrigation.

Work the soil around the roots so that they are not compressed into a tight mass, but spread and are supported by soil beneath them. After each 75 or 100 mm (3 or 4 in) of soil has been placed in the hole, firm the soil around the roots or root ball with your foot, taking care not to tear, bruise, or debark the roots.

After a B & B plant has been set in the hole, remove the rope or twine that holds the burlap. Add soil for 100 to 150 mm (a few inches) around the root ball base to support the plant. Fold back the burlap from around the trunk and the top of the root mass as long as the root ball stays firmly together. Cut off the loose burlap or fold it down to be buried when the rest of the fill is added. If the burlap is sturdy, carefully cut large gashes through it in the lower root ball area so that the roots will not be unduly confined. Complete the backfilling. Be sure no burlap is exposed above the soil, because it can act like a wick, drying the soil below.

The original ground level in the nursery (the soil line on bare root trunks or the top of the root ball) should be 25 to 50 mm (1-2 in) above the finished ground level. Basins should be at least 750 mm (30

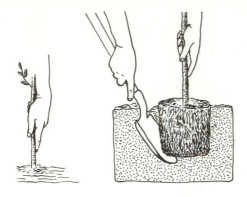

FIGURE 8-7
A bare root tree that has been planted too deeply can be raised to the proper height if you lift on the trunk when the soil is muddy (left). For B & B and container-grown plants, use a shovel to help lift the root ball while you lift on the trunk (right).

in) in diameter. Fill the basin with water to further settle the soil and to provide the plants with water. Attach a 1- to 1.2-m (3- to 4-ft) pipe to a hose and force the pipe, while water is running, through the soil around the roots. This will remove any air pockets and improve contact between the fill soil and the roots. The plant can be rocked slightly to facilitate soil consolidation.

If the plant has settled so that the original soil line is below the soil surface, a bare root tree can be raised by lifting on the trunk while the soil is muddy (Fig. 8-7). It should be raised slightly higher than desired and then settled back into the soil. Use a shovel under the root ball to raise a container plant (Fig. 8-7). This should be kept to a minimum, since the more a plant is raised, the closer the roots are drawn together.

After the soil has drained, the final contour of the basin can be designed to make the base of the plant slightly higher than the bottom of the basin (Harris and Davis 1976). Unless some soil has fallen or been removed from the top of the root ball, do not add soil there to obtain the desired basin contour. This is particularly important if the

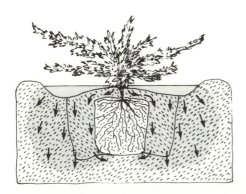

FIGURE 8-8
When a root ball is covered with a layer of soil whose texture is finer than that of the root ball soil, water will tend to flow around the root ball rather than entering it. The root ball will then be wetted only by water that rises from below.

backfill soil is finer in texture than that of the root ball. If so covered, the root ball will be wetted with difficulty or not at all (Fig. 8-8).

Mulch placed over the root ball in the basin and on the surrounding berm will reduce moisture loss, moderate surface soil temperatures, and reduce settling and cracking of the soil and berm surface. Use 50 to 75 mm (2-3 in) of coarse organic material (woodchips, bark, or forest litter), coarse gravel, small rocks, or built-up layers of dry lawn clippings. Around trees in a lawn area, a coarse mulch may interfere with mowing, but lawn clippings will usually work quite well (see Chapter 13).

Pruning

Newly planted trees and shrubs are usually pruned with the thought that the reduced top size will compensate for root loss to bare root plants or the relatively small root ball size of most container-grown plants. If sufficient water is available, plants will usually grow better with little or no pruning than with severe pruning. When water is inadequate, however, prune severely enough to reduce the size or density of a plant's shadow and you will increase its chances of survival. Plants can be pruned severely at planting without suffering serious consequences. Fruit trees are commonly pruned severely at that time so that they will produce low branching structures.

Thinning branches that are close together, crossing each other, or broken can remove considerable leaf area without greatly affecting overall plant size. One-fourth of the leaf area can usually be removed from most plants with only a positive visual effect. If a plant's basic branch structure is not the one desired, remedial pruning should be undertaken (see Chapter 14).

Staking

Before a tree or large shrub is planted, materials for any necessary staking should be at hand. Staking is used to protect or to anchor trees and shrubs and to support young trees (Harris, Leiser, and Davis 1976). The extent of staking for trees will depend on tree strength and conformation, expected wind conditions, the amount of vehicular and foot traffic, the type of landscape planting, and the level of follow-up maintenance. Many young trees can stand alone; others may need support to stand against the wind or to grow upright as desired. In order to decide whether staking is necessary and what type to use, you may want to review the consequences of supporting a tree by either staking or guying.

FIGURE 8-9
Tree ties can injure. If ties are loose, the trunk or branches rub against
the stake (left); if they are tight, a girdle may surround the trunk (center),
restricting translocation in the bark and eventually causing breakage (right).
Note the wire deeply imbedded in the trunk; trunk caliper above the wire
is greater than below it.

Consequences of Staking   Compared to a tree that stands alone
and is free to move, a staked tree will

Grow taller;

Grow less in trunk caliper near the ground but more near the top support tie;

Produce a decreased or even a reverse trunk taper;

Develop a smaller root system;

Offer more wind resistance than trees of equal height (because the top is not
free to bend);

Be subject to more stress per cross-sectional unit at the top support point
(the ground for an unstaked tree);

Be more subject to rubbing and girdling from stakes and ties (Fig. 8-9);

Develop uneven xylem around the trunk if it is closely tied to one stake. The
trunk will grow or bend away from the stake (see Fig. 2-11).

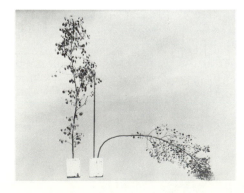

FIGURE 8-10
The cumulative influence of staking is shown
by these two silver dollar gum trees grown
for 11 months in 20-liter (5-gal) cans. One
has grown unstaked with lower laterals on
trunk headed back (left tree); the other has
been tied to a stake with the lower laterals
removed (right tree). The staked tree has
been untied from the stake. (Harris, Leiser,
and Davis 1976)

All of these staking influences make a tree less able to stand without support and, if supported, more subject to injury, particularly if the ties or stakes should break (Fig. 8-10). In addition, staking is expensive and time-consuming and often detracts from tree appearance, especially when improperly done. Even so, staking may be necessary to help young trees until they get root anchorage or their tops become strong enough to stand alone. Staking and other materials may be needed to protect young trees from vehicles, equipment, animals, and vandalism. Necessary and proper staking can overcome most of the problems associated with this procedure.

Types of Staking   No staking is necessary for most shrubs and for many conifers and other trees with limbs close to the ground (Fig. 8-11). Low branches keep people and equipment away from the trunk. These plants are usually short, with root systems or root balls adequate to hold the tops upright.

Protective staking is used primarily to keep mowing equipment, vehicles, and vandals away from young trees that are able to stand alone: most conifers, small trees with upright growth habits, and most trees planted bare root. Even many trees with tops that are large in proportion to their roots can stand alone if about one-third of the branches in the crown are thinned out, reducing wind resistance and top weight.

If vandalism is not a problem, place two or three short stakes (about 50 × 50 mm or 2 × 2 in thick and 1.2 m or 4 ft long) approximately 400 mm (15 in) apart around the tree so that they protrude

FIGURE 8-11
Most trees with laterals close to the ground need no staking (left).
Many deciduous trees need only short stakes to protect their trunks
from lawnmowers and other equipment (center).
Other trees may have sturdy trunks but root development inadequate
to maintain the tree upright, particularly when soil is wet.
Such trees should be anchored by tying them between two low stakes (right).
(Harris, Leiser, and Davis 1976)

FIGURE 8-12
Trees in heavily used areas are frequently protected
with metal grillwork (left) or more substantial barriers (right).

from the ground about 750 mm (30 in) (Fig. 8-11). The stakes should be easily visible, so that people will not walk into them. To minimize vandalism, many public agencies are planting larger trees and using sturdy stakes encircled with heavy wire or metal grillwork (Fig. 8-12). More information is needed on staking that protects trees without unduly restricting top movement.

Anchor staking may be needed to hold the roots or root ball of an otherwise upright tree or shrub until the roots grow into the surrounding soil and can support it. Spring rains or the frequent irrigations given to young plants add to their instability. If the roots are not well anchored, trunk movement could break new roots growing into the surrounding soil. Low anchor staking can hold the lower trunk to keep roots from moving while permitting freedom to the top.

The two or three short stakes suggested for protecting treetrunks usually provide enough anchorage for roots. Place one loop or figure-eight tie between each stake and the treetrunk (Fig. 8-11). Each tie should be secured near the top of its stake and should permit some movement at that level without allowing movement of the roots or rubbing of the trunk against the stake. The tops of some trees may benefit from thinning to decrease wind resistance and weight. You can usually remove ties by the end of the first growing season, but leave the stakes in place to continue protecting the trunk.

Support staking is needed for trees whose trunks are not strong enough to stand without support or to return upright after a wind. Weak trunks often result when trees have been previously staked, grown close together, or had their lower branches shaded or removed (see Fig. 8-10). Weak trunks are common on many container-grown trees. Top support for these trees should be about 150 mm (6 in) above the lowest level at which the trunk can be held and still return upright after the top is deflected (Fig. 8-13). This location will give the top the greatest flexibility while providing adequate support.

Young trees are often tied to a stake higher on the trunk than is necessary. Leiser and Kemper (1968) found on trees with tapered trunks that stress per unit of trunk area increased with increased height of staking (Fig. 8-14). Just the reverse was true when trunks had little or no taper between the ground and the head of the tree. As the height of stake support increases in these trees, the lever arm of the trunk above the stake decreases and the stress at the tie decreases proportionately. On tapered trunks, cross-sectional area decreases proportionately more with increasing height than does the length of the trunk lever arm. The higher the support, the more upright and inflexibly the top is held, so that it must take the brunt of winds instead of taking them more obliquely. In addition, as the leader terminal is approached the trunk is more succulent and less able to withstand wind stress. Leiser and Kemper determined that the most critical stresses occur when staking height for a tapered treetrunk is above two-thirds the height of the tree. Stress per cross-sectional unit increases rapidly as staking height increases above this point. When trees were staked at points closer than 0.75 m (30 in) of the terminal, Leiser and Kemper found that stress per unit was 3 to 4.5 times as great as it was near the ground (Fig. 8-14). They did not obtain stress figures within 10 percent of the height near the tip, but extreme stress would probably continue. The stresses found in the top portion of the trunk exceeded those that most species can endure. It is a wonder that the tops of more staked trees are not deformed by the wind.

Trees with little or no taper might be best suited for staking if they were not to grow after being staked. However, as previously stated, a staked tree grows taller and with less basal caliper than an unstaked tree and is less able to become self-supporting. Also, the stake and tie must withstand the stress of an increasingly larger tree.

Staking stresses vary within the head of a tree, depending on distances from the terminal and on the position and size of individual branches. Tying a leader closer than within 0.75 m (30 in) of the tip, particularly of previously unstaked trees, will subject the leader to maximum stress and maximum likelihood of wind deformation. The onset of damage could be delayed by staking at ever-increasing heights,

FIGURE 8-13
(a) Glossy privet supported by a stake
to which it was tied during container production;
(b) untied, it cannot stand upright;
to determine height of support tie,
(c) trunk is held at about 1.2 m (4 ft)
and the top is bent;
(d) this is the lowest height the trunk can be held
and have the top return upright;
(e) the support tie should be about 150 mm (6 in)
higher than the height determined in "d".
Thinning out the top will reduce wind resistance.
The top of the stakes should be cut to within 50 mm
(2 in) of the top ties to reduce the chance of rubbing.

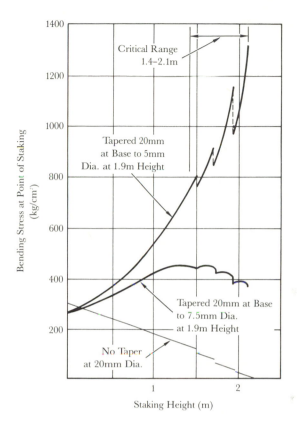

FIGURE 8-14
Bending stress is shown as a function of staking height for staked young trees
(2.4 m or 96 in tall) with different degrees of trunk taper. Staking height was
carried to 2.1 m (84 in). Discontinuities in the curves occur as the point of
staking moves from just below a branch to just above it. A bending stress of
1000 kg/cm^2 equals about 14,000 lb/in^2. (Adapted from Leiser and Kemper
1968)

but practical limits and the weak trunk that would result makes this
solution unattractive. Sometimes small trees are tied to stakes that are
taller than the trees. It should be apparent that, as the terminal grows
beyond the top tie or the top of the stake, the leader will be subjected
to extreme stress from wind unless it is in an extremely protected area.
It would be best in such cases to drop the height of the top tie and
stake top to at least 1.2 m (4 ft) below the leader tip. A flexible
auxiliary stake may be needed to strengthen the leader.

Only a single support stake is commonly used. Two stakes, however, can minimize the difficulties encountered in support staking; some
arborists and landscape architects even recommend three stakes. In any

case, care must be taken to tie or hold the trunk so that it can flex without rubbing against a stake. The ties must not be so tight or inflexible that they are likely to damage tender bark or girdle the expanding trunk. When you decide how many stakes to use per tree, consider the foregoing criteria, along with material and installation costs, expected vandalism, and expected frequency of follow-up maintenance. More latitude can be allowed in selecting the method of staking and tying if the trees will be inspected frequently.

When only one stake is used, the tree is often tied snugly against it at more than one level. Besides the direct effects of trunk immobility, three difficulties may result: (1) The ties may girdle the trunk, although frequent inspection can prevent this. (2) Greater stress will occur at the top tie during a wind if the trunk below is not able to flex in the opposite direction as the top is blown to and fro. The trunk is therefore more likely to break at the top tie or to be seriously deformed (Fig. 8-15). (3) The stake also will shade the trunk, causing the xylem cells to elongate more on the shaded side so that the tree actually tries to grow away from the stake (see Fig. 2-11).

In addition to the many clever methods developed by commercial and public arborists, a number of commercial tree staking and tying devices are available to hold trees properly to single stakes. Black (1978) reports that street trees in Seattle have suffered less than 1 percent loss from vandalism when trees of 50-mm (2-in) caliper are planted and a 16-mm ($\frac{5}{8}$-in). rebar (metal reinforcing rod) 3 m (10 ft) long is

FIGURE 8-15
Bending stress at the top tie of staked trees can cause serious deformation or breakage. *Melaleuca linariifolia* tied to an iron pipe (500 mm or 2 in diameter) (left); a young liquidambar tied to a rigid stake broke at the top of the stake.

driven 1.5 m (5 ft) into the ground close to the trunk. Three ties closely spaced near the top of the stake hold the trunk securely. At that height (1.5 m) it is difficult to break the trunk. The stake is removed after one year. One stake is cheaper and easier to install than two stakes, although any stake is easier to install if it is "planted" with the tree instead of driven afterward.

Two support stakes with one flexible tie near the top of each will hold the tree upright, provide flexibility, and minimize trunk injury and deformation (see Fig. 8-13). If wind is not a problem, the stakes should be placed so as to provide maximum protection from traffic and equipment. In moderate to strong wind situations, though, an imaginary line drawn between the support stakes should be at right angles to the most critical wind direction. The ability of stakes to withstand wind, particularly when the soil is wet, can be increased by connecting them with a 25 × 75 mm (1 × 3 in) wood crosstie partially or entirely below the soil surface. Place the crosstie to the lee (downwind) of the stakes, being careful not to injure roots near the surface. The crosswind orientation of the two stakes lessens the risk that the trunk will suffer rubbing injury and that the stake assembly will jack itself out of the ground in strong winds. One low crosstie keeps the two stakes equidistant along their length but still provides some flexibility in a strong wind.

Some arborists prefer three stakes for greater protection against equipment, vandals, and strong winds. Three stakes will more effectively support protective cylinders of wire or rod which may be desired or necessary in certain situations.

Stake and Tie Materials   Support stakes come in a variety of materials and cross-sectional shapes. Wooden stakes are usually sawn wood or small-diameter poles 50 mm (2 in) in diameter (lodgepole pine is commonly marketed). Poles do not split or break as easily as sawn stakes while in use or while being driven (John Sue 1973, personal communication). Metal pipe, "T" fence stakes, and reinforcing rebars (Fig. 8-16) of various sizes are also used. Metal stakes usually need a flange or plate just below the ground surface for extra stability. A metal flange entangled in roots can make stake removal difficult, however. Availability, cost, strength, durability, ease of installation, and appearance are all factors in selecting staking material.

Several different materials and methods are also used in tying trees to the stakes. Any material used should contact the trunk with a broad, smooth surface and have enough elasticity to minimize trunk abrasion and girdling. Common tie material includes elastic webbing, belting, polyethylene tape, tire cording with wire ties, and wire covered with hose or tubing. A number of patented ties and support devices are available for single and double staking. Although rope, baling wire, covered electrical wire, string, and fishing line are often used to hold

FIGURE 8-16
A single Maxwell tree stake
provides support without shading
or rubbing the trunk, as long as
the tie holds the trunk secure.

tree and stake together, many of these do not measure up to the above criteria. If wire ties are left on too long they girdle the trunk. Especially with a single stake, wire ties may eventually break from flexing in winds. The break is usually between the trunk and stake, so that a circle of wire remains around the trunk. If the trunk is not immediately damaged by losing its support, it may be damaged later by rubbing against the stake and the jagged broken wire or by encountering the wire girdle as it grows. Such a girdle may be obscured by low foliage until it is too late (see Fig. 8-9). Polyethylene or fabric ties that may last only a year will minimize the problem of girdling, although they may allow stake rubbing.

Except for the patented ties and devices, most methods of tying employ a figure-eight loop between the trunk and each stake. The figure-eight tie can hold the trunk secure and yet allow some flexibility; also, two layers of the tie material cushion the trunk against the stake. For two or more stakes, a plain loop between tree and each stake is often used, although these lack the cushioning capacity of the figure-eight if one tie breaks.

Trees with spindly trunks and little or no taper may need extra support along most of their length. A rigid support, however, will delay developing the ability to stand alone. A flexible auxiliary stake tied snugly to the trunk usually will give the needed strength. Its diameter and height should permit the trunk to flex and return upright. It should be slender enough to shade the trunk very little or not at all. A 25×25 mm (1×1 in) stake can deform treetrunks with diameters of 6 to 12 mm ($\frac{1}{4}-\frac{1}{2}$ in) so that they bend away from the stake when untied (see Fig. 2-11) (Neel 1967). Spring-steel wire, fiberglass rods, bamboo, and split wood can be used for trunk support. Steel wire or

224

rods are easily available and uniform. Harris, Leiser, and Davis (1976) have suggested that a rod 3 mm ($\frac{1}{8}$ in) in diameter be used for 20-liter (5-gal) trees. If more support is needed lower on the trunk, a shorter section can reinforce the first rod, which will still provide flexibility at the top. A number of California arborists use 6-mm ($\frac{1}{4}$-in) steel rods for 20-liter (5-gal) trees and 9-mm ($\frac{3}{8}$-in) rods for 60-liter (15-gal) trees. These larger diameters (the cross-section of a 6-mm rod has twice the area that two 3-mm rods have) give more support low on the tree but often lack flexibility at the top.

The rod should not extend within 1 m (40 in) of the terminal of the leader, or deformation of the top may result, as already discussed. The auxiliary stake need not extend to the ground, though it may look more professional if it does. The tip of each rod should be rounded and wrapped with friction or duct tape to lessen the chance of injury to the trunk and to people. Tape on the tip will also keep the top tie from slipping off in a wind. Tie the trunk to the auxiliary stake with polyethylene tape at 200- to 250-mm (8- to 10-in) intervals; the trunk and rod are then treated as a single unit and are tied to the support stake at one level.

The auxiliary stake may be needed through the first season. Remove it as soon as possible. Support stakes will probably be needed through the second year, but the height at which they are tied might be decreased before the second season and the stakes accordingly shortened.

Alternatives to Support Staking

Almost every response to support staking prevents a tree from becoming strong. Support staking represents a last effort to hold a tree upright until it can finally stand alone. For the most part, the sooner a tree can stand alone, the sooner it will become strong.

As an alternative to support staking, a tree smaller than 50 mm (2 in) in trunk caliper may be cut off about 300 mm (1 ft) above the ground or graft union, painted with white exterior latex paint, and allowed to grow a new top without support. I have cut newly planted 25-mm (1-in) caliper Modesto ash and fruitless mulberry 300 mm (1 ft) above the graft union. I selected the topmost vigorous watersprout as the new leader and cut back the other shoots. By the end of the first growing season, the severely pruned trees were almost as tall as others that had been pruned only lightly and left at their current height. The trunk of every severely pruned tree was straight and strong. Similar pruning before growth began was used on eucalyptus of 35-mm ($1\frac{1}{2}$-in) trunk diameter on the Santa Barbara campus of the University of California when these trees could not stand upright if untied from their stakes. One watersprout from each stub was developed into a main trunk. Straight, strong trunks grew, unsupported, into trees that at the end of the first growing season were almost as tall as when they were planted (William Smith 1971, personal communication). Many Oregon nurseries cut deciduous landscape and some fruit trees close to the ground at the end of the first year in the field so that the next year's growth will be straight and vigorous and ready for sale. A few trees may die, but those that grow will be fine, strong specimens requiring less care. Such severe pruning should be done before growth begins.

CHAPTER NINE

Transplanting Large Plants

Large trees and shrubs transplanted into a newly developed landscape give a mature beauty that can be surpassed only by mature plants that were already on the site. There are almost as many ways of moving large plants as there are people who move them. They may be moved bare root or with a root ball that is completely exposed, covered with burlap and laced, wrapped with wire netting, boxed with lumber and banded or bolted, encased in cement, frozen, or dug and moved by special equipment. Care varies from the extremely meticulous (three years of root pruning prior to the move, digging a large root ball, careful wrapping and lacing, loading and moving with large equipment, preparing the site completely, and overseeing follow-up maintenance) to the extremely casual (simply scooping out the plant and moving it with a front-end skip loader or soaking the root area for several days and pulling the plant out with a crane). Any given method has its own advantages, disadvantages, and degrees of success. Transplanting success depends on the plant species, their condition, characteristics of the original and final planting sites, the season of the year, and follow-up care, as well as on the transplanting method itself.

SELECTION OF PLANTS

Species

Experience has shown that some species can be transplanted more easily than others. Plants that have primarily shallow, fibrous roots close to the trunk can usually be moved with greater success than plants with fewer and larger roots. Rooting patterns are determined by soil characteristics and growing practices as well as by species. Transplanting success is usually higher with shrubs than with trees, and with deciduous plants than with evergreens.

Among the 12 genera of trees listed by Himelick (1981) as easiest to move some of the more common include: apple, ash, elm, hackberry, linden, plane tree, poplar, and willow. Also noted for ease of transplanting are honey locust, pin oak, and common pear. The trees of milder climates are not included in Himelick's list; among them, the olive and most palm species can be easily transplanted. Himelick listed hickory and pecan species, walnut and butternut species, sassafras, tupelo, and white oak as most difficult to move.

A survey by Rae (1969) of 24 commercial tree movers in the northeastern and midwestern United States produced the following list of "good risks" when trees are being moved by frozen root ball: crabapple, elm, honey locust, linden, maple, and red, Scotch, and white pine. When moved by frozen root ball, the following are "poor risks": birch, dogwood, hemlock, magnolia, oak, liquidambar, and tulip tree. Soft-rooted trees do not usually survive frozen root balls.

Plant Quality

Smaller plants of a particular species will usually transplant more readily than larger plants. Even so, trees of up to 1.2 m (4 ft) DBH have been moved successfully (Pirone 1978b). The cost and difficulty of moving increase rapidly with plant size. Nursery-grown plants usually have a more fibrous, compact root system and more attractive tops than plants grown without cultivation. Large plants are often available for transplant when land use changes, roads are widened or built, or when plants become too large for their location. Many fine specimens can be obtained in these circumstances, though they may be difficult to dig because they are hemmed in by structures or other vegetation. Wherever large plants are obtained, their health and vigor are important to transplanting success.

SITE CHARACTERISTICS

The root systems of plants are usually more compact and fibrous in fertile, well-aerated soil than in sandy, infertile, droughty soil or in silt or clay soil with a poorly drained subsoil. In these latter situations, roots are more likely to be large, with few laterals or small roots near the trunk; those in sandy soil will tend to be deeper, while those in silt or clay will be more shallow and spreading. Plants are transplanted more easily from sites that are free of stones and other obstructions.

It is particularly difficult to move trees from steep slopes and reset them vertically at a new site.

Planting sites with paved areas or utility wires and pipes may restrict the use of large tree-moving equipment, so that a boxed or B & B tree may have to be moved to the planting hole with great care. Large trees have sometimes been moved into place by helicopter.

A number of urban planting sites are extremely harsh for newly transplanted large plants. Paving and buildings can increase air temperature and radiation intensity. Buildings can create wind tunnels and extreme air turbulence. These conditions can make it difficult for the root system of an exposed plant to supply enough water and to support the top adequately. Frequent watering and sturdy anchor staking may be required.

SEASON OF TRANSPLANTING

The season of the year not only influences the stages of plant growth but in a gross way determines the weather. These factors, in turn, affect the ease with which a particular species can be transplanted. In most cases, however, the time of transplanting is determined by plant availability and the schedule of landscape installation.

In more temperate climates, deciduous plants are most easily transplanted in the fall after the leaves turn color or drop but before the soil freezes, or in the spring before growth begins. Winter planting may be desirable for species that can withstand being moved with a frozen soil ball. Conifers are planted most commonly in early fall or late spring, and broadleaved evergreens as growth begins in spring. In milder winter climates, plants can be moved throughout the winter with equal ease as long as the soil does not become too muddy for easy operation of equipment.

Late summer and fall have the advantage of warm soil to encourage root growth and their shorter and cooler days decrease transpiration. Winter planting takes advantage of cool or cold temperatures, reduced plant activity, and the relative ease of moving frozen root balls over frozen soil. Winter plantings, however, may desiccate or be injured by cold.

Spring planting before top growth begins will avoid most damaging cold weather, allow some root growth before top growth resumes, and ensure ample soil moisture. Most plants, except palms, should not be transplanted in late spring and summer, while they are still making rapid top growth. Most plants survive summer transplanting better after spring growth has matured. Palms, however, are to be moved in late spring and summer, when root growth is usually at a maximum (Muirhead 1961).

Whatever the season, the plant must be protected from freezing and desiccation. It may be wise to avoid moving plants on extremely cold, hot, dry, or windy days. Bernatzky (1978) reports that large trees in Europe are often transplanted at night to protect microorganisms that are thought to promote root growth after transplanting but to be injured or even killed by solar radiation or desiccation.

METHODS OF TRANSPLANTING

Methods of transplanting large plants depend on the plant size and species, soil and site conditions, lead time before moving, time between digging and planting, distance between digging and planting sites, available equipment, personnel, and funds. Needless to say, the larger and more sensitive a plant is, the harsher the weather, and the greater the time and distance between digging and planting, the more protective the method must be.

Bare Root

Bare root transplanting is usually confined to deciduous plants up to 50 mm (2 in) in trunk diameter. This method, however, has been successful with larger trees of many species in regions that have mild winter climates though not necessarily mild summers. Such handling has been confined primarily to plants that grow in sandy soils and are moved short distances. The tree is lifted by a crane with a cable attached to a hardened steel rod inserted through a hole bored in the upper middle portion of the trunk or in large branches. Since the hole is above the center of gravity when no soil is left on the roots, the tree will hang fairly vertical when free (Fig. 9-1). A trench is dug by machine or hand around the root system, similar to that dug for normal balling of roots. Sand is then forked from the roots beginning inside the trench and moving in toward the trunk. Exposed roots are covered or misted to prevent drying out. Sufficient upward strain is applied by the crane so that the tree will lift free when enough soil has been removed. The sandy soil that remains on the roots may be left if the tree is moved immediately to a new planting site; the soil should be hosed off if the tree is to be stored in a shaded area for a time. Well-decomposed organic matter should be worked in around the roots of stored trees, which are held upright and thinned out somewhat. The tops should be misted to reduce transpiration. During planting, moist, sandy soil from the site is worked in around the tree roots, which should be carefully spread. The tree can then be guyed upright, a vertical PVC pipe installed to mist the top, and the soil mulched.

FIGURE 9-1
An *Erythrina caffra* growing in sandy soil at Disneyland is moved bare root for the third time by lifting the tree with a crane. Cables are attached to large metal lifting pins inserted through three main branches above the tree's center of gravity. (Photo courtesy Morgan Evans, WED Enterprises, California)

This method of moving trees has been used successfully at Disneyland for many years.

Ball and Burlap

Moving a plant with a ball of soil wrapped in burlap or other material has been common for centuries, particularly for moving evergreens. The diameter of the root ball should usually be 10 to 12 times the trunk DBH (Fig. 9-2). The higher ratio applies to trees 50 mm (2 in) or less, the lower ratio to larger trees. Since most tree roots are in the upper 750 mm (30 in) of soil, regardless of tree size, root ball depth need not increase in proportion to horizontal diameter (Fig. 9-2). Root ball depth will be about 75 percent of the diameter for small trees and about 40 percent of root balls that are 3 m (10 ft) in diameter. Proper depth is best determined by root density: The ball has been dug deep enough when the root density decreases markedly; going deeper increases the size and weight of the root ball unnecessarily. To lighten the root ball after it is dug, remove surface soil until roots begin to be exposed, then protect them from drying out. The average relation of root ball size to weight is shown in Fig. 9-2. Root balls may be larger or smaller than listed in certain situations, depending on plant species and condition and expected weather.

If there is time, particularly with species that are difficult to transplant, prune back the roots one or two years before the move to promote fibrous rooting within the future root ball. Newman (1963)

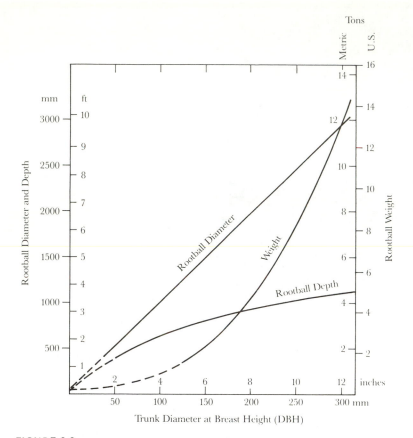

FIGURE 9-2

Estimated weight of root balls of recommended diameter and depth
for standard shade trees of different trunk diameters.
Weights are based on soil weighing 1500 kg per m^3 (110 lbs/ft^3).
(Adapted from Thompson 1940)

describes three root pruning procedures used in Europe. The one
most commonly recommended involves digging a trench around the
plant to leave a root ball slightly smaller than its final size will be. Arcs
equal to half the root ball are dug one year and the remaining arcs the
second year. These are backfilled with a mixture of sandy soil and
organic matter. New, fibrous roots that grow out into the loose soil
are thought to ease the moving process. Alternatively, the trench may
be dug to future root ball size and backfilled with cinders or slag to
encourage fibrous root growth within the soil ball. Finally, the larger
root pruning trench may be left unfilled, although this entails a greater
risk that the root ball will dry out. The advantage of the last two
methods is that new root growth will occur within the root ball, not
out at the periphery where they are easily damaged. As Newman points

out, no experimental evidence favors one method or the other. In few cases, however, is time taken (or available) to prune roots before transplanting.

Sometimes leaves are stripped off or antitranspirant sprays are applied to plants in leaf to reduce transpiration during and after moving. The species sprayed and the concentration of antitranspirant applied must be carefully chosen. Antitranspirants seem to be more toxic to evergreens than to deciduous species. Used with caution, the sprays may reduce water loss not only during the growing season but also during the winter for conifers and other evergreens (see Chapter 12).

To minimize interference with the digging, lower branches should be pruned off or tied up. The root ball should be dug only when soil is moist. A trench is dug around the trunk 75 to 125 mm (a few inches) beyond the finished root ball size, straight down to just below most of the lateral roots and wide enough to be worked in comfortably, usually about 600 mm (2 ft) for large trees. The trench can be rough-dug with a trenching machine or backhoe. Care must be taken to avoid damaging the roots or fracturing the root ball. The ball is finally shaped by shovel so that the base is 200 to 300 mm (8–12 in) smaller than the top diameter. Some arborists begin with a root ball slightly larger than recommended and fork about 300 mm (12 in) of soil from the periphery to form a slightly smaller ball than recommended, thus retaining more roots but decreasing root ball size and weight. The exposed roots will be appressed to the root ball when the burlap is in place and should be kept moist at all times. When planted, the appressed roots must be extended into the backfill soil as it is placed.

When the root ball is covered with burlap or other material, the covering is securely kept in place by rope lacing (Fig. 9-3). Root balls up to 750 mm (30 in) in diameter have also been wrapped with 25-mm-(1-in-) mesh, 20-gauge steel wire netting instead of burlap and lacing (*Amer. Nurseryman* 1970). The ball periphery will dry more quickly in the steel wire netting, but after planting, the wire is said to rust away with time instead of girdling the roots.

FIGURE 9-3
A large root ball has been covered
with burlap, laced with rope,
and then overlapped with chicken wire.
(Photo courtesy W. Rae, Jr., Frost and Higgins,
Burlington, Mass.; Himelick 1981)

After being wrapped and laced, a root ball up to a meter (3 ft) in diameter can be undercut on one side and tipped to break it free. Do not use the trunk to pry the root ball. After the plant is tipped, trim off excess bottom soil and cut projecting roots. Larger soil balls can be cut free if a steel cable is looped around the bottom of the ball and its two ends pulled with a winch. The cable will be pulled under the root ball, cutting the soil and roots.

Plants with root balls up to 600 mm (2 ft) in diameter can be moved by two people using shovels, a heavy canvas or burlap sheet, or a specially designed handcart. A front-end skip loader can be used to lift root balls a meter (3 ft) or more in diameter and move them a short distance. If root balls more than a meter in diameter must be transported some distance, they are usually lifted onto a truck or trailer with a crane or lifted and transported by a truck with a hydraulic boom hoist or a self-loading trailer. The plant can be lifted with a chain or cable attached to the root ball and trunk base (Fig. 9-4). Some arborists insert a steel rod through the treetrunk a short distance from the root ball and use it as a major lifting point. In all of these operations, the trunk should be protected with thick padding where the lifting chain or cable might come in contact with it.

Methods of transporting B & B plants will depend on the plant size, the distance to be moved, and available equipment. Several plants with root balls up to 1 meter (3 ft) in diameter can be transported on one large truck. Special tree-moving trucks with hoists or self-loading

FIGURE 9-4
Lifting trees and root balls. The Newman Frame and Tree Mover lifts a linden
9 m (30 ft) tall and is ready to tilt it by truck winch for moving (left).
The tree and root ball are stabilized by cables attached to a steel pin
through the upper trunk (Photo courtesy C. J. Newman, England).
At the right, a heavily strapped root ball is lifted in a chain sling by the truck-mounted hoist.
(Photo courtesy W. Rae, Jr., Frost and Higgins, Burlington Mass.)

trailers are most suitable for lifting and transporting large plants, one at a time, over reasonably short distances. Limbs should be tied up or in to reduce damage and to reduce the width of the load.

Follow the same general directions as given in Chapter 8 for digging the hole and setting the plant. Some arborists are quite particular about setting the plant with the same orientation it originally had. The main value in this is that the same sides of the trunk and main branches will be exposed to the afternoon sun. Previously protected bark can be killed or seriously injured if exposed to the sun's full intensity. A different orientation, however, may be desirable or necessary. Newly exposed bark can be protected with white exterior latex paint; this will weather so that the bark can develop hardiness as it is progressively exposed. Selectively thin any branches that might be prone to wind damage because of the changed orientation.

Frozen Root Ball
(Himelick 1981, Rae 1969)

Only the outer 100 to 150 mm (4–6 in) of the root ball periphery should freeze. Deeper freezing can cause serious root damage, but less freezing will not ensure a solid enough ball. In areas where the soil freezes to depths of 300 mm (1 ft) or more, trees can be moved with frozen root balls. This extends the transplanting season and can provide work during slow winter months. Frozen root balls require less wrapping and are less likely to be damaged in handling than other root balls. Root balls made of sand or gravel are more cohesive and easier to handle when frozen. Equipment can often be operated more easily and can cause less damage to lawns and thin roadbeds when the ground is frozen.

Serious disadvantages are that frozen root balls can increase costs and the chances of failure for the inexperienced. Work is more likely to be hampered by the weather in winter than at other seasons. Changes in the weather or delays in moving can lead to injurious freezing of the root ball or to a thaw that allows it to break apart. Winter working conditions are difficult for men and equipment; the days are shorter and breakdowns are likely to be more frequent. Many species cannot withstand being moved with a frozen root ball. For trees that can be moved successfully with frozen root balls, certain procedures help ensure success: First, the trees to be moved and the planting sites must be selected in advance so they can be mulched before the ground freezes. When digging the root ball, do not break it free until the move is imminent. If a tree is to be left in a partially dug hole, mulch the excavation to prevent further freezing. The planting hole should be dug shortly before planting the root ball, so that the inside of the hole does not freeze. The soil from the hole may need to be covered or mulched before it is backfilled. If the backfill is not frozen and the temperature is near or above freezing, water only enough to help settle the soil around the root ball and mulch the tree well to prevent further freezing. In the spring, the tree should be watered thoroughly and soil added to the backfill if necessary.

Trees are usually dug and transported only when the daytime air temperature is at least – 7°C (20°F). During transport, the tree and root ball should be completely wrapped with canvas or heavy plastic to prevent drying, further freezing, and exposure to road salt. Other normal practices for plant moving and planting are applicable.

Bare Root Ball

A few hardy species of plants growing in silt or clay soils can be moved with no covering on the root ball. These are usually shrubs and small trees that are replanted soon after digging, although some trees (olive, for example) can be transported several hundred kilometers (300–400 mi) in the summer with no protection of the root ball or top. Although plants handled in this way may survive, they usually are slow to resume growth and regain a healthy appearance. With skill, an operator using a front-end skip loader can successfully move shrubs and fair-size trees up to a kilometer (0.6 mi) or so in a short time (Fig. 9-5).

FIGURE 9-5
A blue oak 8 m (28 ft) tall dug about 2 km (1 mile) away from the planting site in the Sierra foothills in early winter awaits the digging of the hole in which it is to be planted (top); note the small unwrapped rootball. The oak is positioned in the hole by the skip-loader (bottom left) that had lifted and transported it from its hillside site. The same tree six years later (bottom right). About forty blue oak trees were similarly moved with less than a 10 percent mortality. (Left photos courtesy California Dept. of Forestry)

Box

Boxing a square-shaped root ball can substitute for wrapping the root ball with burlap and lacing or wire. Boxing is commonly used on field-grown trees that may be held several months before replanting. Some nurseries grow trees to a large size in boxes. Some tree movers prefer to box trees, particularly large ones, because the ball is held more securely than in burlap and less experienced workers can successfully handle it. Box materials, precut to a particular size, are quickly assembled around a root ball and can be reused.

For boxing, the root ball is dug as if it were going to be balled and burlapped, but it is shaped and trimmed to fit snugly in a square box with sloping sides. Box sides are made of one or two layers of 20-mm (0.75-in) plywood or planks that are 25 to 50 mm (1–2 in) thick and are reinforced with exterior bracing (Fig. 9-6). Two of the sides are wider than the others and have cleats along their vertical edges to hold the other two sides in place. Rods are inserted through and between these cleats to hold the four sides securely against the root ball. The ball can then be undercut and boards inserted to form a bottom for the box. Two beams are placed under and at right angles to the bottom boards; two similar beams are placed across the top of the box and attached to the lower ones by rods. If the soil is loose and the tree is to be transported on its side, plywood can be fitted beneath the top beams to hold the soil ball in place. Heavy metal bands can be used instead of rods to hold the box together. Specially designed slings are used to lift the boxed trees by crane or hoist for loading and transport.

FIGURE 9-6
Ficus microphylla in 1.2 m (48 in) boxes being loaded onto a trailer.
(Photo courtesy Keeline-Wilcox Nurseries, Irvine, Calif.)

Mechanical Tree Movers

Two types of tree-moving equipment are manufactured in a range of sizes to dig, transport, and plant trees up to 200 to 250 mm (8–10 in) DBH; tree size capabilities vary somewhat with the manufacturer. The most common mover, called a tree spade, can be mounted on either a trailer or a truck. It encircles a tree and hydraulically forces four pointed blades into the soil, one at a time, forming a root ball container when the blades meet below the tree (Fig. 9-7). Roots up to 100 mm (4 in) in diameter can be cut fairly cleanly except when they get caught between two blades; any mangled or torn roots extending beyond the blades can be cut back when the tree is lifted. The root ball, encircled by the four blades, is hydraulically lifted out of the ground. The tree can be transported in a vertical position if there are no overhead obstructions, or tilted forward, tied in at the top, and wrapped for longer hauls. Before the tree is dug, the planting hole is dug with the same equipment. On sloping or uneven ground, the planting site may need to be leveled so that the tree will be vertical when planted.

A second type of tree mover, sometimes called a clamshell (no longer manufactured), employs two quarter-sphere cups whose cutting edges are equipped with cutting chains similar to those on a power saw. The cutting cups separate to encircle a tree and lock together before digging begins. The subsequent tree-moving operations resemble those the tree spade performs. Clamshell tree movers come only in larger sizes, for digging large trees, whereas tree spades also come in smaller models.

FIGURE 9-7
A Vermeer Tree Spade is in position to hydraulically force
the black blades diagonally into the soil under the roots (left).
Blades are inserted in sequence so as to not raise the frame.
At the right, a honey locust is being lowered into a pre-dug hole.
When the tree is in place, the blades will be retracted.
(Photos courtesy Vermeer Manufacturing Co., Pella, Iowa)

A variation of the tree spade, called a Tree-Porter, digs a tree mechanically by hydraulically hammering individual blades into the soil to cut the root ball and form the container (Fig. 9-8). The blades are locked together and the root ball lifted with a crane or hoist for transport to the planting site. The equipment, made in England, positions each blade for hammering into the soil. The curved blades come in two sizes, which can be used in various combinations to dig a root ball with one of seven diameters, from 0.8 to 2 m (31–78 in). The equipment weighs about 115 kg (250 lbs) and can be moved short distances by two people or a small vehicle. A tractor or a portable gas-powered compressor supplies hydraulic power. Depending on the number of blades available, any number of trees can be dug before they are lifted for transport. The planting hole is dug by hand, backhoe, or other appropriate equipment, so its size can vary with soil conditions. The root ball and the blades that contain it are planted as a single unit; after the backfill soil has been firmed in place, the blades are unlocked and pulled out individually by hand or leverage.

The clamshell tree mover and the Tree-Porter dig a hemispherically shaped root ball, while the tree spades dig root balls that are more conical in shape. As a result, for a given top diameter, a smaller root ball will be dug with a tree spade, but in the larger sizes the bottom 450 mm (18 in) of the tree-spade root ball adds soil weight without many roots. In other words, for a given diameter, a tree-spade root ball will usually have fewer roots than a root ball dug with one of the other two types of equipment.

A tree spade can move small- to medium-size trees a short distance more quickly than the other equipment. For many trees of varying sizes

FIGURE 9-8
Trees in various stages
of being moved by a Tree-Porter.
The man is hydraulically hammering
one of several blades that will encase
the root ball. Note in the right foreground how
the blades are bolted together.
Chains are attached to the bolted members
for lifting, as is shown in the background.
(Photo courtesy C. J. Newman, England)

and for longer distances, the Tree-Porter would be most economical. If the root ball is firm, both the tree spade and the clamshell can set it on the soil surface or a simple rack so that the ball can be wrapped and laced. Wrapping takes time and skill and does not protect the root ball as well as the Tree-Porter container. The Tree-Porter, since it is smaller and lighter, can dig trees in more difficult terrain than can the tree spade or clamshell. Also, its hammering action digs well in stony and frozen ground where other machines may have difficulty.

If the equipment is used to move a tree larger than it is designed for, the root ball may be insufficient to keep the tree upright and supply it with adequate water and nutrients. It is highly unlikely that bigger tree-moving equipment will be built, because the largest equipment available now is the maximum size allowed on most highways. In addition, too few large trees could be moved in this manner to justify the added equipment cost.

Members of the Forestry Division in Lansing, Michigan, found they could plant a surviving 64 mm ($2\frac{1}{2}$ in) DBH deciduous tree along a street with a tree spade for two-thirds of the cost of planting a surviving 40 to 50 mm ($1\frac{1}{2}$–2 in) bare root tree of the same species (Cool 1976). Survival of bare root trees was 72 percent compared to 99 percent of plantings with the tree spade. Less than 5 percent of the bare root trees were lost due to vandalism and autos. Loss of bare root trees might have been lower had the trees not been so large. Working with 370 trees of eight fruit species in California, Hendrickson (1918) found that trees 12 to 16 mm ($\frac{1}{2}$–$\frac{5}{8}$ in) in trunk diameter grew more than smaller or larger trees. On the streets, smaller trees might be more often vandalized and might require more care than larger trees. An important consideration in Lansing was the favorable response of homeowners when large trees were planted in front of their property.

PLANTING

Planting a large tree, either bare root or with a root ball, involves the same steps and considerations as planting small trees and shrubs. Ideally, the planting site and soil surrounding the hole will be friable (or made so) and moist, in which case the planting hole need only be large enough to allow backfill soil to be easily worked in around the roots.

When holes are dug mechanically for planting, tree spades and augers are more likely than other equipment to glaze the soil surface. The cutting chain of a clamshell is preferable in this respect. Since glazed surfaces can restrict water movement and root growth into the surrounding soil, they should be cut through or broken up. If the soil is layered or compacted, a larger area of the planting site should be loosened and mixed (see Chapter 7). If the planting hole that is to take a root ball dug by a tree spade or clamshell is enlarged to any extent after initial digging, uniform contact and root ball support will be poor and the tree may be planted too deep. This problem can be overcome if the original hole is dug less wide or deep than it would otherwise be dug. Some arborists pour a liquid slurry containing equal parts of peat and soil plus an organic fertilizer into the planting hole before the root ball is planted. As the tree spade or clamshell sets a root ball into the hole, the slurry is forced up between the root ball and the soil, wetting both and improving contact.

Most trees should be guyed after being set in the planting hole, to keep them upright and avoid shifting that will damage the root balls. The guys can be used temporarily until more permanent support is provided.

After a root ball covered with burlap is set in the hole, the lacing can be removed, unless the ball is likely to crumble. The burlap should be rolled back or cut so the outer surface of the ball can be examined. If the ball has dried or been glazed in digging, the hard surface should be slit carefully or scarified with a sharp shovel. Backfill the bottom one-third to one-half of the hole, firmly working the soil around and under the root ball to give it support. If this has not already been done, the lacing can now be loosened and much or all of it removed, particularly if the drum lacing was done with open links. Roll the burlap off the top of the ball and down the sides if the root ball is firm. Cut the burlap above the backfill and remove it. Some burlap can be left to hold the soil together because it will rot in time, but avoid excessive folds of burlap, which could interfere with rooting. If the burlap is left in place, continue backfilling to within several centimeters (2 in) of the surface and then cut the burlap close to the top of the backfill. Fold the burlap down so that none will be exposed when backfilling is complete, because the exposed burlap can act as a wick, drying the soil interface between the root ball and backfill. If wire netting has been used instead of burlap, it can be handled in the same way.

Soil of the same or coarser texture as that in the root ball should be added to the top of the ball to replace the soil removed earlier. The new surface can be feathered to the level of the surrounding soil. The resulting crown on top of the root ball will compensate for any settling.

Anchoring the Root Ball

Anchor staking or guying is needed when trees are so large that the root ball may shift in the soil under windy conditions. Smaller trees should need only protective stakes to keep equipment and people away from their trunks (see Fig. 8-11). Larger (about 50 mm or 2 in. in diameter) and longer (up to 2.5 m or 8 ft) stakes can be used to anchor trees up to 70 or 80 mm (3 in) DBH. In Japan, many street trees are anchored to withstand strong winds (Fig. 9-9).

Guys may be needed to anchor trees larger than 70 to 80 mm (3 in) in diameter. Trees with 70 to 100 mm (3–4 in) DBH should be supported with 2 or 3 guy wires attached to stakes in the soil (Himelick 1981). Trees with 100 to 200 mm (4–8 in) DBH require three guys of 7-strand 3-mm ($\frac{3}{16}$-in) galvanized steel cable, each attached to a deadman. Trees over 250 mm (10 in) DBH should be supported with four guys. Most commonly, the guys, with hose buffers, are looped in a branch crotch about one-third up the trunk. If no crotch will provide the attachment desired, the guys can be attached to the tree by 10-mm ($\frac{3}{8}$-in) screw eyes on trees with up to 150 mm (6 in) DBH and 12-mm ($\frac{1}{2}$-in) screw eyes on larger trees. The screw eyes are usually placed on the side of the trunk that faces the direction of pull. Newman (1963) recommends placing the screw eye on the opposite side of the trunk, so that the pull will be one of compression. A less expensive screw can be used in this case, and will yet provide a "fail-safe" anchor. The trunk must be protected with a hard material where it is in contact with the cable. The screw eyes should be spaced vertically about 200 to 250 mm (8–10 in) apart on the trunk so no point of weakness will develop.

The guys should form about a 45° angle with the trunk and ground. For staking the guys, Himelick recommends stakes that are 50 by 100 mm (2 by 4 in) and 1.2 m (4 ft) long; Newman states that "Good anchorage can be obtained with metal stakes 1.2 to 1.5 m (4–5

FIGURE 9-9
Most street trees,
as this liquidambar,
in Japan are securely anchor staked
to withstand strong winds.

ft) long, made of 38-mm ($1\frac{1}{2}$-in) commercial steel angle sections 5 mm ($\frac{3}{16}$ in) thick.'' Trees in large planters can be guyed to the container corners or sides.

For trees with 100 to 200 mm (4–8 in) DBH, the guy cables should be anchored to deadmen 100 by 150 mm and 1.2 m long (4 by 6 in and 4 ft long) placed 1 to 1.2 m (3–4 ft) below the soil surface. For trees with larger than 200 mm DBH, the deadmen should be 100 by 250 mm and 2 m long (4 by 10 in and 6 ft long) and placed 1.2 to 2 m (4–6 ft) deep (Himelick).

Trenches for deadmen should be dug at right angles to the line of pull, with a channel cut for the cable that will be securely attached to the middle of the deadman. In light or distrubed soil, wide, flat stakes driven on the tree side of a deadman will provide additional anchorage.

There seems to be little agreement on how tight the guys should be. The findings of Jacobs (1939), Fayle (1976), and Leiser and others (1972), however, indicate that some trunk movement, unless the root ball also moves, will increase the size of roots near the trunk and will possibly foster root growth into the surrounding soil. Newman (1963) recommends that the tree be allowed a few inches of freedom at the head of the guys. Because such movement can put considerable strain on a screw eye or a stake when the cable snaps taut in a wind, a compression spring can be inserted in each guy to provide flexibility and increasing restraint as the spring is compressed and closed (Fig. 9-10).

FIGURE 9-10
Large trees may need anchoring until roots grow into the surrounding soil. Compression springs (top inset) in guy wires provide limited flexibility and reduce the strain on the anchor stakes. In turf, guy wires can be attached to removable pins (lower inset) which are inserted into galvanized pipe permanently set flush with the soil surface. (Harris, Leiser, and Davis 1976)

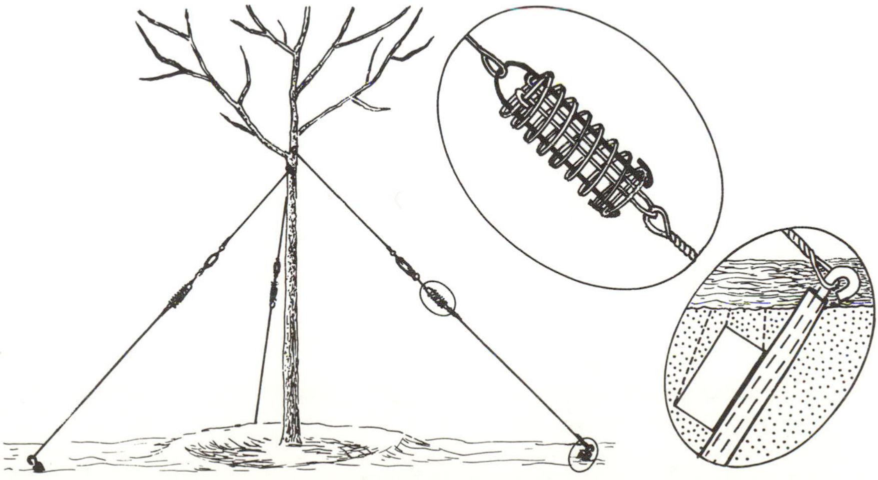

The size of the spring and its compression span will vary with the size of the tree and the flexibility of the trunk. A smaller tree will take a lighter spring with a longer span. For each unit of trunk sway at the height of the cable attachment, the compression span should be 0.7 unit when the guy is at a 45° angle to the trunk; it should be only 0.45 unit when the guy angle is 30°. Should a particular compression spring have a longer span than is wanted, the guy can be tightened until the right amount of span is left in the spring. Turnbuckles (screw tensioners) are usually used to adjust the slack or tension of guys. Guy wires should be marked with streamers to alert people to their presence.

Underground guying is particularly useful for new trees in pedestrian malls and along downtown streets. The root ball is held securely while the trunk and top are free to move, and no unsightly guys will trip pedestrians. Newman (1963) describes a method in which two cables are stretched across the top of the root ball and downward at about 45° to deadmen placed at the bottom of the planting hole (Fig. 9-11). Two boards are laid across the root ball under and at right angles to the cables to distribute pressure on the root ball. The backfill must be well tamped to keep the root ball from breaking as the cables are tightened. The cables can be tightened later by pulling them toward each other. In a variation of this method, the root ball can be cabled

FIGURE 9-11
In heavily used public areas, underground guying may be preferable.
Care must be taken to be sure the backfill soil is firmly in place
before the cables are tightened so they do not break the root ball.
(Adapted from Newman, C. J. 1963. Transplanting Semi-mature Trees.
In *Trees.* R. J. Morling, ed. London: Estates Gazette)

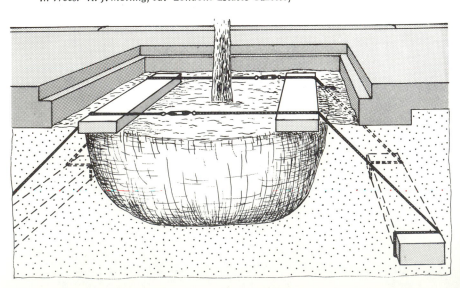

firmly to three or four stakes driven into the bottom of the planting hole alongside and around the ball.

Protecting the Trunk

Wrapping the trunks and sometimes lower branches of newly planted trees with burlap or special paper has been thought to reduce water loss and protect the bark from sunburn, sun scald, and bark borers (Himelick 1981). Pellett (1981) found, however, that traditional wrapping did little to moderate the temperature of trunks of silver maple exposed to full sun; lack of temperature protection may account for Houser's observation (1937) that wrapped trunks of American elm and three maple species were not injured when uncovered and exposed to the sun. On the other hand, trunks painted with white latex or wrapped with Foylon I were respectively 4.5°C (8°F) and 10°C (18°F) cooler than unprotected trunks. Foylon I is an impregnated cloth fabric with a silver-metallic surface; although not as durable, aluminum foil should act similarly. Painting the trunks is easier and cheaper than wrapping them. As the paint weathers and cracks with trunk expansion, the bark is exposed to increasing radiation and becomes more temperature tolerant. Neither white paint nor Foylon I may be particularly attractive in the landscape, however. The more complete protection by Foylon I may sensitize the bark to sun exposure when the wrap is removed.

Tree wrapping paper is available in rolls 100 to 200 mm (4-8 in) wide; Foylon I comes in wide sheets. If used, each spiral of the wrap is overlapped about one-half the width of the material. Wrapping from the ground up will repel rain and sprinkler water, which may be desirable in rainy areas for trees such as pin oak, that are susceptible to fungus cankers (Pirone 1978a). In drier areas and for less susceptible trees, wrapping from the top down will keep the trunk more moist. The wrap is held in place by a cord or heavy string that is wound around the trunk and branches in a direction opposite to that of the wrapping.

Watering and Mulching

Even though some water may be applied during backfilling, more should be added after planting is complete, to further settle the soil, improve contact between soil and roots, and assure adequate initial soil moisture. Construct a berm around the trunk a short distance beyond the root ball. The surface of the basin should slope away from the trunk. If the soil is well drained, fill the basin twice with about 100 mm (4 in) of water, letting the basin drain before the second filling. If the soil is not well drained, more care should be taken to avoid overwatering.

After the basin has drained, it should be mulched 75 to 100 mm (3-4 in) deep. See Chapter 13 for more details.

Pruning

Almost all plants will benefit from pruning after they are planted. Branches that have been broken or seriously damaged in moving should be removed or cut back to a good lateral. Pruning before transplanting should usually be limited to branches that will interfere with digging, loading, and transporting and that cannot be tied in. If branches destined for removal can be left on temporarily, they will protect other branches during transplanting. A general thinning out of the branches, up to one-fourth of the original leaf area, will reduce water use some-what, will reduce wind resistance at the top, and will stimulate growth throughout the plant. See Chapter 14 for more pruning information.

TRANSPLANTING PALMS

Palms are mostly tree-like monocotyledons with one or more un-branched trunks. They have no bark cambium, so that any injury to a trunk is permanent. The roots of most palm species are single strands growing out from the base of the trunk. If a root is cut or broken, it usually dies and new roots must be generated. Large palms flex more easily than do most dicotyledons of similar trunk size. Although palms can be moved at any time of year, the warm spring and summer months are best, when root growth is most rapid. Palms do best in well-drained soil.

Palms for landscape plantings in the United States are grown primarily in nurseries, though some are moved from older landscape plantings, and in Florida a few species are obtained from native plant-ings. A palm tree with trunk sections of varying diameter should be avoided because this indicates previous stress on water or nutrient supplies and could restrict future growth (Himelick 1981).

Field-grown nursery palms in California are commonly boxed, particularly if they are to be held a while before being replanted. The root ball can also be covered with burlap and laced for moving. Muirhead (1961) recommends a root ball of 1.2 m (4 ft) in diameter for large palms. For a number of hardy palms, a small root ball can be dug, undercut, and then lifted with little protection for the roots (Fig. 9-12). One tree-moving firm soaks the soil around the base of the palm for several days and then pulls the tree out with a crane without digging a root ball (Gress 1979).

The general practice is to remove most of the palm fronds before moving, leaving only six or eight per stem. Because each stem has only one bud, it must be protected. The ring of leaves immediately next to the bud is usually removed to prevent undue pressure on the bud when the fronds are tied up. An untreated 4-ply twine is suggested for tying

FIGURE 9-12
A palm is usually moved with a small root ball and only a few fronds left on; these are generally tied together to protect the growing point.

because it will usually deteriorate within a few months and permit the fronds to open up. Removal of all but a few leaves is recommended only for palms that are immediately transplanted from the field, not for palms that have been held in a box for a time after field digging or have been container-grown.

Palms with long, slender trunks (*Phoenix reclinata* and *Acoelorrhaphe wrightii*) should be supported with a long beam attached to the trunk and extending beyond the bud (Himelick 1981). The palm can be lifted with the same range of equipment that is used to move dicotyledonous plants. The trunk surface must be protected to prevent damage from the lifting cables or chains. Nylon slings will provide both support and protection.

For transporting, single-trunked palms can be stacked or shingled on the truck or flatbed trailer. Large, multistemmed palms should be laid down on the truck bed and the individual stems tied to maximize stability and to reduce the width of the load.

Planting palms is similar to planting other trees. The hole is dug, by hand or a backhoe, somewhat wider than the root ball. Along tropical coasts, holes are sometimes augered into coral rock. Pulverized rock from the drilling is used for the backfill (Himelick 1981). The palm should be oriented in the planting hole so that its most pleasing side faces the main viewing perspective. This is particularly important for palms with curved trunks. To encourage new root growth from the trunk base, palms should be set 75 to 125 mm (a few inches) deeper than their original positions.

The backfill used varies from sand or "good soil" to the site soil combined with a planting mix; the important thing is that it be of the same texture as the site soil, or coarser. The backfill should be firmed in around the roots and then well watered. The palm should be guyed for stability, but no nails or screws should be anchored in the trunk. A fender of boards can be held in place around the trunk with steel straps to provide an anchor base for guy wires.

Newly planted palms must be well watered during the first season. Muirhead (1961) recommends that water be slowly applied in a basin around the trunk, running eight hours twice a week for the first two months and then for twelve-hour periods once a week. Obviously, this can only be done in well drained soil. If the weather is particularly hot and arid after planting, palm foliage may benefit from misting, as discussed in the following section.

The fronds of newly planted palms may yellow due to cold weather or manganese and iron deficiencies in alkaline soil. The application of manganese sulfate or iron chelate to the soil should improve the appearance of fronds (see Chapter 11). Nitrogen fertilization will almost always improve growth and appearance.

CARE FOLLOWING PLANTING

In areas where summer irrigation is usually thought to be necessary, deciduous trees and shrubs that are planted bare root and thoroughly watered at planting or afterward should not need irrigation until two to four weeks after growth begins. Plants in other areas may need mid-season irrigation if rainfall has been below normal. Early in the growing season, roots will actively grow into moist soil while the top has few leaves. Overwatering during this time can endanger root growth and function.

On the other hand, container-grown and B & B plants, both deciduous and evergreen, may require rather frequent watering, depending on the soil texture, the weather, and the relative sizes of leaf areas and root balls. Even though most container soils are coarse-textured, their shallow depth while in containers will keep moisture content near the saturation point following irrigation (Davis and others 1974). Depending on plant size and the weather, moisture in the root ball is used fairly quickly by the plant, so aeration is not a problem. When container or B & B plants are transplanted, the water previously retained in the root ball drains into the soil below; if the root ball soil is coarser than the surrounding soil, its moisture content can be reduced below field capacity by capillary forces. The soil of a small root ball can be brought close to the wilting point, even when the plant is using little or no water and the surrounding soil is near field capacity. Costello and Paul (1975) examined a plant newly planted in a loam soil from a

4-liter (1-gal) can and found that 80 percent of the available moisture in the soil mix was lost within 24 to 36 hours after flood irrigation. If a plant in a 4-liter container can last two days between irrigations before wilting, it would thus need irrigation at least every day when first planted in the landscape. The effects of capillarity will be less dramatic in larger root balls, but drainage into the soil below will cause equally great losses down to field capacity.

Daily irrigations may cause the surrounding soil to remain extremely wet. For the first few weeks, water need only be adequate to rewet the root ball and a little of the surrounding soil. A small berm can be established just outside the root ball and high enough to hold water that will rewet the roots. The larger basin beyond need be watered only every two or three weeks. After the first few days, the watering interval for the inner basin can be lengthened until the first signs of wilting appear. An interval one day less than the wilting interval can then be used for irrigation for two or three weeks. Repeat the wilting experiment to adjust the irrigation schedule later. Rain must be taken into account, of course. See Chapter 12 for more irrigation information.

Misting

Large shrubs and trees that are already in leaf or about to be, may not be able to get adequate water through their root systems. Misting such plants will cool and wet the foliage and decrease transpiration. One or more sprinkler heads can be temporarily attached to PVC pipe and hung in the upper part of the tree. The lower end can be joined to others from nearby trees or coupled directly to a standard hose or an automatically timed water source. A light sprinkling twice each afternoon will reduce the transpiration load on the new transplant. If the water is high in soluble salts, the sprinkling should not be continued for many days, or the foliage will become gray with salt. Wet the foliage just to runoff, so that the soil below does not become too wet.

FURTHER READING

BERNATZKY, A. 1978. *Tree Ecology and Preservation*. New York: Elsevier Scientific Publishing Co.

HIMELICK, E. B. 1981. *Transplanting Manual for Trees and Shrubs*. Urbana, Ill.: Revision II—Intl. Soc. Arboriculture.

NEWMAN, C. J. 1963. Transplanting Semi-Mature Trees. In *Trees*, ed. R. J. Morling. London: The Estates Gazette, pp. 29–39.

CHAPTER TEN

Special Planting Situations

People sometimes attempt to grow plants in an extremely difficult situation. Will such a planting be feasible, and what can be done to increase its chances of success? We are continuing to learn how to assess the requirements of the plants in relation to site characteristics and how to adjust the microenvironment. Experiences must be shared if better solutions to many problems are to be worked out. Certain landscape situations are reviewed here, and suggestions offered or approaches described for their use.

PLANTING TREES IN PAVED AREAS

Trees are often planted in paved areas: along downtown streets, in parking lots, and in patios. Provision is not always made for adequate irrigation, drainage, and aeration, particularly as trees become larger. Adequate surface exposure is not always possible when foot or vehicular traffic must be close to treetrunks, and open planting basins can be a hazard to pedestrians. This discussion will concern planting trees in existing or projected paved areas. Placing pavement around existing trees is dealt with in a later section.

Soil, or sand, or loose brick laid on a bed of sand so that its surface is level with the surrounding pavement is often used, but these surfaces make irrigation difficult and usually become dangerously uneven with time. One solution is to build a square planting basin with two half-covers that will extend a solid, smooth surface close to the treetrunk (Fig. 10-1). A standard-size opening (usually 1.2 m or 4 ft square) can be cut in old paving or formed in new. A lip in the paving

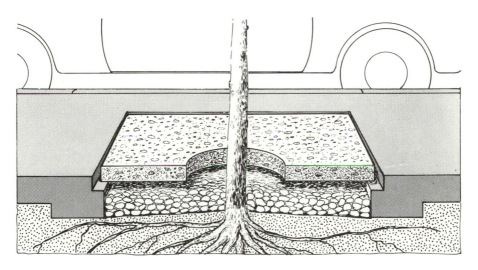

FIGURE 10-1
One method of planting trees in paved areas.
An opening, about 1.2 m (4 ft) square, supports two half covers (one shown).
The basin below provides for irrigation and aeration, while the covers
provide a smooth surface for pedestrians.

edge will hold the two half-covers so that their surfaces are flush with the surrounding pavement. Each cover is made of precast reinforced concrete with a semicircle left on each side for the treetrunk. You can lift out the planter covers to inspect the basin or the lower part of the trunk, to weed, to clean out debris, or to replace them with covers that have larger holes when the trunk gets bigger.

The soil level in the planter hole can be 50 to 200 mm (2–8 in) below the bottom of the covers to provide aeration, a reservoir for water, and additional space between the tree roots and the pavement. Since the early 1970s, the city of Santa Maria, California, has planted many of its street trees with the soil level 500 mm (20 in) below the sidewalk. The planting procedure will be the same as that described in Chapter 9. If tight soil may be a problem, sloping irrigation and aeration wells can be drilled under and out from the root ball and under the pavement to increase the rooting volume that will receive irrigation and to direct root growth downward under the pavement (Fig. 10-2). This technique is described more fully in Chapter 12. The space between the planter soil and the covers can be filled with coarse gravel or small rock to keep the soil from washing when watered by hose or tank truck. Filling this entire space and even the opening at the trunk will minimize trash accumulation and discourage rodents. No special effort need be made to keep the irrigation holes either filled with or clear of gravel.

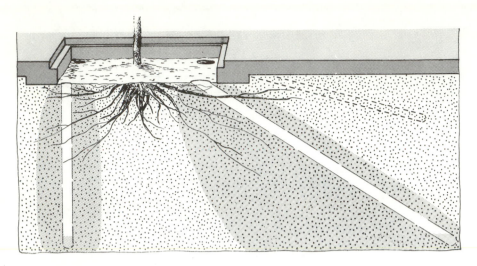

FIGURE 10-2
If needed, aeration and irrigation holes can be drilled
or water-jetted from the tree opening downward under the pavement.
A diagonal hole will serve a larger soil volume than will a vertical one.

In new construction, the pavement around such planting basins should be sloped to carry runoff water to a drain or street gutter rather than into the tree wells. Such drainage will protect the trees from excess water, from any salt that is used to melt ice on streets, and from general increases in alkalinity.

A common problem with covered tree wells is that they often become waste receptacles. San Francisco had a serious rat infestation in tree wells when Market Street was redeveloped. As an alternative to the gravel fill described above, hardware cloth can be used to cover the openings. A solid cover will probably collect less debris than an open metal grate. In heavily trafficked areas, a stake or trunk guard and the planter cover may be an integrated unit (see Fig. 8-12). A whole array of planting holes and covers have been designed.

Trees that will be small to moderate in size when mature can usually be grown safely in pavement planters. The trees will grow according to existing conditions, so they may be healthy even when soil rooting volume is relatively small because of limited aeration and water distribution. Mature tree size may be smaller than typical for the species, but early growth should be good if the trees are given normal care. If full-size large trees are wanted, a gravel subbase should be considered for the surrounding pavement, as described in Chapter 7.

Enlargement of roots near the surface is a problem when trees are close to pavement, walks, and foundations. Roots can be especially

252

destructive if they can reach irrigated open areas beyond the pavement or other sources of air and water. If the soil level is below the pavement as shown in Figure 10-1 and well removed from open areas, root damage to the pavement should be minimized. Damage can be further reduced if small or moderate-size trees are chosen that normally have few or no large surface or buttress roots.

For trees that grow in confined paved areas or between curb and sidewalk and have root access to water and air beyond the paving, arborists have tried various ways of encouraging deeper rooting. They have selected species thought to be deep-rooted, enlarged the opening in the pavement, and planted in large concrete or metal pipe to force roots down; these methods have had varied success, but pipe planting has led to the girdling of roots. A polyethylene planter, wider at the base than at the top, is marketed by the Deep Root Corporation in Beverley Hills, California. Roots are directed downward when they strike the sloping sides instead of in horizontal circles, as often happens with a vertical surface. This planter is open at top and bottom. Its present dimensions are 560 mm (22 in) at the top, 740 mm (29 in) at the bottom, and 460 mm (18 in) from top to bottom. The manufacturers recommend a planting hole 1 to 1.2 m (40–48 in) square and 1 m (3 ft) deep. Their recommendations continue: A mix of organic matter and existing soil should be firmed in place so that the planter will be even with the soil surface. A bare root or container tree (up to 60 liters or 15 gal) is planted in the planter so that the ground level of the tree is level with the soil surface. The space outside the planter is filled to the top with gravel. Roots will enter the surrounding soil 500 mm (18 in) below the surface; this is thought to delay or prevent the heaving of surrounding pavement. After establishment, the tree need be watered only through the gravel; this is thought to further encourage deep rooting and keep the crown roots from becoming too wet.

The success of such a device will depend on the extent of paving, soil texture and structure, tree species, and irrigation practices. Remember the obvious: Roots grow best where conditions favor them. If the surrounding soil is fairly permeable, roots emerging at the bottom of the planter may stay deep or come to the surface only some distance away. Even without the planter, roots may be no problem in permeable soil unless they can reach a water and air source beyond the pavement. In a tight, poorly drained soil, however, roots may grow to the surface quickly, especially if the moist gravel serves as an excellent rooting medium.

These planters have been used for several years, and the manufacturer has established some trial plantings, so a better assessment of their success will be possible in a few years. If you wish to compare this

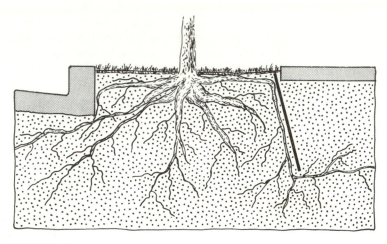

FIGURE 10-3
Tree roots can be deflected downward
if a thin impervious sheet of wood, metal, plastic, or fiberglass is installed
at a slight angle from the vertical near paving
or foundations to be protected.

planter with some other method, prepare and backfill the planting holes of the comparison plantings with the same mix, and choose similar locations and similar trees for each planting.

For plantings between curb and sidewalk, you can direct roots downward at lower cost by inverting a large plastic garbage can, with the bottom cut out, to function in a manner similar to that of the Deep Root planter. Alternately, you can bury an approximately 0.5 X 1.2-m (20 X 48-in) sheet of thin plywood, particle board, metal, or black plastic in a sloping position between the sidewalk and the tree-trunk (Fig. 10-3) and another on the curb side of the tree. The Deep Root Corporation also markets such root barriers. In a paved area, such barriers will probably be needed only if there is a source of air and water close by. It will not matter if the barrier disintegrates with age, because it will be in place long enough to direct the growth of most roots that will become big. Avoid species known to have shallow roots.

LANDSCAPING IN PLANTERS
WITHOUT NATURAL DRAINAGE

Landscape plantings without natural drainage are common on roof gardens, in building foyers, on top of underground garages, basements, or utility lines, in shopping malls, and in hotel lobbies (Fig. 10-4). Almost all of these require plant containers that have no contact

FIGURE 10-4
Landscape plantings can be successfully grown in many situations
with the use of well-designed planters such as these shown in
Constitution Plaza, Hartford, Connecticut.

with parent soil, yet can provide moisture, nutrients, aeration, drainage, and anchorage for successful plant growth. The weight of such plantings is an important consideration when they are placed on a structure, as are the weights of the soil mix, water, plants at maturity, other landscape features, and people occupying the area. Successful planters, whether individual and free-standing or built into a structure, must all be controlled for container size, waterproofing, irrigation, drainage, and soil mix (Davis and others 1974). These must be considered in the initial planning for a structure or project.

Container Size

The container must be deep enough to hold the drainage material and adequate soil mix to provide anchorage and drainage. Minimum container depth should be 150 to 300 mm (6–12 in) for turf or turf substitutes, 0.5 to 1.0 m (1.5–3 ft) for large ground covers and most medium-size shrubs, and 1 to 1.5 m (3–5 ft) for large shrubs and trees

(Davis and others). Bernatzky (1978) recommends a soil depth of at least 450 mm (18 in) for shrubs and small trees. The weight of roof garden plantings can be minimized by varying soil depth with plant size even within individual beds (Fig. 10-4). The planting areas within a bed must conform to the pattern of the structural beams.

Both the texture and the depth of the soil influence the amount of water that will remain in it when drainage is complete. After a container soil is irrigated, a zone of saturation extends up from the bottom. The height of the saturation zone depends on the pore-size distribution (texture) of the container mix. Above the zone of saturation, the soil becomes drier with increasing height until field capacity is reached, providing the soil is deep enough (Fig. 6-7). Note that, regardless of container depth, about the same amount of water will accumulate in the bottom 150 mm (6 in). The next 150-mm layer contains considerably less water than the bottom layer, but each of the soils in that layer contains about the same amount of water. Soil moisture content decreases with height above the bottom until capillary forces are in equilibrium with gravitational forces and the moisture reaches field capacity. It is important to note that even in a coarse, sandy mix (0.5-1 mm), soil moisture is well above field capacity in the bottom 300 mm (12 in), particularly in the bottom 150 mm. In a finer-textured soil mix, moisture will be above field capacity for greater distances above the bottom.

Except for a slight slope (about 2 percent) of the container bottom to assist the movement of excess water to drain lines, the bottom of each bed with a common drainage system should be fairly level. Similarly, the soil surface in each bed should be fairly level, or the high areas will drain to lower moisture contents than the lower areas (see Fig. 6-3). An extreme case would be a sloping surface with uniformly shallow soil (about 450 mm or 18 in) and one side a meter (3 ft) or more above the opposite side. The soil at the top would dry rather quickly, while the soil at the bottom would be excessively wet.

If trees are grown in individual containers that are small enough to move with a forklift, the flexibility of plantings will be greatly increased. Plants can be moved to areas of high visibility when they are blooming or in fall color, and can be replaced with others when they lose their showiness.

Waterproofing

For roof gardens and other landscaping above areas of use, containers must be completely waterproofed (Fig. 10-5). The waterproofing should be planned before containers are built. A double-membrane construction is recommended for most built-in landscape planters.

Irrigation System

A proper and efficient irrigation system will reduce maintenance costs and will help ensure healthy plants. A water line should serve each container or container area of any size. Automatic watering systems with low application rates will improve water use. Shallow, coarse-textured soil leaves little leeway between adequate and insufficient watering for good plant growth.

The irrigation system should be an integral part of the structure supporting the planters. If garden lighting fixtures are placed inside containers for partial concealment, they must be watertight.

Drainage System

Few plants grow well if their roots must function for long without adequate oxygen. No matter how well a soil mix is designed for proper drainage, aeration, and water retention, excess water must be removed from the container. Drainage systems should be designed to handle maximum amounts of excess water, rather than average amounts. The water removal capability for any container surface area should be a minimum of 50 mm (2 in) per hour.

The soil mix must allow the water to drain rapidly to the bottom of the container, and the drainage line and outlets must be large enough to collect water rapidly from the container and move it into building or street drainage systems. Containers should have sloping bottoms with outlets at the lowest point (Fig. 10-5).

Gravel in the bottom of the container will waste valuable space and be a poor medium for plant growth. When sandy soil overlies gravel, the extreme difference in pore sizes of the two materials causes drainage to cease when the sandy soil just above the gravel is still near saturation. Thus, instead of eliminating the water table, the gravel creates an artificial one nearer the surface.

Water moves into drainage tile when soil reaches the point of saturation. In the past, 100-mm (4-in) tile drainpipe was used in landscape planters. While this large size is needed for draining agricultural fields, much smaller plastic pipe can provide adequate capacity in planters at less cost and far less loss of container depth. A pipe 25 mm (1 in) in diameter uses little more than 25 mm of container depth, whereas a 100-mm line, particularly one encased in coarse gravel, may use up to 150 mm of depth.

Plastic pipe has proven adequate for most containers or container areas when 4 rows of holes are spaced evenly around the pipe. Drill 3-mm ($\frac{1}{8}$-in) holes about 75 mm (3 in) apart in each row. Ten- to 25-mm ($\frac{1}{2}$- to 1-in) lines, on 0.5- to 1.0-m (2- to 4-ft) centers connected

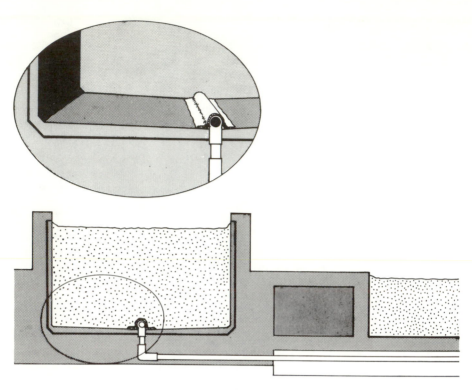

FIGURE 10-5
For landscape planters that must be artificially drained:
slope the bottom to a filter-protected drain line (inset),
waterproof the container, and level each soil surface.
(Adapted from Davis and others 1974)

to larger 25- to 50-mm (1- to 2-in) pipe, will remove water rapidly from the bottoms of planters. Plastic lines are easy to install and add little to the weight load. Copper drainage lines are used between containers and as outlet lines that run to the street.

To keep drainage lines from silting up, wrap them with 12-mm ($\frac{1}{2}$-in) fiberglass filter pads or strips of filter pad laid directly over the drainpipe. Secure pads in place with mastic. After the drainage lines are installed, cover the entire bottom of the container, to a depth of 25 mm (1 in), with coarse sand or with sand used in the parent soil mix. Water will move laterally quite rapidly in this thin layer of sand, which will also help to protect the fiberglass filter pad from damage.

Even with properly installed drainage lines and filter pads, some silt may accumulate, so clean-out standpipes should be connected to each of the main collecting lines and capped. There should be one standpipe for each large container and several in a large container area.

They can also be used to pump excess water from individual containers or container areas, if a major stoppage occurs in the system.

Soil Mix

Weight and aeration are the two most important characteristics of a soil mix to be used for roof gardens and other supported landscapes. Water and nutrients can be fairly easily supplied and large plants guyed if necessary. After the soil is in place, however, little can be done to decrease weight or improve aeration.

Davis and others (1974) examined the container soil mixes recommended by the University of California Manual 23 (Baker 1957) and found them to be quite satisfactory. These mixes combine uniform coarse sand (0.5–1 mm) and organic matter in volume ratios that range from 50:50 to 75:25. Commonly, a little silt or loam soil is added to the mix to increase water- and nutrient-holding capacities, but more than 5 percent of these materials will hamper drainage and aeration.

A coarse sand and organic mix offers a good combination of qualities. When it is saturated with water, a 50:50 sand and organic mix weighs about three to four times as much as when it is dry. The researchers found that the saturated weight was about 1.2 kg per liter (70 lbs/ft^3), but the bulk densities of planting mixes can vary greatly with slight changes in formulation. Therefore, the wet weight of a soil mix can be reliably estimated only by sampling the mix itself. In any planter landscapes of modest or large size, the soil mix should be specified and tested for compliance before it is put in place. The management of high-organic mixes is discussed in detail in the University of California Manual 23 (Baker 1957). The main feature of these practices is close attention to irrigation.

Plant Selection

Outdoor planters are usually focal points, yet the environment for such plantings may be extremely harsh. Wind, direct and reflected radiation, prolonged and varying intensities of light, low humidity, high day temperatures, low root temperatures in winter—all are common in outdoor container landscapes, and all jeopardize the well-being of plants. Generally, woody plants of arid climates will best survive such conditions (Bernatzky 1978). In most cases, trees and shrubs should reach a moderate mature size and grow slowly. Larger plants in windy locations must be supported with guys or anchor stakes. Evergreens are commonly used for their year-round foliage, but many are susceptible to air pollutants.

Determining which species will be suitable for specific micro-climates and will also fulfill other design criteria can be a real challenge, but help is usually available: You can observe which plants grow well in similar situations and can consult local nurseries, botanical gardens, extension agents, public and commercial arborists, and state or regional publications.

Special Problems

In cold winter areas, roots growing in raised individual containers can freeze (see Chapter 4).

INTERIOR LANDSCAPES

Interior landscapes of more than 1000 m^2 (0.25 acre) are common in large buildings and shopping malls (Fig. 10-6). Critical to the success of such landscapes are the selection of plants and the control of irradiation (light intensity), fertilization, and irrigation. An indoor land-scape must be planned as an integral part of the building to ensure

FIGURE 10-6
Interior landscape plantings require good design, engineering, installation, and care if they are to be successful as this planting in the Ford Foundation building in New York City. (Photo courtesy Everett Conklin and Company, Montvale, N.J.)

proper access to electrical and plumbing utilities. Attempts to correct design and specification errors after construction can be costly and unsatisfactory.

Plant Selection

Plant selection depends largely on the irradiance (see Chapter 4) available within the interior site. In most cases, plants are not expected to grow much after installation, only enough to maintain attractive foliage. Tropical and semitropical foliage plants are most commonly used for interior plantings. Conklin (1978) has successfully grown tropical fruit trees: orange, lemon, guava, mango, and loquat. Though temperatures in most buildings will seldom go below 18°C (65°F), a number of temperate-zone trees and shrubs can also be used: southern magnolia, evergreen pear, and camellia, among others (Conklin).

Radiation (Light)

If radiation is inadequate, most plants will grow satisfactorily for only a few months. Artificial light is generally necessary if less than 1600 lux (150 ft-c) of natural light are available. Plants of the dracaena and philodendron families will survive 550 to 800 lux (50–75 ft-c), but greater irradiance will allow for greater leaf retention and growth. Growing plants indoors is primarily a holding operation. Ideally, radiation will be great enough so that plants are slightly above their compensation point, growing slowly but not consuming reserves (Conover 1978). The large, thin, widely-spaced leaves that develop in the shade have thin cell walls and usually only a single epidermal layer of cells at the surface. Such leaves are able to make maximum use of limited radiation. In the nursery shade-grown plants usually grow more rapidly than plants in the sun but often have reduced trunk caliper and lighter root systems. The active absorbant area of the roots will not be substantially reduced, however (Conover).

Conover has found that the chloroplasts and grana of some sun leaves will reorient to shade conditions if the leaves are grown in radiation of intermediate intensity for 4 to 8 weeks. If a sun-grown plant in a nursery is placed directly under low-radiation conditions, however, serious leaf drop usually occurs (Fig. 10-7) before existing leaves become acclimatized or new shade leaves form.

In 1975, about 85 percent of the plants grown for indoor plantings by producers in California, Florida, and Texas were shade-grown and therefore radiation-acclimatized (Conover, Poole, and Henley 1975). The plants that are grown in full sun include: *Ficus, Brassaia, Dracaena marginata*, several palms, *Araucaria*, and *Sansevieria*. Plants of these species should be checked to be sure they have been acclimatized before shipment. The youngest mature leaves of acclimatized

FIGURE 10-7
Ficus benjamina plants grown in full sun and then placed indoors under different
radiation levels for ten weeks; left to right: full sun, 275 lux (25 fc),
800 lux (75 fc), and 1350 lux (125 fc). (Photo courtesy Charles Conover,
Agricultural Research Center, Apopka, Florida)

plants will be larger, thinner, and farther apart than those of plants
grown in the sun. Reliable dealers will correctly inform you if asked
about acclimatization.

Care must also be taken when transporting the plants to see that
they are not exposed to ethylene or kept in boxes or other dark areas
too long. Many plants start to lose foliage after seven days or less of
darkness, and defoliate severely after ten days.

Since windows may be shaded by nearby tall buildings or decrease
radiation by means of tinted architectural glass, irradiance should be
checked throughout the day, particularly in winter. A photographer's
light meter will not measure irradiance accurately (see Chapter 4).
When possible, new buildings should maximize the use of natural radiation for indoor landscapes. When artificial light is needed, it should
be remembered that irradiance decreases by the square of the distance
from its source, so that 100 lux (lumens/square meter) at one meter
(3 ft) from the light source will be only 25 lux at 2 meters. Typical
irradiances in most buildings are quite low (see Table 4-1). Side and
bottom lighting is usually more efficient in conjunction with overhead
radiation and can be more interesting as well.

Light quality is important for plant growth, but growth should
be limited in most interior plantings. The wavelengths (artificial lights)
usually recommended are blue (430–470 nanometers) and red (650–
700 nm). Nevertheless, tropical foliage of excellent quality can be
maintained with predominantly blue light, as from fluorescent fixtures
(Conover, Poole, and Henley 1975). Biran and Kofranek (1976) found
that cool white was the most effective of the fluorescent lights they
used on nine foliage plant species. Fluorescent lamps are about three

times more energy efficient than incandescent lamps, though some incandescent light (150 to 200 lux) may be needed to fill in the red and far red parts of the spectrum.

Fertilization

The proper nutrition regime for indoor plants will depend on plant species and available irradiance (Conover, Poole, and Henley). The level of nutrition must be higher under high radiation. Healthy plants can be maintained indoors on only 10 percent of the fertilizer needed in their nursery production. In fact, if their nutrition level is not reduced before plants are moved to low radiation, they may suffer chlorosis, necrosis, defoliation, general loss of healthy color, and, in severe cases, death. If you are not certain that soluble salts are at a low enough level, leach the containers with at least 150 mm (6 in) of water before taking plants to the planting site.

Conover (1978) recommends using a slow-release fertilizer with a 3-1-2 ratio of nitrogen, phosphorus, and potassium. The rate of application depends on plant species and irradiance (Table 10-1). The low

TABLE 10-1

Fertilization rates for indoor foliage plants under different light intensities for 8 to 12 hours per day (after Conover and Poole undated)

		Grams of Nitrogen per Square Meter of Soil Surface per Year[a]			
Plants	Lux[b] Footcandles	Low 550–800 50–75	Medium 800–1600 75–150	High 1600–2700 150–250	Very High 2700–5400 250–500
Ferns, *Peperomia*		3[c]	7	10	17
Calathea, Dracaena		5	10	17	23
Aglaonema, Cordyline, Dieffenbachia, Scindapsus, Syngonium		7	13	20	26
Palms, *Philodendron*		8	16	22	27
Ficus, Schefflera (*Brassaia*)		10	20	27	33

[a]The nitrogen was introduced as part of a 3-1-2 (N, P_2O_5, K_2O) fertilizer applied three times during the year.

[b]Fluorescent lamps (cool white) used: 100 lux = 1.5 $\mu E/m^2/sec$.
Sunlight equivalent: 100 lux = 1.7 $\mu E/m^2/sec$.

[c]10 g/m^2 is equal to about 2 lb/1000 ft^2. A teaspoon is equivalent to about 5 g or 5 cc.

ratio of phosphorus and potassium is designed to keep soluble salts low. Periodic leaching may be needed to keep the soil solution at safe levels (see Chapter 12).

Irrigation

Practically all of the factors that will influence the frequency and amount of irrigation are determined at the time of installation. These include the water-holding capacity and aeration of the rooting medium, the size of plants, and microclimate factors of temperature, radiation, humidity, and air movement. Most interior landscapes will range from 18 to 27°C (65–80°F), depending on the season. Humidity will usually be lower than that of nurseries and the native environments of plants, particularly in winter. Large, thin shade leaves will be particularly sensitive to moisture stress, which can be seriously aggravated by mite injury. Air circulation is determined largely by the air conditioning system. Sensitive, large-leaved plants should not be planted in the air stream of heating and cooling vents.

A moisture-sensing device will supplement careful observation to help determine an irrigation schedule. Indoor plantings usually evapotranspire less water than do plants outdoors, and overwatering interior plants is more common than underwatering. Most indoor landscapes, once established, will do well with a week or more between irrigations. Plants that wilt first when water is withheld can be a useful guide for irrigation schedules.

Some tropical plants are injured by air temperatures of 5 to 8°C (40–45°F) or by cold irrigation water. If necessary, water temperature should be monitored in winter.

Other Concerns

Compounds used to shine leaves may reduce photosynthesis and can injure a number of species. If such compounds are used, they should not be used on a continuous basis. The foliage of shade-grown plants is particularly sensitive to some common pesticides and to air pollution.

PLANTING
ON STEEP SLOPES

On slopes, watering can be difficult and uneven, erosion can expose roots or bury small plants, and increased exposure to sun can dry plants quickly. Three of many planting alternatives are suggested:

FIGURE 10-8
A plant is set toward the downhill edge
of the basin near the original surface
of the slope. The basin should be deepest
near the back to accumulate water
and any eroded soil.
An overflow is cut into firm soil
so the basin will not be easily
washed out.

Planting Pocket

Digging a planting pocket into a slope is the simplest and most common method of facilitating irrigation (Fig. 10-8). The plant is set well forward in the pocket, establishing a basin to the inside that will retain water and protect the plant from a certain amount of eroding or sloughing soil. An overflow spillway will prevent the pocket from being washed out by all but the heaviest rains. It should be cut into the undisturbed slope at one side of the pocket so that water will flow out of and away from the basin before the berm is breached. You may need to reform the pocket occasionally in the first year or two, until the plant becomes established. For continued irrigation, you may want to consider a drip system (see Chapter 7).

Slope Serration

Slopes are sometimes cut into steps that measure about one meter (3–4 ft) in both the horizontal and the vertical. This has proved particularly useful in certain highway embankments (Fig. 10-9). Instead of

FIGURE 10-9
Slope serration provides growing sites on cut slopes.
Woody plants or grass can be planted, or volunteer plants grow from local seed.

grading cuts to a smooth, steep, sloping surface, as has been so common, some contractors have stair-stepped certain rock or erosion-resistant earth slopes to establish plants and reduce erosion. The steps should slope toward the hill so that water will drain into the soil. Small trees or shrubs can be planted in the middle of the steps and a slight basin dug to the inside of each plant in order to accumulate water. The basin, however, should be kept away from the trunk of the plant.

Wind, water, and sloughing will erode the edge of each step onto the inside of the step below. The speed of erosion will depend on the resistance of the slope material and on the weather. The loose soil and fine rock on each step are favorable germinating sites for seed from nearby plants. In many cases the trees and shrubs planted or those that volunteer from local seed will become established before the serrations are obliterated. Even before the plants develop, the changing light-and-shadow patterns on a serrated slope can provide a more interesting landscape feature than a smooth slope.

Wattling

Wattling has successfully stabilized the surface of fill slopes, reduced erosion, and helped to establish plants on difficult sites (Kraebel 1936, Leiser and others 1974). Recently cut, long, slender branches are tied into elongated bundles that are partially buried in contoured trenches cut across the slopes (Fig. 10-10). The bundles, laid end to end, are staked down to hold them in place. Each row of wattling acts as a sediment trap for materials from up the slope and as an energy dissipater for soil and water movement. The area above a wattling bundle, with its reduced slope angle, is a favorable site for new plants. If the wattling species roots easily, as willows do, the wattling itself

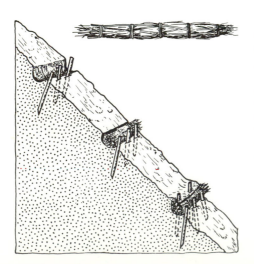

FIGURE 10-10
Wattling in the process of being installed on a slope:
At the top, a shallow trench on the contour is ready to receive the bundles of wattling (center).
Stakes are driven through the bundles for greater stability (bottom).
While workers are installing the row above, the lower row of wattling will be covered by soil and firmed into place.
(Adapted from Leiser and others 1974)

becomes part of the slope-stabilization system. Wattling and stuck cuttings (see Chapter 8) root and grow particularly well in wet sites and can reduce slumping by drying up excessively wet areas. No amount of vegetation, however, will stabilize a basically unstable slope; such slopes must be laid back further or provided with drainage.

Leiser and others (1974) have described the major features of successful wattling installation. Wattling bundles should be prepared from live thicket plants, preferably from species that root easily. The bundles may vary in length but should be tapered for 300 to 450 mm (1-1.5 ft) at either end. Stem butts should be less than 40 mm (1.5 in) in diameter, with half of the stem butts at either end of the bundle. When compressed firmly and tied, each bundle should be 150 to 250 mm (6–10 in) in diameter. Bundles should be prepared within two days of placement and protected from direct sun. If they are covered and kept moist, bundles can be stored up to a week before placement.

Trenches should be dug on the contour to about one-half the bundle diameter and about a vertical meter (3 ft) apart on the slope. Lay the bundles horizontally in the trenches so that their tapered ends overlap. Hold them in place on their downhill side with vertical stakes about 500 mm (20 in) on center (distance between the center of two stakes). Drive stakes on one-meter (40-in) centers through the bundles into and perpendicular to the slope.

Stakes can be made of several materials, including: live wattling material greater than 40 mm (1.5 in) in diameter, 50 by 100 mm (2 by 4 in) lumber cut diagonally, or steel reinforcing bars. In most soils, 600-mm (2-ft) stakes are satisfactory; in loose soil the stakes should be a meter (3 ft) long. In compact or rocky soils where stakes cannot be driven 450 mm (18 in) deep, steel reinforcing bars should be used.

Work from the bottom of the slope toward the top, covering and firmly packing the bundles with soil. If you walk on the wattling while working on the next row above, that alone may cover the bundles and firm the soil sufficiently. The downhill lip of the wattling bundles should be left exposed.

PLANTING UNIRRIGATED
ARID SITES

Most of the considerations for successful low-maintenance land-scapes—site analysis, species selection, and planting methods—have already been discussed. It may be wise, however, to review the more essential considerations for arid sites in particular—those that receive little or no rainfall during the growing season.

The site must be evaluated for its ability to accumulate and store water. Begin by observing uncared-for plants that grow on the actual or a similar site. Find out what has grown there in the past. Have the

soil and native vegetation been disturbed? Have any nearby developments changed the natural surface or underground water flows? What
are the average and extreme rainfalls, seasonal evapotranspiration, soil
textures and depths, depth of the water table, and topography?

If the site is essentially unchanged from its natural state, species
that have grown there before should be suitable, but the range of viable
species may be larger than the range of indigenous species. In most
cases, the smaller the plant when it is introduced into the landscape,
the better the chance it will have of getting off to a good start. Certain
locations within the site may be more favorable than others for plant
survival and growth. Higher moisture supplies and lower evapotranspiration rates are most frequently found on easterly slopes facing away
from the equator; near the bottom of canyons or slopes; in deep, well-
drained soils; near streams; or on the shady side of large boulders.
Individual planting sites can be modified to enhance water accumulation and conservation. One of the simplest methods, if the soil is
well drained, is to slope the soil surface toward each plant or group
of plants, but not so as to accumulate rainfall directly around the
trunks.

Spacing plants farther apart than in irrigated landscapes will provide each with a larger soil-water reservoir. Mulching and weed control,
discussed in Chapter 13, can materially increase water infiltration and
conserve water.

For areas with clear night skies (high radiation heat loss), Jensen
and Hodder (1979) use condensation traps to condense soil water
vapor from a 1.5 to 2 m^2 (16–20 ft^2) area, concentrate it, and direct
it toward plant roots. Their initial condensation traps consisted of a
planting basin, 1.5 m (5 ft) in diameter and 300 mm (1 ft) deep, with a
newly planted tree or shrub on a mound in the center, and polyethylene
sheeting arranged to adsorb evaporating soil moisture.

The researchers found that contractors had difficulty following
the specifications for condensation traps constructed with a basin and
a berm. They therefore designed a prefabricated trap, which is easier
to install and performs more consistently. A frame 1.2 m (4 ft) square,
constructed of 25 by 250 mm (1 by 10 in) lumber, supports the polyethylene sheet (Fig. 10-11). A 20-liter (5-gal) bucket is temporarily
placed in the middle of the polyethylene sheet, whose edges are
wrapped around 1.2-m (4-ft) long laths, which are then nailed to the
outside of the frame. The plant is placed on a slight mound in a shallow basin and protrudes through a slit in the sheet. A few rocks pull
the polyethylene taut, form a seal at the plant trunk, and prevent the
condensate from dropping prematurely at a sag point away from the
plant. Small slits beneath the rocks will drain entrapped precipitation
into the basin. Although clear polyethylene collects about 30 percent
more condensate than black, black polyethylene is used because it will

FIGURE 10-11
The water supply of a new plant is augmented by means of a condensation trap.
Jensen and Hodder found that prefabricated moisture condensation traps
were easy to install and performed consistently.
(Adapted from Jensen and Hodder 1979)

last several years and inhibits weed growth. A small berm of soil seals
the outside of the frame. Available soil moisture is considerably greater
beneath this type of condensation trap than beneath the basin and
berm trap. Moreover, the frame trap does not silt up.

Jensen and Hodder found that the amount of condensate accumulated daily ranged from 16 to more than 200 ml (0.5-7 oz) when a black
polyethylene liner was used in midsummer at 1200 m (4000 ft) elevation in south-central Montana, where annual rainfall was 260 mm
(10.3 in). During that period the maximum air temperatures ranged
from 29 to 38°C (85-101°F); those under the black polyethylene were
39 to 52°C (102-126°F). Air temperatures under clear polyethylene
were 42 to 57°C (108-134°F).

PLANTING IN AREAS
THAT ARE PERIODICALLY FLOODED

Many areas that are subjected to flooding are being planted to
enhance the recreation potential of reservoirs, streams, and canals; to
protect riverbanks and levees; to landscape marinas, drainage-water
basins, and flood plains; and to improve wildlife and fishery habitats.
Some reports identify the species, primarily trees, that have survived
different periods of inundation (Whitlow and Harris 1979). Many trial
plantings support and extend the flooding observations. As would be
expected, species and site are the main determinants of plant survival
and performance.

A number of species will survive water over the soil around them for
more than three months a year (see Appendix 5) (Whitlow and Harris).
A few of these species have been completely submerged for that long
with only reduced growth and trunk rooting of some willows. Some
species that can withstand long periods of water over the soil will lose

269

branches if submerged for more than two weeks. Plants have a higher survival potential on well-drained alluvial soils than on finer-textured shallow soils. Mature valley and blue oaks have grown on alluvial fans in areas with little or no summer rain and are able to survive 100 days of flooding annually. By contrast, interior live oak will die if the soil around their trunks is wet for two weeks. Whitlow and Harris report that trees that had been established for one growing season before flooding were no more tolerant of inundation than trees planted only two months before flooding. Although one would assume that plants are best able to withstand flooding during their dormant period, many are flooded in late spring and summer. In California, plants survive the shift from a flooded condition to an arid one for two to three months in midsummer.

Trees and shrubs on the leeward side of reservoirs may be broken off or seriously damaged by floating debris. These plantings are best located in coves or channels that can be buoyed off to prevent damage during high water. A number of easy-to-root species have been established successfully by pushing or driving stem cuttings, 450 mm (18 in) long and 20 mm ($\frac{3}{4}$ in) in diameter, into the muddy shoreline as the water recedes in spring and summer. An irrigation or two during the first growing season will enhance survival in areas of low summer rainfall. Heavy pruning before flooding can decrease the tolerance of young plants.

PLANTINGS AT LANDFILL SITES

Refuse-landfill sites are often used for parks, golf courses, botanical gardens, and other landscaped recreation areas. The decomposition of organic matter, which usually makes up more than half of the refuse, causes the landfill surface to settle, produces about equal parts of methane and carbon dioxide, and generates high soil temperatures. Each of these conditions, especially the production of methane and carbon dioxide, is detrimental to the well-being of landscape plantings. The amount of organic matter used in landfills, however, will probably decrease in the future as more of it is used for fuel or composting.

Landfills are usually located in abandoned sand and gravel pits, canyons, or excavations made specifically for landfills. The common practice in the United States is to spread and compact the refuse as it is delivered and to cover it at the end of each day with about 150 mm (6 in) of soil, which, in turn, is compacted. As the fill is built up, every fourth or fifth layer of soil may be increased to 300 mm (12 in). The final soil cover is supposed to be at least 600 mm (2 ft) deep, although twice that depth is preferable. Many completed landfills have less than 300 mm of soil over the surface.

Surface Settling

Even though the refuse and soil cover are compacted, decomposition of organic matter causes the fill to settle. Depending on the depth of the landfill, its organic-matter content, and the degree of compaction, settling is likely to be 150 to 450 mm (6–8 in) in places. Not only will the uneven surface hamper activities, but water may accumulate in the depressions, irrigation lines break, and trees and shrubs tilt. The fill may be partly settled by heavy irrigation before the surface soil is brought to final grade, but the practice is of limited value. Excess irrigation can leach refuse pollutants into ground and surface waters. Also, the higher moisture level will increase the rate of refuse decomposition and thus of settling and gas evolution. To minimize damage, main irrigation lines, buildings, and large trees should be placed on firm soil around the edge of the fill.

Decomposition Gases

The rate of decomposition and the volume of decomposition gas produced by a landfill depend primarily on the organic-matter content and the amount of moisture. Most deep landfills in southern California produce between 3 and 6 liters of gas per kg (0.05–0.1 ft^3/lb) of organic waste (Van Heuit 1979). The half-life of solid waste in these relatively dry landfills is estimated to be about 15 years, but such estimates are difficult to make because it often takes 20 to 30 years to complete a large landfill. A gas-collection well in the center of one landfill in southern California produced about 16 million liters (0.5 million ft^3) of methane a day, enough to serve about 2000 homes in the area (Van Heuit). Carbon dioxide is removed from the extracted gas by a molecular-sieve process.

High concentrations of carbon dioxide in the root zone may be directly toxic to plants (Leonard and Pinkard 1946). And even though methane is not directly toxic to plants, the methane and carbon dioxide produced during decomposition will drive out oxygen from the soil or be used by methane-consuming bacteria (Leone and others 1977). Whatever the reasons, oxygen in a landfill can decline to damaging levels and can bring about other reactions in the soil that are injurious to plants.

Polyethylene-lined Planting Holes   A landfill should have a gas-migration control system of collection wells or gravel venting trenches along its periphery for safety, odor control, and reduction of plant injury. Gas may still rise to the upper layers of the landfill to injure or kill plants (see Chapter 18), though a vented planting hole can help vegetation to acclimate. The hole can be installed where plants have been killed by gas in the past or in new plantings where gas injury might be expected. To ensure satisfactory growth of trees and shrubs in land-fill sites, landfill gases must be excluded and water allowed to drain from a planting hole.

One way to accomplish this is illustrated in Figure 10-12. Excavate a hole at least 1.2 m (4 ft) in diameter and 1.2 m deep, into the fill refuse if necessary. Place a 100 to 150 mm (4-6 in) layer of coarse sand on the bottom and sides of the hole to protect the plastic liner that will be installed, to facilitate diffusion of landfill gases toward the surface, and to act as a dry well for drainage water from the planter. The sand base should slope toward the observation well. Line the hole with 6-mil polyethylene and locate a gathered opening adjacent to the observation well. Install a 50 to 75 mm (2-3 in) diameter PVC pipe to drain the planter; the horizontal portion should be perforated, with a trap that can drain into the observation well, and a riser that will surface at the other end. Secure the polyethyelene tightly around the drain line at its point of exit from the planter. The trap will prevent

FIGURE 10-12
A suggested method of protecting plant roots from landfill gases.
The bottom of a hole 1 to 1.2 m (40–48 in) in diameter and equally deep
is covered with 100 to 150 mm (4–6 in) of coarse sand.
The hole is plastic lined and provided with a drain line and a water trap to exclude
landfill gases but allow excess water out.
One end of the drain is exposed to the surface; the other is accessible
in a standpipe (the observation wall) so the drain may be cleaned out.

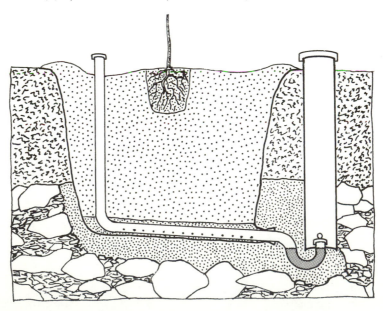

the entry of landfill gases into the planter, and the riser will allow access for cleaning sediment and roots from the drain line. The observation well will allow you to check for water in the trap and to remove debris cleaned from the drain line. Unless the drain line can be cleaned, it may soon become clogged.

Cover the PVC riser, the drain trap outlet, and the observation well to keep them clean, but allow water and air to pass. The exposed covers should be as vandal-proof as possible. Fill the planter with medium-textured topsoil and settle it before planting. Plant shrubs or trees that would normally grow to small or medium size at maturity. Form a berm just inside the fill-soil perimeter to cover the top edge of the polyethylene liner.

Landfill-Tolerant Species Plants tolerant to flooding and poorly drained soils appear to be tolerant of landfill situations where decomposition gas may be a problem (Leone and others 1977). Because the landfill soil may be shallow and unstable, Van Heuit (1979) recommends smaller plants that are known to have shallow roots. Small plants have been found to adapt to landfill conditions more quickly than larger plants of the same species. Plants with an open, flexible habit of growth will best withstand windy conditions. The list in Appendix 6 may help you select species for landfill sites.

Maintenance Suggestions While the plants are small, fertilize and irrigate to keep them vigorous and in good color. Do not overwater, however, because water will drain into the refuse, speeding decomposition and increasing the likelihood of polluting underground water. If individual plants become dense, open them up by thinning out branches so that the wind can pass through freely and be less likely to damage them.

Landfill Soil Temperatures

Soil temperature has reached 58°C (137°F) in the rooting zone of some landfill sites; this is at least 10°C above the temperature that will kill the roots of almost all plants. Soil temperatures in anaerobic soils may be 16°C (30°F) higher than in aerobic soils nearby (Leone and others 1977). In most cases, however, the temperature differences are small, and anaerobic soil is occasionally cooler.

High soil temperatures in a landfill are usually fairly localized. Reducing the amount of moisture that reaches these "hot spots" may slow decomposition and lower the soil temperature. The hot spots can be vented or supplied with extraction wells and suction blowers. On the other hand, these areas can be avoided for plantings.

PLANTING IN SOILS
THAT HAVE HERBICIDE RESIDUES

Where railroad rights of way, parking lots, driveways, vacant lots, roadsides, and canal banks were once treated with persistent or pre-emergent herbicides to prevent weed growth, landscape plantings may now be desirable because aesthetics or land use has changed. Older residual chemicals, like the arsenites, the borates, and the chlorates, were used primarily along railroad rights of way and parking lots, usually in large quantities, and may be toxic many years after application. The more recently developed organic herbicides may persist for less than a week or more than two years, depending on the herbicide, soil texture, and moisture. Organic herbicides can usually be deactivated by adsorption on organic matter, but the inorganic soil sterilants cannot. Occasionally an herbicide may also be spilled, misapplied, or overapplied to a landscape planting (see Chapter 18).

Testing for the Presence of Herbicides

The arborist must first determine what chemical has been applied, in what quantities, and when. If no record or recollection of previous herbicide applications is available, the soil can be analyzed prior to a large or important planting.

In warm, sandy soils that are low in organic matter and exposed to moderate or heavy rainfall or irrigation, organic herbicides will break down or be leached from the surface soil fairly rapidly. Organic herbicides are much more persistent in cold, clay soils that are high in organic matter and exposed to little or no rain or irrigation. Such soils will allow the herbicide to remain effective for a considerable period but will be easy to reclaim because most of the herbicide will be concentrated close to the surface. Preemergent herbicides applied annually at recommended rates are less of a problem than are one or two excessive applications. Even after nine annual applications, simazine at 7.2 kg per hectare (6.4 lb/a; 33 percent above highest recommended rate) was herbicidal only in the top 150 mm (6 in) of a sandy loam California vineyard soil (Leonard, McHenry, Lider 1974).

A bioassay using annual plants can usually determine whether a soil is still toxic enough to injure new plantings (Lange, Elmore, and Saghir 1973). Toxic levels of herbicide in the soil will stunt, twist, or kill test plants. To bioassay a soil for herbicide presence, collect soil in separate paper cups or small pots (about 200 ml or 6 oz) from depths of 0 to 50 mm, 50 to 100 mm, 100 to 150 mm (0-2, 2-4, 4-6 in), and deeper if the herbicide is extremely soluble or if it has been applied repeatedly. Collect samples at several sites in both herbicide-treated and untreated areas. Plant the seeds of at least one grass and of one broadleaved species and grow seedlings under good cultural conditions for at least 20 days; do not overwater or underwater. Compare the growth and foliage condition of plants grown in soil you suspect was treated with those in soil (from the same depth) you know to be untreated. If symptoms are moderate to severe, the soil in planting areas should be replaced or the herbicide deactivated.

Treatment

Soil containing a toxic level of herbicide can be removed, mixed with organic matter, leached, or planted to tolerant plants, depending on the herbicide, soil texture, available drainage, the area involved, the type of planting desired, and the budget available to support it. If the levels of inorganic soil sterilant are toxic or the amounts of organic herbicide excessive, the contaminated soil should be replaced with clean soil. A bioassay will determine the depth to which soil should be removed. If ground-cover plants, turf, or closely spaced shrubs are to be planted, all contaminated soil in the bed should be removed. For individual trees and shrubs, dig holes 1 to 1.2 m (3–4 ft) in diameter and deep enough to remove the contaminated soil. Line the sides of the hole with plastic, insert a 200-liter (50-gal) metal drum with top and bottom removed, or remove the bottom of a large plastic garbage can and invert the can in the planting hole. Fill the hole with clean soil, allow it to settle, and plant the tree or shrub. Do not allow surface water from contaminated soil to reach the clean soil in the planting hole.

If organic herbicides have not been used in large quantities and appear to be in the top 150 to 200 mm (6–8 in) only, it may be easier to dig planting holes 300 to 450 mm (12–18 in) deep and 0.75 to 1.0 m (30–40 in) in diameter. Line the hole with plastic, leaving the bottom uncovered. Mix the soil from the planting hole with an organic amendment and return it to the lined hole. Activated charcoal (carbon) should be thoroughly mixed with the soil at a rate of about 40 g (1 oz) of charcoal for each 150 mm (6 in) depth of soil in a hole 1 m (3 ft) in diameter. This approximates the recommended rate of 150 parts (weight) activated charcoal to one part active herbicide in the soil (Smith and Fretz 1979). Ottoson (1976) reports that a mixture of composted chicken guano and nitrogen-fortified sawdust will be as effective as charcoal, or more so. The amount to be added depends on the nitrogen level of the guano-sawdust mixture; actual nitrogen should not exceed 50 g (0.10 lb) per planting hole. Ottoson credits the improved growth of plants to the increased fertility this soil amendment provides. Any finely divided organic material should adsorb moderate amounts of organic herbicide; the organic matter, however, should not be allowed to deplete the nitrogen in the soil. Tolerance to herbicide-contaminated soil can be increased if roots or root balls are dipped into a charcoal slurry (110 g charcoal/liter or 1 lb/1 gal of water) (Smith and Fretz 1979).

Soluble herbicides in well-drained sandy soils can be leached by heavy rains and irrigation. Herbicides that break down under anaerobic conditions can be deactivated in poorly drained clay soils by several weeks of flooding (Lange, Elmore, and Saghir 1973). Field plantings of corn and grain sorghum are tolerant to simazine and atrazine and can degrade them in the soil (Clyde Elmore 1981, personal communication). Certain landscape plants may have similar abilities to grow in contaminated soil and detoxify it. We need more information on such species.

FURTHER READING

Davis, W. B., and others. 1964. *Landscaping in Containers without Natural Drainage*. Univ. of Calif. Agricultural Sciences Leaflet 2577.

Estevez, M. T. 1976. *From the Plants' Point of View*. Proc. Symposium on the Use of Living Plants in the Interior Environment. Alexandria, Va.: Soc. of Amer. Florists.

Flowers, F. B., E. F. Gilman, and I. A. Leone. 1981. Landfill Gas, What It Does to Trees and How Its Injurious Effects May Be Prevented. *J. Arboriculture* 7(2): 43–52.

Kraebel, C. J. 1936. *Erosion Control on Mountain Roads*. USDA Circ. 380.

Whitlow, T. H., and R. W. Harris. 1979. *Flood Tolerance in Plants: A State-of-the-Art Review*. Vicksburg, Miss.: U.S. Army Engineer Waterways Exp. St. Tech. Report E-79-2.

CHAPTER ELEVEN

Fertilization

Fertilization is one of several cultural practices that may encourage the rapid development and continuing health of landscape plants. Fertilization improves plant vigor, makes leaves grow larger and darker, makes trees less susceptible to certain pests and diseases, and sometimes helps trees overcome a decline. On the other hand, fertilization may increase susceptibility to fall cold or to diseases such as fire blight, *Erwinia amylovora*. Young trees usually grow more rapidly after fertilization; mature trees need little or no fertilization as long as they exhibit good leaf color and reasonably good growth. In fact, if the size of trees and density of foliage are increased unnecessarily, inside foliage or plants under trees may be weakened by heavy shade.

Fertilizer practice has been cloaked in tradition and rules of thumb. Woody landscape plants have not been fertilized widely in either urban or native landscapes. Landscape nutrition is often complicated by plants that show little growth or off-color foliage, typical deficiency symptoms that seem to be nutritional but may be the result of any number of factors: girdling roots, compacted or water-logged soil, root diseases, nematodes, or salt injury. Fertilization has been confounded by the vogue of growing plants "naturally," without so-called "chemical poisons."

As fertilizers become more expensive, as older plantings deplete certain soil nutrients, and as the danger increases that excess nutrients will pollute surface and underground waters, much more attention must be given to wise fertilization practices. An understanding of the interrelationships among soil, nutrients, and water is essential. This chapter reviews these relations and gives suggestions for determining and meeting the nutrient needs of plants. Research on tree nutrition is examined.

RELATIONSHIPS AMONG SOIL,
NUTRIENTS, AND PLANTS

Physical Properties of Soil

Its physical properties influence the amount of nutrients held by a soil and, to a certain extent, the availability of nutrients to plants.

Soil texture, the size distribution of soil particles, directly influences the amount of nutrients adsorbed by the soil. As a soil becomes finer in texture, it will hold more nutrients and water. To adsorb an equal amount of a nutrient, a sandy soil will require more frequent applications of smaller amounts than will a clay soil.

Soil depth can determine the nutrient- and water-holding reservoir available to a plant. As a plant is able to root more deeply and widely, it will reach more water and nutrients.

Soil structure, the arrangement of soil particles, influences root exploration and absorption by plants. A compacted soil, or one lacking an open, granular structure, may restrict root activity by limiting water and air movement and by physically impeding growth.

Chemical Properties of Soil

Sixteen elements have been found essential for plant growth (see Table 6-3):

> In the absence of any one of the elements, a plant will fail to complete its life cycle.
>
> Each element is specific; it cannot be replaced by another.
>
> Each element has a direct effect on plants (as opposed to an indirect effect, such as repelling insects, which might prevent completion of the plant's life cycle).

As shown in Table 6-3, carbon and oxygen are available to plants from carbon dioxide in the air. Oxygen is directly available from either air or soil; hydrogen from water absorbed from soil. The remaining 13 elements are derived from soil and are called plant nutrients. Micronutrients, required in small amounts, are just as essential as macronutrients, required in larger amounts.

Nutrient availability is measured in terms of a soil's capacity to provide nutrients (amount available) and the intensity of these nutrients (concentration). When a soil is unable to meet the nutrient demand of the plant, it is deficient or infertile with respect to the particular nutrients it lacks.

With the exception of urea, which can be absorbed as a compound, almost all elements—whether from organic or inorganic sources—are absorbed as inorganic ions. Until organic forms are mineralized, that is, converted to inorganic ions, they are not commonly utilized by plants.

Plant roots absorb ions selectively. Some ions, such as K^+, NO_3^-, and NH_4^+, are absorbed rapidly and may accumulate in the plant at concentrations much higher than in the external solution. Other ions, such as $H_2PO_4^-$, $SO_4^=$, Ca^{++}, and Mg^{++}, are absorbed less readily. The selective uptake means that nutrient ions are not absorbed in the same proportion as they occur in a soil or soil solution. Absorption rates of specific ions are different for different species, and sometimes even for cultivars of the same species.

A plant does not distinguish between ions originating from inorganic and organic sources. This is not to say that all materials containing the same amount of a particular nutrient will be equally effective. Other factors are important: whether the material contains other essential elements, how it affects soil structure and soil pH, and how long it persists in a soil.

Compared with natural organic sources, inorganic fertilizers usually

Contain greater percentages of a given nutrient

Are easier to handle and apply because they are more concentrated and compact

Are free from unpleasant odors

Are more uniformly available in the soil, not being dependent on the rate of organic decomposition (which is influenced by biological activity, which varies with temperature, moisture, and substrate)

Cost less per unit of nutrient

Contrary to popular belief, plants will grow well on either nutrient source. When nutrients are the primary interest, inorganic forms are usually favored. The principal advantage of natural organic fertilizers is that they improve soil tilth or structure and can meet the nitrogen requirement of plants if applied in sufficient amounts. Manures incorporated in surface soils, for example, reduce crusting and enhance seedling emergence.

Determining Nutrient Needs and Toxicity Problems

Almost all plants in all soils will respond to nitrogen. For woody plants, most soils supply adequate amounts of the other nutrients, although there are some important exceptions. Iron and manganese may be deficient in alkaline soils; phosphorus may be needed for extremely old or acid sandy soils. In semiarid regions currently or formerly used to corral livestock, plants may show severe zinc- and copper-deficiency symptoms. Young trees may not survive their first summer on these sites. Copper deficiency may also be a problem for woody plants on the sites of old Indian camps, even where annual plants grow well. If you suspect this type of nutrient problem, consult a farm advisor of the Cooperative Extension, an agricultural commissioner, a consulting arborist, a farm supply employee, or a nurseryman, all of whom are usually aware of such problems in a given area.

The presence of toxic substances can seriously affect plant growth and appearance; boron, chloride, and sodium can be directly toxic. High salinity and high sodium (alkali) can be problems in areas of low rainfall, poor water quality, or poor drainage. Both nutrient deficiencies and toxic conditions may be diagnosed by soil analysis, plant analysis, and nutritional trials.

Soil Analysis Soil can be tested to appraise pH or the level of elements available before a plant is grown. In areas with high humidity or short growing seasons, testing can indicate the amount of lime needed to adjust the soil pH to a more favorable level (see Chapter 6). Soil analysis can fairly accurately determine the availability of phosphorus and potassium to turf, herbaceous plants, and small shrubs, but it is of doubtful value with regard to trees and deep-rooted shrubs. Only in extremely deficient soils, such as acid sands, have soil-test results correlated with woody-plant response, and then only as regards the level of phosphorus (Pritchett 1979). The amounts of other nutrients in the soil also can be determined, but, to date, are of limited value for determining the fertilizer needs of woody landscape species. Toxic levels of boron, chloride, sodium, and total salt can be determined by soil analysis. Water, too, can contribute toxic levels of these elements to the soil, and should be tested.

A good analysis is dependent upon a careful, accurate, and representative sampling of soil. If you suspect soil problems or do not know the soil situation in the area, it may be wise to have the soil analyzed. Soil samples can be collected by means of various tools (Fig. 11-1). The following sampling procedures will be adequate for most situations (Quick and Rible 1967):

For a fertility assay, apportion the area to be tested into sites that are similar in growth characteristics, general appearance, and soil type. Form each sample by compositing cores or thin soil slices taken from about ten locations in each site. Turf or litter should be removed from the soil surface before sampling. The depth of sampling should be based on soil type and rooting characteristics of the plants you plan to use, as in the examples that follow:

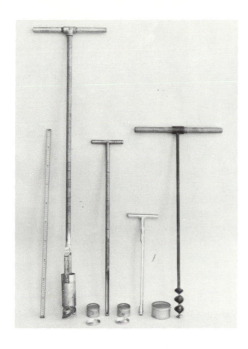

FIGURE 11-1
Soil augers (left and right),
profile tubes (center two),
and sampling containers can be used
to collect soil samples for pH,
nutrient, or moisture analysis.
(Harris, Paul, and Leiser 1977)

Turf: Take samples at depths of 0 to 75 mm (0–3 in) and 75 to 150 mm (3–6 in). For special sites, such as golf greens, take samples at depths of 0 to 50 mm (0–2 in), 50 to 100 mm (2–4 in), and 100 to 150 mm (4–6 in).

Flowers and shrubs: Take samples at depths of 0 to 150 mm (0–6 in) and 150 to 450 mm (6–18 in).

Trees: Take samples at depths of 0 to 300 mm (0–12 in), 300 to 600 mm (1–2 ft), and 600 to 1200 mm (2–4 ft).

To test saline or sodic conditions, take separate (not composite) samples to represent a range of conditions from no or poor plant growth to good plant growth. Depths should be at 0 to 150 mm (0–6 in), 150 to 300 mm (6–12 in), and 300 to 600 mm (1–2 ft), and, when needed, at 300-mm (1-ft) increments to lower depths. For sites you plan to seed, it is desirable to sample the surface 25 to 50 mm (1–2 in) of soil separately. The results of soil and water tests must be interpreted by one who is experienced in evaluating laboratory data in relation to the plants and soils of the given region.

Plant Analysis *Visual symptoms* of nutrient deficiencies are uncommon in most landscapes; exceptions are nitrogen and, in some situations, iron deficiencies. A slight nutrient deficiency may reduce plant growth without producing any noticeable symptoms. This is of concern with young plants only, whose maximum development is desirable. The length of shoot growth (Fig. 11-2), leaf color, color

281

FIGURE 11-2
These shoots of Chinese pistache
exhibit differences in vigor.
Leaves are present on current growth;
bud scale scars (arrows) indicate
starting and stopping points of growth.
Annual shoot growth may be difficult
to determine on species that may have
several growth cycles in a year.
(Harris, Paul, and Leiser 1977)

pattern, and size, and the time of leaf fall can indicate deficiencies. The age of leaves affected and their position on the plant provide additional clues, because some nutrients are more mobile in certain plants. Symptoms for individual nutrient deficiencies and mineral toxicities are given in the next section. Some symptoms have been observed only in nutrient studies under controlled conditions. Because there may be multiple symptoms and symptoms reflecting disease or improper amounts of water, considerable experience is needed to visually determine the nutritional status of a plant.

FIGURE 11-3
Recently matured leaves near
the shoot tip should be selected
for nutrient analysis.
For toxicity analysis, collect leaves
that show symptoms as well as
leaves that appear normal for comparison.
(Harris, Paul, and Leiser 1977)

Tissue analysis can be used to detect certain mineral deficiencies and excesses (Chapman 1960). Leaves are most commonly analyzed because they are easy to collect and the correlation between their analysis and the nutrient status of the plant is relatively high. Collect recently matured leaves from the same relative position (the current season's growth) on both healthy and suspected nutrient-deficient branches or plants (Fig. 11-3). Foliar analysis for woody landscape plants is largely used in research studies and to confirm certain visual diagnoses. We must document more correlation among tissue analysis, plant symptoms, and fertilizer responses before we can develop meaningful standards for determining the nutrient status and needs of woody landscape plants.

Nutritional Trials Nutritional experiments can provide valuable guidelines for fertilization, but they are not as practical for woody plants as may seem at first to be the case. To be accurate and worthwhile, fertilizer trials demand more skill and time than most arborists have or are willing to spend. Test plants and areas must be representative of the species and soils in question. Some plants must be fertilized and other similar plants left as unfertilized controls. Differences in the composition or quantities of nutrient must not be emphasized by the method of application. For example, if the nutrient is sprayed on foliage, control plants should be sprayed with the same amount of water; if the nutrient is injected into the soil, the same amount of water should be injected near control plants. Several comparisons (replicates) between fertilized and control plants must be included. Plant response (shoot growth, trunk increase, leaf size and color) must be measured and analyzed. Even in careful experiments, a difference of 15 to 20 percent may not be statistically significant.

CORRECTING NUTRIENT DEFICIENCIES

Nitrogen is the most commonly deficient soil nutrient. Other nutrients are likely to be deficient only under rather uncommon circumstances. The following text identifies those conditions under which individual deficiencies are most likely to occur. Typical visual symptoms are described and treatments recommended; both are summarized in tabular form, along with the relative mobility of each nutrient in plants and soil. Nutrient mobility within a plant will tend to determine where symptoms first appear and whether nutrients applied to the foliage will be translocated to foliage that develops later. When nutrients are mobile in plants, deficiency symptoms are likely to appear first on older foliage; less mobile nutrients tend to appear on newer leaves.

To a large measure, the mobility of a nutrient in the soil determines how it should be applied and how vulnerable it is to leaching. Nutrients that move readily in soil can be applied to the surface in relatively small amounts, but those that are less mobile must be incorporated into soil and applied in larger quantities or sprayed on foliage. Nutrient mobility increases loss from leaching. Mobility in soil is strongly influenced by soil texture and pH. As soil particles become coarser and the soil more acid, more nutrients are likely to be mobile. Organic matter and certain soil compounds may decrease nutrient mobility. The extent of nutrient mobility and its effect should be considered tendencies, not absolutes.

In the sections that follow, nitrogen, phosphorus, and potassium are considered first because they are among the first nutrients that were found lacking in soils, particularly for herbaceous plants. The rest of the macronutrients, and then the micronutrients, are presented in alphabetical order. Several nutrients may be present in toxic concentrations, which must be reduced if plants are to grow satisfactorily. Toxicities are also discussed individually.

Macronutrients

Nitrogen   Plant response to nitrogen (N) fertilizer is almost universal. Nitrogen becomes available to plants through

> Mineralization of organic matter
> Addition of fertilizers
> Fixation of atmospheric nitrogen by bacteria and actinomycetes

Nitrogen in soil is rendered unavailable to plants due to

> Absorption by weeds or other nontargeted plants and by organisms during
>   the decomposition of organic matter low in nitrogen
> Denitrification by soil organisms
> Leaching

These processes, except direct fertilizer application, are influenced by the environment (temperature, moisture, soil aeration, soil pH, and soil flora and fauna). Due to the transitory nature of soil nitrogen, soil and foliar analyses are not particularly useful for woody plants in the landscape.

Nitrogen fertilizers come in different chemical forms. Nitrate (NO_3^-) and ammonium (NH_4^+) ions are the most common of the inorganic forms. Nitrate ions, being negatively charged, are not adsorbed by soil colloids; they move with soil water, available for absorption by plant roots and other organisms. If excessive water is applied, some of the nitrate may leach below the root system and be lost to the plants.

Ammonium ions, being positively charged, are adsorbed on soil particles and do not move readily with the soil water. Many plants have shallow feeder roots in the surface soil and can absorb ammonium ions directly, particularly if the soil is protected by low branches or mulch. Ammonium ions, however, are commonly transformed to nitrate by

---

Deficiency Symptoms: NITROGEN (Nitrogen is mobile in plants.)

Broadleaf: Leaves are uniformly yellowish-green, the color being more pronounced in older leaves; leaves are small and thin, leaves have high fall color, and drop early; compound leaves have fewer leaflets. *Shoots* are short and small in diameter, and may be reddish or reddish brown. *Flowers* bloom heavily but may be delayed. *Fruit* set is light; fruit is small, highly colored, and early to mature (Harris, Paul, and Leiser 1977).

Conifer: Needles are yellowish, short, and close together. Young seedlings may remain in the primary needle stage with little or no branching. Older plants exhibit poor needle retention. Lower crowns may be yellow, while upper crowns remain green (after Powers 1979).

Treatment: NITROGEN DEFICIENCY (Nitrogen mobility in soil depends on chemical form.)

Application	Chemical Form	Amount to Apply	How to Apply	Years Effective
		$kg/100 \ m^2$ [a]		
Soil	Several (N)	1–2	On surface	1–2
		$kg/100 \ liters$ [b]		
Spray	Urea	1	Foliage[c]	1

[a]One kilogram per 100 m^2 is about 2 lb/1000 ft^2. To convert kg/100 m^2 to g/cm diameter of treetrunk: Multiply by 20 when trunks are less than 15 cm in diameter; multiply by 40 when trunks are more than 15 cm in diameter. To convert lb/1000 ft^2 to lb/in. diameter of treetrunk: Multiply by 0.05 when trunks are less than 6 inches in diameter; multiply by 0.1 when trunks are more than 6 inches in diameter. When rainfall or irrigation is heavy, this amount might be split into two applications.

[b]One kilogram per 100 liters is about 8 lb/100 gal.

[c]Wet to runoff.

soil microorganisms before they move down below the surface soil. Depending on the temperature and microbial activity of the soil, most ammonium ions will be converted to nitrate in about two weeks. Therefore plants will respond more slowly but for a longer period of time to ammonium nitrogen. Ammonium ions are also less readily leached from the soil.

With the exception of urea, organic nitrogen is generally unavailable for plant use. Urea, though organic, is water soluble and moves with water into the soil, where it is converted first into ammonium nitrogen, then into nitrate. In practice, urea can be irrigated into the soil and made available to roots as urea, ammonium ions, or nitrate ions. If a second irrigation takes place when most of the nitrogen is still in ammonium form, it will not leach nitrogen out of the root zone, because ammonium ions do not move readily with soil water. Other organic forms of nitrogen, whether natural or synthetic, must be decomposed or transformed into more soluble forms before they move through the soil to plant roots. The rate of decomposition or transformation is usually dependent on the form of the material, temperature, soil moisture, and soil microorganisms.

Bernatzky (1978) discourages the use of both ammonium and urea, arguing that too much oxygen in the soil is required to disintegrate them. Both forms of nitrogen are used widely in the United States, however, with no such problem reported. In fact, ammonium is retained in surface soil, where oxygen is ample, until it is converted to nitrate.

Tattar (1978) advises against fertilizing declining trees with nitrogen because it primarily stimulates foliage production, apparently at the expense of root growth. Proebsting's (1935) experiments in unirrigated California almond orchards showed just the opposite: Where lack of water curtailed growth and yield of mature trees, nitrogen fertilization increased growth and yield; other nutrients were without effect. Root growth must have been stimulated sufficiently to supply the tops of the trees with enough water.

In recent years, several nitrogen products that release nutrients over a longer period have become available. These "slow-release" fertilizers include

> Slowly converted organic forms
>
> Inorganic materials of low solubility, such as magnesium ammonium phosphate
>
> Inorganic materials coated with plastic or sulfur or enclosed in plastic bags that govern their rate of release to the soil

The organic forms may release larger amounts of nutrients if the temperature is high. Inorganic forms, which release nutrients more uniformly, are therefore superior in certain situations.

Slow-release fertilizers are usually more expensive than normal nitrogen sources, but their use may be justified where fertilizer application is difficult or required at frequent intervals, since they can be applied in larger quantities and less often. They provide a less concentrated but more continuous source of nitrogen with less likelihood of loss by leaching. On the other hand, plant growth may not be controlled as easily by the slow-release fertilizers as by the more soluble forms. Slow-release nutrients will be available at a fairly uniform level for an entire season or more, while a soluble form applied at the correct rate in the winter will be readily available when growth begins and will drop to a low level in late summer. This may be desirable for winter hardiness or fall leaf color.

Even when the particular form of nitrogen supplied does not have a pronounced effect on plant response, it may influence soil reaction. For example, ammonium ions tend to acidify a soil (see Table 6-6). Also, as noted previously, forms vary in the ease with which they are leached from the soil.

Surface application is easiest and quickest and is just as effective as placement of nitrogen in the soil (Neely, Himelick, and Crowley 1970, Smith and Reisch 1975). If urea or an ammonium-containing fertilizer is applied on the surface of sandy alkaline soils, however, some nitrogen may be lost due to ammonia volatilization within the first two or three weeks. Laboratory tests estimate losses to be about 5 percent, but they may be as high as 50 percent on sandy alkaline soil that is alternately dried and wetted (Connell and others 1979). Volatilization loss is not a problem with nitrate or on acid soils. Ammonia loss from urea can be virtually eliminated if a good rain or irrigation dissolves and leaches the urea into the soil. Watering in ammonium fertilizers reduces volatilization by increasing adsorption of ammonium ions on soil particles. A mulch reduces ammonia loss by providing surface for adsorption and by maintaining a uniform and fairly high moisture level in the surface soil, which can absorb some of the ammonia that is volatilized. Incorporating ammonium-containing fertilizers or urea into the soil will practically eliminate the volatilization loss of nitrogen, though the savings may not be worth the effort.

Phosphorus In almost all soils, phosphorus (P) occurs in amounts that are adequate for trees and large shrubs. Phosphorus fertilization research has focused much more extensively on fruit and forest trees than on landscape trees and shrubs. Trees and large shrubs are able to grow well even when soil phosphorus is available at relatively low levels. In certain California soils, satisfactory cover crops can be grown only when phosphorus is added; in these same soils, however, fruit trees have not responded to phosphorus applications (Proebsting 1958). Dickey (1977) reports similar response differences between woody species and field crops in Florida. Boynton and Oberly (1966) find no conclusive evidence that phosphorus applications affect the growth or fruiting response of apple trees under field conditions. By means of mycorrhizae, woody plants are apparently able to exploit larger volumes of soil than are annual crops and may use less available soil phosphorus (Pritchett 1979).

Phosphorus availability to plants is low in most soils. Some clay soils, clay loam, red soils high in iron oxides, and highly organic soils have high phosphorus-fixing capacities. Very old soils (as in Australia and New Zealand) and some coastal sands and peat lands have extremely low levels of phosphorus. Availability is also influenced by soil pH: Phosphorus is most available to plants between pH 5 and 7 and becomes less available as the pH deviates further from this range. At pH 7, phosphorus combines with calcium and becomes increasingly unavailable to plants. Below 5, it is precipitated out of the soil solution, forming aluminum and iron compounds. Phosphorus deficiency occurs on many surface-mine spoils, which are often extremely acid.

Deficiency Symptoms: PHOSPHORUS (Phosphorus is mobile in plants.)

Broadleaf: Leaves are green to dark green; veins, petioles, and lower surfaces may become reddish (dull bronze) to purple, especially when young; foliage may be sparse, slightly smaller than normal, and distorted; leaves drop early. *Shoots* are normal in length unless the deficiency is severe, but they are small in diameter. *Flowers* are few. *Fruit* is sparse and small (Harris, Paul, and Leiser 1977).

Conifer: Needles turn purple in young seedlings, starting at the tips of lower needles and progressing inward and upward. Few or no secondary needles may appear. Needles die, starting in the lower regions and spreading upward through the tree. Buds may set early or seedlings remain dormant longer than usual. Older trees take on a dull blue- or gray-green color. Roots are sparse with no evidence of mycorrhizae (after Powers 1979).

Treatment: PHOSPHORUS DEFICIENCY (Phosphorus is relatively immobile in most soils.)

Application	Chemical Form	Amount to Apply $kg/100\,m^{2\,a}$	How to Apply	Years Effective
Soil	Several (P)	1–2 on sandy loam	Incorporate	3–5
		2–4 on clay loam	Incorporate	3–5

[a]One kilogram per 100 m^2 is about 2 lb/1000 ft^2.

Pritchett (1979) reports that young slash pine in the coastal plains of the southeastern United States respond more to phosphorus fertilization when soils are poorly drained. Conversely, in well-drained soils there is little or no response to phosphorus additions. Applying nitrogen to soil that is severely deficient in phosphorus may actually suppress plant growth until the deficiency is corrected (Maftoun and Pritchett 1970).

Phosphorus does not move readily in soil. It tends to concentrate near the surface and is often fixed in the humus layer there. Grading soils that are low in phosphorus may aggravate phosphorus deficiencies in young or shallow-rooted plants.

Newly planted trees and shrubs in soil that is low in phosphorus may respond to phosphorus fertilization, but their response usually decreases as plants increase in size. The roots of larger plants are able to explore more soil and become mycorrhizal, thereby increasing their phosphorus supply. This may account for the differences between reports by Wyman (1936) and Chadwick (1941) on the one hand, who observed responses from newly planted trees to phosphorus fertilization, and by Neely, Himelick, and Crowley (1970), Smith and Reisch (1975), and Dickey (1977) on the other, who worked primarily with more mature woody plants. In fact, when Pridham (1938) observed the trees originally fertilized by Wyman, he found that after seven years the trees not fertilized with phosphorus were as large as those that had been fertilized. In forest soils where phosphorus has been found to be deficient, it has been applied only to young stands of trees (Pritchett 1979).

Phosphorus fertilization is sometimes recommended to enhance the root growth of trees growing under difficult urban conditions (Bernatzky 1978, Pirone 1978b) or where soil has been filled around them (Tattar 1978). I have been unable to find experimental results to confirm that a deficiency or an excess of phosphorus will affect root growth more than a deficiency or an excess of any other nutrients. Epstein (1972), Halfacre and Barden (1979), and Kramer and Kozlowski (1979) make no mention of such rooting responses to phosphorus in their reviews of plant nutrients.

It is reasonable to apply phosphorus to young woody landscape plants if a deficiency is expected. One should ask appropriate authorities, however, whether mature woody plants have responded in the past to phosphorus fertilization on a particular soil. Pioneer species in natural forest successions generally have lower nutrient requirements than succeeding species (Powers 1981a). It may be worthwhile to test leaves for phosphorus content or to conduct a fertilizer trial. Phosphorus should be applied if the level in the current season's foliage of conifers and broadleaved plants is below 0.1 percent. If additional phosphorus is needed for good plant growth, it needs to be applied only once every three to five years. When phosphorus is not needed, application will increase soil salinity and can pollute surface water if erosion occurs.

Except in well-drained acid soils, phosphorus does not move well. If needed, the mineral should be placed among the roots or close to them. For young trees, mix 5 g of phosphorus for each 10 liters (0.3 oz/ft^3) of backfill soil, or ring the bottom of the planting hole (500 mm or 20 in. in diameter) with 10 g (0.2 oz) of phosphorus. The hole should be 100 mm (4 in) deeper if phosphorus is added, so it can be covered with unfertilized backfill to minimize toxicity.

In northern California forests, Robert Powers (1981b) has obtained good response when he has sprayed phosphoric acid on the foliage of young white fir and sugar pine planted in phosphorus-fixing soil.

Potassium   Most soils contain enough potassium for woody plants, which absorb this nutrient in relatively large amounts, concentrating it in rapidly growing shoots. When leaves and brush are left on the ground, potassium is quite effectively replenished in surface soil, since it leaches readily out of organic matter.

Potassium deficiencies occur primarily in soils that are acid, sandy, low in organic matter, and low in total cation exchange capacity (Leaf 1968). Deficiencies are most likely to develop in soils of alluvial and blown-sand origin, are somewhat less common on heath lands, peat, and muck soils, and are rarer yet in heavier soils. On sandy soils in northwestern Florida, Dickey (1977) has identified potassium deficiency symptoms in wax-leaf privet, eastern dogwood, and red maple, but not in other plants. Potassium deficiencies have been identified on many agricultural and forest lands. Surface soil is usually higher in potassium than the soil underneath, so in areas where potassium is low, grading may leave cuts deficient and fill areas adequate in nutrient content. Similarly, cultural practices that encourage rooting in surface soil will decrease the likelihood of potassium deficiency symptoms (see Chapter 13).

If you suspect potassium deficiencies in a certain area, consult agricultural extension agents, extension foresters, or knowledgeable arborists. Leaf analysis may be unreliable because potassium is mobile in a plant and is readily leached from leaves during rain or sprinkler irrigation. Visual symptoms may be the best guide.

The amount of potassium required depends on the type of soil. In fine-textured soil that is neutral or alkaline and has little or no shallow rooting, apply potassium in larger amounts in the rooting zone; less is needed in sandy soil. A positively charged ion, potassium tends to be adsorbed strongly on clay particles. Apply it as a liquid by injection or dry in holes bored in the soil or in a trench dug near the drip line of the tree.

In sandy soils with an organic mulch cover to encourage shallow rooting, correct deficiencies by applying 2.5 kg of potassium per 100 m^2 (5 lb/1000 ft^2) to a surface area that extends outward from the trunk for one and one-half times the radius of the plant canopy. In New York forests on sandy acid soils, White (1956) obtained response from conifers with only 1.2 kg/100 m^2 of potassium applied by air.

Fertilizer potassium comes from a number of sources. In areas where salinity (due to arid conditions or salting winter roads) is not a problem, potassium chloride, or muriate of potash, (KCl) is generally used. Chloride, however, can be toxic, particularly in arid regions. Potassium sulfate (K$_2$SO$_4$) is a satisfactory source of potassium, particularly where salinity may be a problem. Potassium nitrate (KNO$_3$) is usually more expensive than other potassium sources and will cause a buildup of excess nitrogen if it is used to supply large amounts of

---

Deficiency Symptoms: POTASSIUM (Potassium is highly mobile in plants.)

Broadleaf: Leaves exhibit marginal and interveinal chlorosis, followed by scorching that moves inward between the main veins to the entire leaf (older leaves are affected first); leaves may crinkle and roll upward. *Shoot* tips die back late in the season; shoots from lateral buds result in zigzag growth that is short and brushy. *Flower* buds are few. *Fruit* is small and poorly colored. (Harris, Paul, and Leiser 1977).

Conifer: The oldest foliage takes on a dark blue-green color that progresses to yellow and reddish-brown; finally, necrosis occurs at needle tips. Needle retention is poor and needles are often stunted; seedlings have short, thick, abundant buds; frost injury is frequent (after Powers 1979).

Treatment: POTASSIUM DEFICIENCY (Potassium is fairly immobile in soil.)

Application	Chemical Form	Amount to Apply kg/100 m^2 [a]	How to Apply	Years Effective
Soil	Several (K)	2–8 on sandy loam	Incorporate	5–10
		8–15 on clay loam	Incorporate	5–10

[a]One kilogram per 100 m^2 is about 2 lb/1000 ft^2.

potassium. At rates above 5 kg/100 m^2 (10 lb/1000 ft^2), one application should last for five to 10 years.

Wood ashes have been used as a source of potassium for a very long time. The term *potash* is derived from *pot ashes*, which were obtained by leaching wood ashes and evaporating the solution to dryness. Unleached hardwood ashes contain from 8 to 30 percent potassium, mostly in carbonate form (Branson 1980). Softwood ashes contain less potash, and ashes from trunk wood are poorer in potash than are those from twigs and small branches. Rain can leach about 90 percent of the potassium from ashes. Hardwood ashes also contain 15 to 25 percent calcium but less than 2 percent phosphorus. Potassium and calcium carbonates are quite alkaline: Although on acid soils they are beneficial for both pH adjustment and potassium, on alkaline soils they can raise the pH to critical levels (see Chapter 6).

Potassium is said to provide a number of benefits to plants: It increases the root growth and drought resistance of trees in urban areas (Bernatzky 1978), overcomes succulence and brittleness (increases stem strength), hastens plant maturity, intensifies flower color (Pirone 1978b), and enhances a tree's resistance to disease and cold (Murphy and Meyer 1969). I have been unable to find any experimental results to suggest that potassium affects these responses any more than other nutrients do. Again, Epstein (1972), Halfacre and Barden (1979), and Kramer and Kozlowski (1979) attribute no such responses to potassium in their reviews of the roles of plant nutrients.

All of these responses (except for the increased drought resistance and root growth) may be attributed to potassium because the early use of fertilizers assessed quantities by total weight, not by the proportions of individual nutrients. By such a method of calculation, one nutrient increases in a fertilizer only when one or more of the other nutrients decrease. Because phosphorus and potassium are adequate for woody plants in most soils, the significant component in any fertilizer containing nitrogen, phosphorus, and potassium will be the nitrogen alone. As differing proportions of these three nutrients are used, varying amounts of nitrogen will be much more important to plant response than the presence or absence of potassium and phosphorus. Thus the responses attributed to increased amounts of potassium may in reality be due to the concomitant, and more significant, decrease in nitrogen.

Complete Fertilizers "Complete" fertilizers are still recommended widely for improving the growth, appearance, and health of landscape plants. A complete fertilizer contains nitrogen (N), phosphorus (P), and potassium (K). Early trials found these three nutrients deficient for field and vegetable crops in most soils. Complete fertilizer use was therefore adopted for almost all crops, for fruit trees, landscape plants, and, more recently, in some forests. Even though field trials have shown that most soils contain phosphorus and potassium in sufficient levels for fruit, forest, and landscape trees, the inclusion of these minerals in fertilizer is still commonly recommended for landscape trees, shrubs, and vines, regardless of circumstances.

On a phosphorus deficient soil in Tennessee, five of six five-year-old broadleaved tree species grew more rapidly when fertilized with a slow-release 14-14-14 fertilizer than with ammonium nitrate or a soluble 20-20-20 fertilizer, but on soils of moderate phosphorus levels, the same species responded most vigorously to ammonium nitrate and to a 20-20-20 fertilizer (same amounts of N) (van de Werken 1981). It is not clear that any of the trees responded to the phosphorus. Phosphorus and potassium fertilization, however, neither increased growth nor improved the appearance of landscape trees and shrubs tested in Illinois, Ohio, and Florida (Neely, Himelick, and Crowley 1970, Smith and Reisch 1975, Dickey 1977). Neely, Himelick, and Crowley worked with 20 deciduous and two evergreen species at five sites in Illinois to determine the response of established young trees to different nutrients and to various methods and schedules of application. Nitrogen was the only nutrient to which the trees responded (Fig. 11-4). Those results have been confirmed by similar trials throughout the

FIGURE 11-4
Trunk growth in 1963 and 1964 of pin oak. Trees grown in sod at Lisle, Illinois, were fertilized in the spring of 1963 and 1964. Fertilizers containing nitrogen were applied at 3 kg of nitrogen per 100 m^2 (6 lb/1000 ft^2).
(Adapted from Neely, Himelick, and Crowley 1970)

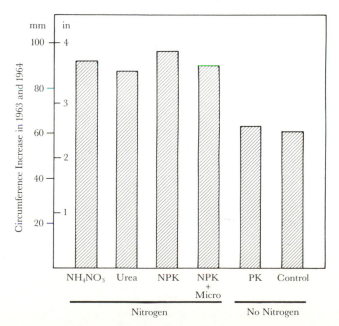

world. It is the exception, not the rule, when woody plants respond to additions of phosphorus and potassium.

Large amounts of potassium can cause magnesium deficiency, particularly in sandy soil. Even when potassium is needed, a complete fertilizer will supply excessive amounts of nitrogen and phosphorus along with the required amounts of potassium. It is best, therefore, to apply phosphorus or potassium separately. Applications once every three to five years should be adequate even in soil originally deficient in one of these nutrients (Neely and Himelick 1968). In forest soils deficient in phosphorus or potassium, a single application of the deficient nutrient during or shortly after planting will usually produce sufficient growth (Pritchett 1979). Nitrogen, which is more soluble, must usually be added at least every two years if it is to continue to benefit plants. For certain shallow-rooted woody plants, a complete fertilizer may be most effective, but ought to be applied only once every few years, with nitrogen applied in the remaining years.

Phosphorus, potassium, and complete fertilizers are usually placed in holes drilled or punched in the soil or are injected into the soil in a solution under pressure. Air pressure has also been used to force dry fertilizer into soil through drilled holes. By contrast, nitrogen fertilizers are most easily applied on the surface and washed into soil by rain or irrigation water. In tight, poorly drained soils, plant growth may improve following phosphorus or potassium treatment because of the increased aeration provided by fertilizer holes, rather than as a direct response to the nutrient itself.

Since complete fertilizers serve a genuine purpose for growing vegetables, herbaceous landscape plants, and turf, it is advisable to understand how they are labeled. In the United States and many other countries, a fertilizer container must have a nutrient analysis printed on its label. The analysis gives the percentage by weight of the three nutrients, always listed in the same order: nitrogen (N), phosphorus (P or P_2O_5), and potassium (K or K_2O). Percentages of the three nutrients are often prominently displayed on the label (as in 10-6-4 or 10-10-10), with the understanding that their order is N-P-K. Attempts have been made to standardize the elemental forms of all three nutrients, but the older oxide designations are still common. Be sure you know whether phosphorus and potassium recommendations and label analyses are given in terms of the elemental or oxide forms. Conversions can be made in this way:

1 gram* P = 2.294 grams P_2O_5

1 gram P_2O_5 = 0.436 grams P

1 gram K = 1.205 grams K_2O

1 gram K_2O = 0.830 grams K

*Other units of weight may be handled in the same manner.

Calcium   Most soils contain more than enough calcium for plant use.   Some soils low in calcium become so acid that deficiencies of nutrients or excesses of chemicals other than calcium become more serious to plant health than the calcium deficiency alone.   However, even when vegetative growth is normal, low calcium can result in fruit disorders such as bitter pit of apple.   In soils that are about pH 7 (neutral), calcium represents 60 to 85 percent of the total exchange capacity (Chapman 1966a).   In well-drained soils with an annual rainfall above 750 mm (30 in), hydrogen and aluminum ions replace calcium and other bases.   If the acidity becomes high enough, manganese, aluminum, copper, and other elements reach toxic concentrations. Increasing sodium, on the other hand, displaces calcium, increases the soil pH, and reduces the solubility of remaining phosphorus, manganese, calcium, and, in certain cases, zinc, boron, and iron (Chapman

---

Deficiency Symptoms: CALCIUM (Calcium is extremely immobile in plants.)

Broadleaf: Leaves become chlorotic and/or necrotic; young leaves are small and distorted with tips hooked back. *Shoots* are stunted with terminal dieback. *Fruit* of apple trees may have bitter pit. *Roots* are usually affected first, with dieback of root tips severely reducing growth (Chapman 1966a).

Conifer: Primary needles are usually normal, but secondary needles may be stunted or killed. Terminals are stunted and needles may hook at tips. Symptoms are most severe in the youngest foliage in the upper crown (after Powers 1979).

Treatment: CALCIUM DEFICIENCY (Calcium is relatively immobile in soil.)

Application	Chemical Form	Amount to Apply $kg/100\ m^2$ [a]	How to Apply	Years Effective
Soil pH <6.0[b]				
pH >7.0	$CaSO_4$[c]	40–75 on sandy loam 75–150 on clay loam	Incorporate	5–10

[a]One kilogram per 100 m^2 is about 2 lb/1000 ft^2.
[b]Apply lime to bring pH to about 6.0 (see Table 6-4).
[c]Twenty percent as much sulfur can also be used; it will act more slowly.

1966a). Soil structure deteriorates with high sodium content, leading to decreased water movement, permeability, and aeration. Low levels of calcium commonly occur in acid soils, in sandy soils where the annual rainfall is above 750 mm (30 in), and in highly acid peat soils. Alkali or sodic (high sodium) soils are quite alkaline, so that any calcium present is extremely insoluble. Soils derived from serpentine rock are calcium deficient.

Serpentine soils, generally sterile and unproductive (Whittaker 1954), have been reported in isolated areas on all continents but South America. On these soils, plants are usually sparse or stunted and composed of many species restricted to the serpentine habitat. Molybdenum deficiency and heavy metal toxicities have been reported as the cause of serpentine infertility, but the basic cause is low calcium. In the United States, scattered serpentine outcroppings occur along the Appalachian Mountains from western Massachusetts to Georgia. The Pacific Coast states have even more extensive serpentine areas in their mountains and valleys.

Add lime to acid soils to increase calcium and pH (see Chapter 6). In alkaline soils, add gypsum or elemental sulfur to decrease pH and make more calcium available. The amount of gypsum needed depends on the soil, its sodium and magnesium contents, and the change in pH desired. A simple soil test can provide this information (Chapman 1966a). Gypsum or sulfur should be worked to 100 or 150 mm (4–6 in) below the soil surface. If heavy amounts are needed, apply them before you plant and incorporate the material to deeper levels.

Serpentine soils will require about the same amount of gypsum or lime to supply calcium, depending on the soil pH. It is important to incorporate material into the soil because movement of calcium can be slow, and root growth in soil that is extremely deficient in calcium is almost nil (Martin, Vlamis, and Stice 1953).

These drastic treatments should only be needed once if follow-up management maintains the new situation.

Magnesium   Although magnesium is sufficient in most soils, it is readily leached from sandy acid soils. In calcareous soils, it is tied up in relatively unavailable forms (Dickey 1977). Magnesium may be deficient in soils in which potassium is deficient.

Magnesium deficiency in woody plants is often difficult to control in the field. Raising the pH of acid soils to between 5.5 and 6.5 by liming (see Table 6-4) with dolomite limestone (magnesium and calcium carbonates) and supplementing the organic matter content will increase the availability of magnesium and its retention in the soil (Dickey 1977). These treatments should be carried out before planting. Lowering the pH of calcareous soils by applying acidifying materials will also increase the availability of magnesium.

Magnesium deficiency in established plants can be particularly difficult to overcome; response is often slow. Dickey recommends adding magnesium sulphate to magnesium-deficient plants on acid soils in Florida in early spring and early summer, at rates of 5 to 10 kg per 100 m^2 (10–20 lb/1000 ft^2). Substituting dolomite for up to half of the magnesium sulfate may improve response. Fertilizer should be spread evenly under the plant canopy and water used to soak it in; magnesium sulfate will move with the water into acid soil. In lawns,

---

Deficiency Symptoms: MAGNESIUM (Magnesium is mobile in plants.)

Broadleaf: Leaves are thin and brittle, and drop early. Older leaves may show interveinal and marginal bright chlorosis (reddening) with interveinal necrosis late in the season. *Shoot* growth is not reduced until deficiency is severe. *Fruit* yield is reduced when deficiency is severe; apples may drop prematurely (Harris, Paul, and Leiser 1977; Boynton and Oberly 1966).

Conifer: Needle tips are orange-yellow and sometimes red. Primary needles remain blue-green in young seedlings, but in older plants, older needles and the lower crown show symptoms first. In affected needles, the transition to green may be sharp (after Powers 1979).

Treatment: MAGNESIUM DEFICIENCY (Magnesium is mobile in acid soils and fairly immobile above pH 6.5.).

Application	Chemical Form	Amount to Apply	How to Apply	Years Effective
		kg/100 m^2 [a]		
Soil	$MgSO_4 \cdot 7H_2O$[b]	12–25 on sandy loam	Incorporate	5–7
		25–50 on clay loam	Incorporate	5–7
		kg/100 liters		
Spray	$MgSO_4 \cdot 7H_2O$[c]	2.5	Foliage[d]	1

[a]One kilogram per 100 m^2 is about 2 lb/1000 ft^2.

[b]Use magnesium sulfate (Epsom salts) in neutral or alkaline soils and dolomitic limestone in acid soils.

[c]Dickey (1977) reports that citrus growers in Florida use 1.25 kg/100 liters each of magnesium sulfate and calcium nitrate.

[d]Foliage application is not usually recommended, but can determine whether a plant will respond to the nutrient.

the material must be injected into the soil. For magnesium-deficient sandy soils, Dickey recommends that applications at this rate be repeated annually until symptoms disappear, then continued at a rate of 2 kg/100 m^2 (4 lb/1000 ft^2) annually. For magnesium-deficient soils in California, Proebsting (1958) recommends magnesium sulfate for neutral or alkaline soils and dolomite for acid soils. Bernatzky (1978) cites Ruge as recommending that 2 percent MgO (about 1 kg MgSO$_4$/100 m^2 or 2 lb/1000 ft^2) be regularly added to soils in Germany.

A foliage spray of magnesium sulfate alone or magnesium sulfate and calcium nitrate can be used to determine plant response to magnesium. This is a temporary measure at best, however, and has produced inconsistent results.

Sulfur   Irrigation water, rainfall, decomposing organic matter, fertilizers, and some fungicides provide enough sulfur for normal plant growth in most soils. In the past, domesticated plants received sulfur from the decomposition of manure and other organic matter in the soil. Later, fertilizers began to contain inorganic materials such as ammo-

---

Deficiency Symptoms: SULFUR (Sulfur is mobile in plants.)

Broadleaf: Leaves are entirely pale yellow-green in both young and old plants; they are small on some species and exhibit other symptoms associated with nitrogen deficiency. *Shoots* are stunted (Harris, Paul, and Leiser 1977; Eaton 1966).

Conifer: Symptoms are similar to those associated with nitrogen deficiency; needle tips may be yellow, red, or mottled, particularly on older needles; necrosis may follow; needle retention is poor (after Powers 1979).

Treatment: SULFUR DEFICIENCY (Sulfur is somewhat immobile in soil.)

Application	Chemical Form	Amount to Apply kg/100 m^2 [a]	How to Apply	Years Effective
Soil	CaSO$_4 \cdot$ 2H$_2$O[b]	5–8 on sandy loam	Incorporate	5–7
		8–12 on clay loam	Incorporate	5–7

[a]One kilogram per 100 m^2 is about 2 lb/1000 ft^2.

[b]Chemical treatment should not be needed if sulfur-containing materials are used to remedy other deficiencies or to control pests.

nium sulfate, potassium sulfate, and superphosphates. Since the 1950s, the trend has been toward fertilizers that contain high concentrations of primary nutrients, but little or no sulfur. This trend could lead to sulfur deficiencies if it were not for the increased amounts of atmospheric sulfur dioxide produced by burning coal and oil. Maugh (1979) reports that Noggle of the Tennessee Valley Authority found that as much as 40 percent of a plant's accumulated sulfur had been absorbed directly as sulfur dioxide from the air. Compounds that are washed from the air by rain and settle as particulates supply more sulfur. Dickey (1977) estimates that each centimeter of rain deposits 0.45 kg of sulfur per hectare (or 1 lb of sulfur per acre from 1 in. of rain). Although sulfur dioxide is a serious air pollutant that can damage plants and fish, at lower concentrations it may be an important source of sulfur for plant growth. In regions where sulfur is not added to the plant environment in one or more of the ways described, deficiency symptoms may appear.

Sulfur can be added by increasing the decomposition of organic matter as part of a sulfate-containing fertilizer or other sulfur-containing compounds. Elemental sulfur is commonly applied to increase acidity in soil that is close to the neutral range (see Table 6-5) or as a corrective for alkaline soils (Proebsting 1958). Sulfur is readily available in gypsum, but it may take several weeks for microorganisms to oxidize elemental sulfur.

Micronutrients

Boron Even though plants require only extremely small amounts of boron, that nutrient is deficient in many parts of the world. Boron excess, while not as widespread, is more serious than a deficiency, which can be easily corrected. Excess boron can be difficult to overcome.

Boron is a rather unusual nutrient because of the narrow range between deficiency and excess in a plant. For example, peach trees can be deficient at leaf concentrations less than 20 ppm, normal between 20 and 80 ppm, and in excess above 90 ppm (Bradford 1966). Roses growing in solution culture are deficient at leaf levels below 5 ppm and toxic at levels above 20 ppm. Most trees (primarily fruit) analyzed are deficient at leaf levels below 20 ppm, but toxic levels vary considerably with species, ranging from 80 ppm to 500 ppm. Woody perennials are more sensitive to boron excess than are most annuals, particularly field crops.

The boron content of soils is determined by geological origin and degree of weathering. Boron deficiencies occur most commonly in coarse sandy soil, in acid soils, soils derived from acid igneous rocks, soils from freshwater sediments, and alkaline soils, especially those

Deficiency Symptoms: BORON (Boron is extremely immobile in plants.)

Broadleaf: Leaves are occasionally red, bronzed, or scorched, and young leaves are affected first. Leaves are small, thick, brittle, and, on some species, distorted. *Shoots* exhibit rosetting, discoloration, and dieback of new growth, which becomes zigzag, short, brushy, thick, and stiff. *Flowers* may be few. *Fruit* set is light and deformed, with cracked, necrotic, spotty, corky surfaces. Fruit may drop before it is mature (Bradford 1966; Harris, Paul, and Leiser 1977).

Conifer: Shoot tips are bent (J topping), and meristematic tissue of the main leader may split. Necrotic blotches are visible on magnified cross-sections of buds and cause the death of terminal and some lateral buds. Plants become more like shrubs than trees (after Powers 1979).

Treatment: BORON DEFICIENCY (Boron moves readily in soil.)

Application	Chemical Form	Amount to Apply	How to Apply	Years Effective
		$kg/100\ m^{2}$ [a]		
Soil	Borax	0.2–0.5 on sandy loam	On surface	5–7
		0.5–1.0 on clay loam	On surface	5–7
		$kg/100\ liters$		
Spray	Boric acid[b]	0.125–0.250	Foliage	1

[a]One kilogram per 100 m^2 is about 2 lb/1000 ft^2.

[b]Boric acid is more soluble in cold water than is borax. One and one-half kg of borax equals about 1 kg of boric acid in boron content.

Toxicity Symptoms: BORON

Broadleaf: Leaves are yellow along margins and tips, later turning brown or black; dark necrotic spots may appear interveinally and on undersides of the mid-rib and petiole. Symptoms worsen progressively from the tip to the base of shoots. *Shoots* exhibit dieback with swelling and cracking below lateral buds; shoots gum and have short internodes (Bradford 1966; Harris, Paul, and Leiser 1977).

Conifer: Shoot dieback is found in most pine (*Pinus patula* is an exception). The leaders of some pine may curl on drying. Resin flow is common in some pine species (Stone 1968).

containing free lime (Bradford). As of 1968 (Stone), all reported boron deficiencies in cultivated forests occurred in sandy or highly weathered soil. No native forest stands have been found to be boron deficient. In the United States, areas of known boron deficiency are located in the eastern third of the country and portions of the Pacific states. In Lake County, California, boron-deficient soils are within 10 km (6 miles) of soils with excess boron.

Boron deficiency can be aggravated by certain cultural practices. Irrigation water that is very low in boron and high in calcium will increase the need for boron. Symptoms of deficiency or excess can be intensified by drought. Adding lime to soils that are low in boron will inhibit boron uptake and utilization.

The correction of boron deficiency is relatively easy. One surface application of 0.25 to 1 kg of borax (sodium tetraborate—11 percent boron) per 100 m^2 (0.5-2 lb/1000 ft^2), depending on soil texture and pH, should last several years. Other boron fertilizers can be used to add similar amounts of boron. Soil applications during summer and fall will usually produce growth response the following spring.

Many species will respond if borax is sprayed on foliage during the early growing season, at a mixture of 120 g/100 liters (1 lb/100 gal) of water. Borax sprays must be applied each year, while soil applications last several years. Sprays are usually more difficult in landscapes than are surface applications. Borax is fairly soluble and moves into the soil with rain or irrigation. Boron can be injected into treetrunks, but soil application is much easier. Care should be taken to avoid applications that are too large or too frequent because it is easy to turn a deficiency into an excess.

Excess boron can be lethal to plants and is most likely to occur in soils that originate from marine sediments or geologically young deposits, or in parent material that is high in boron minerals or has developed in arid climates (Bradford 1966). Certain cultural practices can also increase boron toxicity: irrigating with water above 0.75 ppm boron, acidifying neutral or alkaline soils already high in boron, applying heavy amounts of boron fertilizers, and applying heavy amounts of potassium to high-boron soils.

High levels of boron may be reduced by leaching soil with water containing less than 0.5 ppm boron. The soil must be well drained (see Chapter 12). Some plants are more tolerant of high levels of boron than others (Table 11-1). The liberal application of nitrogen fertilizers, especially calcium nitrate, sometimes helps, as does moderate liming (Bradford). Plants accumulate more boron when rates of transpiration are high. At the same boron level in the soil, a plant in a hot, dry area may show toxicity symptoms, while the same species in a cool, humid area will be normal.

TABLE 11-1

Boron tolerance in selected species of woody plants
(Francois and Clark 1979)

Common Name	Botanical Name
Tolerant[a]	
Natal Plum	*Carissa grandiflora* 'Tuttlei'
Indian Hawthorn	*Raphiolepis indica* 'Enchantress'
Chinese Hibiscus	*Hibiscus rosa-sinensis*
Oleander	*Nerium oleander*
Japanese Boxwood	*Buxus microphylla* var. *japonica*
Bottlebrush	*Callistemon citrinus*
Ceniza	*Leucophyllum frutescens* 'Compactum'
Blue Dracaena	*Cordyline indivisa*
Semi-tolerant	
Brush Cherry	*Syzygium paniculatum*
Southern Yew	*Podocarpus macrophyllus* var. *Maki*
Oriental Arborvitae	*Platycladus orientalis*
Rosemary	*Rosmarinus officinalis*
Glossy Abelia	*Abelia* X *grandiflora*
Sensitive[b]	
Yellow Sage	*Lantana camara*
Juniper	*Juniperus chinensis* 'Armstrongii'
Chinese Holly	*Ilex cornuta* 'Burfordii'
Japanese Pittosporum	*Pittosporum tobira*
Spindle Tree	*Euonymus japonica* 'Grandifolia'
Pineapple Guava	*Feijoa sellowiana*
Wax-Leaf Privet	*Ligustrum japonicum*
Laurustinus	*Viburnum tinus* 'Robustum'
Thorny Elaeagnus	*Elaeagnus pungens* 'Fruitlandii'
Shiny Xylosma	*Xylosma congestum*
Photinia	*Photinia* X *fraseri*
Oregon Grape	*Mahonia aquifolium*

[a]Species are listed in order of decreasing tolerance. Tolerant species were affected little, if at all, by 7.5 mg of boron per liter of irrigation water.

[b]Sensitive species were severely damaged or killed by 7.5 mg of boron per liter and moderately damaged by 2.5 mg of boron per liter of irrigation water.

Chlorine   Although chlorine has been shown to be an essential nutrient, deficiencies of the element under field conditions for woody plants have not been reported. Chloride is highly soluble and widely distributed in soils and natural waters (Eaton 1966). If a chlorine deficiency occurs or is suspected, the equivalent of 0.25 kg of chloride per 100 m^2 (0.5 lb/1000 ft^2) can be included in a fertilizer mixture.

Chloride is much more likely to be toxic than deficient, particularly in irrigated arid regions, near seacoasts, and adjacent to roadways treated with salt during the winter. Symptoms include small leaves with marginal and tip scorch. Such leaves will yellow and drop early. Leaching is the most effective way to reduce the level of chloride and other salts in soil. The effects of excess chloride, sodium, boron, and total salts are discussed in Chapters 12 and 18.

Copper   Although not as common as boron or zinc deficiency, copper deficiency is fairly widespread on most continents. It is not likely to occur on soils that are sandy, organic, alkaline, or calcareous (Reuther and Labanauskas 1966). In the United States, old livestock corrals and Indian burial sites are commonly deficient in both copper and zinc; these elements are apparently tied up by the combination of certain organic matter and soil compaction. Copper deficiency can be aggravated by alkaline irrigation water and by nitrogen or phosphorus accumulation. The woody plants most likely to show deficiency symptoms are apple, apricot, camellia, citrus, jasmine, olive, peach, Japanese pittosporum, plum, wax-leaf privet, rose, and tung (Dickey 1977; Reuther and Labanauskas 1966).

Most soils deficient in copper respond readily to copper fertilization. The amount and frequency of fertilization should vary with soil texture, pH, and organic matter content. No additional copper should be added until and unless deficiency symptoms reappear. Do not exceed recommended amounts, because copper can become toxic at higher rates. A 5 percent copper sulfate solution was one of the first sprays used to control weeds chemically (Reuther and Labanauskas). In fine soil, where copper is slow to move into the root zone, you may have to drill or inject copper sulfate into soil (Proebsting 1958) or spray it on foliage.

Copper-deficient plants usually respond quickly and satisfactorily to foliage sprays such as the Bordeaux mixture (copper sulfate and lime) (Reuther and Labanauskas). In fact, many fruits and vegetables grow in copper-deficient soils without exhibiting deficiency symptoms because they have been sprayed with copper fungicides. Soil applications are easier and last longer than foliage sprays. Trunk injections are not recommended because copper is quite toxic to bark, cambium, and young sapwood. Proebsting (1958) reports that growing alfalfa in fruit orchards helps correct mild copper deficiency and greatly improves plant growth in severe cases. It is not known whether turf or other ground covers will improve copper availability.

Excess copper often occurs in soils derived from or influenced by copper ore deposits and those on which crops have been heavily fertilized with copper or sprayed with copper fungicides for many years

(Reuther and Labanauskas). Copper excess commonly induces foliage symptoms similar to those associated with iron deficiency (Stone 1968). In solution cultures that receive excess copper, roots will be stunted and top growth noticeably reduced. A given level of copper will be most toxic in sandy soils of pH 5.0 or below.

Little can be done to reduce toxicity when an entire soil profile has excess copper. The soil will need to be replaced for planting pockets and a barrier created to prevent rooting into the surrounding

---

Deficiency Symptoms: COPPER (Copper is relatively immobile in plants.)

Broadleaf: Leaves may be small and may develop necrotic spots and brown areas near the shoot tip. *Shoots* exhibit a rosetting of leaves near the terminal, followed by dieback, producing a bushy, stunted appearance. *Fruit* gum appears on the surface of citrus fruit (Reuther and Labanauskas 1966).

Conifer: The needles of young pine show tip burn. The shoots of douglas fir are weak and drooping and form crooked stems; needles at the tips of shoots may discolor and drop during winter (Stone 1968).

Treatment: COPPER DEFICIENCY (Copper mobility is greater in acid than in alkaline soil.)

Application	Chemical Form	Amount to Apply	How to Apply	Years Effective
		$kg/100\ m^2$ [a]		
Soil	$CuSO_4 \cdot 5H_2O$	0.5–1.5[b] on sandy loam	Incorporate	5–7
		1.5–5.0 on clay loam	Incorporate	5–7
		$kg/100\ liters$		
Spray	$CuSO_4 \cdot 5H_2O$	0.4–0.8[c]	Foliage	1[d]

[a] One kilogram per $100\ m^2$ is about 2 lb/1000 ft^2.

[b] To reduce the possibility of copper toxicity in sandy soils, Reuther and Labanauskas (1966) recommend applying only 0.125 to 0.250 kg of copper sulfate per $100\ m^2$ (0.25–0.5 lb/1000 ft^2) annually for several years until the total of 0.5 to 1.5 kg/$100\ m^2$ has been added.

[c] One-half kilogram of hydrated lime should supplement this amount.

[d] This is not a long-term solution, but will determine the plant's response to copper.

toxic soil. If the excess copper is in the surface soil, as will result from heavy fertilization or spraying with copper, liming acid soils and spraying plants with iron chelate should reduce copper toxicity and encourage renewed rooting in surface soil (Reuther and Labanauskas). If the soil is alkaline, large amounts of well-decomposed organic matter may be worked in to reduce copper toxicity. Replacing contaminated soil may be the only way to assure satisfactory plant growth, however.

Iron Iron deficiency is the most common micronutrient deficiency. In many areas, its frequency is second only to that of nitrogen deficiency. Because symptoms occur most frequently in alkaline soils and those high in lime, the deficiency is sometimes called *lime-induced chlorosis*. Iron deficiency also occurs on particularly sensitive plants in acid soils. Symptoms are usually more pronounced in early spring, when the soil is cold and wet. Iron deficiency is more severe on poorly-drained wet soils. Some species will show symptoms more readily than others, and symptoms may be quite spotty: A single limb may be affected, while others appear normal.

Iron deficiency usually occurs not because iron is lacking in soil, but because it is unavailable. The availability of soil iron to plants decreases rapidly as the soil pH rises above 7.0. In acid soils, an excess of heavy-metal nutrient elements (such as copper, zinc, or manganese) can also produce iron deficiency. The heavy use of copper fungicides has caused iron deficiency in acid sandy soil (Dickey 1977).

For more than 100 years, people have been trying to correct iron chlorosis with various iron compounds, soil amendments, and methods of application. These efforts have often been disappointing, and it may be worthwhile to explore other possibilities. In many situations, plants that exhibit severe chlorosis are growing on soils that are improperly drained and high in salt content. These problems must be corrected before long-term iron nutrition can be maintained. If large areas are involved, the best long-term solution is to use plants that grow well even in alkaline or high-calcium soils. Attempts to acidify large areas of alkaline soils with existing plants have generally not been successful, and such a process requires extremely large amounts of sulfur or acid over a number of years (see Table 6-5). For small planting areas, it is feasible to acidify soil or bring in new soil before planting. Excessively wet soil commonly results in chlorotic plants, especially during cold springs and when plants are young. Attention to surface and internal drainage should alleviate some of the problem. When young trees and shrubs are located in flower, ground cover, or turf areas that are frequently watered, iron deficiency is further aggravated.

According to Dickey (1977), adequate organic matter and organic mulch will reduce the possibility of iron deficiency. Dickey recommends that backfill soil that is low in available iron be amended with ample organic matter. He also cautions that soil around new buildings—

Broadleaf: Young *leaves* are yellow with contrasting narrow green veins; older basal leaves remain darker green. Exposed leaves are bleached and will eventually exhibit apical or marginal scorch. Leaves may be small; symptoms will be severe in cold, wet springs. *Shoot* length is usually normal, but diameter will be small; twig dieback and defoliation will occur when the deficiency is severe. *Fruit* has poor color and some species, such as citrus, will drop fruit heavily (Harris, Paul, and Leiser 1977; Wallihan 1966).

Conifer: New growth will be very stunted and chlorotic; older needles and the lower crown will remain green. In seedlings, cotyledons remain green (after Powers 1979).

Treatment: IRON DEFICIENCY (Iron mobility in soil decreases with increasing pH.)

Application	Chemical Form	Amount to Apply	How to Apply	Years Effective
		$kg/100\ m^{2\,a}$		
Soil	Chelate[b]	0.2–0.5 on sandy loam	On surface	1–3
		0.5–1.0 on clay loam	On surface	1–3
	$FeSO_4 \cdot H_2O^{c}$	12 on sandy loam	Incorporate	3–5
		18 on clay loam	Incorporate	3–5
		$kg/100\ liters$		
Spray	Chelate	0.12–0.2	Early foliage	1
	$FeSO_4 \cdot H_2O$	0.5	Early foliage	1

[a]One kilogram per 100 m^2 is about 2 lb/1000 ft^2.

[b]A number of iron chelates are available: for acid soil, FeEDTA (Sequestrene NAFE® or Versene Iron Chelate®); for neutral, alkaline, and most lime soils, FeHEEDTA (Perma Green Iron 135® or Versenol Iron Chelate®), FeDTPA (Sequestrene® or Chel 330®), or FeEDTAOH (Versenol®); and for difficult soils, FeEDDHA (Sequestrene® or Chel 138®) (Dickey 1977; Wallihan 1966). Manufacturer's directions should be followed for effective treatment.

[c]Apply one half as much during the growing season. For liquid incorporation in soil, dissolve 1 kg of ferrous sulfate in 10 liters (1 lb/gal) of water and inject into holes 250 to 300 mm (10–12 in) deep and 750 mm (30 in) apart around the drip line of the tree. In the dormant season, inject 5 liters in each hole (half that amount during the growing season). For upright trees, space the holes 600 mm (24 in) apart. In a shrub area, apply 35 ml (2 oz) at sites 750 mm (30 in) apart.

block, brick, and particularly stucco—is often contaminated with lime when mortar is spilled from the walls, foundations, interior plaster, or wallboard. This contaminated soil should be removed and replaced with clean soil that is more acid in reaction.

Iron compounds can be applied on the soil surface, injected into the soil, sprayed on leaves and bark, and injected or placed in trunks and branches. Iron added as a simple salt is quickly made insoluble in alkaline soils, and little or none is available for plants. Heavy amounts of iron sulfate applied on the surface and worked into the soil have given inconsistent results. If iron sulfate or iron-sulfate solution is concentrated in holes or trenches, however, it will maintain localized zones of availability (Locke and Eck 1965). Dickey suggests that a mixture of three parts dusting sulfur and one part iron sulfate be applied once or twice a year at 5 kg per 100 m^2 (10 lb/1000 ft^2) and worked into the soil for 150 to 200 mm (6–8 in) under the surface, if possible. Dickey has found this particularly successful for local areas that are alkaline or have been excessively limed, but not for extensive areas of calcareous coastal soil.

In the 1950s, the synthetic chelates improved the continued solubility of iron even in alkaline soils. Iron chelates are more mobile in soil and plants than are other forms of iron. The differences among the various iron chelates principally involve their relative stabilities at high pH values (Wallihan 1966). Iron chelates can be applied to the soil surface and watered in or, for more rapid response, can be dissolved and applied as a soil drench. For more concentrated iron sources in large trees, Wallihan suggests applying the solution at 10 to 50 spots under the crown and then irrigating to leach into the root zone. Neely (1976) recommends injecting the chelate solution into soil to a depth of 300 mm to 375 mm (12–15 in). He obtained the best results by injecting chelate in the spring and found that treatments remained effective for two or more years.

Iron sulfate or chelate solutions can be sprayed on foliage. Response may be irregular among species and nonexistent in citrus and gardenia (Dickey 1977). Iron sprays have certain disadvantages, particularly for landscape plants, because

The sprays are difficult to apply.

They stain brick and cement.

They leave an undesirable residue on the foliage.

Iron does not move to new foliage.

Chelates can burn certain plants.

Iron sprays have a place when symptoms must be corrected immediately or the user wants to determine whether chlorotic symptoms can be corrected by iron treatments.

When organic salts, such as iron citrate or iron tartrate, are placed in holes in trunks, they can correct iron deficiencies for up to three years (Bennett 1931). Gelatin capsules containing ferric ammonium citrate (sold as Medicaps®) or ferric citrate can be tapped into holes drilled into the trunk. The number of capsules and the amount of iron salt in each should be varied with tree size. Solutions of iron sulfate or one of the chelates can also be injected into holes bored into treetrunks.

Smith (1976) reports that trunk implants of ferric ammonium citrate are more effective in correcting chlorosis in Ohio pin oak than are either foliar sprays or soil applications of various compounds. In Illinois, Neely (1976) has also successfully treated pin oak with iron implants. Neither Smith nor Neely tried injections of iron solutions.

When iron chlorosis symptoms are corrected, applications of iron can be discontinued until symptoms reappear. If soil acidification is part of the treatment, it can be continued at a more modest level or with the aid of acid-forming fertilizers (see Table 6-6).

Manganese   Although it is deficient less often than iron, manganese is commonly deficient under similar conditions and in a wide variety of cultivated plants. Reports of manganese deficiency in forest trees are rare, but in shade trees and shelterbelts on calcareous soils and in orchards are quite common (Stone 1968). The solubility of manganese decreases as soil alkalinity increases, and it is not readily available to plants above pH 6.5 (Labanauskas 1966). At pH values above 6.5, manganese is converted from the manganous to the manganic form, which is less soluble. Manganese deficiency is more likely to occur in poorly drained soils that are high in organic matter. Calcareous soils are noted for their alkaline reaction, poor drainage, and their low availability of iron and manganese. Manganese deficiency can also occur in sandy, acid, mineral soils that are low in native manganese or have been heavily leached. Cultural practices that increase soil alkalinity—such as liming, irrigation (in most soils), and burning (in organic soils)—increase the possibility or severity of manganese deficiency (Labanauskas). Manganese deficiency often occurs when soils are alternately well-drained and waterlogged. Deficiency symptoms are more likely to occur under drought conditions than when moisture is sufficient.

Labanauskas notes that apple, cherry, and citrus trees are especially sensitive to manganese deficiency. Dickey (1977) lists more than 30 woody species in Florida landscape plantings that have exhibited manganese deficiency. It is interesting to note that eight species of palm are thought to exhibit manganese deficiency symptoms, but no palms are thought to exhibit iron deficiency symptoms. Fruit trees may exhibit pronounced chlorotic leaf symptoms of manganese deficiency before there is any reduction in growth or fruiting (Boynton and

Oberly 1966).   More severe deficiencies cause necrotic leaf spotting and margins, twig dieback, and reduced growth and fruitfulness.

Mild leaf symptoms of manganese deficiency may not be worth correcting, because such correction is unlikely to increase growth or fruitfulness.   To restore dark green foliage, however, corrective steps must be taken.

Trunk injections, foliar sprays, and soil applications of manganese have corrected manganese deficiency in trees, primarily fruit trees, with varying success.   In general, foliar applications of manganese sulfate have been more successful than soil applications (Labanauskas 1966), but soil application, where effective, is easier and lasts longer (Dickey 1977).   Even so, Dickey had no success with soil applications of manganese sulfate to camphor trees, crape jasmine, and citrus on calcareous sandy soil.   The landscape plants Dickey examined probably received minimum cultivation and had roots near the surface.   Soil applications

---

Deficiency Symptoms: MANGANESE (Manganese is immobile in plants.)

Broadleaf: Leaves become yellow, with wide green bands along veins. Later, necrotic interveinal spots will appear and leaves may seem limp. *Shoot* growth may be reduced.   *Fruit* is often smaller than normal (Harris, Paul, and Leiser 1977; Labanauskas 1966).

Conifer: The symptoms are difficult to distinguish from those associated with iron deficiency; new growth is stunted and chlorotic; older needles and the lower crown remain green (after Labanauskas 1966; Powers 1979).

Treatment: MANGANESE DEFICIENCY (Manganese mobility in soil decreases with increasing pH.)

Application	Chemical Form	Amount to Apply	How to Apply	Years Effective
		$kg/100\ m^{2}$ [a]		
Soil	$MnSO_4 \cdot 2H_2O$	2–10	Incorporate	1–5
		$kg/100\ liters$		
Spray[b]	$MnSO_4 \cdot 2H_2O$	0.25–1.0[c]	Foliage	1

[a]One kilogram per 100 m^2 is about 2 lb/1000 ft^2.

[b]The preferred method is spray application to foliage.

[c]Also apply 0.6 kg (5 lb/100 gal) hydrated lime or soda ash.

of manganese sulfate will be more effective for these plants than they would be in cultivated orchards with few or no roots in surface soil. Except in acid or very sandy soils, little or no manganese sulfate applied to the surface reaches the deeper roots of large woody plants. Manganese sulfate can be placed in holes or injected into the soil, however. Although it is more expensive, manganese chelate may be more successful where soil applications of manganese sulfate have failed.

Foliar sprays have been almost universally successful in correcting manganese deficiency symptoms, but they must be applied nearly every year. Lemon foliage has been injured by manganese sulfate sprays, while orange and grapefruit foliage was not (Parker and Southwick 1941). Parker and Southwick found that soda ash (anhydrous sodium carbonate) is preferable to hydrated lime for mixing with manganese sulfate because the resulting spray leaves less visible residue on foliage and fruit. Sprays of manganese chelate are also effective. A combination of foliage and soil application might provide both a short- and a long-term solution. Once manganese deficiency has been corrected, additional applications should be made only when symptoms reappear.

Although trunk injection of manganese sulfate or chelate has been successful, it is time-consuming and causes wounds that are unsightly and can lead to decay. Acidification of alkaline or calcareous soil with sulfur will usually correct the deficiency, but the slow response and the cost make it unattractive.

Manganese toxicity frequently causes a mottled chlorosis of leaves, and necrotic spots develop as the deficiency worsens (Halfacre and Barden 1979). Manganese toxicity symptoms often resemble those associated with iron deficiency, and, in fact, the two minerals can be competitive. Of the fruit plants, apple, citrus, and pineapple are most seriously affected by manganese excess. Apples, particularly Red Delicious, develop corky measles.

Manganese toxicity can occur in strongly acid and poorly aerated soils (Labanauskas 1966). In conifers, toxicity may occur in high elevation fir sites with leached, exposed, sandy loam soil that is low in calcium and pH (Powers 1981b). Sprays or soil amendments that acidify soils high in manganese will chemically reduce manganic manganese to the manganous form (which is much more soluble), thereby increasing toxicity. The effects of manganese toxicity can be reduced if lime is applied and drainage improved. Because the effects of toxicity appear to be more severe in young trees, it is important that soil be limed to obtain a pH of 6.0 to 6.5 before new landscapes are planted (Boynton and Oberly 1966).

Molybdenum   Among the woody plants, molybdenum deficiency has been reported only in Chinese hibiscus and citrus in Florida (Dickey 1977), rose in Australia (Johnson 1966), and apple in New Zealand

(Boynton and Oberly). This deficiency is more common in vegetable and field crops.

Molybdenum deficiency has occured in soils derived from a variety of parent materials. Its incidence depends more on soil formation and leaching. In contrast to most of the other micronutrients, molybdenum is less available at lower pH values. Deficiencies can occur in heavily leached, well-drained calcareous and serpentine-derived soils, highly podsolized soils, and old soils with extensive secondary mineral formation (Johnson). Molybdenum deficiency commonly occurs in soils that are extremely low in phosphorus and sulfur. Because of its role in the biochemistry of nitrogen fixation, molybdenum may be deficient when

---

Deficiency Symptoms: MOLYBDENUM (Molybdenum is mobile in plants.)

Broadleaf: Leaves are similar in color to those deficient in nitrogen; they exhibit marginal scorching and rolling and reduced width (strapping). *Shoot* internodes are short when deficiency is severe. *Flowers* are few and small when deficiency is severe (Johnson 1966).

Conifer: No description is available.

Treatment: MOLYBDENUM DEFICIENCY (Molybdenum is mobile in soil.)

Application	Chemical Form	Amount to Apply $g/100\,m^{2}$ [a]	How to Apply	Years Effective
Soil pH 5.5[b]				
pH 5.5	$Na_2MoO_4 \cdot 2H_2O$ or $(NH_4)_2MoO_4 \cdot 2H_2O$	2–20[c]	On surface with other fertilizer[d]	3–7
		$g/100\,liters$		
Spray	Either of the above	10–100[c]	Foliage	1+

[a]One gram per 100 m^2 is about 0.03 oz/1000 ft^2.

[b]Apply lime to bring pH to 5.5 or above (see Table 6-4). If this does not work, apply molybdenum.

[c]Note that the amount specified is in grams; excess molybdenum may poison forage.

[d]Molybdenum is usually added to phosphate fertilizer.

nitrogen-fixing species, such as acacia, alder, and locust, show symptoms of nitrogen deficiency (Johnson).

Molybdenum deficiency in soils can be treated in one of three ways: soil application of sodium molybdate at low rates, usually in conjunction with superphosphate; foliage spray of sodium molybdate; or liming of acid soils to above pH 5.5 when total molybdenum is thought to be adequate but unavailable because of low pH (Johnson). Dickey (1966) suggests affected plants be sprayed to runoff with 8 g sodium molybdate per 100 liters (1 oz/100 gal) of water and the soil drenched with the excess solution, up to a total (on plant and soil) of 10 to 20 g of sodium molybdenum per 100 m^2 (0.5–1 oz/1000 ft^2). This treatment should be effective for at least one year and should be repeated when symptoms reappear.

Molybdenum excess or toxicity in plants rarely occurs in the landscape, but ruminant animals can be seriously affected by forage from soils that contain excess molybdenum (Johnson 1966). This danger must be kept in mind when fertilizers are applied or soils are limed.

Zinc   Zinc deficiency is fairly common among cultivated trees and large shrubs, and its effects on growth can be quite marked. However, when Viets (1960) and Schutte (1966) published their papers on this subject, South Africa was the only country whose native vegetation had responded to applications of zinc (Stone 1968). The level of zinc available in surface soil is often double that in the soil below (Chapman 1966b). ˉGrading that removes surface soil and other practices that reduce rooting in surface soil will aggravate zinc deficiency.

As with many other nutrients, zinc is more likely to be unavailable in the soil than low in total amount. As with several other micronutrients, zinc compounds decrease in solubility as alkalinity increases. It follows that zinc deficiency is more likely to occur in alkaline and calcareous soils like those of the western United States. Zinc availability may also be low in some organic soils, in clay soils with low silicon-to-magnesium ratios, and on the sites of old livestock corrals and Indian burial grounds (Chapman 1966b). Total zinc may be deficient in soils derived from granites and gneisses, and in acid, sandy soils that have been severely leached.

Zinc deficiency can be aggravated in high-phosphate soils, in those that have received heavy or prolonged applications of phosphate fertilizers, and in acid soils that have been limed. Zinc deficiency is more likely when topsoil is removed and when soil is cultivated frequently, an act that restricts rooting near the surface. Nitrogen fertilization has been said to increase zinc deficiency, but this may be due to an increase in pH or to cation effects rather than to nitrogen itself (Chapman 1966b). Sodium nitrate has been found to decrease zinc uptake, while ammonium nitrate and ammonium sulfate have increased it.

The symptoms of zinc deficiency were known long before the cause was determined in 1931. Disorders included: *rosette* (the formation of extremely short internodes) of pecan and walnut, *little-leaf* of peach and pear, *frenching* or *mottle* of citrus, and *bronzing* of tung. Dickey (1977) lists a number of Florida landscape plants commonly afflicted with zinc deficiency, including orange jessamine, loquat, wax-leaf privet, surinam cherry, carambola, barbados cherry, and silk oak. Fruit trees are particularly susceptible to zinc deficiency; vegetable and field crops are less often affected (Halfacre and Barden 1979).

Deficiency Symptoms: ZINC (Zinc is mobile in plants.)

Broadleaf: Leaves are uniformly yellow, sometimes mottled with necrotic spots. Leaves are small (*little-leaf*), very narrow, and pointed; older leaves drop. *Shoots* of small diameter have tufts (*rosettes*) of leaves at their tips, which may die back. *Fruit* set is light with small, pointed, highly colored fruit (Harris, Paul, and Leiser 1977).

Conifer: Branches and needles are extremely stunted; foliage yellows. Trees lose all but their first- or second-year needles; terminals die back (Powers 1979).

Treatment: ZINC DEFICIENCY (Zinc mobility in soil decreases as pH increases.)

Application	Chemical Form	Amount to Apply	Where to Apply	Years Effective
		$kg/100\ m^2$ [a]		
Soil	Chelate	1 when soil pH <6.0	On surface	1–3
		2 when soil pH >6.0	On surface	1–3
		$kg/100\ liter$		
Spray	Chelate	0.125–0.25[b]	Early foliage	1
	$ZnSO_4 \cdot 7H_2O$	1.25–6.0 during dormancy	Dormant twigs	1
		0.5–0.75 during growing season	Foliage[c]	1

[a]One kilogram per 100 m^2 is about 2 lb/1000 ft^2.
[b]Also apply 0.125 liter of detergent.
[c]Also apply 0.4 kg of hydrated lime.

Depending on the plant species, soil, and climate, zinc deficiency can be corrected by spray and soil applications, injections into tree-trunks, and by zinc-coated nails or pieces of galvanized iron driven into trunks and limbs of woody plants (Chapman 1966b).

Soil application is usually preferred because of its ease and its longer-lasting effects. Surface applications of zinc sulfate are primarily successful in soils that are low in total zinc and have a pH lower than 6.0. At higher pH values, zinc sulfate can be more effectively applied in holes or injected into soil. Zinc chelate has been successful in a number of soils in which plants did not respond to zinc sulfate, but the variable fixation and movement of zinc in soils and the danger of toxicity have discouraged soil applications in some areas. In alkaline soils or when soil applications are ineffective, zinc sprays will correct deficiency symptoms in most plants; sweet cherry and walnut are two reported exceptions (Proebsting 1958). For many deciduous species, an annual dormant spray of zinc sulfate will control deficiency symptoms in all but young and vigorous plants. A foliage spray of zinc sulfate and lime, zinc oxide, or zinc chelate can be used on evergreen plants or when deficiency symptoms appear during the growing season. Zinc oxide, however, can injure thin-skinned fruit.

Metallic zinc points, or pieces of galvanized iron driven into the trunks and branches parallel to the grain will correct deficiency symptoms in most species for several years. This is the most satisfactory method for sweet cherry and walnut, and perhaps for other plants that do not respond to soil or foliage applications (Proebsting). An area around each piece of metal will die, and if affected areas merge, the trunk or branch will be girdled. To prevent this, the metal pieces should be placed in a spiral at least 50 mm (2 in) apart with 1 to 2 pieces per 100 mm (4-6 in) of circumference. This placement is most effective when the trunk or branches are less than 250 mm (10 in) in diameter. Zinc solutions injected into trunks or dry salts placed in bored holes in trunks have also been successful. Deficiency symptoms have been corrected for three years or more when one gram of zinc sulfate per hole is placed in holes about 100 mm (0.04 oz/4 in) apart around the trunk. Because of the labor involved and the danger of decay, metal, salt, or solutions should be placed in trunks only if soil or spray applications are ineffective or will cause toxicity or drift problems.

Alfalfa and cover crops can reduce or prevent zinc deficiency in fruit orchards (Chapman 1966b), and a buildup of organic matter on the surface will improve the growth and appearance of plants deficient in zinc. Both of these practices tend to accumulate zinc in surface soil. If soil is fumigated or sterilized before planting, zinc availability will frequently be improved.

Initial symptoms of zinc toxicity resemble those of iron deficiency, and may in fact be a sign of iron deficiency, because the two nutrients compete with one another. Plants can be severely injured and killed by toxic levels of zinc in soil. This has occurred in some acid peat soils, soils contaminated by mine spoils or seepage, and soils derived from rock or materials high in zinc. Tree seedlings can be stunted and, in some cases, killed when they are grown in galvanized iron tubes (Chapman 1966b). Liming soil and applying phosphate fertilizer may help mitigate zinc toxicity.

Application Methods

Fertilizers can be applied in a variety of ways; they can be

Broadcast on the soil surface
Placed in holes in the soil
Injected into the soil in solution under pressure
Sprayed on foliage
Injected into or placed in holes in treetrunks

The appropriate method will depend on the nutrients applied, equipment available, other plants in the area, nature and slope of the soil surface, and, in some cases, the species to be treated.

Surface Application   Surface application is the easiest method and can be maximally effective for nitrogen and most chelated micronutrient fertilizers. Nutrients like phosphorus and potassium, which do not move readily into and through the soil, should not be applied in this manner. Lawn fertilizer spreaders or cyclone seeders can be used to apply nitrogen fertilizers (Fig. 11-5). For trees that are growing in

FIGURE 11-5
Lawn spreaders and cyclone seeders can be used to apply nitrogen
and chelated micronutrient fertilizers. (Harris, Paul, and Leiser 1977)

a lawn, fertilizer should be applied in winter before the turf begins rapid growth. Grass blades should be free of moisture. After spreading the fertilizer, sprinkle the area thoroughly to wash fertilizer from the grass and into the soil. A second irrigation the following day will move the material further into the root zone, will minimize injury from high concentrations of fertilizer, and will reduce volatilization of nitrogen from urea and ammonium fertilizers.

Soil Incorporation   Phosphorus, potassium, and other nutrients of low solubility can be effectively applied in lawn areas if they are injected or placed in holes dug in the soil. Placing such nutrients in the root zone will make them more readily available to plants. In addition, the holes will increase aeration and water penetration in most soils. In poorly drained silt and clay-loam soils in Ohio, Smith and Reisch (1975) found that young crab apple, linden, and maple trees benefited as much (they produced 20 percent more caliper growth) when holes were drilled 300 mm (1 ft) deep and no fertilizer was added as they did when 3 kg (6 lb) each of nitrogen, phosphorus, and potassium per 100 m^2 (1000 ft^2) were applied in holes or on the surface. Trees and large shrubs seldom need phosphorus and potassium fertilization. Therefore, to determine plant response to these nutrients, which are incorporated into soil, the soil around half of the control plants (those receiving no nutrients) should also be similarly disturbed.

Holes can be punched in soil with a bar (which may seriously compact the soil) or drilled with an auger by hand, with an electric drill, or with the power takeoff on a tractor (Fig. 11-6). To minimize compaction around each hole, the moisture content of the soil should be at or below field capacity before holes are made. Drill holes 250 to 300 mm (10–12 in) deep and 0.6 to 1 m (2–3 ft) apart, beginning away

FIGURE 11-6
Holes for fertilizers whose nutrients
do not move readily in the soil
can be punched, cored, or drilled,
either by hand, as shown, or mechanically.
(Harris, Paul, and Leiser 1977)

from the trunk to avoid injury to main roots and extending up to one-fourth of the radius beyond the drip line of the tree. The number of holes will range between 100 and 275 per 100 m² (110–250/1000 ft²) with such spacing. Apportion the fertilizer among the holes.

Another method of incorporating fertilizer is to dig a circular trench 250 to 300 mm (10–12 in) deep at the drip line of the tree. Place the fertilizer in the trench and cover it with soil.

Fertilizer can be effectively injected into the soil if proper equipment is available (Fig. 11-7). Fertilizer materials must be soluble in water so they can be applied uniformly, even though the nutrients may later be adsorbed or reduced in solubility by the soil. Water-soluble fertilizers containing phosphorus, potassium, and other nutrients are usually more expensive per unit of nutrient than are fertilizers not readily soluble in water. Ammonium phosphate, potassium phosphate, and potassium nitrate are water-soluble sources of nitrogen, potassium, and phosphorus. Potassium chloride is water-soluble, but should not be used where salinity may be a problem.

A high-pressure hydraulic sprayer is probably the most economical piece of equipment to use for injecting fertilizer, because it can be used for other maintenance practices. Fertilizer solutions are corrosive, however, and the equipment must be cleaned thoroughly after being used for fertilizer application. Stainless steel or plastic-lined tanks reduce maintenance, but the pump and lines must be cleaned carefully to prevent corrosion.

Because the fertilizer is in solution, the injections can be spaced somewhat farther apart than can holes for dry fertilizer. The probe

FIGURE 11-7
A needle can be used to inject nutrient solutions into the soil when the elements involved are adsorbed by and do not move readily in the soil.
(Harris, Paul, and Leiser 1977)

should penetrate the soil about 300 mm (1 ft), applying 100 to 150 kg/cm^2 (150–200 psi) of pressure to force the solution into the soil. Inject at least 800 liters (200 gal) of solution for each 100 m^2 (1000 ft^2) treated (5 liters or 1.2 gal/injection). The concentration of the solution will depend on the nutrients and the desired rate of application.

Foliage Sprays   Sprays can quickly overcome deficiency symptoms involving micronutrients like iron, zinc, and manganese. Small-scale spray applications of a single micronutrient will quickly determine a plant's response; this can be a very useful preliminary to a longer-term corrective program (Fig. 11-8). A detergent spreader (a dishwashing detergent, for example) added to the water at a rate of 60 to 120 ml/100 liters (0.5–1 pint/100 gal), will increase the effectiveness of sprays.

Spraying to alleviate deficiency symptoms can be expensive unless it is included in a regular pesticide program. A number of micronutrients have been incorporated into the pesticide spray programs of many horticultural crops, but few landscape plants are sprayed so routinely. Spraying any material on large shrubs and trees is difficult in most landscapes and may involve injury to nontargeted plants, to other property, and to people. Spray may leave unattractive whitish film or spots on foliage. If fertilizers and pesticides are applied together, they must be compatible.

FIGURE 11-8
Applying foliar sprays of an individual nutrient may be a quick way to determine plant response. (Harris, Paul, and Leiser 1977)

Trunk Implants and Injections   Trunk implants and injections should be used only when other methods prove to be ineffective or too difficult. Boring the necessary holes, implanting or injecting fertilizer, and sealing holes takes a great deal of time, disfigures trunks, and can lead to decay around insertion sites. On the other hand, these methods may be the only effective way for plants to obtain adequate nutrition.

Fertilizer implants and injections are almost entirely restricted to micronutrients whose availability is reduced by soil conditions. Implants contain small amounts of a soluble nutrient in a gelatin capsule, and are placed in holes drilled into trunks or branches. Holes should usually be 10 to 13 mm ($\frac{3}{8}$–$\frac{1}{2}$ in) in diameter and deep enough to allow the implant and a plug to be placed entirely in the xylem. A wood, cork, plastic, or asphalt plug inserted deep enough will lessen the chance of injury to the cambium and favor rapid callus coverage of the wound. Some commercially available capsules, such as Medicaps®, seal the hole when inserted properly, so no plug is needed. Follow the manufacturer's directions for the number of implants. If several implants are to be placed in a tree, they should spiral around the trunk to minimize injury.

Trunk injection holes can usually be smaller than those necessary for implants. A hollow, tapered 6-mm ($\frac{1}{4}$-in) wood screw with a quick coupler attached can be used effectively for injecting orchard and landscape trees with pesticides and nutrients (Reil 1979). Nutrient solutions can be injected at pressures of 7 to 14 kg/cm^2 (100–200 psi). Some species may gum severely after injections at more than 7 kg/cm^2, and pressures higher than 14 kg/cm^2 will not appreciably decrease injection time. You may use a mechanical high-pressure sprayer: Reil and Beutel (1976) report on portable pressure-injection equipment that uses bottled nitrogen gas or compressed air to force the solution into the plant, usually at three injection sites (Fig. 11-9). A convenient feature of this equipment is a two-way hydraulic cylinder with 1 liter (1 quart) capacity, which measures the amount of nutrient solution for each injection. One liter of solution can be injected in 30 seconds to 5 minutes, depending on the tree and the time of year. Reil and Beutel have injected many fruit and landscape trees successfully, including broadleaved trees, conifers, and palms. Some maple, fir, and pine trees, however, accept injection very slowly.

With regard to implanting iron in trees, Neely (1976) observes that although injection and implantation have consistently provided the most prompt and thorough correction of iron chlorosis, some authorities are reluctant to recommend the method. Objections center on the numerous holes that must be routinely drilled (every 3 to 4 years) into the trunk, sap leakage from holes, toxicity to cambium where salts are inserted, and toxicity to leaves when too much nutrient

FIGURE 11-9
A portable pressure-injection machine operated by compressed air or nitrogen.
Cylinder, valves, and pressure gauge of injection machine; the coil spring recharges
the system automatically with solution to be injected after air pressure is released
(left). Worker connects quick coupler on distribution line to an injection screw
(right). (Reil and Beutel 1976)

is applied at the wrong time to sensitive plants. In his experiments on
pin oak, however, Neely observed no injury to treated trees regardless
of the time of treatment, source of iron, or amount implanted. Both
Simon (1976) and Neely report that implantation holes were callused
closed within one year.

Working with red maple in West Virginia, Shigo, Money, and Dodds
(1977) found negligible amounts of discolored wood and cambial dieback (8 mm or 0.3 in) above and below injection wounds made one
year before; no nutrients had been injected into these holes. Similar
holes that had received Mauget injections of iron, magnesium, manganese, or zinc were more seriously affected. The cambium died back
20 mm ($\frac{3}{4}$ in) above and below the holes injected with magnesium,
while 50 to 60 mm (2-2.3 in) of the cambium was killed around holes
injected with any one of the other nutrients. New wood formed after
treatment was healthy around all wounds. One year after red maple
trees were injected with Bidrin, the dead cambium extended 350 mm
(14 in) above and below wounds; two years after injection, it extended
only 110 mm (4.3 in). The wounds apparently close fairly rapidly, but
there is always the possibility of decay, especially in trees of low vigor.
The researchers did not try implants or other fertilizer methods.

Time of Application

The timing of fertilization should depend on the nutrient, the
formulation, the method of application, soil texture and drainage,
climate, and nutrient levels in the plant. Nutrients should usually be
available to the plant during the period of rapid growth in the spring.

Deciduous oak and many conifers, whose growth is limited by the number of preformed initials in their buds, may show only improved leaf color the first year and no increase in growth until the following year. For macronutrients other than nitrogen, the time of application is not an important factor for perennial plants.

In most situations, nitrogen should be applied annually during the late dormant season so that it is available for spring growth. Heavy amounts of nitrogen applied in the dormant season, or moderate amounts applied during the growing season, can extend late-season growth and make plants liable to injury by fall and winter cold. Low nitrogen levels in the plant in late summer and fall favor development of color in the leaves and fruit of many species. On sandy soils with heavy winter rainfall, nitrogen should be applied two to three weeks before growth begins to minimize leaching.

Where sandy soils receive heavy summer rains or where irrigation must be heavy to control salt buildup, plants may respond best if nutrients are applied more than once a year. This is particularly true for young trees to encourage rapid growth. In these situations, "slow-release" fertilizers may most efficiently reduce cost of application while maintaining plant performance.

Time chelate applications of iron, zinc, and manganese, which are soluble in the soil, as you time applications of nitrogen. Since these nutrients can be applied as foliage sprays, however, spray application of them should begin in early spring, after the first few leaves attain full size. Additional sprayings may be needed during the growing season if deficiency symptoms appear.

Fertilizing Young Plants

Recommendations for fertilizing newly planted trees and shrubs vary considerably: Some experts suggest that a complete fertilizer be mixed in backfill soil; others advise no fertilizer during the first growing season. Landscape trees did not respond to differences in soil nitrogen levels during the first three years after planting (van de Werken 1981). I am aware of no experimental evidence showing that nitrogen is more effective if applied in the planting hole than in equal amounts on the surface. Nitrogen is the only nutrient needed in most situations and can be applied on the surface after planting more easily and with less likelihood of toxicity to the plants.

The slow-release fertilizers that have become available in recent years can supply low concentrations of nutrients continuously for two to 18 months. These fertilizers are particularly useful for plantings that are inconvenient or expensive to fertilize regularly, or for those in sandy soils receiving heavy rainfall. Slow-release fertilizers should be placed deep in the planting hole and backfilled with 100 to 150 mm

(4–6 in) of soil before planting, to minimize fertilizer absorption by weeds and turf.

The quantity recommendations for nitrogen application also vary considerably: from 10 to 50 g (0.02–0.10 lb) of nitrogen in the planting hole or backfill soil, and from 10 to 75 g (0.02–0.15 lb) of nitrogen per surface application in the planting basin. In Florida, Meskimen (1970) made monthly surface applications of 8 g (0.18 lb) of nitrogen per square meter (as 6-6-6) to newly planted red gum in a fine sandy soil. In California, Harris (1966) made two applications of 125 g (0.25 lb) of nitrogen per square meter (as NH_4NO_3) to newly planted southern magnolia 'Saint Marys' and sawleaf zelkova in a silt loam soil. Both field experiments produced significant growth responses and no injury to trees.

TABLE 11-2

Analysis and relative cost of some nitrogen-containing fertilizers for woody plants

	Analysis on a percentage basis			Cost $ /	Weight kg	Cost per kg of N
Fertilizer	N	P_2O_5	K_2O			
Calcium nitrate	15.5			7.70 / 36	(80)[a]	1.36 (0.62)[a]
Ammonium sulfate	21			8.25 / 36	(80)	1.08 (0.49)
Ammonium nitrate	33.3			11.60 / 36	(80)	0.97 (0.44)
Urea	46			14.00 / 36	(80)	0.84 (0.38)
Ammonium phosphate	16	20		12.10 / 36	(80)	2.07 (0.94)
Complete—local mix	6	20	20	15.25 / 36	(80)	7.00 (3.18)
Complete—local mix	12	12	12	12.00 / 36	(80)	2.75 (1.25)
Complete—local mix	16	16	16	14.00 / 36	(80)	2.40 (1.09)
Complete—brand[b]	16	16	16	17.00 / 36	(80)	2.93 (1.33)
Slow-release fertilizers						
Urea formaldehyde—brand	38			28.50 / 22.5	(50)	3.32 (1.50)
Slow-release—brand	19	6	12	44.00 / 22.5	(50)	10.17 (4.63)
Slow-release—brand	7	40	6	55.00 / 22.5	(50)	34.55 (15.71)
Special fertilizers						
Spikes—trees and shrubs	16	8	8	11.29 / 3	(7)	22.18 (10.08)
Spikes—evergreens	12	6	8	11.29 / 3	(7)	29.57 (13.44)
Water soluble[c]	23	19	17	9.95 / 2.25	(5)	19.03 (8.65)
Water soluble[c]	20	20	20	2.69 / 0.45	(1)	29.59 (13.45)

[a]The weight of a bag of fertilizer in lbs and the cost per lb of nitrogen are given in parentheses. These are 1981 retail costs from a farm supply store. Commercial and public accounts would get about a 20 percent discount; buying by the ton would save about 10 percent.

[b]National brand of fertilizer.

[c]These fertilizers should be diluted for application to the soil.

Nitrogen (25–100 g or 0.05–0.20 lb) can be applied in the form of ammonium nitrate or ammonium sulfate to an area one meter square (3 ft by 3 ft) around each tree after planting and again six weeks after growth begins. Slow-release fertilizers should be placed in the planting hole according to the manufacturer's directions.

Estimating Fertilizer Costs

The cost of fertilizer nutrients is important not only as you budget for landscape maintenance but as you determine which fertilizer to buy. Fertilizers are available in a number of formulations and mixtures; they vary in the amount of nutrient as a percentage of weight (Table 11-2) and can vary greatly in price. Fertilizer is usually applied by area or individual plant. Be careful to consider what amounts of a nutrient are to be applied and how large an area or how many plants this amount will cover. The following equation may be helpful:

$$\frac{\text{Area to be}}{\text{fertilized}} \times \frac{\text{Rate of}}{\text{application}} \times \frac{100}{\text{Analysis of fertilizer}} \times \frac{\text{Price of}}{\text{fertilizer}} = \text{Cost}$$

$$\text{Area} \times \frac{\text{Weight}}{\text{Area}} \times \frac{100}{\text{Nutrient analysis on a percentage basis}} \times \frac{\text{Price}}{\text{Fertilizer weight}} = \text{Cost}$$

Example: What is the cost of applying 2 kg of nitrogen per 100 m^2 to a grove of trees in an area 300 by 400 m? Ammonium nitrate ($33\frac{1}{3}\%$ nitrogen for ease of calculation) costs $9.00 per 40-kg bag.

$$300 \text{ m} \times 400 \text{ m} \times \frac{2 \text{ kg N}}{100 \text{ m}^2} \times \frac{100\% \text{ NH}_4\text{NO}_3}{33.3\% \text{ N}} \times \frac{\$9.00}{40 \text{ kg NH}_4\text{NO}_3} = \text{Cost}$$

$$120{,}000 \times \frac{2}{100} \times \frac{100}{33.3} \times \frac{\$9.00}{40} = \$1620.00$$

Note that all of the units except the price cancel out to give the cost.

Pollution Caused by Fertilizers

As concentrations of nitrogen and phosphorus increase, plant life, particularly algae, will increase in water. Lakes with high concentrations of these nutrients may bloom with algae in summer, resulting in low oxygen levels that may be fatal to fish. Swimming and boating under such conditions will be unpleasant.

Nitrogen will leave soil and enter bodies of water primarily through leaching. Phosphorus loss is almost entirely due to erosion, though phosphorus can also be leached from sandy, coarse-textured soils.

The following measures, designed to minimize pollution, will also increase the efficiency of fertilization programs (Gowans 1970):

Do not continually over-irrigate. Large amounts of water seeping below the root zone leach out nitrates.

Take care to apply only the amount of nitrogen fertilizer needed; apply at optimum times and in such a manner that the fertilizer is most readily available to roots.

Keep erosion to a minimum, particularly near streams and lakes.

Avoid using phosphate to fertilize trees or large shrubs on sites that are susceptible to excessive leaching, water runoff, or erosion. If plants require phosphate for satisfactory growth, carefully control the amount used and the method of application to minimize transport of phosphate from the site.

Use plants that are suited to conditions in the area and will require only minimal fertilization.

FURTHER READING

CHAPMAN, H. D. 1966. *Diagnostic Criteria for Plants and Soils*. ed. H. D. Chapman. Riverside, Calif.: Chapman.

HAUSENBUILLER, R. L. 1978. *Soil Science: Principles and Practices*. 2nd ed. New York: John Wiley.

NEELY, D., E. B. HIMELICK, and W. R. CROWLEY, JR. 1970. Fertilization of Established Trees: A Report of Field Studies. *Ill. Natural History Survey Bull.* 30(4):235-66.

TISDALE, S. L., and W. L. NELSON. 1975. *Soil Fertility and Fertilizers*. 3rd ed. New York: Macmillan.

TENNESSEE VALLEY AUTHORITY. 1968. *Forest Fertilization: Theory and Practice*. Symposium on Forest Fertilization. Muscle Shoales, Ala.: Tenn. Valley Authority.

CALIFORNIA FERTILIZER ASSOCIATION. 1980. *Western Fertilizer Handbook*. 6th ed. Danville, Ill.: The Interstate.

CHAPTER TWELVE

Irrigation

Landscape plants probably suffer more from moisture-related problems than from any other cause. For them, it is either feast or famine, flood or drought, air or suffocation, acceptable or saline water. Life as we know it is not possible without water. Water is a primary constituent in the photosynthetic production of organic matter. It is the solvent for nutrient and food transport within plants. Transpiration cools plants. Roots extend into soil and shoot tips grow only by absorption of water and the turgor it provides. But although water is vital to plants, excessive water is often responsible for their decline and death.

As a source of water for plants, irrigation is more reliable than rain, but irrigation has drawbacks with which horticulturists must reckon. Irrigation without adequate drainage can create aeration and salinity problems that are almost as serious as drought. It is essential to understand the relations among soil, plant, and water in order to grow healthy, attractive landscape plants. The retention and movement of moisture in soil are discussed in Chapter 6. Some methods of improving soil-moisture conditions in the landscape are presented in the last portion of Chapter 7. This chapter will discuss factors that affect water use by plants: the timing and quantity of irrigation, soil salinity, drainage, and antitranspirants.

WATER USE BY PLANTS

When you are managing irrigation, you may want to determine the rates at which water is lost by transpiration from plants and evaporation from the soil (these losses are called *evapotranspiration* or ET).

ET is usually expressed as water loss (millimeters or inches) per time unit; the area need be considered only when the total volume of water must be known. The term ET usually refers to potential ET, that which occurs when there is adequate water in the root zone. Actual ET will be less than potential ET when soil moisture is inadequate. When characterized for an area without reference to a specific crop or planting, ET usually is calculated for a perennial planting (such as turfgrass or pasture grass) that covers all or almost all of the soil surface. Because of differences in climate, ET varies according to place, season, and weather.

You can estimate potential ET by using a lysimeter, a device that measures the water loss of plants growing in a container when total weight (of container, soil, plants, and water) can be determined periodically. Approximations of ET can also be calculated according to a plant's incoming, outgoing, and net radiation. It can also be estimated by determining the evaporation rate from an open water surface under prescribed conditions. In a number of irrigated agricultural regions, weather reports include ET measurements to assist growers in regulating the timing and amount of irrigation more accurately. The use of these measurements will be described shortly.

Other environmental factors that influence ET are sunlight, temperature, humidity, and wind. The combination of factors that creates the lowest vapor pressure around leaves and the highest vapor pressure within leaves will result in the highest ET. Intense solar radiation invariably produces higher temperatures, particularly in exposed leaves. High air temperature is often accompanied by low humidity, which tends to increase transpiration. Transpiration is usually higher on windy days, but it does not increase in direct proportion to wind velocity. The wilted, dull appearance of some leaves after strong winds is probably due to a combination of low humidity, high temperature, intense sunlight, and mechanical injury. These factors determine the rate at which plants will use water under different conditions and in different seasons. Of course, ET varies considerably from seacoast to desert (Fig. 12-1), but it can also vary greatly within a single landscape.

As mentioned, the calculation of ET is usually based on the ET of turfgrass or pasture grass. Larger plants, usually depending on their shape and exposure to the sun, will use more water than does grass. Solid or nearly solid plantings of shrubs or trees often use 10 or 20 percent more water than would turf. Because of its greater exposure to the sun and wind, a large, solitary shrub or tree may use two to three times as much water as a comparable area of turf. It should be remembered that the reported ET applies when soil moisture is not a limiting factor. All plants can get by with less than the potential ET; many woody plants can survive on half that rate.

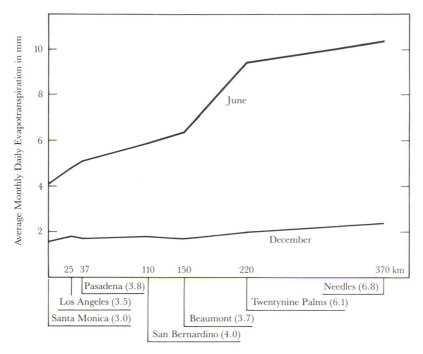

FIGURE 12-1
Evapotranspiration (grass surface) from the Pacific Coast (Santa Monica)
to the California-Arizona border (Needles) for June and December.
The numbers in parentheses indicate the average annual daily ET in mm per day.
(Data courtesy W. O. Pruitt, University of California, Davis)

Moisture use is strongly influenced not only by climate but also by soil and the plants themselves. Plants vary in size, shape, and the extent of their root systems, depending on species, plant density, and soil texture and depth. Some plants have a larger soil-moisture reservoir than do others. Leaf and branching characteristics can govern the rate of transpiration. Different surroundings will subject plants to different radiation loads. A large shrub surrounded by concrete on the sunny side of a tall building will require more water than a similar plant surrounded by turf on the shady side.

ET varies greatly during the year for evergreens and turf, but it varies even more for deciduous plants (Fig. 12-2). In addition, it can be difficult to estimate ET for deciduous plants early in the season. Deciduous plants should be near their maximum ET as soon as the shadow cast by their foliage is complete or as dense as it will get. For a plant nearing maturity, full foliage density should occur within two months after the beginning of spring growth. In the summer, deciduous

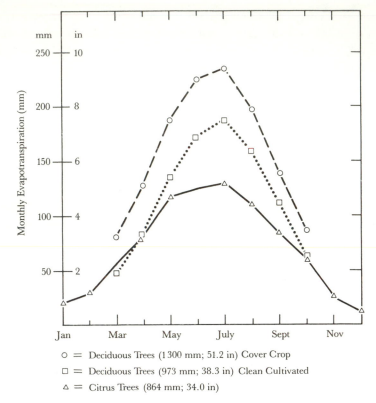

○ = Deciduous Trees (1300 mm; 51.2 in) Cover Crop
□ = Deciduous Trees (973 mm; 38.3 in) Clean Cultivated
△ = Citrus Trees (864 mm; 34.0 in)

FIGURE 12-2
Monthly evapotranspiration for deciduous fruit trees,
with and without cover crop, and for citrus in the San Joaquin Valley, California.
Numbers in parentheses indicate the annual ET.
(Source: Interagency Agr. Info. Task Force, Calif. Dept. Water Resources 1977)

plants may use 40 percent more water per day than broadleaved evergreen plants of similar size. Annually, however, deciduous and evergreen plants may use virtually the same quantities of water (Fig. 12-2).

DETERMINING WHEN
AND HOW MUCH TO IRRIGATE

Variations in the weather and length of day, adequacy of previous
irrigation or rainfall, depth and spread of rooting, and size of top affect
the moisture requirements of plants. If you know when and how much
to irrigate, you will be able to balance plant performance and water use.

Observe the Plants

Most plants wilt noticeably when too little water is available. Leaves that were once shiny become dull, and bright-green leaves fade or turn gray-green. Leaves fall early and young leaves sometimes die. Plants must be watched carefully as the time nears for the first irrigation of the season, when the amount of moisture in the soil and the rate of use are difficult to assess. A few plants wilt before others because of their species or location and can be used as indicators.

Feel the Soil

With experience, you can roughly estimate moisture adequacy by the feel of the soil. The soil sample must be representative, however. Collecting a sample with a soil probe (see Fig. 11-1) can be difficult and time-consuming, particularly when the roots of large shrubs and trees range far and deep. For shallow-rooted plantings, soil samples can be gathered more easily. To estimate moisture adequacy, roll or squeeze a small sample of soil into a ball. If the soil will not mold into a ball, it is too dry to supply adequate water to plants. If the ball formed will not crumble when rubbed, the soil is too wet. If the soil can be molded into a ball that will crumble when rubbed, the moisture is probably about right. Sandy soils, however, will crumble even when wet. More details on this method of assessing soil moisture are given in Table 12-1.

Use Tensiometers

In both production agriculture and landscape maintenance, tensiometers are being used more widely to determine irrigation needs, particularly with turf. A tensiometer is a closed tube filled with water. A hollow ceramic tip is sealed to the bottom, and a gauge or device for measuring vacuum is attached to the other end (Fig. 12-3). The tube is placed in the soil so that the ceramic tip is located at the spot to be measured. As the soil dries, water is sucked out through the porous wall of the ceramic tip, creating a partial vacuum inside the tensiometer. This power of the soil (soil suction) to withdraw water from the tensiometer increases as the soil dries and the gauge reading rises. When soil is irrigated, soil suction is reduced and water is drawn back into the tensiometer, lowering the gauge reading.

Most tensiometer gauges are calibrated with graduations from 0 to 100 centibars (100 cb = 1 atmosphere). Within a range of 0 to 80 cb, a tensiometer is more accurate than any other practical means of evaluating the relative wetness of a soil (Marsh 1975). Even though the permanent wilting point (PWP) has a tension of 1500 cb, for plants

TABLE 12-1

Practical chart for determining soil moisture (Harris and Coppock 1977)

Sand (gritty when moist, almost like beach sand)	Sandy loam (gritty when moist; dirties fingers; contains some silt and clay)	Clay loam (sticky and plastic when moist)	Clay (very sticky when moist; behaves like modeling clay)	Amount of Moisture Readily Available to Plants
		Feel or appearance of soils		
Dry, loose, single-grained; flows through fingers.	Dry, loose; flows through fingers.	Dry clods that break down into powdery substance.	Hard, baked, cracked surface. Hard clods difficult to break; sometimes loose crumbs on surface.	Close to 0%. Little or no moisture available.
Still appears to be dry; will not form a ball with pressure.	Still appears to be dry; will not form a ball.	Somewhat crumbly; will hold together with pressure.	Somewhat pliable; will ball under pressure.	50% or less. Approaching time to irrigate.
Same as sand above.	Tends to ball under pressure but will seldom hold together.	Forms a ball; somewhat plastic; will sometimes slick slightly with pressure.	Forms a ball; will ribbon out between thumb and forefinger.	50% to 75%. Enough available moisture.
Tends to stick together slightly; sometimes forms a very weak ball under pressure.	Forms weak ball; breaks easily; will not become slick.	Forms a ball and is very pliable; becomes slick readily if high in clay.	Easily ribbons out between fingers; feels slick.	75% to field capacity. Plenty of available moisture.
Upon squeezing, no free water appears but moisture is left on hand.	Same as sand.	Same as sand.	Same as sand.	At field capacity. Soil won't hold any more water (after draining).
Free water appears when soil is bounced in hand.	Free water is released with kneading.	Can squeeze out free water.	Puddles and free water forms on surface.	Above field capacity. Unless water drains out soil will be waterlogged.

FIGURE 12-3
Two tensiometers, one (dial visible)
installed in the soil
with its ceramic tip at the depth
soil moisture is to be measured, can be used
to determine when to irrigate.

with roots 750 mm (30 in) or more deep, up to 90 percent of the available water is held between 30 (field capacity, FC) and 80 cb (see Fig. 6-8). Shallow-rooted plants and turf may begin to experience water deficiency as tension increases above 25 cb. A reading of zero indicates a saturated soil; between zero and 5 cb a soil is too wet for most plants. Readings between 10 and 25 cb represent ideal water and aeration conditions (Marsh). With observation and some experimentation, you should be able to schedule irrigations according to tensiometer readings.

Areas with varying soil, plants, and water use should be measured with tensiometers at two depths; these will also give a good indication of the moisture level between the two depths. Two tensiometers can be useful in determining when to stop applying water. Insert tensiometers so that the porous tip is in the active root zone, in good contact with the soil, and in a place where irrigation water is sure to wet the soil. Do not allow a plant or other object to block water from a sprinkler. When drip irrigation is used, tensiometers should be placed 300 to 450 mm (12–18 in) from a water emitter. For trees and shrub beds in soil at least one meter (3 ft) deep, place one porous tip 300 mm (12 in) deep and a second 750 mm (30 in) deep. Tensiometers should be unobstrusive and protected from damage and vandalism. Gauges may be placed below the soil surface and protected by a box or cover.

Farmers and landscape managers have reported considerable savings in time and water and considerably improved plant growth when tensiometers are used to determine when and how long to irrigate. Tension measurements can be especially helpful in spring, when variations in weather and plant growth can affect water use markedly, and in summer, when fog, rain, or clouds can make irrigation need uncertain. In one study in southern California, the irrigation interval for turf, as determined by tensiometer, varied from two to six days within a two-week period in the spring (Morgan 1965). Some tensiometers have electric contacts so that they can be used with timed controls to almost completely automate irrigation. Needless to say, these tensiometers must be monitored closely to ensure that they continue to work properly.

Use Electrical Resistance Meters

Soil moisture can be measured by a bridge or meter that determines the electrical resistance between electrodes embedded in a small block of gypsum, fiberglass, or nylon. The block is buried in the soil where moisture readings are wanted. Insulated wires connect the block to the surface, where they can be attached to a meter. Resistance blocks are most accurate at soil moisture tensions between 100 and 1500 cb. None of the resistance units, however, is as sensitive as the tensiometer when tensions are less than 100 cb (Haise 1955). Most available water will be absorbed by plants before the tension increases to 100 cb. Hand-held electrical resistance probes may help monitor moisture in shallow containers, but a buildup of salinity will cause the probes to indicate more water than is actually available.

Estimate Evapotranspiration (ET)

Irrigation can be better planned if the available water-holding capacity (in millimeters or inches) of the root zone to be irrigated is known and if there is access to ET reports. The amount of water available to a plant depends on soil texture (Table 12-2) and depth of roots (see Chapter 6). In shallow soils, depth of roots may be determined by soil depth; in deeper soils, root depth can be roughly estimated as follows:

Trees and large shrubs	1.2 m (48 in)
Medium shrubs and vines	1.0 m (40 in)
Small shrubs and ground covers	0.6 m (24 in)

In many farming regions, ET data are included in the weather reports. These reported figures may have to be adjusted somewhat to allow for differences in mesoclimates between the reporting station and the planting area. The ET level in any area will usually be consistently higher or lower than that at the reporting station. Whoever reports the ET should be able to estimate the ET correction needed for a particular nearby area.

The first irrigation in spring or early summer is the most difficult to schedule because of the relatively extreme weather (temperature and rainfall) variation and the increasing leaf cover. According to one approach, daily adjusted ET levels should be collected when spring rains are nearly over and deciduous plants are fairly well leafed out. As the accumulated estimated ET levels approach the amount of avail-

TABLE 12-2

Estimated irrigation requirements for a planting with roots occupying a 12×12 m (40×40 ft) area to a depth of 1.2 m (4 ft) (adapted from Booher and Houston 1958)

Soil	Amount of Available Water		Duration of Available Water $\left(\dfrac{available\ water}{ET}\right)$[a]	Suggested Interval between Irrigations $\left(\dfrac{75-80\%\ available\ water}{ET}\right)$	Amount of Water to Apply (75–80% available water)[b]		Time Required to Replenish Water[c]
	mm	(in)	Days	Days	mm	(in)	Hours
Sandy loam (Hanford)	70	(2.8)	11	9	60	(2.25)	5.3
Silt loam (Yolo)	150	(6.0)	24	20	130	(5.0)	16.9
Clay loam (Auburn)	240	(9.6)	38	32	200	(8.0)	36.2

[a]ET = 6 mm (0.25 in) per day.

[b]Amount required to replace water lost by ET. The leaching requirement, if applicable, is not included.

[c]Calculated at a typical infiltration rate for each soil with cover.

able water in the soil, plants should be watched for wilting. Begin irrigation when the first plants begin to wilt, adding only enough water to equal the accumulated ET level or the estimated amount of available water in the root zone, whichever is less. The accumulated ET will probably be greater than the estimated amount of available water, because plants usually root and obtain water from beyond the edge of the irrigated area. Rooting beyond the irrigated area is less of a factor if the area is large. The accumulated ET can also be greater than the available water in the root zone, because either level may be estimated inaccurately. Whatever the reasons for discrepency between ET and available water levels, adding more than the lesser amount will probably cause the soil to become waterlogged or will allow excess water to move below the root zone.

The depth to which water has penetrated can be checked by pushing a 3-mm- ($\frac{1}{8}$-in-) diameter spring-steel rod (with one end sharpened) into the soil. This probe can be pushed fairly easily through wet soil but only with difficulty through dry soil. If water movement is not yet complete, the wet soil may be above field capacity. Probing the soil in representative locations two to three days after irrigation will provide a rough check on the estimated available water capacity of the root zone. If soil is not wetted to the full depth of the soil sample (expected rooting depth), you can estimate how much water would have been needed to do so. Even with a well-engineered and carefully operated irrigation program, it is necessary to add an extra 10 to 15 percent more water in order to wet most of the soil. Uneven water application and variations in soil surface and texture make it almost impossible to wet a soil evenly without wasting water and overwatering some areas. This must be taken into account when you probe or take samples to determine the depth that irrigation water has reached.

For succeeding irrigations, the same essential procedure should be followed. Collect daily estimated ET levels since the date of the last irrigation. As the ET level approaches the amount of water applied in the previous irrigation, you may irrigate again or wait until the first sign of wilting appears, adding either enough water to equal the accumulated ET level or an amount based on the first irrigation plus any additional water that would have been needed to wet the entire root depth; use the lesser of these two amounts. The accumulated ET should be much closer to the amount of water applied at the second irrigation because the roots will have found much less or no water beyond the irrigated area. Check the depth of water penetration after this irrigation also.

If the soil irrigated is fairly uniform over the planting area and plants are similar, future irrigation can be scheduled according to the accumulated ET levels between irrigations. If soil varies within the planting or plants vary (particularly in their root depth), you may have to irrigate while much of the soil still has available water. You can estimate the magnitude of variation by spot-watering the plants that wilt first and keeping track of daily ET levels until most of the plants have begun to wilt. If the area involved is large and the differences in irrigation intervals are significant, you have to change the planting or the irrigation method.

In the second year, accumulated ET levels can be a fairly reliable guide to irrigation. Either the ET level or the amount of available water in the root zone will indicate how much water should be applied. Irrigations can be made at shorter intervals (at lower accumulated ET levels), so that none of the plantings will wilt and the irrigation will not interfere with other field operations or activities.

These procedures are not as complicated as they may seem at first, and even though they are not as precise as many would like, they can save considerable water and improve plant growth. Experiences during the California droughts of 1975-1976 and 1976-1977, and other observations, have indicated that many landscape plantings are over-irrigated to the detriment of the plants and maintenance budgets.

Measure Foliage Temperatures

Water stress in a plant reduces transpiration (evaporative cooling), which in turn raises the temperature of leaves exposed to the sun. This suggests that leaf temperatures could be measured to determine when healthy plants should be irrigated. Such a determination would be useful for mature plants in areas where water conservation is important. Exposed leaves of plants with sufficient soil moisture will have temperatures that equal or are slightly below air temperature. As soil moisture becomes limited, however, transpiration is reduced and leaves exposed to the sun become warmer. Landscape plants in dry soil will have leaf temperatures 5 to 15°C (10–25°F) above ambient temperature, often as high as 40°C (104°F). For most landscape plants, little or no permanent leaf injury will occur at these elevated temperatures (Sachs, Kretchun, and Mock 1975).

Various infrared sensing instruments (Fig. 12-4) are able to mea-

FIGURE 12-4
By indicating temperature differences between exposed leaves and those not exposed to the sun, infrared sensing instruments can be used to evaluate a plant's need for water.

sure differences in the leaf temperatures of exposed and unexposed leaves at some distance from a plant if they are pointed at the surfaces in question. This method of measuring a plant's moisture level is in a developmental stage but potentially offers a substantial new tool for managing plant water regimes. It can also be used to detect plant stress from other causes when soil moisture is adequate.

ESTIMATING
HOW MUCH WATER
IS APPLIED

The amounts of water in the soil that evapotranspire or are applied as rainfall or irrigation are most easily expressed as the equivalent to the depth (in millimeters or inches) of an equal amount of water on the surface of the ground. That measure is useful even when the area involved is not known. The depth of rainfall and sprinkler application is measured easily. The amount applied by other means must be measured by volume (in liters or gallons), though some irrigation distribution systems have a meter to record the amount of water applied. You can estimate the rate of water application by measuring how long it takes to fill a large container of known volume (about 50 liters or 10 gal) when the rate of application is similar to the irrigation rate. For a drip system, catch the flow of each of several emitters in a calibrated cylinder for 1 minute. The rate of discharge will vary with changes in water pressure and in the length and diameter of the distribution line beyond the point of measurement. The amount of water applied can be determined by the time and rate of discharge. The amount of water needed can be calculated by multiplying the area to be irrigated by the depth of water wanted. Table 12-2 records the effects of soil texture, soil depth, and ET on available water, irrigation intervals, and time required to replenish the water.

MANAGING
WATER APPLICATION

In most situations, enough water should be applied at each irrigation to replace water that is used. If enough water is applied to penetrate deeply and to be absorbed before subsequent irrigations, deep rooting will be encouraged and crown-rotting organisms minimized. If the soil is allowed to dry between irrigations, it will shrink and swell maximally, a process of internal cultivation that improves soil structure. Shallow rooting is usually caused by irrigating so lightly that only surface soil is wetted or by irrigating so heavily that soil is waterlogged and only the surface soil has adequate aeration. Shallow-rooted plants are more likely to be blown over by wind and are more subject to water stress than deeply-rooted plants.

If the correct amount of water is evenly distributed over a uniform, deep soil, it should produce the best plant growth for the least amount of water. But uniform distribution also requires that the water infiltrate near the site of application. The way in which water is applied greatly affects its distribution; soil topography and the rate of application greatly affect infiltration. When water is applied faster than it can infiltrate, soil topography becomes important. Reshaping the soil surface or creating small catch basins will sometimes allow water to enter the soil more uniformly. See Chapter 7 on Irrigation Systems.

Basin Irrigation

Water can be applied in basins by fixed bubblers (small sprinkler-like heads that spray or dribble water onto the soil), hoses, sprinklers, soakers, or tank trucks (Fig. 12-5). Even in nearly level basins, too much water will be applied near the water inlets if the rate of application is too slow or the number of outlets too few for the texture of the soil and the size of the basin. This can be corrected by creating smaller basins, adding more water outlets around the basin, or increasing the rate of water flow. Basin irrigation may wash lightweight mulch to low areas or away from water inlets. A gravel or rock mulch, at least around the water inlets, will reduce this problem if the flow of water itself cannot be sufficiently reduced.

FIGURE 12-5
Basin irrigation can be used in areas that are relatively level and will be used little for recreation or thoroughfare.

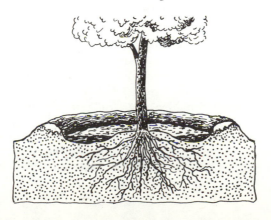

FIGURE 12-6
A series of crosschecks in a three-furrow system provides
for fairly uniform water distribution even on a slope.

Furrow Irrigation

Uniform water application along furrows depends on the infiltration rate, the furrow grade, and the rate of water flow. The infiltration
rate and furrow grade are not easily changed in existing plantings. A slow
infiltration rate can probably be increased slowly over years if organic
matter is incorporated in the surface soil of the furrows, if cultivation
is kept to a minimum, and if irrigation is adequate but infrequent. If
more water enters the soil at the lower end of the furrow than at the
head, the rate of water flow should be decreased. However, the furrow
could be steep enough or infiltration slow enough that decreasing water
flow may not be adequate. If space along the furrow run and under the
plants will allow a three-furrow system with alternating crosschecks,
they can be added to slow water flow and provide better infiltration
(Fig. 12-6). On the other hand, if more water is applied at the upper
end of the furrow, the rate of water flow should be increased. This
may cause erosion and upset work scheduling unless there is an automatic controller. The length of furrow run can be shortened if one or
more water inlets are introduced along the furrows.

Sprinkler Irrigation

Uniformity of water sprinkling depends on the sprinkler system
and its design, water pressure, and wind conditions. Penetration of
water into the soil depends on application and infiltration rates, topography, and soil cover. Most sprinklers apply water faster than the soil
can absorb it. Mulch, turf, or ground cover will slow water movement
on sloping surfaces so that more will penetrate. Sprinkler heads with

slower application rates can be installed, but they may adversely affect the distribution pattern.

Sprinklers are best used in the early morning, when water pressure is usually greater, there is little or no wind, irrigation will not interfere with other activities, and foliage will dry more quickly. This last factor may be important for plants that are susceptible to water-related diseases. With sprinkler irrigation of full canopy plantings, time of day makes little difference in the total water loss by evaporation and transpiration under relatively calm conditions. Even on a hot, dry afternoon, due to foliage wetting, the increase in ET during sprinkler application to a large planting will be less than 3 percent (Heermann and Kohl 1980), because evaporation during sprinkling and from the wet foliage reduces transpiration by an almost equal amount.

Sprinkler irrigation can pack the surface of bare soil and reduce infiltration. Wind can result in uneven distribution. Water can damage flowers and knock them down or bend tall plants. Frequent light sprinklings with highly saline water may cause an unsightly and even toxic buildup of salt on foliage; longer, less frequent applications usually correct or reduce this. Some believe that sprinkler application favors mildew and rust diseases on susceptible plants, but observations do not bear this out. Mildew and rust are particularly unlikely in arid regions, or when water is applied in early morning so that foliage dries quickly.

To determine the rate of application and the distribution pattern of a sprinkler system, distribute several cans evenly between two adjacent sprinkler heads in the same row and diagonally between two sprinklers in adjacent rows. After running the sprinklers for a specified length of time under the pressure at which they will usually operate, compare the depth of water in different cans. Many systems in use today may distribute more than twice as much water to some areas as to others, but the timing of irrigation must be based on the area receiving the least water. Many inadequate sprinkler sytstems can be redesigned to improve distribution, plant performance, and water conservation. If a planting is irrigated by a movable sprinkler, variations in the distribution pattern can be minimized if the distances between sprinkler settings are equal to the radius of the area covered by the sprinkler.

Drip Irrigation

The goal of drip irrigation is to wet only about half of the soil in the root zone of mature plantings or those whose canopies shade most of the ground. Water is applied slowly for long periods to only a portion of the soil. When plants are young or farther apart, water can be applied so little or none wets soil without roots. Studies indicate that 50 to 65 percent of the root zone of plants should be wetted. The amount of soil wetted depends on soil characteristics, irrigation operation time, and the number of emitters used. The number of emitters will range from less than one per plant to eight or more around individual large trees (Marsh 1977; Hamilton 1978) (Fig. 12-7).

In a vertical section of soil the shape of the wetted zone under an emitter is ovoid, or egg-shaped. The zone is elongated vertically in well-drained soil and quite flat in fine-textured or stratified soil. Usually the maximum wet zone below the surface is 0.3 to 1.2 m (1–4 ft) in radius and 0.6 to 1.2 m (2–4 ft) deep. The wet zone in sandy soil might be 0.3 m in radius and 1.2 m deep, and, in clay, 1.2 m in radius and 0.6 m deep. To wet about half of the soil volume in a large planting, space emitters at distances up to 2.5 times the radius of wetted surface in heavy soil and 3.5 times the radius in sandy soil.

FIGURE 12-7
Small-diameter tubing (3 mm or 0.12 in) on the soil surface supplies water
from an underground distribution line (top);
the length of the tubing controls the amount of water discharged.
A 125-mm (0.5-in) distribution line with an emitter and tubing
to control water flow to individual plants (bottom);
the lines and tube have been exposed for ease of viewing.

Emitters should be placed at least 300 mm (1 ft) from the trunk of any plant. You may need to loop a lateral (secondary distribution line) around large trees (just inside the drip line) to space emitters evenly. Emitters should not be placed under paving, where their operation cannot be checked easily. On uneven terrain, pressure-compensating emitters, pressure regulators in the lateral, and emitter tubing of different lengths will help to establish even discharge over changes in elevation. Shorter emitter tubing may be used at high elevations and longer ones at lower elevations; the manufacturer should supply information needed to make adjustments. On sloping plantings, pressure control is simplified if laterals are on the contour and if an adjustment valve is placed where each lateral connects to the main.

If laterals are buried, the emitters or emitter-tube outlets should be on the surface or within 50 to 75 mm (2–3 in) of the surface, so that you can check their operation easily by looking for wet spots. Laterals covered with mulch should have emitter outlets at or above the mulch surface. If emitters are placed below the surface, they should be designed so that they do not plug.

For optimum growth, the amount of water applied should equal water lost through evapotranspiration. This can be determined with a tensiometer or by trial and error. When drip irrigation is installed, the plants and irrigation system must be watched closely, but drip systems are easy to automate if water application is controlled by tensiometers. Place the control tensiometer in the active root zone 300 to 375 mm (12–15 in) from the nearest emitter in a representative part of the planting (Marsh 1977). Set the controls to operate between 10 and 20 cb. Locate a second tensiometer at the lower boundary of the feeder roots (300–600 mm below the first) to provide a check on water use and the adequacy of application. A few tensiometers at other representative locations can be used to adjust the control settings for the most efficient use.

In contrast to other systems, drip irrigation must be frequent; waterings should occur daily or every two days during the main growing season (Kenworthy and Motes 1975). Plants deplete the water more quickly than if the entire root zone were wetted. Capillarity, which accounts for most water movement between emitters, decreases rapidly as soil dries and is reduced to zero if soil cracks, but a mulch will slow the drying of surface soil. To restore the water used by plants since the previous irrigation, application time may range from one to 16 hours, but should not be continuous. If more than 16 of each 24 hours are needed regularly, the number of emitters should be increased. The formation of ponds at emitters is undesirable and can be avoided if application is interrupted periodically to allow percolation.

Monitor the system to be sure it is operating properly. Check emitters visually each week for correct flow. If performance is satisfactory, the intervals between checks can be increased. Take precise measurements at least twice each year by catching the flow from a number of emitters. Measurements can reveal problems of emitter performance or pressure control in lateral lines.

The most common problem is that drip-irrigation lines and emitters become plugged. Even self-cleaning emitters may become clogged. Some emitters can be flushed manually. If all else fails, the emitter or discharge tube must be removed from the line and cleaned. The most common causes of plugging are soil particles, calcium and iron precipitates, bacterial slimes, and algal growths. The usual treatments are acidification for chemical precipitates, and chlorination for bacteria and algae. To rid lines of calcium and iron plugging, fill the system beyond the filter with water acidified to pH 2 with either sulfuric or hydrochloric (muriatic or pool) acid and let it stand for an hour. Flush thoroughly and repeat the process. Emitters that are completely plugged must be removed and cleaned individually. Once they are cleaned, maintain the lines by keeping irrigation water acidified to pH 6.5 for most calcium problems or pH 4.5 for most iron problems (Rible and Meyer 1979). Most managers, however, prefer to clean the lines as necessary rather than inject acid continuously.

Bacterial slime and algae can be cleaned from lines if hypochlorites are injected beyond the filter to fill the lines with 30 to 50 ppm chlorine, which is left to stand for an hour. Measure chlorine residual at the end of the line; it should be at least 1 ppm. If it is not, repeat the chlorination procedure until the chlorine level is 1 ppm at the end of the line after an hour. Then flush out the lines and clean the filter. Do not add chlorine to the system if the water contains iron compounds. You can check for iron compounds by putting the planned chlorine concentration in 1 liter (1 quart) or more of water and letting the water stand overnight. If a considerable amount of precipitate forms, acidify the water to pH 4.5 and repeat the test. If there is still a problem, consult a water specialist. A swimming pool maintenance person may be able to help because both acid and chlorine are used in pools.

Filters and screens must be cleaned periodically by hand or by built-in backflushing if drip systems are to operate properly. Cleaning frequency may vary from twice a week to twice a month, depending on water quality and filter size. The clamp at the end of each lateral line should be released once a month to flush out accumulated sediment.

Some Common Irrigation Problems

Table 12-3 summarizes solutions to some common irrigation problems.

TABLE 12-3
Solutions to common irrigation problems

| | | Solutions | | | |
| | | | Surface Flood Systems | | |
Problem	Sprinkler Systems	Level	Graded Basin	Furrow	Drip Systems
1. Surface runoff	Decrease application by: A. replacing worn nozzles, B. reducing nozzle size, C. reducing pressure. Decrease set time. Repair system leaks. Increase intake rate by: A. installing ground cover or turf, B. surface mulching, C. applying amendments.	Repair border leaks.	Decrease stream size. Install tailwater return system. Increase run length. Decrease slope.		Repair system leaks.
2. Ponding at lower end	See (1) above.	Level basin.	See (1) above.		See (1) above.
3. Slow advance of water stream	Not applicable	Not applicable	Increase stream size. Decrease basin width. Decrease basin length.	Increase slope. Decrease run length.	Not applicable
4. Rapid advance of water stream	Not applicable	Not applicable	Decrease stream size. Increase basin width and/or length. Decrease slope.	Decrease stream size. Increase furrow length. Decrease slope. Use cross-checks.	Not applicable

TABLE 12-3 (cont.)

Solutions to common irrigation problems

| Problem | Sprinkler Systems | Surface Flood Systems | | | Drip Systems |
		Level	Graded Basin	Furrow	
5. Uneven distribution.	Check sprinklers for proper operation. Repair leaks. Check operating pressure for pump and system. Shut down during high wind. Change: A. sprinkler spacing, B. sprinkler head, C. nozzle size.	Adjust basin boundary to soil types.	Regrade. Increase stream size. Install tailwater return system. Use basin checks.		Check proper operating pressure. Check for clogged emitters. Change emitter spacing and/or location.
6. Erosion	See (1) & (5) above.	Decrease inlet velocity.	Decrease inlet velocity. Decrease stream size. Decrease cross slopes. Decrease irrigation slopes.	Decrease stream size. Decrease slope. Change furrow shape.	See (1) above.
7. Muddy tailwater	See (1) & (6) above.	Not applicable	See (6) above.	See (6) above.	Not applicable

Problem					
8. Saline water supply	Replace system with a more corrosion resistant material. Irrigate at night to reduce leaf burn.	Maintain high soil moisture content by irrigating more frequently. Apply adequate water for leaching.	Use special furrow shapes for salinity control.	Use special furrow shapes for salinity control.	Not applicable
9. Salt deposits on plants	Irrigate at night to reduce leaf burn.	Not applicable	Not applicable	Not applicable	Not applicable
10. Salt deposits on soil surface		Change to a more salt tolerant crop. Apply adequate water for leaching. Improve surface drainage.		Use special furrow shapes for salinity control.	Apply adequate water for leaching using other irrigation methods.
11. High water table	Improve water management. Improve subsurface drainage.				
12. Uneven plant growth	See (2), (5), (10) & (11) above.				
13. Distribution system losses	Repair pipeline leaks.	Replace open ditch with pipeline. Line open ditches or ponds with concrete, plastic, bentonite, or chemical sealers. Install interceptor subsurface drains parallel to canal, ditch, or pond.			Repair pipeline leaks.

Adapted from Interagency Agricultural Information Task Force. n.d. *Drought Tips: Common Irrigation Problems: Some Solutions.*

MINIMUM IRRIGATION

Regions subject to summer rainfall have little need for irrigation except during occasional dry spells. In regions with Mediterranean climates (with hot, dry summers), irrigation practices that replace evapotranspired water will provide the amount of water that plants need. Many plants, however, will function satisfactorily in a landscape even if they get less water than would be evapotranspired under adequate soil moisture conditions. Minimum irrigation of mature landscape plantings can reduce not only the cost of water and its application but also the need for other care, such as pruning and weed control. When plants have reached their effective size in a landscape, shoot growth is not as important as appearance and longevity. The minimum leaf area consistent with acceptable appearance and shading or screening functions should be satisfactory in most situations. In contrast to production agriculture and young landscape plantings, mature landscape plantings will withstand reduced water and fertilizer applications without showing undesirable effects in most cases. In arid regions, of course, particularly when species are indigenous to humid regions, you may need to irrigate once a month or more frequently during the dry growing season.

In experiments by Sachs, Kretchum, and Mock (1975) on deep California soils in San Jose and Santa Ana, one supplemental irrigation in midsummer adequately maintained the appearance of six out of eight mature shrub and ground-cover species. *Cotoneaster pannosa* and

FIGURE 12-8
The growth of established Monterey pines was influenced by the frequency of irrigations during the growing season at San Jose, California.
All pines were sheared to a height of about 1200 mm (4 ft) each winter.
Tree heights at the end of a growing season reflect the irrigation frequency:
no irrigations (left), bimonthly irrigations (center), and biweekly irrigations (right).
(Photos courtesy of Roy Sachs, University of California)

mirror plant required two irrigations during the summer to maintain health and appearance. Oleander and surinam cherry required one irrigation. English ivy, two ice plant species, and shiny xylosma maintained adequate or excellent appearance with no irrigation. Plants in these experiments were vigorous species commonly grown in California, which were used for several years in growth-regulator studies before being differentially irrigated. Even though soil moisture in the root zone sampled was at the permanent wilting stage for several weeks and leaf temperatures were 6 to 15°C above ambient, few species were injured seriously. Growth was reduced by the less frequent irrigations, but appearance was quite acceptable for most species (Fig. 12-8).

Most landscapes in California were in satisfactory condition after two years of drought (1975–1977), when winter rainfall was less than the winter ET and only half the normal amounts of water were available for summer irrigation. Many trees and landscapes had no supplemental water. Some plantings did not show stress until two years after normal watering was resumed. It is obvious that the amount of water applied to most irrigated landscapes could be reduced to the benefit of plants and owners.

Several factors should be considered in minimum irrigation situations:

> The transition from a regime of adequate soil moisture at all times (and even excess or too frequent watering) may require care to make sure that plants are not severely injured or killed because they have an inadequate root system or moisture reservoir.

> Some plants tolerate infrequent irrigations because they have extensive root systems. In drought years, the water reservoir may be low, thus reducing a plant's supply of water. It may be necessary to apply more water than normal to keep the plant alive.

> Minimum irrigation regimes are designed for maintaining plants, not for promoting vigorous growth. New plantings require frequent irrigation because of their limited root balls. Even after they become established, the plants usually need frequent watering until they reach optimal size.

> Under saline conditions, irrigations must be more frequent to avoid salt toxicity.

> Plants differ in their water requirements, so irrigation schedules may have to differ accordingly. In mixed plantings, irrigation schedules should be determined according to the needs of species that are the least tolerant of heat or desiccation. Alternatively, plants with similar water requirements can be grouped in the same areas, so that each group can be watered as needed.

> Minimum irrigation regimes are not restricted to so-called "drought-tolerant" species; most plants can survive on less water than they receive in the landscape. A minimum irrigation regime merits serious consideration in the design and maintenance of landscape plantings, particularly areas such as freeways, regional parks, and open spaces.

SOIL SALINITY

Soil salinity is a problem in many arid and coastal regions, where annual rainfall is commonly less than 750 mm (30 in). Soils contain a mixture of salts. A white crust on the soil surface is usually a mixture of sodium, calcium, and magnesium salts. Most moist, dark, oily spots on the soil surface indicate an excess of calcium chloride. These salts dissolve easily in water, thereby increasing salinity. Other salts, such as lime and gypsum, are only slightly soluble and do not appreciably increase soil salinity.

Excess salt can also be a problem along roads and streets where sodium or calcium chloride is used to melt ice in winter (Carpenter 1970). Sodium chloride has been shown to be more toxic than calcium chloride, whether applied to the soil or sprayed on plants. When salts are applied to roads, the salt in the spray generated by traffic may be more damaging to plants than the salt that drains into the soil. Plants that are the most tolerant to soil salinity may not necessarily be the most tolerant to salt spray or mist on foliage, and vice versa (see Chapter 18).

Saline soils are not desirable for several reasons. Salinity decreases growth of sensitive plants, kills leaf margins, and may cause death. Sodium and chloride can cause specific injury to certain plants, including leaf burn, leaf drop, and stem dieback. High sodium content also adversely affects soil structure, resulting in poor aeration and extremely low rates of water infiltration. The availablility of water to plants is also decreased in saline soils because of increased osmotic tension, by which water is held in the soil. In such circumstances, irrigation must be more frequent. If salinity is suspected, soil samples should be analyzed. A list of commercial laboratories is usually available from the local cooperative extension or soil conservation services.

Salinity has usually been expressed in millimhos per centimeter of soil extract (mmhos/cm), a unit for measuring electrical conductivity (EC). The metric unit for electrical conductivity is decisiemens per meter (dS/m). One dS/m equals one mmho/cm. The higher the soil salinity, the greater will be the conductivity of the extracted soil solution (EC_e). A soil is saline when its total soluble salt content is high enough to affect plant growth adversely. This could be as low as 3 dS/m for sensitive plants. A few plants will grow satisfactorily in soils with levels up to 10 dS/m. Most plants tolerate salinity of 5–7 dS/m (Bernstein 1964).

Water Quality

Salt may accumulate in the soil from irrigation water, high groundwater tables, or existing salt deposits. If irrigation water does not penetrate the soil deeply, most of the salt brought in by that water will remain near the surface. Ground water generally contains more salt than does surface irrigation water. When the ground-water level is near the surface, some water moves upward because of evaporation and plant use. The salt content of the surface soil increases as this happens and plant growth can decrease. A good soil can become saline in one season.

There is increasing use of secondary-treated sewage effluent for landscape irrigation. The main concern is the salinity of the effluent. Check the range of effluent quality before beginning effluent irrigation. In addition, monitor regularly effluent quality because incoming sewage can vary considerably. Close cooperation with the treatment facility manager is essential to alert you to changes in effluent composition that might adversely affect plants.

Assessment of water quality is based upon three main considerations: the total salt content, the proportion of sodium in salt, and the boron concentration (Table 12-4). Water from the Colorado River contains up to 75 kg of salt per hectare cm (2000 lb/acre ft) of water (800 ppm, or an EC of 1.25 dS/m). With reasonable irrigation practices, however, water with an EC of 0.75 dS/m or less should present few or no salinity problems. An EC greater than 3.0 will cause severe problems except in areas restricted to only a few salt-tolerant species (Stromberg 1975). Salt-sensitive species can tolerate an EC no greater than 2.0 dS/m.

Even though the total content of irrigation water may be low to moderate, if the sodium constitutes more than 60 percent of the total of calcium, magnesium, and sodium, soil permeability to water and air is usually adversely affected. It is worthwhile to remember this

TABLE 12-4

Guidelines for interpretation of water quality for irrigation (Ayers 1977)

	Water Quality Guidelines		
	No Problem	Moderate Problems	Severe Problems
Salinity EC_w in decisiemens[a] per meter	0.75	0.75–3.0	3.0
Boron Parts per million (ppm)	0.5	0.5–2.0	2.0

[a]Electrical conductivity expressed as decisiemens per meter (dS/m); one decisiemen per meter equals one millimho per centimeter (mmho/cm).

expression: Hard water (high in calcium and magnesium) makes soft soil, and soft water (high in sodium or very low in total salt) makes hard soil." Before you can assess the salinity hazard of any irrigation water, you must know how much salt the plants can tolerate and how much leaching is needed to maintain the desired salt level in the soil water.

Removing Salt by Heavy Irrigation

Each area or basin to be irrigated should be fairly level to obtain uniform leaching with the least water. Adequate drainage is also important. If natural drainage is inadequate, you may have to install tile or open drains. If poor soil drainage is caused by excess sodium, you can first reduce sodium by adding gypsum ($CaSO_4$) and then leach salt out of the root zone.

Planters, containers, and beds with shallow soil are quite vulnerable to salt buildup. Soils of limited volume and depth need frequent and excess watering to keep salinity from increasing to injurious levels.

Ordinary irrigation methods should produce some leaching, which reduces but does not eliminate the accumulation of salts in the soil water. The amount of water that must be applied to reduce soil salinity to a safe level depends primarily on the salinity of the water and the sensitivity of plants. Generally, for each 100 mm (4 in) of medium-textured soil

50 mm (2 in) of irrigation water leaches out $\frac{1}{2}$ of the salt

100 mm (4 in) of irrigation water leaches out $\frac{4}{5}$ of the salt

200 mm (8 in) of irrigation water leaches out $\frac{9}{10}$ of the salt

After soil salinity has been reduced to a level plants can tolerate, water must be applied in excess of the quantities used by the plants in order to maintain soil salinity at safe levels. The saltier the irrigation water and the more sensitive the plants, the more water must be applied (Table 12-5). For instance, a moderately salt-tolerant plant should be irrigated with 125 to 166 percent of the water that transpires if the salinity of the water is 2.0 dS/m. If 30 mm of water has transpired, the plant should be irrigated with 38 to 50 mm of water (30 mm times 1.25–1.66). This would maintain soil salinity at a safe level (below 5–8 dS/m.

It has generally been assumed that the effects of saline water can be offset by increasing the amount of leaching, so that the *average* salt content of the root zone will not be increased. The U.S. Salinity Laboratory has demonstrated that yields of alfalfa (and probably other crops) are governed not by the average soil salinity but primarily by

TABLE 12-5

The depth of water required for maintaining a safe level of soil salinity
(Bernstein 1964)

Salinity of Irrigation Water EC_w (dS/m)[a]	Amount of Water (Percentage of Evapotranspired Water) Required to Maintain Safe Salinity Levels for Plants of Different Salt Tolerances		
	Tolerant (10 dS/m max)	Moderately Tolerant (5–8 dS/m max)	Sensitive (3 dS/m max)
0 (salt free)	100	100	100
0.5	105	106–111	120
1.0	111	115–125	150
2.0	125	125–166	300
4.0	166	200–500	

[a]Electrical conductivity expressed as decisiemens per meter (dS/m); one deci-siemen per meter equals one millimho per centimeter (mmho/cm).

the salinity of irrigation water (Bernstein and Francois 1973). Experimental yields were reduced as the salinity of the water increased, no matter how much leaching was done. With this in mind, the practice of blending saline drain waters with low-salt irrigation water must be reassessed. In native landscapes where yield is not a factor, reduced growth due to saline water may be desirable as long as plant appearance is satisfactory.

Irrigation water may be more saline than sensitive plants can tolerate. Irrigation water with a conductivity of 4 dS/m can never lower the soil salinity below a lethal level for certain trees, such as the Lombardy poplar. Additionally, if the irrigation water is strongly saline, the amount of water needed by sensitive plants may make leaching impractical. In fact, the high amount of water necessary for adequate leaching may be as harmful as the salinity.

Other practices may reduce salt buildup in soil. High-analysis fertilizers provide adequate nutrition with the least increase in soil salinity. For example, an application of 2 kg of nitrogen per 100 m^2 would require 2000 kg of 10-6-6 fertilizer per hectare but only 600 kg of urea (45% N). Surface mulch reduces evaporation and salt buildup in the surface soil. If the mulch is of wood chips, bark, or other organic matter, it should be at least 50 mm (2 in) deep. Sprinkler irrigation may be necessary to keep the mulch from floating to low areas.

Salt buildup on surface soil may be encouraged to minimize salt concentration in the root zone of shallow-rooted plants. Since salts usually accumulate at the high spots in a planting, berms and ridges used in basin and furrow irrigation can be formed so that salt accumu-

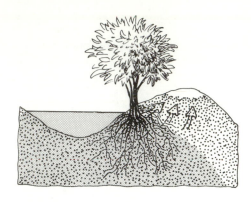

FIGURE 12-9
Establish small shrubs at the bottom
or side of a furrow to minimize
salt buildup in the root zone
since much of the excess salt
will accumulate at the top of the ridge.

lates away from the root zone (Fig. 12-9). However, since light rains may carry this salt into the root zone, heavy irrigation may have to be used if heavier rains do not follow. In some landscapes, it may be necessary to select relatively salt-tolerant plants for satisfactory growth and appearance with reasonable maintenance (Bernstein 1964; Maire and Branson 1971).

ANTITRANSPIRANTS

Antitranspirants are chemicals capable of reducing transpiration. They have been used for a number of years by arborists and nurserymen to reduce water loss in plants, particularly during and after transplanting. Antitranspirants have also been used to reduce the water requirement of plants during drought, to reduce the frequency of irrigation in arid regions, or to protect conifers against windburn in winter. Since plants lose most water through their stomates, antitranspirants are usually foliar sprays. The idea of coating foliage to curtail transpiration, particularly in transplants, is not new, although research on the subject increased in the 1970s.

When a plant wilts, growth is retarded. By slowing down the rate of water loss, antitranspirants can help to prevent or delay wilting. Dendrometer measurements have shown that the trunks of trees usually shrink during daylight hours, when water uptake lags behind transpiration. A film-forming antitranspirant sprayed on five-year-old almond trees can reduce daytime trunk shrinkage by more than 50 percent, indicating that the water balance of an entire tree can be improved if water loss from the leaves is curtailed (Davenport, Hagan, and Martin 1969).

Foliar antitranspirants may reduce transpiration in one of three ways:

By the use of reflecting materials that reduce absorption of radiant energy, thereby lowering leaf temperatures and reducing transpiration

By applying emulsions of wax, latex, or plastics that dry on foliage to form thin, transparent films, which hinder escape of water vapor

By applying certain chemical compounds that affect the guard cells around stomatal pores and prevent stomata from opening fully, thus decreasing the loss of water vapor from leaves

To be effective, reflecting materials must coat the exposed leaves with a light-colored film, but a plant so colored may not be acceptable in most landscapes, except when needed to keep the plant alive. Another drawback is that the film restricts light penetration into leaves and reduces photosynthesis. Spray emulsions that form transparent films not only reduce water loss but reduce the carbon dioxide intake needed for photosynthesis. Chemicals that inhibit stomata from opening fully appear promising because the smaller openings restrict water vapor movement more than they restrict the passage of carbon dioxide (Kramer and Kozlowski 1979). Three such chemicals are phenylmercuric acetate, abscisic acid, and silicone. Although phenylmercuric acetate has been more widely tested than the others, its use will probably be limited because it poses the danger of phytotoxicity and mercury contamination of the environment.

While antitranspirants of the reflecting type cause a reduction in leaf temperature, the film-forming and stomata-closing types tend to increase leaf temperature by curtailing transpiration and thus reducing evaporative cooling. Under normal conditions, however, the increase in leaf temperature induced by antitranspirants is usually not serious, because thermal emission rather than evaporative cooling becomes the more important means of heat dissipation.

The effectiveness of antitranspirants seems to depend on species, the plant's stage of development, and atmospheric conditions (Gale and Hagan 1966). Even when antitranspirants are effective, the duration of effectiveness will vary with the efficiency and durability of the material, the accuracy of application, environmental conditions, and the amount of new foliage produced after spraying. Cracking of the film and new growth both limit the duration of reduced transpiration. Repeated application can retard growth by seriously reducing photosynthesis. Even so, film-forming materials are the most commonly used antitranspirants and have a place when reduced transpiration for a few days to two weeks is crucial. Antitranspirants have been helpful in transplanting plants that are in leaf, particularly during the growing season. Though they may reduce water loss initially by 40 percent, the film materials have limited application for longer-term reduction of water loss. In tests by Turner and DeRoo (1974), antitranspirants did not reduce

winter-burn injury to conifers when air temperatures were below freezing, but they seemed to be helpful in the spring when air temperatures
were warmer but the soil was still cold or frozen.

Antitranspirants can be used occasionally on small plantings or
container plants to extend the interval between irrigations when it is
not convenient to water on schedule. Cut foliage, including Christmas
trees, will stay fresh longer if they are sprayed with an antitranspirant
before cutting. Antitranspirants can also reduce salt damage to conifers
along highways that are salted during the winter months (Gouin 1979).

In general, the use of antitranspirants will be limited until more
effective materials are found and benefits are more specifically
documented.

MANAGING DRAINAGE

In an established landscape, little can be done to facilitate drainage
short of installing a drainage system (see Chapter 7). The main concern
is to prevent excess water from reaching the rooting zone. Surface water
from rain or irrigation on soil, pavement, and buildings should be
diverted before it reaches the planting or planter. On shallow or poorly
drained soils, overwatering should be avoided.

When standing water remains more than several hours after a rain
or irrigation, soil infiltration is quite obviously slow, but poor drainage
may be more difficult to detect. The first indication may be poor
growth or even wilting of plants (see Chapter 18). You can identify
the problem by probing the soil with a soil tube or auger. For an ongoing problem, you may insert a perforated pipe 25 to 50 mm (1–2 in)
in diameter into the soil to monitor the level of the water table. It
should be capped between observations.

If water is slow to infiltrate an otherwise well-drained soil, infiltration may be improved by working some slowly decomposing organic
matter into the tight surface soil (see Chapter 7). A surface mulch will
reduce future compaction, and its organic leachates will improve soil.
In a planting, traffic should be kept to a minimum or restricted to certain paths or areas, particularly when the planting is wet. If the planting
is irrigated, the soil should be dried almost to the permanent wilting
point between irrigations. Sprinkler or drip irrigation should not be
applied faster than water can infiltrate.

Water penetration and aeration can be further improved if holes
are drilled or water-jetted into the soil around plants. Aerating turf
around trees will usually improve the soil for the tree roots as well.
Holes drilled or injected for fertilizer placement also improve water and
air movement; in many cases, the resulting aeration may be more beneficial than the fertilizer. Deeper holes affecting a wider volume of soil

FIGURE 12-10
A water jet is used to drill (wash) a hole to improve future
water penetration and aeration. Jetting holes at an angle will have a greater effect
than a vertical hole (Fig. 10-2).

can be washed with a water jet (Fig. 12-10) made from a 20-mm ($\frac{3}{4}$-in)
galvanized pipe 1.5 to 2 meters (5–7 ft) long with a beveled hard-steel
tip on one end, and a valve and hose connection on the other. Water
is best supplied by a high-pressure sprayer at 10 to 20 kg/cm^2 (150–
300 psi) pressure. A garden hose will work, but its low pressure makes
the job a slow one. The water jet is to be used for making holes, not
for irrigating.

To use the water jet, push its tip into the soil at an angle, turn
the water on, and push the pipe into the soil. The water washes the
soil ahead of the tip and carries it to the surface around the outside of
the pipe. A channel about 50 mm (2 in) in diameter can be washed
fairly easily in all but the tightest soils. Holes are usually started near
the trunk of a large shrub or tree and slanted down and away from it.
Three or four holes per plant are usually sufficient; for large trees,
additional holes near the drip line may be desirable if they do not
interfere with other plants. Holes may have to be plugged while the
remaining ones are bored so mud and water do not flow into them.
After all holes have been dug, they are usually left open or filled with
gravel.

These holes not only improve drainage but increase water distribution and aeration. Roots usually grow into the channels and eventually
impair their effectiveness. After two to three years, new channels can
be made in between the previous holes. Water-jetting young trees in
pavement planters (which are open to the soil around) and along streets
should encourage deeper, wider rooting, which improves tree growth,
increases the interval between irrigations, and minimizes later damage
to sidewalks and pavement (see Fig. 10-2).

CHAPTER THIRTEEN

Soil Management

A number of cultural operations (other than fertilization and irrigation) may be employed to protect the soil, maintain or improve its physical condition, and provide an attractive surface for landscape use and enjoyment. Most of these operations involve management of the surface soil, but some involve the tops of plants as well as soil and roots.

Development and management of the soil surface should be integral to any landscape plan. Paved surfaces are used for thoroughfare and recreational activities. Lawns provide aesthetic protection for the soil surface. Areas not covered by pavement or lawn are usually treated in one of three ways:

With ground covers
By clean cultivation or chemical weed control
With organic or inorganic mulches

Proper handling of the soil surface will depend on the function of the landscape, the type of soil, and the kinds of plants that are grown.

TURF AND GROUND COVERS
AROUND TREES

Turf and ground-cover plants present few problems for existing trees. When turf or other low plants are irrigated, however, care must be taken so that trees and shrubs are not over- or underwatered. Overwatering may suffocate roots and even allow crown rot fungi to infect treetrunks. On the other hand, light, frequent irrigation may keep turf healthy but may fail to supply enough water to trees and shrubs.

Turf

Turfgrass can greatly reduce the growth of young plants if it is planted too close to them. Adequate irrigation and fertilization will help overcome some of the growth inhibition, but not all of it. A number of turf species and some broadleaved plants have allelopathic effects on young trees and shrubs (see Chapter 2). Keeping turf and other plants at least 300 mm (12 in) away from the trunks of young plants will eliminate almost all such retarding effects (Fig. 13-1) (Harris, Paul, and Leiser 1977; Meskimen 1970; and Whitcomb 1979b).

Clear areas near treetrunks or protective stakes will also minimize trunk damage caused by mowing equipment (see Fig. 18-8). After turf growth has ceased in late fall, nitrogen fertilizers for trees can be spread on the lawn surface under them. If rain is adequate, most of the nitrogen will be carried into the root zone of the tree without excessive stimulation of the turf.

FIGURE 13-1
The influence of alta fescue turf on young southern magnolia trees.
The photograph shows growth after two years: 0, 0.36, and 17.6 m^2 (0, 4, and 196 ft^2) of soil (front to back) were kept free of turf around each tree. (Harris, Paul, and Leiser 1977) The three 35-year-old ginkgo trees show the influence of turf on tree growth: The one on the right was surrounded, while the other two grew at the edge of the turf in a shrub bed. The smallest was so low in vigor that the bark exposed to the afternoon sun was killed.

Ground-cover Plants

Where foot traffic is unimportant or undesirable, low-growing shrubs, vines, and herbaceous perennials can often be used and will require considerably less maintenance than turf. Dense plantings will quickly minimize weeds, but physical or preemergent weed control will usually be needed until the ground cover blankets the soil. Irrigation may be needed until the plants become established. Most ground covers, particularly vines, must be pruned to keep them in bounds and attractive (see Chapter 14).

Edging The spread of turfgrass, weeds, and other ground-cover plants can be controlled with herbicides, growth regulators, mechanical edgers, and mulches. The development of string trimming equipment has greatly facilitated edging around plants and isolated objects, and along walks and fences (Fig. 13-2). A short length of flexible cord rotating at high speed cuts or breaks leaves and small-diameter stems but supposedly does not hurt larger items. The flexible cord allows trimming in difficult-to-reach places. Caution should be exercised, however, when trimming around the base of plants, particularly in the spring when the bark of young plants is easily separated at the cambium.

FIGURE 13-2
A string trimmer is effective in edging turf
around shrub beds and trimming around trees.
In the spring when the cambium is active, bark can be damaged or torn loose
by the impact of the flexible cord.

BARE SOIL AROUND PLANTS

Cultivation

Seedbed preparation, weed control, and incorporation of organic matter are the primary purposes of cultivation. There is reason, however, to minimize cultivation in landscape plantings. The process increases costs, cuts and injures shallow roots, and makes soil on slopes more subject to erosion. Even though cultivation can destroy weeds, it brings buried seeds to the surface to propagate the species. In plantings, weeds can be controlled with chemicals, mulch, or surface hoeing, which disturbs little of the soil.

Some people would insist on the necessity of stirring the soil around existing plantings in order to improve aeration and water infiltration. Cultivation may appear to improve tilth, but in fact it breaks down soil structure; the surface of cultivated soil can quickly become sealed from splashing rain or sprinkler drops. Cultivation can, however, break up a tight soil, so that wetting and drying of the fractured soil (large clods) from rain (irrigation) and plant extraction can bring about an internal cultivation that will improve soil structure.

Chemical Weed Control

Chemical control of weeds came into its own with the development of 2,4-D (2,4-dichlorophenoxyacetic acid) in the early 1940s, and later with the introduction of preemergence herbicides. Wise use of herbicides can minimize weed problems and greatly reduce the cost of weed control. Herbicides are classified in a number of ways. *Contact* herbicides kill only those plant parts they contact, while *translocated* herbicides can affect the entire plant even though only a portion of it has been treated. A translocated herbicide that is particularly effective against perennial broadleaved weeds, 2,4-D is translocated from the leaves throughout the plant, including the roots. Contact sprays are primarily used against annual weeds; once their tops are killed, few will sprout again. At the right concentrations, 2,4-D is also a *selective* herbicide, one that can selectively kill broadleaved plants without damage to surrounding grass. Contact herbicides are usually not selective and kill everything they strike.

Fumigants can be used before planting to rid confined planting areas of weeds and weed seeds. This *preplanting* treatment will greatly reduce the need for future weed-control measures, particularly if the treated planter is isolated from other weed-seed sources.

Other chemicals can be applied to the soil and landscape plants to kill any germinating weed seedlings. These *preemergence* herbicides are extremely valuable in keeping ground cover and shrub plantings free of weeds. A selective herbicide or a herbicide applied so as to avoid contact with the landscape plants can kill certain plants while leaving desirable ones uninjured. This would be a *postemergent* use of herbicide.

Like all poisons, herbicides must be used with caution. The margin between killing weeds and seriously injuring or killing landscape plants may be fairly narrow. You cannot give too much attention to correct concentration and rate of application. Avoid spraying desirable plants or applying herbicides too heavily, because they may be further concentrated by being washed to the bottom of the basin around a tree or shrub. Applying sprays with large droplets at low pressure will minimize drift, which can cause plant injury, particularly under windy conditions. Weed oils, which have a disagreeable odor, will stain concrete, but kerosene and other contact sprays will not.

Bare Soil around Plants

Whether weeds are controlled by hand pulling, cultivation, or herbicides, the soil surface cleared of them is left unprotected from water and wind erosion. The soil surface on slopes can be shaped to trap water so it will soak in near the site of application, however. Bare soil under plant canopies can protect plants, particularly trees, against grass fires in unirrigated, arid landscapes.

MULCHES

Material placed on soil to cover and protect it is called *mulch*. Organic mulches include bark, wood chips, leaves, conifer needles, lawn clippings, straw, corn cobs, peanut hulls, and numerous other organic byproducts. Inorganic or synthetic mulches include crushed stone, black polyethylene, asphalt, pavement, and aluminum foil. Although it is not recommended now, in the early 1900s, loose, dry surface soil, called a *dust mulch*, was thought by many to reduce water loss from deeper soil. Spread organic mulches evenly over the soil in layers 50 to 150 mm (2–6 in) thick, taking care to keep them at least 150 mm (6 in) away from the trunks of trees and shrubs. Inorganic mulches may be applied in comparable thicknesses or in thin sheets.

Benefits of Mulching

The plant microenvironment is so extensively modified by mulching that it is seldom possible to isolate specific causes for each of the benefits of a mulch. It is helpful, however, to consider each of the possible benefits so as to better understand how mulches can be used or modified to be most effective.

Soil moisture is conserved by mulches because they reduce evaporation from the soil surface and reduce weeds that use water. In 1939, Russel conducted a laboratory study with soils at field capacity and found that a layer of straw mulch 40 mm (1.5 in) thick reduced evaporation primarily by obstructing solar radiation (about 50 percent), protecting against wind (about 20 percent), reducing soil temperature, and obstructing vapor escape (Fig. 13-3). Mulches are most effective in reducing evaporation when the surface soil is near field capacity, unless there is a water table within 5 to 10 m (15–30 ft) of the surface (see Chapter 6) to provide a continuous water supply. Once the water level of surface soil is much below field capacity, evaporation is very

FIGURE 13-3
Water loss from four soil columns, initially at field capacity,
as influenced by soil surface protection during four summer days.
The straw mulch did more to reduce evaporation than mere protection
from sun and wind would account for. (Adapted from Russel, 1939)

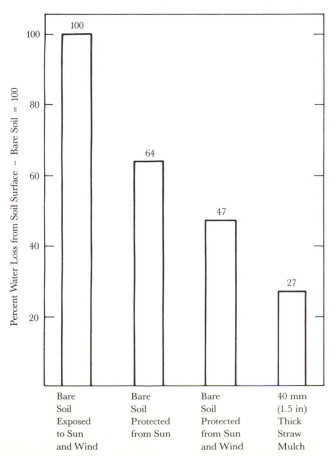

slow.   Weeds can use as much or more water than landscape plants, depending on their size and the area they occupy. A cover crop or a dense stand of weeds in a mature deciduous California orchard will increase evapotranspiration about 25 percent, or 1.5 mm (0.07 in), per day in the summer. A thick mulch (100–150 mm or 4–6 in) can suppress all but the most persistent weeds, such as bindweed and quackgrass.

A mulch will also hold water near where it falls, so that more of it soaks into the soil before flowing away. In addition, moisture can accumulate under a mulch because water vapor in the soil condenses on the cold mulch at night. This moisture may be equivalent to 0.1 mm of rain per day.   Under high humidity and radiation cooling conditions, moisture from the air also condenses on cold mulch surfaces. For a peat-mulched soil, Jacks, Brind, and Smith (1955) estimate that dew equivalent to 1.25 mm (0.05 in) of rain per day condenses on the mulch and wets the root zone. The mulch surface must be exposed to clear sky if that much condensation is to occur.

Soil erosion and water loss are reduced because mulch breaks the impact of rain and sprinkler drops, slows down the movement of water, and keeps water in contact with soil that has a faster infiltration rate than bare soil. Borst and Woodburn (1942) have shown that raindrop impact is a much more important erosion-causing factor than is flowing water (Fig. 13-4). On a soil that had been puddled and sealed by long exposure to rain, they found that a straw mulch 15 mm (0.6 in) thick only reduced water runoff when rain fell at a rate of 57 mm (2.25 in) per hour by 43 percent but reduced soil erosion by 86 percent. When a similar mulch was placed on wire mesh 25 mm (1 in) above the soil, water runoff increased 25 percent, but soil erosion did not increase. More water would probably have infiltrated the soil if the mulches had been left in place long enough to improve soil structure.

Soil fertility is usually increased by nutrients from mulch, either by direct leaching or by decomposition. The surface soil under a mulch is favorable for microorganisms, which increase the availability of nutrients. Jacks, Brind, and Smith (1955) cite results from both farm and forest studies in which almost all soils showed increased nutrients under organic mulches.   In short-term studies, nitrogen may be depressed by the use of mulches, while in studies more than five years long, nitrogen levels are usually enhanced after the early years. The more easily a mulch decomposes, the lower the initial nitrogen levels. A fine-textured mulch that is high in cellulose can cause a nitrogen deficiency because decomposition requires an increased number and activity of microorganisms.   Mulch protection also allows increased rooting in the surface soil, which is usually the most fertile soil.

Weed competition is reduced by mulches, so more water and nutrients are available for the landscape plants and any allelopathic

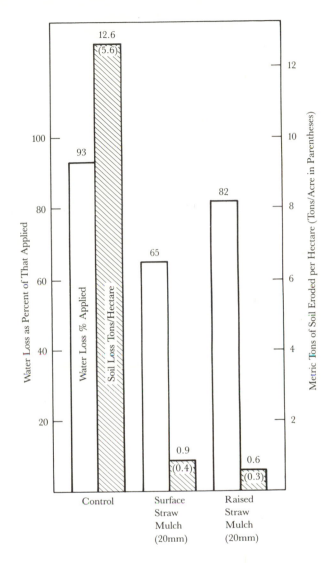

FIGURE 13-4
The influence of mulch in reducing rainfall runoff and erosion from
a sloping soil that had been puddled and sealed by prolonged exposure to rain.
Erosion was greatly reduced by mulch, whether it was in contact with the soil
or not, demonstrating that the main function was to reduce the impact
of raindrops. (Adapted from Borst and Woodburn 1942)

effects on young plants are minimized. A thick or impervious mulch
will reduce weed seed germination and keep weeds that do germinate
from growing toward the light.

Soil structure is improved when soil is covered with a thick mulch, particularly if the mulch is organic: The size of soil aggregates and the total porosity of soils increase (Jacks, Brind, and Smith). Improved aeration, temperature, and moisture conditions near the surface encourage rooting and other biological activity that enhances soil structure. Just the absence of cultivation and the low amount of compaction that even a thin inorganic mulch provides will allow soil structure to improve through wetting and drying cycles and through biological activity. Improved soil structure increases the infiltration rate and allows more uniform water distribution and less soil erosion, all of which favor plant growth.

Soil compaction is reduced when a protective covering reduces rain and sprinkler impact on the soil surface and disperses the weight of vehicles, people, and animals. Bare soils can be seriously compacted so that water and air movement is extremely slow and root growth is impeded. If surface soil becomes puddled from the impact of rain and sprinkler drops (Fig. 13-5), water infiltration and air exchange are limited.

Mulches provide an improved surface for traffic during wet weather or after irrigation. A coarse organic mulch reduces dust when soil is dry and mud when soil is wet. Many mulches improve the appearance of landscape plantings and are also good sound absorbers.

FIGURE 13-5
The impact of a raindrop on wet soil splashes water drops and soil particles in all directions and leads to soil crusting and slow water infiltration. (Photo courtesy of U.S. Soil Conservation Service)

Clay soil is less likely to crack when protected by mulch. Some clay soils shrink upon drying and form large cracks in the surface that increase moisture loss from the soil below, break small roots, and make furrow and basin irrigation difficult.

Mulches reduce salt buildup in surface soil by reducing evaporation of water that may be high in soluble salts. This is particularly important when summer rainfall is low and water tables are shallow and saline.

Mulches moderate soil temperatures so that surface soil is cooler in summer and warmer in winter than it would be otherwise. Moderate soil temperatures and good aeration encourage rooting in the fertile surface soil; this can be quite important in shallow soils. In areas with extremely cold winters, soil under a mulch is less likely to freeze; if soil does freeze, the mulch will minimize the alternate freezing and thawing that can injure the roots of young plants. Snow can serve as a mulch to effectively insulate the soil from cold and changing air temperatures. Mulch that has highly concentrated, tiny air spaces can be the most effective insulation. When they become soggy, compressed, or thin, mulches lose much of their insulating properties. Thin, dark mulches or those that are made of dirty paper or polyethylene, increase soil temperatures because they absorb more heat than they dissipate. In fact, summer temperatures under clear plastic can be lethal to plant roots to a depth of 300 mm (12 in). On the other hand, aluminum and white paper mulches reflect radiation and keep the soil cooler in summer than if it were bare or covered with clear or dark films.

Coarse mulches reduce the reflection and reradiation of heat, thus increasing human comfort and plant well-being in summer. Rough pieces of mulch diffuse heat in many directions instead of concentrating it, as a smooth, bare surface or pavement may do. The mulch surface, however, can become quite hot and damage emerging plants.

The incidence of some diseases is reduced when mulch minimizes the splashing of soil and pathogenic organisms onto plants during rain or sprinkler irrigation. Avoid mulch materials, of course, that may harbor diseases to which the landscape plants are susceptible. Though it is probably not appropriate for landscape plantings, aluminum foil has been used to reduce thrips and aphids on certain vegetables.

Problems Caused by Mulches

Most problems associated with mulches can be avoided or greatly minimized. The potential problems are discussed in this section; and the methods of avoiding or minimizing them are also discussed here and more generally in the section on mulch management. Mulches may not be advisable in some situations. Many of the advantages just discussed have associated disadvantages, so the latter are presented in the same order as were the former.

Nitrogen deficiency may develop when mulch materials, such as sawdust and straw, decompose rapidly, particularly if materials are partially incorporated in the soil. This problem can be overcome fairly easily if you use mulch that decomposes more slowly, add nitrogen fertilizer, or minimize incorporation of the mulch in the soil.

Excessive moisture may occur in fine-textured and poorly-drained soils, resulting in poor aeration, particularly below a compact organic mulch. Rickman (1979) found that mulches can aggravate anaerobic conditions in slowly draining soils, leading to soil denitrification or loss of nitrogen fertilizer. Even in better-drained soils, a number of studies (Jacks, Brind, and Smith 1955) report slower nitrification in mulched soils than in bare soils. This is thought to be due to the higher moisture content and lower temperature of mulched soil during the warmer part of the year. When such a problem occurs, the use of mulch should probably be discontinued and drainage improved (see Chapter 7).

Insufficient water may be available under plastic mulch because large sheets of unbroken plastic greatly reduce the amount of rain or sprinkler irrigation reaching the root zone. Openings around the plants, at the edges, and along the overlapped sheets usually allow enough water to flow under the plastic and wet the soil. Pricking holes in the plastic will also overcome this problem and will not encourage weeds. The problem would not arise, of course, under flood irrigation.

Air temperature extremes are accentuated above a mulch. Because mulches insulate soil from radiation and air temperatures, temperatures are more extreme just at and above the mulch. As mentioned, a mulch surface can become so hot in summer that young plants growing through it will be injured. A more frequent threat to most plants is the increased cold above a mulch under radiation frost conditions (see Chapter 4). A mulch has low heat absorption and storage characteristics, and its surface cools quickly under night radiation. In Maryland, Creech and Hawley (1960) found that minimum air temperatures in mid-fall (Oct. 4 to Nov. 12) averaged 3°C (5.5°F) colder 50 mm (2 in) above a hay mulch than above bare soil; temperatures above the mulch were as much as 4°C (7°F) colder on clear nights. They attributed the winter injury of buds and bark-splitting of two-year-old azaleas during the first fall frost to the prolonged growth and the abnormally low air temperatures brought about by continuous mulching. If cold is a hazard to plant tops, mulch should be removed before cold weather arrives so that the heat capacity of the soil can moderate air temperatures.

Mulched plants usually grow more vigorously and longer in the summer, thereby delaying development of hardiness. Whitcomb (1980) examined winter dieback of mulched one-year-old Chinese pistache in Oklahoma. Of trees mulched with black plastic and 50 mm (2 in) of bark mulch, an increasing number experienced top dieback as the fertilizer level increased from zero to 3.5 to 7 kg of actual nitrogen per

100 m² (0 to 7.5 to 15 lbs/1000 ft²). Trees mulched with only 50 mm of bark were killed back only at the highest nitrogen level. Unmulched pistache trees were not injured, nor were any of the other three species tested (sawtooth oak, dwarf burford holly, and pfitzer juniper), regardless of the mulch and fertilizer treatments. Around the pistache trees mulched with black plastic and bark, soil moisture had been higher than around the other pistache during a dry period the previous summer. Where bark or plastic and bark were present, growth was 30 percent and 21 percent higher, respectively, than that of trees in bare soil. The most vigorous late top growth was most susceptible to winter injury. Therefore, you can avoid injury by not mulching, by protecting the plants from the cold (see Chapter 4), or, as the three uninjured species testify, by planting hardier species.

Some diseases and pests may be more serious on mulched plants. A mulched soil favors the increase and development of most microorganisms, some of which may be infectious. Armillaria root rot (*Armillaria mellea*) (see Chapter 19), a soil-borne disease, is encouraged by the moist soil under a mulch. Crown rot is more likely if the trunks of susceptible plants are kept moist by mulch piled against them. Proper mulch selection and placement will greatly reduce the likelihood of infection. In certain situations favoring disease, a mulch should not be used.

Because rodents often burrow and live in loose mulches, particularly if snow is common, mulch should be kept away from the trunks of plants. Trapping or poison may be necessary if the problem becomes serious. Mulches also harbor slugs and some insects that can occasionally be troublesome. These pests can be eliminated by specific control measures. Some mulches, such as straw and manure, contain bothersome weed seeds and should be avoided.

Some mulches increase fire hazard, but most can be used with little or no danger. Except for the surface centimeter ($\frac{1}{2}$ in) or so, most mulches are quite moist and resistant to fire. Many federal and state agencies use wood chips as ground cover to control dust and reduce soil compaction in picnic areas and campgrounds, even in the arid Southwest. That type of coarse mulch works down into the soil and stays fairly moist. Even mulches around stoves and campfires have not proved hazardous. Vapor from the soil below condenses in the colder surface mulch at night and keeps it moist even during dry periods. Rain and sprinkler irrigation, of course, are even more effective in moistening mulches.

On the other hand, a loose mulch of straw or other dry, lightweight organic matter can be a fire hazard even in irrigated areas. Straw dries quickly and is highly inflammable. Such materials are usually unattractive in landscapes, although they are frequently used in conjunction with vegetable and fruit production where summer rains are frequent.

A dry, compact organic mulch can catch fire from a discarded match or cigarette and smolder for days. In a peat-moss mulch plot in Davis, California, two days of continuous sprinkler irrigation wetted only the mulch surface and failed to extinguish a smoldering fire under the surface. The mulch had to be broken apart so that a fine spray of water or fire retardant could be applied directly to the burning material. Wyman (1957) reports a similar experience. Although a smoldering mulch fire may not endanger most plants, wind or nearby inflammable material can cause it to grow to dangerous proportions.

Nearly all fire problems can be eliminated by proper mulch selection and management.

Some mulches may be toxic, particularly to young plants. The bark, sawdust, and foliage of a number of tree species have been reported to be toxic to other plants (Jacks, Brind, and Smith 1955; Del Moral and Muller 1970; and Green 1978). Toxic tissues include eucalyptus sawdust and leaves, redwood and cedar sawdust, Douglas fir, larch, and spruce bark. Reported toxicities occurred with fresh mulch and young plants in almost all cases. Toxic substances can be removed or inactivated by leaching or composting (Hoitink and Poole 1977).

Mulches may affect depth of rooting. Halfacre and Barden (1979) recommend that transplanted trees and shrubs be mulched only after the first year, so that a deeper root system is initially encouraged. Whitcomb (1979a) has found that roots develop in a thin layer beneath a mulch of black plastic and bark, and he stated that plastic restricts depth of root development. This belief that mulched plants develop shallower root systems than unmulched plants, however, is not borne out by other studies. There is no doubt that more roots grow near the soil surface in mulched than in unmulched soil. Several papers, however, report that the roots of mulched trees penetrate as deep or deeper than those of unmulched trees. Jacks, Brind, and Smith (1955) cite a Japanese study in which mulched peach seedlings developed almost 50 percent more total root weight than did unmulched seedlings and double the weight between 0.6 and 1.2 m (2 and 4 ft) deep. In a mature Michigan apple orchard, Kenworthy (1953) found that moisture depletion was uniform at all depths to 1.06 m (42 in) under a hay mulch on an orchard grass sod, while there was no appreciable depletion under an unmulched sod below 0.6 m (24 in), indicating that mulch promoted root penetration. Beckenbach and Gourley (1932) examined the root distribution of an apple orchard that had been mulched for 35 years and found that mulched trees in the lower soil horizons had root systems at least as dense as those of unmulched trees. Whitcomb modified his earlier observation (1979a) in 1980 to say that although distinct differences in the location of fine, fibrous roots are apparent, both mulched and unmulched plants have larger roots penetrating to a depth of 0.3 m (12 in) or more.

It appears that mulches encourage roots to grow deep as well as close to the surface. Mulches should not jeopardize plant stability, though increased early surface rooting may later damage nearby curbs and sidewalks.

Mulch Materials

The range of mulch materials is great. Most are organic and are usually byproducts of industry and agriculture. The following considerations should guide selection of a mulch.

Availability, including price and delivery, is probably the most important factor in determining which mulch to use. Several mulch materials may be free, but transportation and handling make delivery expensive. Large quantities of potential organic mulch materials are now being used as alternate sources for energy, building products, and nursery soil amendments. Plant litter obtained near the site and used directly or after composting may be the most economical source.

Lawn clippings and the leaves of some plants are considered poor mulch materials because they compress and mat together, restricting water movement. They can also become slimy and unattractive. This often happens with fresh leaves that are applied so thickly that they are slow to dry. If applied in thin layers, they dry more quickly; shredding leaves with a mulcher before application will quicken drying and reduce matting. Composting will also reduce the problem.

In Nebraska mulch trials, dried turf clippings proved as effective as dried alfalfa in moderating soil temperatures, conserving soil moisture, and improving growth and nitrogen content of plants (Shearman and others 1979). Of six turfgrass species tested, five were about equally effective and all were much better than creeping bentgrass. Bentgrass clippings were little better than no mulch at all, because they were difficult to dry, tended to compact, and had a crusted surface. Alta fescue and buffalograss should form a more open mulch, since they are coarser and usually heavier when dry than other species.

Ease of application and maintenance is an important consideration, particularly in existing plantings. Wood chips are easy to apply and maintain, while plastic sheeting can be difficult and expensive.

Mulch appearance may be a key element in the design of a landscape (Fig. 13-6). A mulch should enhance the appearance of a garden and conform to the design of a landscape. Many organic materials are quite versatile. Gravel, crushed colored rock, cobblestones, and sand are sometimes used, usually underlaid with black plastic for weed control. The scale of the landscape should suggest the appropriate mulch texture.

Mulch stability under windy and wet conditions is essential. A number of materials, such as rice hulls, blow easily and are unsatis-

FIGURE 13-6
A wood-chip mulch is being renewed in an area of a park
that had originally been planted to turfgrass (upper left).
Two colors of crushed rock on black plastic edge a shrub
and tree border along a sidewalk (upper right).
An entire front yard is mulched with gravel on black plastic;
specimen roses and other shrubs are planted through the plastic (lower left).
Weeds are growing (left foreground) in an old gravel mulch on plastic (lower right).

factory for mulching in most locations. Organic mulches may be washed away during heavy rains, just when they are most needed. Coarse or heavy materials in close contact with the soil are less likely to be blown or washed away. Surface water flow may have to be diverted to protect the landscape as well as the mulch.

Freedom from contamination in a mulch will protect plants and ease maintenance. Mulch material should be free from weed seeds, harmful insects and diseases, and toxic chemicals. Any one of these can create maintenance problems and inhibit the success of a planting. Weed seeds are most likely to occur in mulch material made from the entire tops of mature plants, such as hay, straw, and weeds, or from leaves of small-seeded fall-fruiting plants, such as hackberry and privet. As mentioned, mulch material made from plants that are susceptible to serious diseases such as Armillaria root rot (*Armillaria mellea*) and fire blight (*Erwinia amylovora*) should be selected with caution. Some organic mulch materials, when fresh, may be toxic to plants, particularly young ones. These materials should be leached or composted (see next section) before being applied.

Slow decomposition will prolong the effectiveness of a mulch and reduce the chance of nitrogen deficiency. Organic materials that are high in lignin, are partially decomposed (composted), or have large-sized particles are slower to decompose.

Low fire hazard characterizes many coarse organic materials that work into the soil. Rock and plastic pose no problem, but straw and peat moss can be hazardous when dry.

Permeability is necessary so that rain and sprinkler water can penetrate the mulch and enter the soil. Peat moss does not rewet easily after it dries. Thick thatches of some turfgrasses may repel water so effectively that little gets into the soil.

Mulch Management

Mulches can be most effective and most easily maintained if the appropriate guidelines are followed. A mulch may be used primarily for appearance, weed control, water conservation, erosion control, improvement of soil structure, protection of soil from cold or heat, provision of a firm, clean surface for pedestrians, or for disposing of garden litter. These purposes are not mutually exclusive, but the choice of a mulch and maintenance methods should depend on the primary purpose.

Soil Preparation Installing a mulch is usually one of the last steps in developing a planting bed or landscape. The site should first be brought to grade to provide surface drainage. Weeds should be killed and large ones removed, especially if they have gone to seed. If surface soil is not in good physical condition, loosen it to increase water and air infiltration. Unless plastic is used, apply a preemergent herbicide such as Treflan® or Surflan® before the mulch is placed to reduce weed growth. In small planting areas isolated by buildings and pavement, it may be wise to fumigate soil to destroy most weeds and their seeds, as well as any disease organisms and other pests.

Mulch Decontamination   The bark, wood, and foliage of some plants contain substances toxic to other plants, particularly young ones. Leaching will remove water-soluble toxicants, and composting will remove or inactivate toxicants and kill most disease organisms. The toxic substances can be leached from redwood and cedar sawdust by heavy sprinkling or several months of rain. Water should drain into a sewer or special settling basin so that surface and underground water and nearby plants are not polluted. Pollution-tolerant fish have been killed by as little as 10 ppm of leachate extracts from western red cedar (Scroggins 1971).

Even though they may contain substances toxic to other plants, pieces of freshly chipped wood and bark (with an average diameter of about 20 mm or $\frac{3}{4}$ in) are commonly used to mulch mature plants with no apparent ill effects. A mulch containing toxins, however, is most likely to cause injury if the mulch particles are small (as in sawdust), if the mulch is particularly deep (over 100 mm or 4 in), if heavy rains or irrigation follows application, or if a high proportion of the plant roots are in the surface soil. When there is concern about the toxicity of fresh chips, they should be spread thinly (in layers 20–30 mm or 1 in thick) under the plants. The toxins will be adsorbed on the soil particles and inactivated by other soil constituents. If injury should appear, the mulch can be removed from underneath affected plants or the soil leached heavily to move toxins below the major part of the root zone, or both.

Compost Preparation   Bark and sawdust can be safely used as soil amendments, provided they are composted first to minimize later nitrogen draft and to eliminate toxins. Leaves, grass clippings, and other plant refuse can be composted along with sawdust and chips of wood and bark. Decomposing microorganisms need nitrogen, in amounts that increase with the speed of decomposition. Wood decomposes more quickly than bark, softwoods more quickly than hardwoods, fine particles more quickly than coarse, succulent material more quickly than woody, fresh tissue more quickly than dry. For each cubic meter of compost composed primarily of woody material, 0.5 to 1.5 kg of nitrogen (1–3 lbs/yd^3) should be used, the larger amounts for more rapid decomposition. The total carbon to nitrogen ratio of the materials in a compost pile should be about 25 to 1 (Raabe 1974).

Methods of handling compost depend on the amount of material and availability of space or facilities, equipment, and labor (Fig. 13-7). Bins can be used to store compost for small landscape operations, while large stacks or long windrows can be used for larger quantities of compost. A compost bin is usually 1 to 2 m (3–6 ft) wide and about 1 meter deep, with a variable length. The sides of the bin can be built of wire, wood, or open brick or cement block; if the compost will be

FIGURE 13-7

After the 21-day composting period, a pile of wood-chip and sewage-sludge compost is moved to a holding area ready for use (left); 100-mm (4-in) diameter perforated flexible hose was used to force air through the compost pile.
(Photo courtesy N.H. Experiment Station, University of New Hampshire)
Compost material is forked from the right to the left bin (left bin cover visible above the person's arm) (right). Each bin (about 1 m^3 or 1 yd^3) has a cover to confine heat and maintain optimum moisture.
(Photo courtesy Robert Raabe, University of California, Berkeley)

turned frequently, the sides should be solid. Large amounts of compost should be handled with equipment and formed into stacks or windrows about 5 m wide and 2.5 m high (15 × 8 ft); the amount of material and space will determine length.

Those who use bin composting are commonly advised to fill the bin with alternating layers of organic matter 150 to 300 mm (6–12 in) thick and of topsoil about 25 mm (1 in) thick. To each layer of organic material add an appropriate amount of nitrogen fertilizer. Recommendations for composting, however, vary a great deal: Some people use organic matter and nitrogen alone, while others add alternating layers of soil and manure, more exotic fertilizer combinations, and lime. Each compost enthusiast seems to have a favorite recipe.

C. G. Golueke of the University of California Richmond Field Station found that garden wastes could be converted to compost in three weeks by turning the pile every second or third day if the compost material had a carbon-to-nitrogen ratio of about 25 to 1 and enough moisture to favor decomposition (Raabe 1974). A desirable carbon to nitrogen ratio is not difficult to obtain. Fresh plant material (green leaves, grass clippings, and weeds) has a ratio of about 12 to 1; dried plant material (dried leaves and straw) has a ratio between 50 to 1 and 100 to 1; woody plant materials range from a low ratio in small, soft material to 200 to 1 in sawdust. Mixing equal volumes of green and dried (straw or fallen leaves) material will give approximately the correct ratio; no nitrogen should need to be added (Raabe).

Compost materials should be placed in solid-walled bins to hold the heat. Compost turning is most easily and completely accomplished by transferring the compost from one bin to another. Material toward the outside of a bin should be placed near the center of the second bin to equalize temperatures because temperatures are higher in the center. Frequent turning aerates the compost to aid decomposition and prevents the compost from becoming so hot that most of the decomposing organisms are killed. At the beginning, the pile should not be turned until it has become very warm (several days). If the pile is slow to heat, nitrogen fertilizer can be sprinkled on the material as it is moved to the second bin. If the pile smells of ammonia, sawdust (carbon) should be added during the next turning to absorb the excess nitrogen. Raabe (1974) has found that turning a compost pile daily will result in finished compost in two weeks.

Compost piles in the open should be packed more at the edges than at the center so that the center will settle lower than the sides. Rain or water applied will then wet the compost rather than run off. Compost windrows are built in about 300-mm (12-in) layers with nitrogen fertilizer spread between the layers. The top of the windrow should be fairly level to retain water.

Frequent turning and mixing of the compost of woody material (once every two to three weeks) hastens decomposition and raises compost temperatures to between 50 and 80°C (120 and 160°F) (Hoitink and Poole 1977). Compost mixing can be a slow, tedious procedure with a pitchfork, but a skiploader can turn large quantities of compost quickly. In warm weather, frequently-turned composts high in hardwood bark should be ready to use for mulching in 10 to 12 weeks, and those high in softwood bark within six weeks.

Many cities are composting wood chips and sewage sludge as an alternative to incineration or landfill disposal of sludge. A system of windrowing and forced aeration is simple, easy to operate, and adaptable to different situations (Fig. 13-7) (Leighton, Harter, and Crombie 1978). Wet sewage sludge (about 20 percent solids) is mixed with wood chips at a ratio of three parts chips to one part sludge by volume, then the pile is covered with about 250 mm (1 ft) of composted sludge for insulation purposes. Fans draw air through the pile for the first 10 to 14 days and exhaust it into a small pile of composted sludge for odor absorption. After this period, the fans are reversed, and air is blown through the pile for another 10 days. The resulting compost is stable, practically odorless, and essentially devoid of pathogens.

Leighton, Harter, and Crombie found that composting costs, about $72 per dry ton, were competitive with other methods of sludge disposal: $115 to $134 for trenching, $57 to $93 for incineration, and $62 to $115 for heat drying. Composting appears to be a viable and satisfactory method of sludge disposal and compost production for communities with a low concentration of heavy metals in the sludge.

Plant toxins and pathogens will normally be eradicated during composting. The material will not be completely composted, but will be safe to use around plants. During cold weather, detoxification and composting will be slower. When compost is to be used as a soil amendment, composting should continue until the desired degree of decomposition has occurred.

Mulch Application   Loose organic mulches are usually applied in a layer 100 to 150 mm (4-6 in) deep, but even a thin layer of these mulches (10-20 mm or 0.5 in thick) can quite effectively reduce surface puddling, erosion, and mud splashing from rain. It can also moderate summer soil temperatures and reduce evaporation from soil. Mulches, however, must be thicker to effectively control weeds, minimize soil compaction from traffic, and provide firm footing under wet conditions. The insulation value of an organic mulch, particularly against winter cold, depends on its thickness.

Fresh leaves and grass clippings should be spread in thin layers so they can dry before more are added. Coarse leaves are less likely to mat, but wind can often blow them away. Composting or shredding large leaves might be the best solution.

Black plastic is sometimes used in conjunction with bark or rock to mulch shrub plantings that are free of traffic. Plastic is most effective in controlling weeds and surface evaporation but is easily torn. The soil surface should slope away from the trunks of the plants and drain to an outlet that removes excess water. When placing plastic around a plant, slit the plastic in a single line to a hole cut 25 to 50 mm (1-2 in) larger in diameter than the trunk of the plant, depending on plant spacing and size. When in place, the slit edges can be slightly overlapped to give complete cover. Adjacent plastic sheets should be overlapped about 100 mm (4 in). If you wish to assure that enough rain will enter the soil, you can prick the plastic with holes to allow water to penetrate without encouraging weeds. Usually, enough air exchange will take place at the trunk holes, seams, and edges.

After the plastic is in place, apply bark, gravel, crushed rock, cobblestones, or other material thickly enough to cover the plastic. The border of the planting should be high enough to contain the mulch. The plastic should last several years if it is not disturbed by traffic.   Problems may include dust, plant litter, and small pieces of trash that accumulate in the mulch on top of the plastic. After a few years, seeds will also germinate in the mulch and lower its effectiveness (see Fig. 13-6). It is tedious to remove leaves and litter from a mulch; such litter would just blend in with most organic mulches. To replace plastic and bark or rock mulches, you must remove the mulch covering, install new plastic, and replace the mulch cover. Most other mulches can be simply upgraded if you add more mulch material.

Fertilization   Improved growth from mulching may decrease or even eliminate the need for fertilizing. Whitcomb (1980) reports winter injury to young Chinese pistaches that were invigorated after being mulched with plastic; the number of injured plants increased with increasing nitrogen application.   In some instances, however, nitrogen deficiency may occur after installation of an organic mulch; most other nutrients will increase under mulching. A fine-textured, rapidly-decomposing material in close contact with a relatively low-nitrogen soil will be most likely to cause nitrogen-deficiency symptoms in mulched plants. Close observation of plant growth is the best guide to fertilizer needs.

If growth is vigorous and leaf color dark, no fertilizer is needed. On the other hand, if growth is less than desired and leaves are yellowish, nitrogen should be applied.   If the nitrogen-deficient plants are small, the mulch can be raked from beneath them, 25 to 100 g of nitrogen per square meter ($0.05-0.20$ lb/yd^2) spread on the surface around each plant, and the mulch replaced. Fertilizer should not be concentrated in basins. If the plant canopy is more extensive, apply a nitrate or urea fertilizer on the mulch (assuming there is no plastic) at a rate of about 1.5 kg N/100 m^2 (3 lbs/1000 ft^2) and sprinkler irrigate the fertilizer (with about 25 mm or 1 in of water) through the mulch into the soil below. If experience indicates that mulching may cause nitrogen deficiency, nitrogen can be applied before the mulch is put in place (see Chapter 11). Alternatively, fertilizer can be mixed with the mulch before application, at a rate of 1 kg N/100 kg or 600 liters of mulch (1 lb/100 lb or 10 ft^3). The fertilizer, whether mixed with the mulch or applied on top and watered through, will quicken decomposition of the mulch but should also correct or prevent nitrogen deficiency.

If deficiency symptoms persist two months after fertilization, the soil may be poorly aerated because of high moisture under the mulch. The mulch may have to be thinned in such a case.

Irrigation   If irrigation is necessary, mulching can reduce its frequency.   Plantings treated with organic mulches respond best when they are drip-, soaker-, or sprinkler-irrigated. Flood and furrow irrigation will carry light mulch to low areas or the downwind portions of a basin. Sprinkler irrigation keeps a mulch moist and thus reduces fire danger. Plantings treated with plastic mulch can usually be irrigated satisfactorily by flood, sprinkler, or drip irrigation.

Fire Protection Straw and peat moss are highly inflammable and should normally not be used for landscape mulches. As mentioned, wood chip mulches can be used extensively in campgrounds without posing serious fire problems. Wyman (1957) has also found that spent hops burn slowly. Sprinkler irrigation and rain will keep most organic mulches resistant to fire. Where straw is used on top of ground-cover plants to protect them from winter cold, it should be removed in spring to allow for the best growth and appearance of plants, as well as to reduce the fire hazard.

In arid regions, where organic mulches are more liable to catch fire, they can be sprayed with a solution of ammonium sulfate or diammonium phosphate (1-1.5 kg/40 liters or 2-3 lb/10 gal of water per 100 m^2 or 1000 ft^2). These effective fire retardants will also enhance soil fertility and mulch decomposition as they soak in below the mulch surface. They are not used in irrigated landscapes or areas with summer rain, where mulches are already slow to burn, and chemicals would wash off the mulch surface.

Inorganic mulches offer most of the benefits of organic mulches and in addition will not burn or decompose.

Disease and Pest Control Do not use mulch materials that might be infected with disease organisms to which your plants are susceptible. Even though composting would probably kill any pathogens, it is better not to compost diseased material. Do not mulch plantings where *Armillaria mellea* is present, because the higher soil moisture under a mulch will encourage root rot.

If rodents damage the trunks of trees and shrubs during the winter, place gravel or crushed rock 100 to150 mm (4-6 in) deep around the trunks. Rodents find it difficult to burrow in gravel. Keeping mulch away from the base of plants will also decrease rodent damage, even without the gravel barrier. If roots are also being damaged, the rodents must be trapped or poisoned; see a pest control officer in the local agricultural commissioner's office for specific recommendations.

Protection from Cold Most mulches protect against heat loss by reducing radiation and conduction from below. Low-growing plants can be protected from cold if they are covered with a loose mulch, such as straw, from soon after the first frost until spring growth. On a calm, clear night, temperatures above a mulch will be colder than those above bare soil. This difference in temperature can sometimes be the difference between severe injury or death and safety to the upper parts of plants. In these situations, the mulch should be removed so the soil is warmed by the sun during the day and in turn protects the plants at night. Under a full canopy, however, a mulch will make little difference, because exposure to the sun and the cold night sky will be limited in any case.

Mulch Incorporation   Although most organic mulches will improve the physical and chemical properties of soil, that is usually not their primary purpose. In most situations, mulches should remain on the surface and should not be incorporated into the soil.

CHAPTER FOURTEEN

Pruning

Pruning is the removal of plant parts: usually shoots and branches, but sometimes buds, roots, and even flowers and fruit. By pruning, one can control the growth of plants to enhance their performance or function in the landscape. Pruning can increase the productive capacity of fruit trees, the trunk quality of lumber trees, the quality and size of flowers and fruit, the structural strength of fruit and landscape trees, and the aesthetic appeal of many plants.

GENERAL PRINCIPLES
OF PRUNING

Purposes of Pruning

Plants grow in many shapes and sizes. Some have central leaders with tall, straight trunks (an *excurrent growth* habit) (see Fig. 2-12). Others have several main branches with spreading crowns (a *deliquescent*, *diffuse*, or *decurrent* growth habit) (see Fig. 2-13). Between these extremes, intermediate forms occur. The natural characteristics of different plants can be exploited through landscape use and maintenance practices. Pruning can enhance plant appearance and safeguard plant health and well-being. Depending on its extent, pruning can affect a plant from root to crown. Growers should therefore be familiar with pruning techniques and plant responses to pruning. You should be able to determine the growth habit of a plant by observing its growth and its response to previous pruning and thereby be able to properly prune even unfamiliar species. This discussion will cover pruning techniques and plant responses in general, but will offer little advice on pruning particular species.

Plants may be pruned for a number of reasons:

Compensation for Root Loss   This is often given as a reason for pruning back the tops of newly planted plants. It is true that considerable root loss occurs between the nursery field and landscape planting, but removing a large portion of the top may not necessarily improve a plant's ability to survive and grow (Whitcomb 1979a). Such pruning may instead delay the initiation of root growth (see next section).

Training Young Plants   The arrangement, attachment, and size of scaffold branches can be controlled to produce vigorous and mechanically strong plants. Pruning should take advantage of the plant's growth habit, accentuating its natural tendencies, seldom modifying them greatly. You may create unusual plant forms through pruning, including topiary, espalier, bonsai, pleach, and pollard forms (discussed later), but you must be familiar with the individual plant's responses to pruning.

Maintenance of Health and Appearance   Pruning can remove dead, diseased, injured, broken, rubbing, and crowded limbs. A dense top may be thinned to allow for the passage of light and air. Light is needed by the interior parts of the plant and by other plants beneath it; air circulation reduces the incidence of certain diseases and allows sprays to penetrate more effectively. Proper thinning of the top reduces wind resistance, which can create deformities or break the trunk and branches.

Control of Plant Size   Pruning can reduce shade, the danger of wind-throw, and interference with utility wires, can simplify pest-control spraying, and can prevent the obstruction of views and traffic. If you choose plants that will be an appropriate size at maturity, this will minimize the need for pruning. If a plant must be pruned more than every five to seven years to control its size, it is the wrong plant for the particular location or use. Withholding nitrogen fertilizer and growing lawn under trees and shrubs will slow growth and reduce the need for pruning when a plant reaches the desired size.

Influencing Flowering, Fruiting, and Vigor   Pruning can change the balance between vegetative growth and flower bud formation. If young plants that flower on one-year-old wood are pruned, the development of flower and fruit may be delayed. On mature plants, pruning helps maintain vigor, minimizes overcropping (which results in small blossoms and fruit and broken limbs) and encourages annual flowering and fruiting. Pruning plants that flower on current season's growth stimulates shoot growth and usually produces fewer and larger flowers, particularly those that flower only from terminal buds.

Invigoration of Stagnating Plants   When plants are doing poorly but show no symptoms other than extreme lack of vigor, they may be pruned in a "kill or cure" operation.

Increasing the Value of Conifers   Nursery stock and Christmas trees may be sheared to improve their shape.

Pruning Responses

Young Plants   Pruning removes leaves and buds that would develop into leaves. Two seemingly opposite effects occur when young plants and those that do not have a heavy flower and fruit load are pruned:

Invigoration of individual shoots is the universal response to pruning (Fig. 14-1). Pruning off foliage and buds that would develop into leaves allows the root system (which is not immediately affected) to supply each remaining shoot, leaf, and bud with more water and nutrients than before. Individual shoots grow more rapidly and later into the season; leaves become larger and darker in color. Even with larger leaves, though, total leaf area may be reduced on more severely pruned plants because there will be fewer shoots. A pruned plant will transpire less water than will an unpruned plant if its shadow is reduced in size or density. The reduction in transpiration, however, will be less than the reduction of foliage.

Dwarfing results from pruning young plants and those that do not flower and fruit heavily. Fewer leaves and buds are left. Though individual leaves may be larger, total leaf area will be less. Shoots will grow later in the season, using foods produced by the leaves. After shoot growth stops the pruned plant has less time to produce food for other growth and to store food for the next season. Less total growth is the result. This can be observed and even measured by the relative sizes of trunks on plants that have been pruned more and less severely.

A young plant that has been pruned will usually exhibit the following characteristics at the end of the first growing season:

The top and root systems are in balance.

The top and roots will be smaller than if the plant had not been pruned.

There will be less stored food because the plant had a smaller leaf area for photosynthesis, and its leaves were active for a shorter time.

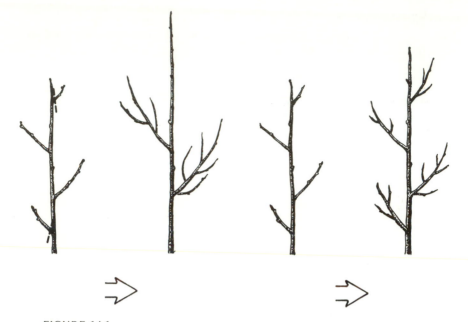

FIGURE 14-1
The effects of pruning.
The trees to the left of each arrow
were exactly the same before two of four lateral branches
were pruned off the tree on the left.
The two remaining laterals and the terminal of the pruned tree
grew more than the corresponding laterals on the unpruned tree,
but by the next dormant season the unpruned tree (extreme right)
will have made more total growth.

The extent of invigoration and dwarfing will depend on the severity of pruning. A branch that needs subduing must be pruned more severely to reduce its relative growth. A branch that needs encouraging, however, must be pruned lightly or not at all; branches that shade or compete with the branch to be encouraged must be pruned more severely. Removing dead, weak, and heavily shaded branches has a much smaller effect than removing healthy, well-exposed branches.

Mature Plants Pruning does not always result in dwarfing. Plants, such as peach and pyracantha, that produce flowers and fruit heavily on one-year-old wood can attain a greater leaf area by the end of the growing season if they are pruned when dormant. Both leaf and flower buds are removed by pruning. No new flower buds will form and bloom in the spring following pruning, but shoots from the leaf buds that remain can form more leaves than they would have without pruning. Thus, shoots and leaves will form in greater quantities than flowers and fruit. It is possible to prune severely enough that the leaf area developed will not only increase fruit size but also enhance total growth of the plant. Pruning a stagnated plant can stimulate it to grow more vigorously and also more in volume.

Top Pruning   Removing branches of a young plant reduces not only the food supply to the roots but also the flow of auxin formed in developing buds and leaves. The spring root growth of deciduous plants may be delayed by the removal of shoot terminals, which provide the first auxin needed to stimulate root initiation. Richardson (1958) found a close correlation between the initiation of spring root growth in sugar maple and the presence of at least one physiologically active bud. The roots of sugar maple whips began to grow just before the terminal bud began to expand; when the terminal bud was removed, root growth slowed to zero in three days and did not resume for another five days, when the two uppermost lateral buds began to swell. This response may explain Whitcomb's (1979b) finding that pruning bare-root trees at planting (designed to balance the top with a reduced root system) improved neither survival nor growth. On unpruned trees, expanding buds and new leaves may stimulate enough additional root growth to more than compensate for any increased transpiration due to a larger leaf surface. It is not known which plants respond to pruning as the sugar maples do. The best established reasons for pruning plants at planting are to remove damaged branches and to begin to develop tree structure. It may be desirable, however, to leave as many terminals as possible in order to stimulate root growth.

The Influence of Pruning on Flowering and Fruiting   In most young plants, particularly those that produce flowers on one-year-old wood (Fig. 14-2a), pruning will delay flowering. It will invigorate the growth of individual shoots, so less food and fewer hormones will be available during the early summer, when the buds for next season are differentiating into leaf or flower buds. If you keep pruning to a minimum and withhold nitrogen fertilization, plants will be more floriferous and will flower at a younger age. Plants of low vigor may produce few flowers because of a lack of nitrogen and associated compounds; pruning channels available nutrients into fewer shoots, thereby increasing the nitrogen supply of each. Plants that flower laterally on growth of the current season (Fig. 14-2b) will produce more flower buds along invigorated shoots. Plants that flower terminally will have larger flowers or flower clusters. Measures that increase flowering will usually increase fruiting.

FIGURE 14-2a
Spring-flowering plants bloom from buds
formed the year before (extreme left).
Flowers usually open before new shoot growth is extensive (left center).
New shoots grow after the setting of fruit and fading of bloom (right center).
Dormant buds on newly formed shoots overwinter
to bloom the next spring (extreme right).

FIGURE 14-2b
Summer-flowering plants bloom from buds on the current season's shoots.
Overwintering twigs have dormant vegetative buds
and often remnants of the previous summer's blossoms (extreme left).
New shoots grow from lateral buds in the spring (left center),
and flowers or flower clusters form on current shoots,
often terminally (right center).
The following winter, flower stalks may remain
on the previous season's shoots (extreme right).

Root Pruning   As would be expected, root pruning has essentially the opposite effect of top pruning. It decreases the supply of nutrients and water to the top, thus producing less vigorous growth of individual shoots and less total growth. On a young, vigorous plant, root pruning will encourage flowering and fruiting at a younger age. The roots of established trees are sometimes pruned to obtain more fibrous rooting within the root ball (see Chapter 9); this is thought to enhance transplanting success. When surface roots threaten to damage walks, streets, paving, and foundations, they are commonly pruned back and efforts are made to stop or divert their growth. Root pruning can be used to decrease the vigor of severely top-pruned, vigorous shrubs. Root pruning is hard, dirty work, and the results are not always obvious, so it is usually discussed more than it is practiced.

The Time of Pruning   The appropriate time to prune will depend on the type of plant, its condition, and the results desired. Light pruning can be done any time. Unwanted growth is most easily removed while it is small, and early removal will have less of a dwarfing effect. Broken, dead, weak, or heavily shaded branches can be removed with little or no effect on a plant, no matter what the timing.

Rapid plant development can be best maintained when the required pruning is done before the period of rapid growth that usually occurs in spring. Most deciduous plants can be pruned any time during the dormant period between leaf-fall and spring growth with similar results. Evergreens will be set back the least if they are pruned just before spring growth starts. A few broadleaved evergreen plants grow most rapidly after the weather warms up later in the season. If these plants are pruned just before the period of most rapid growth, leaves will be kept productive for the longest time and pruning cuts will be concealed more quickly by new growth. The growth of young plants can be directed during the growing season itself. Branches in desired positions can be encouraged if you pinch back or remove competing shoots.

Plant development can be slowed and plant size maintained if pruning takes place soon after growth is complete for the season. Such pruning should not be so severe or so early as to encourage new shoot growth. If maximum dwarfing is desired, most plants should be pruned in the period from spring to midsummer. This will reduce leaf area for the longest period. Pruning cuts should be discretely placed for minimum visibility.

Corrective pruning may be easier during the growing season. Branches that hang too low from the weight of leaves or fruit can be thinned. Dead and weak limbs can be easily spotted for removal.

If you wish to maximize flowering, timing will depend on the flowering habit of the plant (see Fig. 14-2). Those that flower in summer or fall on the current season's growth, such as crape myrtle, jacaranda, and rose, should be pruned during winter, before growth begins. Moderate-to-severe pruning will favor the growth of fewer but larger blossoms or blossom clusters. Plants that flower in the spring from buds on one-year-old wood, particularly flowering fruit trees, should be pruned near the end of their blooming period. You can enjoy the blossoms and then remove them before they produce fruit that competes with new shoot growth. This will encourage vigorous growth to support the next year's bloom. If pruning is delayed until the end of spring bloom, it will have little or no debilitating effect on growth. Blossoming utilizes food in the bud and its shoot and makes little or no demand on other food reserves. This is well demonstrated by the branches that will bloom and initiate shoot growth after they are cut and brought indoors to flower.

Callusing should be somewhat more rapid if a wound is made a few weeks before or after growth begins, assuming that bleeding is not a problem. In Illinois, Neely (1970) found that wounds on ash, honey locust, and pin oak callused about equally rapidly in the next growing season whether cuts were made in the spring, summer, or winter, but about 20 percent more slowly when they were made in the fall. Summer wounds, however, callused much less than spring wounds during the current growing season, because they had less time to do so. This difference did not carry over into the following season.

Shigo (1981, personal communication) has suggested that pruners avoid the spring flush of growth so that there will be less chance of tearing loose bark, which is particularly vulnerable when the cambium is active. He also suggests that the fall, when most decay fungi are sporulating, is not a good time for pruning. This, along with Neely's (1970) findings, should discourage you from pruning in fall without good reason.

Bleeding can be minimized if predisposed species are pruned in the fall and early winter instead of late winter and early spring. Wounds on mature trees, particularly deciduous trees such as birch, elm, and maple, can bleed heavily (Brown 1972). On susceptible trees, bleeding can be minimized if only small cuts (less than 75 mm or 3 in) are made. Bleeding is usually not harmful to plants, but if it is heavy and persistent it can cause bark injury below the pruning cut and can retard callusing in the lower portion of the wound. Brown (1972) describes a procedure for tightly binding small wounds to stop bleeding.

Cold injury may be increased if substantial pruning takes place before growth begins in the spring. Some plants, such as roses and subtropicals, can be stimulated into new growth if they are pruned in the fall and early winter. A pruned plant may begin growth during a warm period in the winter, only to be injured when the weather turns cold again. When winter temperatures go below $-20°C$ ($0°F$), the hardiness of tissue near pruning cuts may be reduced even though growth is not stimulated by pruning. This is particularly true of some conifers (Brown). If cold injury is a danger, it is best to delay pruning until just before growth begins in the spring.

The incidence of some diseases can be affected by the season of pruning. Ceanothus that is native to California is much more susceptible to fungus dieback (*Sclerotinia fructicola*) if pruning in the winter or early spring is followed by rain, which spreads the fungus; light annual pruning in the summer is best. Fire-blighted shoots and branches (those affected by *Erwinia amylovora*) should be pruned back heavily, at least 300 mm (12 in) into healthy wood, as soon as possible to minimize further infection. Himelick and Ceplecha (1976) were able to stop or greatly retard the development of Dutch elm disease (*Ceratocystis ulmi*) by pruning out limbs that showed symptoms (yellowing and wilting of leaves) before 5 percent of the tree was affected. Pruning during or just before beetle flight, however, greatly increases the chance of infection, because fresh pruning wounds attract beetles. Check the influence of timing on infection and development of diseases before you prune species that are susceptible to serious vascular and foliage diseases.

Solar heating capacities will be enhanced if you prune in the early fall to open up trees and allow more sunlight to reach exposed windows, solar collectors, and patios (see Chapter 10). The bare branches of deciduous trees can block up to 70 percent of the incoming radiation.

Types of Pruning Cuts   The type of pruning cut not only affects the initial appearance of a branch or plant but also to a large measure determines growth. *Heading* and *thinning*, two types of pruning cuts, produce quite different plant responses.

Heading or heading-back is cutting a currently growing or one-year-old shoot back to a bud, or cutting an older branch or stem back to a stub or a tiny twig (Fig. 14-3). There are several variations of heading to a bud: *Pinching* is the removal of the terminal 20 to 50 mm (1–2 in) of a growing succulent shoot. *Tip pruning* is the selective heading of shoot terminals that may or may not still be growing. *Shearing* is tip pruning without selecting individual laterals or buds, as when a hedge or topiary form is maintained. Heading a large branch or trunk is often called *stubbing*.

In response to heading young branches and leaders, new growth develops from one or more buds just below the cut; lower buds usually do not grow (Fig. 14-4). Depending on pruning severity, the new growth is usually vigorous, upright, and dense. New foliage and branches may

FIGURE 14-3
Heading back is pruning to a stub (lower
branch), a small lateral (trunk), or a bud
(terminal on small lateral).

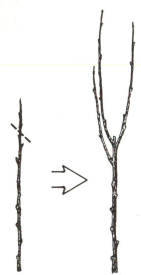

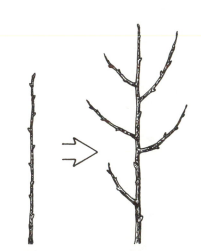

FIGURE 14-4
Heading a one-year-old shoot (left pair above)
will force two to four buds just below
the cut into vigorous upright growth.
Growth from a similar but unpruned shoot
will be more uniformly distributed along
the shoot and will be less vigorous
(right pair). The London plane tree
in the foreground of the photograph was
headed back (stubbed) shortly before the
photograph was taken (right); the tree
immediately to its left was headed back
the winter before. Note the vigorous
upright shoots stimulated below the
pruning cuts.

be so thick that lower leaves, and nearby plants, are severely shaded. Shoots that grow from the trunk or large branches (epicormic shoots) after stubbing come from latent buds and are attached by only a thin layer of new wood formed by the cambium (Fig. 14-5). Branches from these shoots, especially when young, are weakly attached and can break off easily.

FIGURE 14-5
Laterals that begin growth while a tree is young usually are well attached (left).
New (epicormic) shoots forced on older limbs are weakly attached
and can split out easily (right). Note the lack of union (between arrows)
of the lateral branch that grew after this coast redwood was headed back (bottom);
until the stub is completely engulfed with growth from this and other laterals,
the branch will not be securely attached.

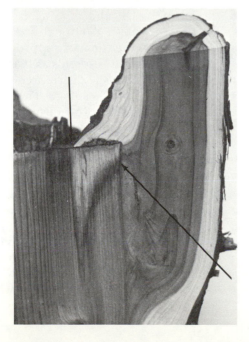

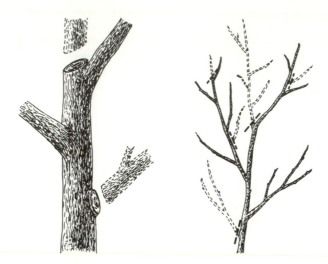

FIGURE 14-6
Thinning-out is removing a branch at its point of origin or shortening
a branch or trunk by cutting to a lateral large enough to assume the terminal role.
This applies to mature (left) as well as young (right) trees.

Thinning or thinning-out is the removal of a lateral branch at its point of origin or the shortening of a branch by cutting to a lateral large enough to assume the terminal role (Fig. 14-6). Reducing the height of a tree or branch by thinning the terminal to a large lateral is called *drop-crotching*. The lateral to which a branch or trunk is cut should have at least one-third the diameter of the cut being made (N.A.A. 1979). "Heading" is sometimes used to denote this cutting back to a large lateral, but "thinning" seems to be the more appropriate term, because little or no stub is left and the plant responds as it would to other thinning cuts. The response to thinning is distributed more evenly through the plant than is the response to heading. The plant becomes more open but retains its natural form. More light will penetrate a plant that has been thinned, and foliage will grow more deeply inside it. For a given severity of pruning, shoot growth will be less vigorous after thinning than after heading.

Location of Pruning Cut The type of pruning cut not only influences subsequent growth, but its location in relation to the branch attachment determines the size of wound and also affects the callusing of the wound, exposure to decay, and the possibility of ring shakes (circumferential separation of xylem along an annual ring) (Shigo and others 1979). The closeness of a cut also affects the amount of regrowth from the base of a cut and the strength of attachment when a branch is thinned to a lateral.

From the evidence to date, most pruning cuts should be made close to but beyond the branch bark ridges (shoulder rings) and the collar at the base of a branch (Fig. 14-7). Branch bark ridge is synonymous with shoulder ring, but a collar may extend further, particularly on a weak or dead branch. When ridges or collar are not visible, cut slightly beyond an imaginary line joining the vertices of the upper and lower angles of the branch attachment.

Branch bark ridges are areas, rings, or lines of bulging bark that may be rough and darker in color than surrounding bark. When a branch attachment is cut longitudinally somewhat similar ridges are seen where the xylem of each growth ring of the trunk or branch and that of the lateral meet (see Fig. 16-12). Shigo and coworkers (1979) point out that this zone is usually a strong physical barrier to decay between the lateral and its mother branch and can become a more effective chemical barrier should the lateral die or be injured. If this be the case, it would be wise not to cut the branch bark ridges that indicate the barrier zone in the xylem between the branch and the trunk without good reason.

A pruning cut outside of the branch bark ridges and collar differs from a "flush" cut (which is made as close as possible). Several authors recommend that the final cut should be flush with or as close as possible to the trunk or mother branch (Bartlett 1958; Baumgardt 1968;

FIGURE 14-7
In order to preserve the branch bark ridges, thin lateral branches as shown by the single dashed lines and leaders back to a lateral as shown by the double dashed lines.

Wittrock 1971; Grounds 1973; Bernatzky 1978; and Pirone 1978a), but flush cuts are unnecessarily large. Shigo and coworkers (1979) observed that some flush cuts on black walnut were more than three times as long (vertical) as the diameter of the limb removed. The long cuts often resulted in multiple ring shakes and opened trunk tissue to decay.

The location of pruning cuts given by Davey (1967) and the N.A.A. (1979) should be modified slightly outward. They recommend the final cut be made through the center of the collar or shoulder rings. Although they did not mention shoulder rings or collars, Brown (1972) and Bridgeman (1976) agree that the final cut should make a smaller wound than will most flush cuts.

Although the foregoing remarks pertain primarily to large pruning cuts, Shigo and coworkers (1979) indicate that branch bark ridges should not be violated even on young trees. An exception might be with vigorous, young trunks on which a close flush cut would reduce protruding buds and thereby reduce the number of shoots that might be stimulated to grow. Such wounds would be small and callus quickly. Some species in cold winter areas might be subject to radial shakes (xylem separation along a radial plane) that could lead to frost cracks (see Chapter 4).

Leave a short stub and cut upward when you thin the terminal of a young plant to a lateral (Fig. 14-13). The new terminal is less likely to split out. Preserving the branch bark ridges will decrease the likelihood of decay.

Making the Pruning Cut

Pruning shears come in a variety of sizes and shapes, but are essentially of two types. One has a curved blade that cuts by passing close to a curved or hooked anvil (Fig. 14-8). The other has a straight blade that cuts against a flat anvil. The straight shears are commonly available only as hand shears (secateurs). The curved shears make closer cuts and are less likely to crush stem tissue, particularly if the shears are dull.

Hand and power saws are used to cut branches that are usually too large for pruning shears (Fig. 14-9). Chain saws are used to remove large branches (Fig. 14-10). They have greatly eased the effort needed in pruning, but often the chain saw blade is dropped into a branch crotch with little regard for the location of the branch bark ridges. The resulting large flush cut opens trunk tissue to decay and usually takes longer to callus than a smaller proper cut.

Other mechanically powered pruning tools, particularly those on poles (Fig. 14-10), greatly increase the versatility of the arborist. Even though most pruning cuts will not be accurately placed, the cuts cause few or no problems because they are usually made in the smaller branches. These tools can be powered by pneumatic, hydraulic, elec-

FIGURE 14-8
One-hand shears (secateurs)
may be of the hook and blade (top center)
or the anvil (bottom center) type.
Long-handled shears also come
in varying lengths and are of both types
though the hook and blade
is more common (right).
Hedge shears have two blades
of similar design (left).

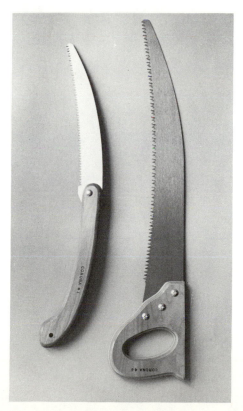

FIGURE 14-9
A hand saw that folds (left)
is convenient when working with young trees
because the saw can be folded
and put in a pocket when not in use.
A larger saw with larger teeth
is more effective for more frequent sawing
of larger limbs (right).

FIGURE 14-10
Power tools greatly add to the versatility and speed
with which pruning cuts can be made.
Gasoline-engine-powered chain saw (top)
and hydraulically operated circular and chain saws and fopping shears (bottom);
photos are not shown at the same scale. (Photo of three tools
courtesy of Fairmont Hydraulics, Fairmont, Minnesota)

FIGURE 14-11
A boom-mounted power saw
is able to cut and hold limbs,
so it is particularly suited for work
near utility lines or in areas that make it
difficult to lower large cut limbs.
(Photos courtesy of Dorsey Trailers,
Montgomery, Alabama)

tric, and gasoline-driven equipment adapted to be used with aerial-lift equipment. A boom-mounted power saw with a limb clamp now available can cut limbs up to 170 mm (6.75 in) in diameter, hold the cut portion, and remove it from the tree (Fig. 14-11). This equipment is particularly adapted for utility-line clearance and other situations in which limbs would be difficult to remove and lower safely.

Small Branches You can make a close cut by placing the blade against the branch bark ridge of the limb to be removed; an even closer cut can be made by placing the blade against the trunk or branch that is to remain (Fig. 14-12). A close cut reduces the number of shoots that might grow near the pruning cut. Conversely, if a short stub is desired, the anvil should be placed against the trunk or branch to

FIGURE 14-12
Make a close cut by placing the blade of hook-and-blade shears
just to the outside of the branch bark ridge of the branch to be removed (left).
Less effort will be required if the blade cuts up (left)
or diagonally (center) instead of down.
Similarly, less effort is needed to make a diagonal heading cut (right);
the branch is less likely to be crushed with a diagonal cut.
The top of the diagonal cut should be about 5 mm ($\frac{1}{4}$ in)
above the topmost bud left.

remain. Leaving a short stub and cutting upward when thinning a terminal to a lateral (Fig. 14-13) will lessen the chance of the new terminal splitting out.

Less effort will be required to remove a branch and less tissue will be crushed if the blade cuts up or sideways (Fig. 14-12). A heading cut can be made with less effort when the cut is made diagonally instead of at right angles to the branch cut. It is also easier to cut a limb when you take a deep bite and place the fulcrum of the shears near the limb to be cut. If you cut through a limb too large for the shears or twist them while cutting, you can strain the shears and permanently damage them.

Where appearance is important, you can hide pruning cuts on shrubs and low-branched trees somewhat by angling cuts away from the direction of most frequent viewing (Fig. 14-14). The plants will appear more natural if they are cut to a lateral arising on the top of the branch; if the pruning cuts are horizontal, or parallel to the ground, they will be hidden from view. If you prune just before growth begins, the cuts will be covered by new growth most quickly.

FIGURE 14-13
When cutting back to a lateral place the
anvil of the shears in the crotch (left)
and cut up parallel to the direction of the lateral;
leave a short stub so the branch bark ridge of the lateral is not cut (center).
If you place the blade in the crotch and cut down, however,
you will usually cut into the branch bark ridge and often
split out the selected lateral (right).

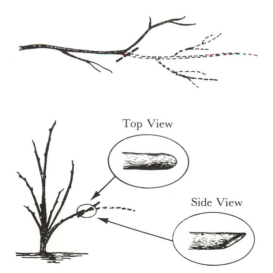

FIGURE 14-14
On low-growing shrubs, you can often hide pruning cuts by cutting back
to a horizontal lateral growing from the top of the branch (top)
or cutting to a bud so the cut surface is toward the ground
and away from a viewer's angle of vision (bottom).

Large Branches   Limbs larger than 25 mm (1 in) in diameter must usually be cut with a saw.   Branches much larger than 50 mm (2 in) should be cut in three steps to avoid splitting back the branch and tearing the bark (Fig. 14-15).   The first cut should be made on the underside of the branch 300 to 600 mm (1-2 ft) from the crotch.   Cut the branch about one-fourth of the way through or until the saw begins to bind.   Begin the next cut on top of the limb within 25 to 50 mm (1-2 in) of the first cut and saw until the limb breaks off.   Many arborists believe the second cut (on top of the branch) should be farther from the crotch than the first so that the cut branch will "jump" from the crotch when it splits and the lower cut surfaces bump together. The branch should jump equally well and the limb split cleanly between the two cuts as long as the second is reasonably close to the first.

Place the third cut at the crotch, making as small a wound as possible, but do not leave a stub that will be slow to callus.   As described earlier, this is best done by cutting just beyond the branch bark ridge or the branch collar.   Observations of old pruning cuts usually reveal that the most uniform callusing is on those that were cut in or just beyond the branch bark ridge area.

If the branch stub is heavy, it should be removed with two cuts: the first from the bottom and the second from the top.   Until cut, a heavy stub should be supported by a rope sling which can later be used to lower the stub safely (Bridgeman 1976).

Many books and articles on pruning give directions for paring the edges of large pruning wounds with a sharp knife or chisel.   Paring means cutting the thick bark around a wound to give a smoother, more even surface.   The process is thought to speed callusing.   However, although the wound will look neater, paring does not seem to be worth the effort.

FIGURE 14-15
Remove a large limb by making three cuts.
Make the first cut on the bottom of the branch
300 to 600 mm (12–24 in) from the branch attachment (left).
Make the second cut on the top of the branch within 25 mm (1 in)
of the under cut. Make the final cut just beyond the outer portion
of the branch collar or branch bark ridges (right).

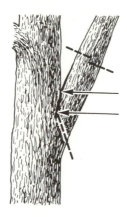

FIGURE 14-16
When removing a large branch
with a sharp branch attachment,
angle the third cut upward
toward the top of the actual union
of the branch with the trunk.
Although they touch,
the branch and trunk are not united
between the two arrows.

Cambium may be damaged by a mallet and chisel and will be subject to greater desiccation if too much bark is cut. If you do pare bark around a wound, pare only thick bark and cut no more than one-half its thickness at the wound edge. Feather out the pared bark to the bark surface around the wound (Keith Davey 1981, personal communication).

In seeming contradiction to Davey (1967), Brown (1972), and Bridgeman (1976), Thompson (1961), Bernatzky (1978), and Pirone (1978a) recommend that pruning wounds be traced. This requires a flush cut and removal of any shoulder rings. An elliptical tracing further removes bark above and below the wound. This bark would supposedly die because of inadequate sap flow, but such is not the case; observations support that wound closure is most rapid when the cut is through the shoulder rings. I am unaware of any study that supports the practice of tracing pruning wounds.

Remove a branch with a sharp "V" crotch in a similar three-step process (Fig. 14-16). Be aware, however, that the actual union of the two branches is often much lower than the apparent junction. The cut should slope downward from the point of attachment at a 30° to 40° angle from the horizontal.

Dead Branch Stubs Advice conflicts somewhat on how to remove a dead branch stub on which a collar has formed. Bartlett (1958) states the wound will heal more quickly if the cut is made into the collar of the callus. Shigo and co-researchers (1979) dissected the trunks of seven 36-year-old black walnut trees from which dead branch stubs had been pruned 11 years earlier. Ring shakes (tangential and longitudinal separation in the wood) were associated with 14 of 17 stubs that had been flush-cut; they were not associated with four stubs whose branch collars had not been removed from the tree. In addition, more discolored wood was associated with the flush cuts and was subject to decay. The authors consequently recommend that when pruning

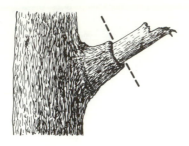

FIGURE 14-17
A dead branch stub that has a collar
of live wood should be cut just at the
outer edge of the collar.

is done late in the life of a tree, care must be taken to preserve branch collars that form around the bases of dying and dead branches (Fig. 14-17). When this zone is removed, the stub is vulnerable to infection.

Protecting Pruning Cuts

It is doubtful that pruning cuts can be protected by an asphalt emulsion or other materials (Neely 1970; Shigo and Wilson 1977; Harris and others 1981). The purpose of these coverings is to protect the cut surface from wood-rotting organisms and to reduce surface checking. Upon exposure to the sun, however, all but the thinnest coverings may crack. Moisture from rain, sprinklers, or dew can then enter the cracks and accumulate in pockets between the wood and the wound covering. These circumstances are even more favorable for wood-rotting organisms than an uncovered wound. If you wish to paint pruning or bark wounds for the sake of appearance, bonding of the paint will be strongest if you allow the wound to dry before applying a thin coating. Examine the wound several times in the first year and retreat it if the coating cracks.

A growth retardant, NAA (naphthaleneacetic acid), has been added to some asphalt emulsion and aerosol paints for application to pruning cuts. This will reduce the number of water sprouts, as well as the vigor of those that grow, by about 50 percent (Ashbaugh 1968) (see Chapter 15). Chapter 16 further discusses the treatment of wounds and cavities.

PRUNING TREES

Most trees grow quite well with little or no pruning; they have done so for centuries. But if trees are taken from their natural settings or if their natural settings are changed, a number of new requirements for growth and form are imposed. Many trees are no longer protected by other trees in a grove or forest, but are exposed to the elements. Low branches may hamper activities that take place under and around trees. Trees may grow into utility lines, block views, obstruct the sun's rays, grow too close to buildings, become deformed or destroyed by strong winds, grow too large, or assume an unattractive shape. Growth patterns started in the nursery may be undesirable; structure may be weak. All of these developments are reasons for pruning.

Many tree species naturally develop a crown with desirable branch spacing and characteristic structure. Even though many more branches initially grow than will be needed, competition and shading allow some branches to develop more rapidly than others. Many smaller, weaker branches die and drop, in a natural pruning process. Trees allowed to develop with minimal pruning will often require only the correction of obvious structural faults, such as poorly positioned or strongly competing limbs, weak branch attachments, or limbs that are damaged or dead.

Structural Strength

Certain features contribute to the structural strength of the trunk and main branches of a tree:

Tapered Trunks   Trees with tapered trunks can withstand greater stress from wind and vandals (Leiser and Kemper 1973). A tapered trunk decreases in diameter with height, and when it bends, the curvature is fairly even throughout its length, allowing for a uniform distribution of stress. The tops of well-tapered trunks bend farther under the wind than those with less taper, reducing the danger of broken trunks and other deformation. During the growing season, the tip of a leader may bend far enough to be nearly parallel to the wind load, relieving almost all stress on the immature wood of the tip.

Temporary branches on the trunk will strengthen and protect it, and the trunk will increase in base diameter more rapidly if laterals grow along it. The leaves and growing points provide food and auxins for rapid trunk growth. Branches along the trunk also shade it and reduce the likelihood of sunburn injury to the bark and cambium. Leiser and Kemper (1973) found that stress is most effectively reduced if at least one-half of the foliage grows on branches originating on the lower two-thirds of the trunk. Branches along the trunk will increase total tree growth, even though height will not increase quite as rapidly as it would if no branches grew along the trunk (Chandler and Cornell 1952; Larson 1965; Harris and Hamilton 1969). This slight reduction in total height is a definite advantage in developing a structurally strong tree.

Branch Attachments   For a strong attachment, a branch must be smaller than the trunk or limb from which it arises (MacDaniels 1932;

Ruth and Kelley 1932; Miller 1959). Relative branch size is apparently more critical than the angle of attachment. When Miller subjected 60 apple branches of the same variety, age, and size to stress until they split, there was no correlation between the angle of attachment and the force at which the branch split. Their angle has little direct effect on the strength of branch attachments, but a branch will grow more upright and vigorously when the angle of attachment is more acute. Not all upright-growing branches have acute angles of attachment, but almost all branches attached in that way are upright. Upright branches are usually larger in relation to the trunk than are more horizontal limbs. With time, the height and weight of vertical branches increase the stress at the crotch, which is weaker than if the trunk were larger in relation to branch diameter.

MacDaniels (1932) states that the angle of branch attachment does not affect strength except when there is embedded bark. In his studies, embedded bark weakened the branch attachments of apple trees. Miller (1959), however, found that embedded bark does not influence crotch strength consistently. Bark-embedded crotches that have split out certainly appear to be structurally weak compared with those that have connective wood in the crotch (Fig. 14-18). Bark-embedded crotches are easy to detect, and it seems wise to eliminate them as soon as possible if large, permanent branches would otherwise grow from them. A branch with a more desirable angle of attachment can often be found. If not, a second branch may grow from the same node and with a wider angle of attachment; it will also be smaller in diameter than the first branch, permitting the attachment to be stronger (Fig. 14-19). Most broadleaved plants produce more than one bud at a node, but usually only one develops unless growth is quite vigorous. The wider angle of attachment when a branch develops from a second bud is particularly evident when the buds are superposed, as in ash and walnut (Harris and Balics 1963).

These interpretations of branch attachments do not materially change how a young tree should be pruned. Unless bark is embedded, the growth of large, potential scaffold branches should be slowed to assure strong branch attachment and to lessen competition with the leader. Wherever a leader or branch forks, one of the branches should be substantially larger than the other. If there is a choice among permanent branches, those with wider angles of attachment should be selected; their less vigorous upright growth and smaller ultimate size will usually be advantageous.

In choosing permanent branches, estimate the diameter of trunk and lateral just beyond the point of attachment. If the lateral is too large in relation to the trunk, prune it to remove some of its leaves or leaf buds. Any laterals on the branch can be thinned; if the branch

FIGURE 14-18
Wide angles of branch attachment (top left)
and narrower angles with bark ridged up in the crotch (top right)
are stronger than sharp angles of attachment with embedded bark (bottom left).
As limbs increase in spread and weight, they can split at this point of weakness
(bottom right); note the lack of attachment in the upper part of the tight crotch.

FIGURE 14-19

On young trees, shoots with sharp angles of attachment
should be pruned off when they are about 150 to 200 mm (6–8 in) long.
A second bud will usually form a shoot that grows at a wider angle of
attachment; in the photograph, a short stub indicates the position of the
shoot removed (left). A seven-year-old Modesto ash whose first laterals
were removed during its first year in the landscape and later laterals
thinned for vertical spacing is pictured on the right.
(Compare with Fig. 14-20.)

has few laterals or none at all, it should be headed. In many decurrent
tree species, the lowest branch outstrips the growth of the trunk and
the upper branches. When developing a trunk, therefore, examine
lower laterals regularly during the growing season to ensure that they
do not outgrow the leader. Branches from latent buds, even though
they are much smaller than the trunk, are usually weakly attached
(see Fig. 14-5). If such branches are to be kept, prune them to decrease
growth and permit the attachment to strengthen.

Branch Spacing When the branches of broadleaved trees are
evenly spaced on the trunk, both vertically and around the circumference, they are more likely to have strong attachments than when
several branches arise at about the same level (Fig. 14-20). Vertical
branch spacing is more critical in large trees than in smaller ones.
If the main leader is headed in the nursery or at planting, close branch
spacing will develop in many trees; in other trees, the leader will tend
to form several branches close together at the base or near the tip of
one season's growth.

FIGURE 14-20
This Modesto ash was headed
in the nursery or at planting;
the many shoots that grew from
below the heading cut were not thinned.
Most of these branches are weakly attached
and are in danger of splitting out
as the tree grows larger
and continues to spread.

Some tree species have branches with extremely narrow angles of attachment, acute enough to form indentations in the trunk. Certain trees, such as Lombardy poplar, have been selected for their erect branching (fastigiate) habit. These branches will remain relatively small in relation to the trunk, and there is little or no need to remove them.

Training Young Trees

A plant may need to be pruned to achieve a desired form or structure. Little or no pruning is needed on young trees that have lateral branches on current growth of the leader (weak apical dominance). Such trees usually form a strong central leader and a conical shape, as do liquidambar, pin oak, and most conifers. On the other hand, trees with no laterals on current growth (strong apical dominance) may have quite irregular growth habits, as does Chinese pistache, or may form vigorous laterals in the second year, as do many flowering fruit trees. Poor branch attachments may also be common, as in Modesto ash. These trees have decurrent growth and usually need considerable pruning while young in order to develop a central trunk and a strong branch structure. *Scaffold* or *permanent* main branches are those that make up the framework of a decurrent tree. In contrast, most excurrent trees naturally have strong central trunks with many laterals that are much smaller in diameter.

The irregular, zigzag growth of some treetrunks can be minimized by generous applications of nitrogen and prudent watering, which will increase tree vigor. You may also head the trunk back to its vertical portion so that new growth will be straighter. In Oregon nurseries, a number of deciduous tree species are headed close to the ground after the first season to develop straight, vigorous trunks.

In the nursery, most trees are trained to central leaders so that they have one main trunk. It is a common practice to remove laterals on the lower half of the developing leader of broadleaved trees or cut them back to stubs (100-150 mm or 4-6 in) before they become 250 mm (10 in) long. If stubs are left during summer training, they are often removed the following winter (Brown 1972), but a stronger, more tapered trunk will develop if lateral shoots grow on the lower portion throughout nursery production. These shoots must be cut back so they remain small in caliper and do not take up too much space. If these temporary shoots are not kept within bounds, height growth

FIGURE 14-21
Container-grown trees are often headed in the nursery (extreme left)
to encourage branching and to develop proportions that will be attractive
at the time of sale (two center sketches).
The laterals are usually headed a second time to ensure
another set of branching (right center).
Even for small-growing trees most of the branches are too low;
for large trees, they will be too close together also.
Once the tree is in the landscape, you may select the top branch as the leader,
prune another high branch so that it may become the lowest permanent scaffold,
and remove or severely cut back the lower laterals on the trunk (extreme right).

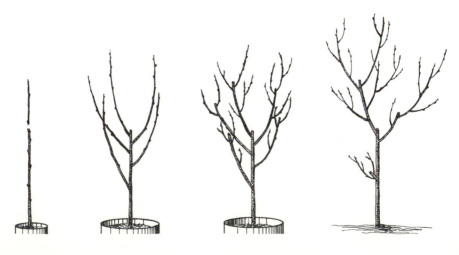

of the leader can be unnecessarily slowed. These laterals along the trunk are cut back (but not removed) up to the height of the lowest permanent branch. Those above the lowest permanent branch can be thinned so that they are about 150 mm (6 in) apart vertically. Temporary laterals should be spaced radially around the trunk, with no more than one per node, and none should compete with the leader.

Many small-growing deciduous trees are headed in the nursery when they become 1 to 1.5 meters (3–5 ft) tall; this forces laterals below the cut. Unless one branch is selected as the leader and most of the others are removed or cut back, these branches will form a compact head, with seemingly good proportions when the tree is small, but with no leader (Fig. 14-21). These branches will be too low and close together for most landscapes.

Trees must be given enough room in the nursery to develop lower laterals and must be somewhat buffeted by wind; otherwise they will grow exceedingly tall, with little basal trunk caliper and poorly developed laterals (see Fig. 4-11) (Harris and others 1972). Close spacing is often accompanied by staking, which further encourages height growth at the expense of trunk caliper (Chapter 8).

Pruning at Planting

Pruning at planting may provide little or no improvement in growth and survival, particularly if plants have vigorous root growth and were planted well in advance of new spring top growth. Richardson (1958) found that removing the terminal bud of a one-year-old sugar maple seedling whip (single stem) just as it was beginning to grow delayed the initiation of root growth until another bud developed. When pruning removes most terminal buds, therefore, the reduction in potential leaf area may be accompanied initially by a smaller root system. When six species of dormant bare root trees were pruned when spring planted in Oklahoma, Whitcomb (1979b) found that nearly all the trees survived, but none benefited from the pruning. Whitcomb examined trees whose tops had been pruned back (probably headed) 0, 15, 30, and 45 percent, respectively. Those pruned back 30 or 45 percent did not have a growth form typical of the species.

In most cases, pruning of bare root and B & B plants after planting should be restricted to the removal of damaged branches and the correction of structural weaknesses. If you remove about 25 percent of the leaf area or potential leaf area of container-grown plants by removing entire shoots and leaving others with terminal buds intact, the plant will appear to be full size but will have fewer leaves, which might reduce transpiration somewhat. The unpruned terminals may favor more rapid root development.

Training in the Landscape

The first three to five years in the landscape are critical in the training of most trees. Pruning to achieve desired forms and strong structure will greatly enhance tree performance and the ease of later maintenance. An arborist should have a clear understanding of the landscape function of each tree in order to train it to the desired form. Pruning should be only extensive enough to direct a tree's growth and correct any structural weakness. Trees that are pruned most lightly experience the most total growth. A young tree should be left with more branches than will ultimately be wanted; some of these will grow more and dominate the others. Dominant branches are usually fairly well spaced. Severe pruning of young trees may remove potentially dominant limbs, leaving some that will not develop as readily. For this reason, it is best to remove or prune back only those branches that are clearly unwanted and those that compete with the leader or other desirable limbs. Alternately, the growth of these branches can be retarded by pruning.

Branches to be removed or retarded on young trees during the first two to four years in the landscape are those that are too low to be permanent, those that grow from the same node as or within 100 to 150 mm (4–6 in) of a potential scaffold branch, those that outgrow the leader or other potential scaffolds, and those that have sharp angles of attachment with imbedded bark in the crotch (see Fig. 14-18). Occasionally a tree will be so vigorous that the pruning needed to retard growth of less desirable branches would excessively stimulate potential scaffold branches. Extremely vigorous young branches are susceptible to breakage, particularly in windy sites. In such cases, about one-quarter of the current growth of temporary branches should be headed back. In order to provide some support for the future main branches, leave most of the temporary branches on the tree unless they severely compete with potential scaffolds. Do not fertilize the trees; in extreme situations, withhold irrigation to curb excessive growth.

Temporary Branches Branches below the lowest permanent branch will strengthen and protect the trunk (Fig. 14-22). When young trees have not yet reached the desired height of the lowest scaffold, treat laterals as temporary branches. At planting and at each dormant pruning, select laterals of weak to moderate vigor to remain as temporary branches. Remove vigorous low-growing laterals if less vigorous ones can be selected. Short, horizontal laterals can be left unpruned. More vigorous laterals, if not removed, should be headed back to two- or three-bud spurs during dormant pruning.

Although their angle of attachment and spacing along the trunk are not particularly important, temporary branches should be 100 to

FIGURE 14-22
Temporary branches, particularly on the side exposed to the afternoon sun,
can shade and nourish a trunk (left);
unprotected bark, particularly on young trees of low vigor,
is often killed by the sun (right).

300 mm (4–12 in) apart because closer spacing may unduly retard over-all increases in height. Temporary branches on the western side of the trunk, which is exposed to the afternoon sun, will reduce the chance of sunburn. During the growing season, pinch the tips of vigorously growing temporary branches to keep them in bounds and to reduce competition with the leader and potential permanent branches. Examine vigorous trees at least two to four times during the growing season. Pinch back new shoots about 50 mm (2 in) when they become 150 to 200 mm (6–8 in) long. Vigorous shoots may have to be pinched more severely to keep growth within bounds. Pinching back temporary growth requires little time and provides an opportunity to observe any other problems that may develop.

As a young tree develops a sturdy trunk and a top that shades the trunk, temporary branches can be reduced in number and eventually eliminated. Three or four years after planting, when the trunks of small trees (such as crape myrtle and Japanese maple) are 50 to 75 mm (2–3 in) in caliper, and those of large trees (such as elm and sycamore) are 100 to 150 mm (4–6 in) in caliper, the number of temporary branches can be reduced over a two- to three-year period. If, at each pruning, you remove the largest temporary branches, you will minimize the total surface of the pruning wounds.

Unfortunately, many young trees are grown with no laterals left along the lower trunk. Thus, any shoots that do begin growth in this region should be encouraged. If a trunk is spindly and could benefit from the nourishment and protection of temporary branches, it may be possible to stimulate their growth by notching the trunk. The removal of a small slit of bark above or below a bud will influence its differentiation into a leaf or flower bud (accomplished soon after bud formation) or will stimulate a dormant bud into growth. Removing a slit of bark 1 to 2 mm (0.04-0.08 in) wide around 15 to 25 percent of the trunk circumference and 10 to 15 mm (0.4-0.6 in) above a bud a month before expected growth will usually assure the growth of that bud. In Europe and the Orient, the technique has been used on young fruit trees, particularly apples and pears, that sometimes branch irregularly. The notch reduces the flow of bud-inhibiting substances from buds higher on the stem.

A less precise method of encouraging shoot growth is to bend the trunk so the side where the shoots are wanted is uppermost. The terminal portion of the leader is held nearly parallel to the ground. This should take place about two weeks before buds are expected to grow and should continue until buds break, possibly within three to six weeks. The trunk must be protected from sunburn with white latex paint or wrapping. After the buds break, the tree can be returned upright. Container-grown trees may be laid on their side before planting to encourage new shoots along the upper side of the trunk; again, the plant must be protected from the sun. Inclining the trunk from the vertical decreases bud-inhibiting substances on the upper portion, thereby releasing the buds.

Notching and bending are not so much recommended here as used to exemplify the manner in which a basic knowledge of plant responses can be used to develop growing techniques. Temporary branches have often been pruned off by uninformed people for a variety of misguided reasons. More education is needed.

Height of the Lowest Permanent Branch   Height of the lowest branch is usually prescribed by the function of the tree in the landscape. The lowest permanent branch can be several centimeters (or inches) or more than 4 meters (12 ft) from the ground, depending on how the tree is to be used. A certain clearance is needed over streets or patios, but limbs may be lower if they will not interfere with traffic or the use of ground underneath the tree. The position of a limb on a trunk remains essentially the same throughout the life of the tree. In fact, as a branch increases in diameter, the distance between it and the ground actually decreases (Fig. 14-23).

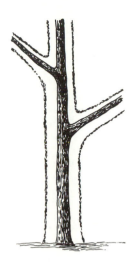

FIGURE 14-23
As a tree grows,
branches retain their position
on the trunk and at the same time
increase in diameter,
becoming closer to the ground.

Vertical Branch Spacing   In many decurrent species, branch spacing is important for future leader dominance, structural strength, and appearance of the tree.   Two or more vigorous branches arising at or near the same level on the trunk are apt to "choke out" the leader and limbs above.   This is especially true in "fast-growing" trees whose laterals grow from buds formed during the previous season, such as flowering fruit trees, mulberry, and zelkova.   On mature trees, closely spaced scaffolds break out more easily than those with wider spacing. Closely spaced scaffolds usually have fewer laterals and thus develop as long, thin branches with little or no taper and little structural strength (Fig. 14-24).   These branches are particularly susceptible to snow and

FIGURE 14-24
Closely spaced scaffolds
on large trees are usually
long, thin, and unbranched.
If one should break out,
the remaining branches
are particularly vulnerable.

FIGURE 14-25
Occasionally a young vigorous branch
(watersprout) will grow more upright
than the others and will compete
with the leader.
Unless the sector in which it grows
is devoid of limbs, the upright branch
should be removed (broken line).

ice breakage; if one is lost, the others are more likely to break out as
well.

Vertical spacing between permanent branches should be greater
on a large-growing tree with large-diameter branches than on a tree of
smaller mature size. Major scaffold branches on large trees should be
spaced at least 300 mm (12 in) vertically, and preferably 450 to 600 mm
(18–24 in) or more. Many attractive mature trees have branches 1 to
4 meters (3–12 ft) apart. On excurrent trees, the vertical spacing of
branches is much less critical because the more vigorous branches are
naturally well-spaced, while others become relatively weak. Little or
no pruning is needed except when an occasional limb becomes overly
vigorous and competes with the leader (Fig. 14-25).

Radial Branch Distribution   You can usually select five to seven
scaffolds to fill the circle around a trunk without undue crowding (Fig.
14-26). This can be done in one or two rotations around the circum-

FIGURE 14-26
On large-growing trees,
the permanent scaffold branches
should be well spaced
vertically and radially.

FIGURE 14-27
Proper radial distribution
will allow limbs adequate space
for development.

ference. Although an ascending spiral may appear more symmetrical and pleasing to the eye, branches seem to grow equally well when they do not arise from the trunk in a spiral. Avoid allowing one limb to grow directly over another, which prevents both from developing properly (Fig. 14-27): The lower branch will be shaded and develop few or no ascending branches; the upper one will be less vigorous in the presence of the lower, which competes with it for water and nutrients.

Selecting Scaffold Branches   Examine the vertical and especially the radial distribution of potential scaffold branches.   One sector of the tree may have few branches from which to choose. If so, choose a scaffold in that sector first and choose others in relation to it. This assures the best radial symmetry and selection.   Depending on the tree's growth rate, scaffold branch selection may take two to three years to complete.

Pruning during the Growing Season   Growth should be directed when the tree is active as well as when it is dormant.   Pinching the growing point (heading) or complete removal of a shoot (thinning) will direct growth into the leader and remaining shoots. Pruning during the growing season is usually confined to temporary shoots and branches. On a young tree, the leader or a scaffold will only occasionally need substantial pruning.   Shoots that are too low, too close, or too vigorous in relation to the leader and selected scaffolds should be pinched or removed.

In many species, shoots do not branch during the season in which they form.   Even in the second year, some trees may develop few or no laterals, except near the previous season's terminal. On vigorous leaders it is possible to obtain branches during the growing season by pinching the growing point when it has grown a little above the desired height of

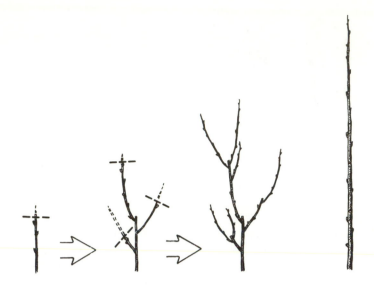

FIGURE 14-28
Vigorous young trees with strong apical dominance may grow 2 m (6-7 ft)
with no lateral branches (extreme right). On the other hand,
if the leader is pinched during the growing season
when the terminal is 50 to 100 mm (2-4 in)
above the height desired for the first scaffold (extreme left),
two or more shoots can be encouraged to grow.
Select the most vigorous and upright shoot as the leader;
select a second one as a lateral and head it lightly if necessary;
head the others more severely (left center). Repeat this process
as long as a vigorously growing leader can be selected (right center).

a lateral (Fig. 14-28). Remove about 50 mm (2 in) of the tip, and buds below the pinch will begin to develop. One will usually grow more vigorously than the other shoots, and this can become the leader, although it may need encouragement. Choose a second developing shoot as a lateral by pinching the tips of the other shoots that were forced. The new leader should arise on the pruned trunk above the selected lateral; if it does not, the selected lateral may not develop into a vigorous scaffold. On a vigorous tree, the new leader may in turn be pinched when it reaches a height that is suitable for another lateral branch. It may be possible to force as many as three well-spaced laterals in one season. Without such pinching, the leader would require severe heading during the dormant season, usually to the height at which the lowest lateral is desired.

You can keep the development of scaffold branches in balance with the rest of the tree either by thinning laterals or by pinching the tips of the most vigorous laterals during the growing season. Growth will be channeled where it is most effective. For a tall, upright trunk, keep the leader dominant by preventing laterals from outgrowing it.

Uncontrolled laterals will commonly outstrip the leader in decurrent species.

Low Branches   Nursery trees with low, large laterals can be a problem in many landscapes. Such branching may be satisfactory for small trees in areas with little activity, but not for large trees or in active areas. As mentioned, many landscape trees are headed once or twice in the nursery to force low branching and a well-proportioned small tree. The result is many low branches with little or no vertical spacing. At planting, it may be possible to select the most upright and vigorous branch to become the leader. A second branch may be chosen as the first scaffold if it is high enough above the ground for the site. In some cases, only a leader can be selected. Other branches should be thinned and those remaining treated as temporary branches.

The sooner corrective pruning is done, the less dwarfing it will cause. In some cases, however, the pruning needed will be so severe that it should be done over at least two years. This is more true for older nursery trees than for young ones.

Upright vs Horizontal Branches   An upright branch will usually be more vigorous than one that is less upright, and you may want to use it as a permanent branch if its position is desirable. Because the branch may, however, compete with the leader, a more horizontal branch should usually be selected. As a tree matures, even one of excurrent form, the leader decreases in apical control and the tree becomes more round-headed. Occasionally, a branch will grow vigorously upright in a tree that normally has subdued horizontal limbs, as do most conifers, pin oak, and liquidambar (see Fig. 14-25). Remove these limbs or cut them to an outward-growing lateral as soon as you see them. Otherwise, they will upset the symmetry of the tree.

In contrast to upright branches, those growing more horizontally are usually of low vigor. Horizontal branches will seldom compete with the leader and are desirable as temporary branches to protect and nourish the trunk. Unless they become too long, the smaller ones can be left unpruned. Horizontal or drooping limbs may be a problem in some young trees, however. If they droop because of excessively vigorous growth, buds behind the top of the bend will often grow, and the new shoots will usually be more upright. Select the well-placed shoots from among these by thinning the lateral back to the selected shoot (Fig. 14-29). Thin out other new shoots that might compete or interfere with the one selected. If the horizontal or drooping limb has no well-placed upright laterals, head the branch to an upward-pointing bud slightly behind the top of the bend or to a point where you wish a lateral to form. Certain trees (such as weeping willow and Chile mayten) are chosen for their drooping branches, and the characteristic is exploited.

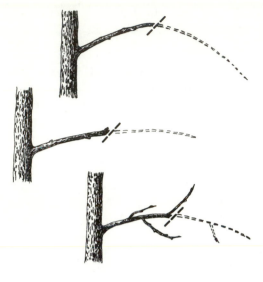

FIGURE 14-29
To cause a horizontal branch
to grow more upright,
prune it back to
an upright lateral (bottom),
cut it back to
an upward facing bud (center),
or cut it back near
the top of an arch (top).

Windy Situations Prevailing winds can deform trees so that most of the growth is on the downwind side. (Planting in windy locations is discussed in Chapter 5.) Depending on wind conditions and the type of tree, the leader may or may not be bent by the wind. Many trees, such as conifers, liquidambar, and plane tree, resist deformation by moderate prevailing winds.

In windy locations, open up the top of the tree by thinning out moderate-size branches; this cuts the tree's wind resistance. Laterals

FIGURE 14-30
A tree deformed by the wind
can be made more symmetrical
if branches on the windward side
are headed back to a bud
pointing into the wind,
if the leader is thinned
to a more upright lateral,
and if some of the downwind
branches are lightly headed.

will develop more extensively on the downwind side. In certain situations, such a condition may be picturesque and desirable. If it isn't, thin out branches on the downwind side to laterals to keep the tree more symmetrical (Fig. 14-30). Head curving branches on the windward side near the point at which they begin to bend with the wind; prune them to a lateral or a bud pointing into the wind. Repeat this each time the endmost new shoot starts to bend from the wind. Branches so pruned will be stockier and more resistant to bending. In some locations and for some species, windbreak protection may also be needed.

Maintaining a Leader   Sometimes a leader may lose its control and become outgrown by a vigorous lateral, which may make a better leader than the original one. If so, remove the original leader by pruning back to the new leader (Fig. 14-31), which should be the topmost lateral on the trunk. Do not leave part of the original leader above the new one. It will seldom amount to much because new growth will go into the new leader. If caught young enough, laterals can be headed to keep the leader dominant (Fig. 14-32).

In species with moderate to strong apical dominance, several buds may begin to grow near the tip of an otherwise branchless leader or scaffold branch late in the season. They may grow 50 to 150 mm (2-6 in) in length and as thick as the terminal. Unless these branchlets are thinned out on young trees, they and the terminal will grow weakly the next season. It is best to head to a bud below this tuft of branchlets or to thin the tuft, leaving one branch and the terminal.

FIGURE 14-31
When a leader has been outgrown
by one or more laterals,
thin the leader back
to one of its most vigorous
and upright laterals
which in turn
will become the leader.

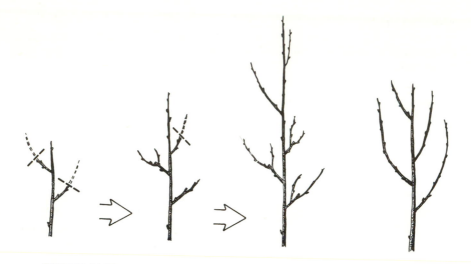

FIGURE 14-32
You can maintain a leader by heading back any laterals that may
compete with it (extreme left); prune these laterals fairly severely
if they are temporary (left center). The tree will grow taller (right center)
than if it had not been pruned (extreme right).

Other Tree Forms   Although the previous discussion emphasizes
the development of a well-tapered single trunk with well-spaced scaffolds, this does not mean that other tree forms are not recommended.
The comments focus on developing a structurally strong tree and can
be applied to whatever structure is desired. Trees pruned according to
these guidelines should perform well and long. Plants can become fine
sculptures in the landscape either naturally or through the skill of a
horticulturist (Fig. 14-33).

FIGURE 14-33
A tree can be a sculpture
in a landscape:
The leader of this thirty-year-old
California sycamore was staked
to the ground
for the first three years after
planting.

FIGURE 14-34
A lateral branch that is too upright (arrow, left) can be inclined more
if you insert a *spreader*, cut from a pruned branch of similar caliper,
between the branch and the leader.
The branch should be smaller than the leader and have a strong attachment.

Training Is More Than Pruning   Although pruning is the primary method of training young plants, other procedures may be used.
Staking may be used to encourage a straight trunk or a more upright
growth habit, though support staking has limits for most species if a
sturdy trunk and well-proportioned top are to develop (see Chapter 8).
Branches may be tied to supports to create special forms, cover a wall,
or form a screen; espalier is a prime example.

The vigorous branches of young fruit trees, particularly pear and
apple, are sometimes bent and tied so their tips are at or below the
horizontal to encourage flowering and fruiting at a younger age. Such a
practice also increases the spread of trees that might otherwise be quite
narrow. An upright limb can be made to spread if a length of branch
pruning is wedged between the trunk and the limb (Fig. 14-34). First
ascertain that the branch attachment is strong and the limb smaller
than the trunk. The spreader can be left in place for two or three
years and will be more effective than pruning to an outside bud, which
in turn will most likely grow upright.

The scaffold limbs and the main structure of a decurrent tree can usually be selected by the third or fourth year, depending on the type of tree and its growing conditions. If scaffolds are well placed, the tree may need little or no pruning for several years. Temporary branches along the trunk will probably have been thinned out or reduced in number by this time. Staking to support a tree should not be necessary, but short stakes can be used to protect the trunk from mower, auto, or other damage. Place stakes so that the trunk or branches do not rub against them.

Inspect young but mature trees annually to ensure that the main scaffolds are growing well and in balance with each other and the trunk. Remove low, broken, interfering, and diseased branches. A densely headed tree and the plants growing under it may benefit if you open up the tree to allow for the passage of light. Depending on tree species and growing conditions, inspections can be less frequent as the tree matures. Mature trees usually need pruning for one or more reasons. The National Arborist Association (N.A.A.) in the United States and the British Standards Institution (BSI) in Great Britain have each attempted to standardize terms for the various pruning operations. N.A.A. standards describe four classes of pruning, based primarily on the size of the smallest pruning cuts normally made (N.A.A. 1979). For example, in Class III (coarse pruning) undesirable branches larger than 50 mm (2 in) are usually removed, but smaller branches are not pruned out. In Class I (fine pruning) this dividing line is 13 mm (0.5 in). The BSI standards base pruning categories on pruning purpose, such as removing low branches or opening up a tree (BSI 1966).

Cleaning Out (BSI) Cleaning out or "dead wooding" consists primarily of removing broken, diseased, dying, and dead limbs, those that cross, are weakly attached, or are of low vigor. Water sprouts and suckers (epicormic shoots) can be quite a problem on some species and on trees that have been severely pruned; they should be cut as close to their base as possible in order to reduce the possibility of renewed growth. Treating these cut surfaces with a growth retardant also minimizes new growth (see Chapter 15). Climbing plants should be removed. Wire, rope, nails, and other foreign material should be eliminated. A tree can also be examined for defects that might require additional attention.

Pruning can quite effectively reduce infection and stop or slow the spread of infection within a plant. Brown (1972) describes pruning treatments to guard against 20 diseases and pests. Take care in pruning, because a few diseases can be spread by contaminated pruning tools. When you prune plants that are infected with a canker-forming disorder, disinfect tools after each cut to avoid spreading the disease to healthy wood. Alcohol (70 percent), Lysol® (5 percent), or Amphyl® solutions are effective disinfectants. Household bleach (sodium hypochlorite) can be used, but it corrodes metal tools.

Raising the Crown Lifting the crown (BSI 1966), raising the head, raising the canopy, and lift-pruning: All these terms refer to the removal of lower branches from the trunk or lower limits of a tree (Bridgeman 1976). Lower branches may block pedestrian and vehicular traffic, may obstruct a view, may grow too close to buildings, or may eliminate sunlight or breezes. As most trees increase in size, their branches bend lower under increased weight. Lower branches may also tend to grow downward because light intensities are higher below the tree canopy than within it. Lower branches should be removed only for valid reasons, however, because they have certain advantages if allowed to grow near the ground.

Plan the height of the lowest permanent branch when training a young tree so that you do not need to remove large branches later on, when pruning will leave large wounds. Some arborists perfer to prune for clearance in late summer, when branches are heaviest with foliage and will therefore reveal the extent of the problem.

When raising the crown, thin back to a more upright large lateral, or remove the branch entirely. All too often, low branches are cut back to small, more upright laterals (Fig. 14-35). These low branches

FIGURE 14-35
If there is a suitable branch above, completely remove a low, spreading branch (arrow) instead of pruning it back to less drooping laterals.

are usually heavily shaded from above. The loss of foliage and the shade combine to weaken a low branch so that it becomes unattractive and needs further pruning. If there is a limb above, the low one should usually be removed to provide long-term clearance and to improve appearance.

If the findings of Leiser and Kemper (1973) about young trees can be applied to young mature trees, the removal of low branches might increase stress along the lower trunk, immediately below the new lowest branch, and on the root system. Particularly when you raise the head of a young mature tree, it may be wise to remove low branches over several seasons (Brown 1972); keep in mind that at least one-half of the foliage should be on branches originating on the lower two-thirds of the trunk (Leiser and Kemper 1973). In order to leave a tree well balanced, both in appearance and weight, some arborists remove branches opposite those removed in raising the head. This is primarily a matter of aesthetic preference. Most trees can withstand considerable asymmetry without hazard, but removing lower branches, even to balance the tree, may increase trunk stress.

Crown Thinning (BSI 1966)   Opening up the top of a tree permits deeper light penetration, which benefits inner leaves and branches. Moderate to high light intensity is necessary for active and productive leaves. If branches arise close together along the trunk, they will grow long and slender with few laterals (see Fig. 14-24). Thinning out some of these branches will give the remaining ones more room, help initiate new laterals, and increase the taper of remaining branches. Wind resistance and weight at the top are reduced by crown thinning, which can be especially important for trees with weak branch structure or insecure roots.

The structural features of a tree can be emphasized by moderate thinning that opens the trunk and branches to view. Pittosporum, dogwood, olive, and ginkgo are particularly suited to this treatment.

The first branches to remove in crown thinning are those that would normally be removed in cleaning out the crown. After broken, weak, crossing, diseased, and dead limbs have been removed, it is easier to see what further thinning is required to open up the crown. On a medium to large tree (12–18 m or 40–60 ft tall), moderate (25–50 mm or 1–2 in. in diameter) thinning cuts of limbs are effective. Somewhat smaller cuts are appropriate on smaller trees. Make these in the top and around the periphery. Remove branches that are close to others. In some large trees, you may remove limbs of up to 150 mm (6 in) in diameter. Such large cuts, however, should not be needed unless the tree has been improperly pruned or its use in the landscape has changed.

Brown (1972) points out that crown thinning, which may involve the removal of one-third or more of the branch system, is generally confined to deciduous trees. It is seldom necessary to prune hardy evergreens or conifers in this manner. Most arborists start thinning in the top and work their way down the tree. Prunings can be cleared as the work progresses downward; if falling branches damage those below, it is not too late to modify the selection of permanent branches.

Crown Reduction (BSI 1966)   Many trees become larger than is desirable or safe: They may grow into overhead wires, block views, grow into buildings or other trees, shade solar collectors or other areas where sunlight is wanted, or become hazardous because of size or condition. You may control tree size by pruning, but it will be a continuing task. You may also slow growth without greatly affecting tree appearance by stopping or reducing nitrogen fertilization and irrigation. Size can be maintained most effectively if you prune as a plant begins to reach the maximum acceptable size. If you delay pruning until the tree is much larger than wanted, pruning will be more difficult, cuts will be harder to hide and slower to callus, and renewed growth will be encouraged. Species vary in the severity of pruning mature trees can withstand if they are to remain healthy (Table 14-1).

Thinning-out can reduce the height and spread of a tree while retaining its natural shape (Fig. 14-36). Prune branches back to lower laterals that are at least one-third the size (diameter basis) of the portion removed (N.A.A. 1979). Pruning cuts will be less obvious and the tree less subject to water sprouts than if you cut to smaller laterals or headed (lopped) the branches. Observation indicates that a thinned tree will usually take longer to grow back to the critical height than a headed tree. The finest compliment an arborist can receive after materially reducing the size or density of a tree is when passers-by fail to notice it has been pruned. Thinning-out requires greater skill and time than heading, but in most situations it is worth the effort: It will retain a tree's characteristic form and will minimize the problems of decay and regrowth.

TABLE 14-1

Relative tolerance of species to severe pruning of mature trees (Arboricultural Association 1976)[a]

High	*Intermediate*	*Low*
Horse Chestnut	Acacia	Beech
Elm	Maple	Birch
Lime tree	Ailanthus	Hornbeam
Oak	Alder	Eucalyptus
Mulberry	Ash	Walnut
Poplar	Cherry	Most conifers
Willow	Catalpa	
	Sycamore	

[a]Virtually all broadleaved trees withstand heavy pruning during their formative years.

FIGURE 14-36
You can reduce the height
and spread of a tree
and yet maintain its natural shape.
Branches thinned are outlined
by broken lines.

FIGURE 14-37
This plane tree, in the Royal Botanic Garden at Kew,
has a natural lower crown (broken line) to which the tree could be pruned
should one want to lower the height of the tree.

Even when severe pruning is needed, thinning can usually accomplish the task more satisfactorily than heading. Thinning to a large lateral is often called "drop-crotching." Tall, decurrent trees, such as elm, hackberry, and eucalyptus, occasionally branch in a pattern that provides a natural lower top to which the tree can be pruned (Fig. 14-37) (Keith Davey 1980, personal communication). This second, lower top usually occurs between two-thirds and three-fourths up the tree. Except for the large thinning cuts to the branches that form the lower head, the tree reduced to such a lower head will appear unpruned. When the height of a tree must be reduced even though a natural lower top is not readily apparent, you can usually find appropriate lower branches to accomplish the same results. Treating the large cuts with NAA should keep regrowth to a minimum (see Chapter 15).

Although most excurrent trees can be reduced in width and still retain an unpruned natural form, it is difficult to reduce tree height without hastening development of a round head. Bridgeman (1976) cautions that mature beeches and birches do not respond to crown reduction and may die back from cuts (Table 14-1).

Crowded trees can become misshapen and their branches weakened by shading and rubbing. In many situations, the smaller, more deformed, or less desirably located trees should be removed. The remaining trees will grow into the new space with improved health and appearance. Trees are often planted closer together than will be desirable when they reach mature size. As these trees begin to crowd, the less desirable ones should be pruned back more severely each year until they are removed (Fig. 14-38). This will allow proper development of

FIGURE 14-38
The center tree has been pruned back every few years to allow more room for adjacent trees and can soon be removed with little or no loss.

the more permanent trees while retaining the value of the temporary trees. Tree removal decisions may be difficult, and many people oppose the practice, but its omission will be to the detriment of a planting.

Old trees that are of low vigor and have failing branches can often be kept healthy and attractive for several additional years by severe pruning. Wise fertilization, irrigation, and pest control may also be beneficial.

Heading (also called stubbing, dehorning, or lopping on older trees), unfortunately, is often used by well-intentioned but ill-informed people to reduce tree size. In such pruning, main branches are cut to stubs with little regard for their location. Regrowth from below the cuts is dense, vigorous, and upright. New shoots form a compact head, cast dense shade, and are weakly attached to the older branches (see Fig. 14-5). They are held only by the surface layers (rings) of wood formed after the shoots begin to grow; these layers are similar in thickness to the outer layer of a sheet of plywood, but are not attached as securely. Shoots usually develop more rapidly than the strength of their attachment to the older wood. Branches from such regrowth are weakly attached, particularly if the heading cuts are large, and can be hazardous throughout their life. Regrowth near a heading cut on a vigorous small branch (less than 50 mm or 2 in) may, however, cover over the cut surface and unite around the wound to form an intact cylinder of new wood in direct contact with the older wood. In such cases, there is little or no weakness and the actual lopping line may disappear (Brown 1972).

The unprotected surfaces of large heading cuts are also vulnerable to decay. Exposed branches may sunburn and be vulnerable to decay. If a branch has already begun to decay, cutting into it will speed the process. Large stub cuts seldom callus. Some arborists recommend that a large heading cut should be made at a slant (about $30°$ from the horizontal), with the upper edge above a small branch or in the direction in which regrowth is most desired. Field observations, though not verified by research, indicate that a slanted surface will remain drier, will decay less easily, and will callus more quickly.

Pollarding is a training system used on some large-growing trees that are severely headed every few years to hold them to modest size or to give them and the landscape a formal appearance. Species that grow quickly and in difficult conditions are usually used for this purpose. Most commonly pollarded species are London plane, linden, black locust (Free 1961), and fruitless mulberry. A pollarded tree can be trained to an excurrent form, with short laterals distributed around the upper third of a central leader, or to a more decurrent form, with fewer laterals arising closer together on the upper trunk (Fig. 14-39).

FIGURE 14-39
A pollarded London plane tree at the University of California, Berkeley.
Enlarged knobs develop at the branch ends to which the shoots
are cut back each year (inset).

Pollarding can be started soon after the lowest permanent branches are selected or at a subsequent dormant period during the first few years. For a decurrent shape, the laterals are vertically spaced within one meter (3 ft) or so at the top of the trunk. Every one to three years, one to three shoots that grow from each of the headed laterals are in turn headed back to within 10 to 15 mm (about 0.5 in) of the earlier pruning cut; the other shoots are removed close to their origin. After a tree has been pruned this way for several years, a knob of cut stubs and callus forms at the end of each branch. As would be expected with such severe heading, the regrowth is vigorous, upright, and dense. Anthracnose (*Gnomonia veneta*) on American sycamore is sometimes minimized by pollarding because it allows less disease inoculum to be carried over to the next spring, but the dense growth often causes early fall of interior leaves because of heavy shading.

Crown Renewal   Bridgeman (1976) uses this term to refer to the practice of reshaping a tree that has been severely headed into one that will have a more natural form, improved health, and greater structural strength.   *Restoration* (BSI 1966) and *corrective pruning* are other terms for this practice.   A tree is probably worth saving if the main scaffolds and the trunk are sound or can be cut back to sound wood. Branches that have grown from headed scaffolds should be thinned until only one to three remain on each scaffold.   Even though thinning out branches opens the top so the tree has less wind resistance, the remaining individual limbs are more exposed to wind damage. Therefore, the remaining branches should be thinned back to large, low laterals.   The reduction in number and size of the branches helps develop their attachment to the main scaffolds, particularly in relation to their size.

Such severe pruning might best be done over two to four years to minimize its side effects, particularly the vigorous regrowth.   Pruning during the growing season should reduce excessively long growth, strengthen branch attachment to the scaffolds, and dwarf total growth somewhat.   In areas subject to fall frosts or winter cold, pruning should not be done so as to prolong growth and the beginning of cold hardening.   Fertilization, irrigation, and other practices should be adjusted to minimize excessively vigorous growth on healthy trees.   Pruned trees must be examined annually for structural development, presence of decay in framework branches, and general health.   The safety of pedestrians and property is paramount.

Pruning near Energized Lines   Energized lines can be hazardous to arborists who work near them.   Any power line can be lethal, depending on the contact that completes the circuit between an energized conductor and the ground.   Because it is not possible to anticipate when a power line may energize other lines because of a storm or accident, the American National Standards Institute (ANSI) recommends that "All overhead and underground electrical conductors and all communications wires and cables shall be considered to be energized with potentially fatal voltages and shall never be touched either directly or indirectly"(cited in Haupt 1980, p. 94).   This includes fire alarm, cablevision, telephone, and house drops, as well as utility transmission and distribution lines.

The United States Occupational Safety and Health Act (OSHA) and ANSI have established minimum distances between electrical conductors and working arborists.   Workers whose primary responsibility is pruning trees for line clearance must have special training in working near conductors, and they are permitted to work closer to power lines than are other arborists.   Line clearance workers must stay at least 0.6 m (2 ft) from a primary conductor carrying up to 15 KV (15,000 volts), at least 4.5 m (15 ft) from a conductor carrying 552 to 765 KV.

Arborists who do general tree work must stay at least 3 m (10 ft) from conductors carrying up to 50 KV. Since standards may be changed, be sure of the current ones.

Before beginning work, the tree worker and supervisor should inspect individual trees if an electrical conductor passes through the trees or within reaching distance of the tree worker. The utility company should be notified before any work is performed within 3 m (10 ft) of an energized conductor. Work should be started within 3 m of an electrical conductor only after a representative of the utility company has declared the condition safe. The utility may move the conductor, de-energize the line temporarily, or send a line-clearance tree-trimming crew to do the work near the conductor. During and following storms, telephone lines, wire fences, and metal guard rails within 1 km (0.6 mile) of damaged lines may be lethally energized. Workers should be alert to the possible dangers.

Tree crews should be trained to work safely near energized conductors, whether they climb in trees or use an aerial lift. Lowering cut branches near conductors requires planning and special care. As during any work in large trees, a trained person should be on the ground at all times to assist the person in the tree and to act in emergencies. In general, the previously described methods of pruning mature trees can be used to keep utility lines clear and to direct tree growth away from the conductors.

CONIFEROUS TREES

Most conifers have an excurrent growth habit, particularly while young, and their training and pruning needs are similar in many respects to those of excurrent broadleaved trees. Since most conifers have strong central leaders, they need little or no training unless you desire some atypical effect. Conifers are pruned primarily to control the density of branching, the shape of young trees, and the size of older ones. Double leaders should be thinned to one; dead, diseased, crowded, and structurally unsound branches should be removed. See that branches are also smaller in diameter than the trunk and have wide angles of attachment.

Conifers have several characteristic, though not unique, growth patterns, which should be kept in mind during pruning (Table 14-2). The main leader of a conifer is seldom subdued by the several branches that arise at or near one level on the trunk. Therefore, branches arising in whorls or close together along the trunk can be left. Even though laterals may be close, vertical spacing between branches will be adequate in most conifers. Branches may be thinned, however, to reduce wind resistance or enhance appearance.

TABLE 14-2

Pruning guidelines for conifers (Harris and others 1981)

| Genus and Branching Pattern | Distribution of Latent Buds or Dormant Growing Points | Type of Growth | Method of Pruning for a Given Response[a] | | Comments |
			Reduce Size, Direct Growth	Increase Density	
Usually whorled					
Pine	Almost none. See comments for exceptions.	From preformed initials in most species. Some have more than one flush of growth.	Tip or cut to lateral shoots only. Do not cut below terminal buds.	Pinch candle when it expands in the spring. This may be done on each flush of growth.	Canary Island pine and a few others have latent buds and may be pruned more severely if necessary.
Spruce Fir Douglas fir	Some.	New growth from preformed initials.	Tip or cut to lateral shoots or visible dormant buds.	Pinch lateral shoots in spring when they expand. The terminal (leader) may or may not form multiple leaders if pinched.	These usually require little pruning. Remove bottom limbs for clearance over several years.

Random-branching					
True cedar Larch	Some latent buds on short spur-like shoots.	Long shoots have new growth from preformed initials but continue growth with favorable conditions.	Cut to lateral on long shoots. Short shoots may break into long shoot growth.	Pinch expanding shoots.	These usually require little pruning. Remove bottom limbs for clearance over several years.
Podocarpus	None or few.	More or less continuous under favorable conditions.	Thin to laterals or visible foliage.	Pinch expanding shoots.	Removal of leader may produce a multistem crown.
Bald cypress Dawn redwood Cryptomeria Giant sequoia	May have some growing points capable of new growth.	More or less continuous under favorable conditions.	Thin to laterals or visible foliage tufts.	Pinch expanding shoots.	These species have both persistent and annual (deciduous) shoots. Prune to persistent shoots.
Arborvitae False cypress Incense cedar Cypress	No true buds. Numerous dormant growing points where foliage persists.	More or less continuous under favorable conditions.	Thin to laterals or within visible foliage.	Tip prune.	Do not prune into bare wood. Cypress are very slow to make new growth after pruning.
Juniper Yew	Numerous dormant growing points in foliage areas, some on older wood.	More or less continuous under favorable conditions.	Thin to laterals, tip prune.	Tip prune.	Those with needle-like foliage may not respond to pruning as well as those with awl-shaped or scale-like foliage.
Coast redwood	Numerous latent buds.	More or less continuous under favorable conditions.	Thin to laterals, tip prune.	Tip prune.	Remove multiple leaders.

aFor maximum dwarfing effect, prune in any of the suggested ways in late summer.

FIGURE 14-40
Some conifers, such as pines (left), have whorled branching;
others, such as cedar (right), have random branching along the trunk
(see Table 14-2).

Laterals on conifers radiate from the trunk either randomly or in whorls (Table 14-2 and Fig. 14-40). In species that branch in whorls, shoot elongation is usually determined by the number of preformed initials in the terminal buds; those that branch randomly continue growth as long as conditions are favorable. To a large measure, the distribution of latent buds or growing points limits the severity with which conifers can be pruned (Table 14-2). Few conifers have latent buds below the foliage area in older wood; if branches are headed back to older wood with no foliage, the branch stub usually dies. Pruning to control size or direct growth is easiest when you can cut within foliage alone. More severe pruning can be done by cutting back to active laterals while retaining a semblance of natural form. On the other hand, when conifers are able to form new shoots from older wood, pruning is much simpler and trees are better able to recover from injury by storm or fire.

Whorl-Branching Species

These should be selectively pruned back to a bud or the branch from which the pruned portion originated. In vigorous trees, branch whorls are sometimes farther apart on the trunk than will be desired. You can create a denser tree by heading current growth back to a bud. This can easily be done with pine trees, whose buds elongate into candles before starting their period of greatest elongation (Fig. 14-41).

FIGURE 14-41
You can easily reduce the growth of pine shoots by pinching
their candles back proportionately to the growth reduction desired.

As candles approach full length, needles begin to elongate; this is the
time to pinch them (USDA 1969). Growth will be inhibited according
to the severity of the pinch. Break or pinch candles, if feasible, because
cutting the needles with shears causes them to develop brown tips. If
you must shear, shorten candles after they have hardened.

You can shape a pine by removing most of the candles on shoots
you wish to discourage and leaving most or all on those you wish to
encourage. Reduce the size of whorl-branching conifers by thinning
a branch tip back to a laterally growing shoot. Spruce, fir, and other
whorl-branching trees whose buds elongate into shoots can be kept
dense and small if you pinch the expanding shoots in the spring. Trees
that have short internodes, as do firs, may need little or no pruning.
In fact, Scarlett and Wagener (1973) warn that the topmost whorl of
firs should not be pinched because regrowth may be poor.

Pinching shortens the distance between the topmost branch
whorl and the next one to be formed, but excessive distance between
existing whorls cannot be decreased. Bare trunks between widely
spaced whorls can be unattractive on vigorous pines. Heading to
shorten a long distance between whorls usually results in a stub that

does not resprout. Some arborists and Christmas tree growers induce branching between whorls by girdling, or removing needles between the whorls. To girdle a trunk, they remove a narrow (3 mm or $\frac{1}{8}$ in) band of bark at the height at which branching is desired. One or several branches may grow below the girdle and help fill the void. Alternately, one can remove the needles 25 to 50 mm (1-2 in) above the level at which branching is wanted; this may stimulate shoot growth from the band of needles just below the exposed area (Francis, Breece, and Baldwin 1974). Branches developed in this way have a relatively late start and will not grow as large as older limbs. The younger the tree and the closer to the terminal girdling or needle removal is done, the more likely success will be.

Random-Branching Species

These can usually be sheared or pinched to control size, branching, and form. When latent buds occur along older branches as well as on younger shoots, pruning cuts are usually made near a latent bud, which will then become active and develop a new growing point.

To control the spread of most random-branching conifers, prune back new growth of side branches to their halfway point in late spring. The branches can be taken back more severely and still appear natural if each lateral is pruned back to a shorter shoot growing on the top of it. The small shoot will hide the pruning cut and give a tip-end appearance to the shortened lateral. When conifer branches reach 250 to 300 mm (10-12 in) in length, new growth can be cut back to 20 to 30 mm (1 in). Small side shoots will develop, making the tree more dense. You can then maintain the tree at about that size by cutting back new growth. Williamson (1972) suggests that even tall-growing deodar cedar can be kept as low as 2 m (6 ft) if you cut back the leader each year and tape a side branch near the cut into a vertical position to form the new tip.

Pruning Severity

As mentioned, response to pruning severity is largely determined by the presence or lack of latent buds on older wood. Pruning severity can also be based on the duration of growth during the season. If conditions are favorable, some species (including some pines) with pre-formed shoot initials in their buds may experience more than one growth flush during a season. Young, expanding shoots can be pruned during any or all of these flushes. If there are no visible latent buds, however, pruning into old wood will usually produce a stub from which no new growth will arise.

At the other extreme are species with buds or dormant growing points (with no bud scales) whose shoots continue to elongate even after their preformed initials are fully expanded. These species grow almost continuously or in successive flushes during the growing season. Their usually abundant latent buds produce new growth even when pruning is severe and extends into old wood. Branching is usually spiral or random, and growth may be either excurrent or decurrent, at least when the trees mature. Even though plants in this group can be pruned severely, thinning will produce the most attractive results.

As with all natural things, there are intermediate forms. Many conifers continue growth in a series of flushes under favorable growing conditions or the stimulation produced by pruning or the removal of adjacent trees. These conifers usually have latent buds randomly spaced along stems; they may retain active laterals or short shoots for many years on older wood. If pruning is needed on such conifers, it should be moderate. These species may either maintain or develop a decurrent growth habit with age.

Columnar Conifers

A number of upright conifers, such as arborvitae, cypress, false cypress, and yew, whose branches arise from the ground to the top, are often used in formal settings. Many are clipped into "unnatural, spectacular ugliness" (Chandler and Cornell 1952, p. 29), while most can be pruned to have a natural, textured appearance. The branches of columnar conifers that start near the ground may become quite tall, growing up on the outside of the branches above; in such circumstances, the inside foliage is shaded and may be lost. Outer branches that bend away from the column are unattractive. Most of these conifers can be trained or retrained to have short branches if upright branches are cut back to short, spreading laterals. These spreading laterals can be spaced to provide pleasant shadows and glimpses of the trunk. Weak and dead foliage can be washed out with a hose.

Outside branches are often held close by wires wrapped around the tree at several elevations to keep them from bending out of the columnar form. Even though some of the outside shoots will be sheared, the plants accumulate many dead twigs and leaves between the inner stems. As such trees mature or lose vigor, the multitude of bare, slender branches becomes quite unattractive.

Yew and podocarpus may be too slender or sparse if branches are cut back to spreading laterals. In these cases, the branches should be headed at varying lengths above their laterals; the branches of yew might be left 300 mm (1 ft) or more long, and those of podocarpus cut to stubs with six to 10 leaves and as many buds, several of which will grow and increase the number of branches (Chandler and Cornell).

Nursery-Grown Trees

Young nursery-grown conifers may develop asymmetrical or decurrent growth in a normally excurrent species, as well as particularly compact branching. Unless you want these features, do not select such trees for planting. Such growth is the result of vegetative propagation from lateral shoots (buds or cuttings) of species that have strong radial symmetry. A compact conifer, trained by severe pruning or shearing in the nursery, may look well-proportioned when young, but unless similar pruning is continued when the tree grows in the landscape, the new branches will be more widely spaced, giving the appearance of a tree growing on top of a shrub.

To correct asymmetry in a small tree, choose the most upright and vigorous shoot to produce a new leader, so that the growth habit will be changed from decurrent to excurrent. This shoot should have strong two- or three-year-old wood below it and a cluster of buds at the tip. If the chosen branch is not the longest, remove competing branches or reduce them in size to direct major growth into the new leader. If you must plant an asymmetrical conifer, partially tip (rotate) the root ball in the planting hole to keep the selected leader vertical. You may also have to stake the selected leader into a vertical position. Continue follow-up pruning and staking until radial symmetry is obtained.

You can correct excessive and closely spaced laterals by selectively thinning to match the more normal later growth. Thin crowded laterals lightly until tree growth in the landscape indicates what the new spacing between whorls or branches will be. Even then, extend the thinning over two to three years. The tree will usually be most attractive if this thinning is done early in its life, so that the selected branches develop without crowding. If remedial pruning is delayed, much of the interior foliage may die from shading.

Young Conifers in the Landscape

Conifers are most attractive when the tips of low branches almost touch the ground, but if clearance is required under the canopy, remove lower limbs more gradually than you would with broadleaved trees, about 250 to 350 mm (10–14 in) at a time on young trees. Some pruning back of laterals that are scheduled for removal will keep them relatively small so that later pruning wounds will also be smaller than they would be otherwise.

Conifers that normally have an excurrent growth habit often form multiple leaders; all but one of them should be removed. Some conifers, however, including a number of pine, some cypress, cedar, juniper, false cypress, and arborvitae species, form decurrent crowns. In these conifers, you can allow multiple stems to develop into their characteristic form, unless the branches are too low or their angles of attachment weak. These main stems should be differentially pruned, however, so they develop unequally to increase structural strength and present an interesting shape.

Conifers occasionally lose their leaders. When this occurs, a new leader may develop from one of the uppermost branches or from a latent bud near the top. In many conifers, when the top dies, is cut, or breaks out, branches in the topmost whorl will bend upward as reaction wood forms. One branch usually dominates to become the leader, and the others return to approximately their original orientation. Alternately, a latent bud may grow into a new leader. When a leader is lost, latent buds are more likely to grow if a 50- to 70-mm (2- to 3-in) stub is left above the top whorl (Scarlett and Wagener 1973). A leader from a latent bud should usually be selected over one developed from a lateral; the former will usually be more vigorous and more symmetrical. Shorten laterals in the topmost whorl to reduce competition with the developing leader. If no leader develops naturally, tie one of the topmost branches upright to induce it to become the new leader.

PALM TREES

Even though palms are pruned primarily to remove old and unsightly fronds and fruiting clusters, several of the most commonly planted species must be pruned quite frequently and are consequently some of the most expensive trees to maintain. The terminal growth is never removed because single-trunked palms have only one growing point. Entire stems are sometimes removed in clustering species that produce more than one stem.

Except to minimize disease or fire damage, you need not prune established palms for their own well-being. In public landscapes, however, many palms must be pruned frequently to keep them safe and attractive (Fig. 14-42). A palm normally maintains only a certain number of fronds, depending on the species and growing conditions. As new leaves develop at the terminal, the lower, older fronds die. Some palms shed their old leaves; others retain dead fronds until eventually they rot away, are blown off by the wind, or are removed. Fronds can sometimes cover the trunk to the ground (Ledin 1961). In some palms the leaf blade will break off, leaving the petiole base or boot remaining on the trunk for many years. Dead fronds or their bases harbor insects and rodents and can be a fire hazard. Falling fruit and fronds can be a serious nuisance. Low-growing palms may have spines on their trunks or spiny leaflets that can be dangerous.

FIGURE 14-42
Mexican fan palms are pruned frequently in intensively maintained
landscapes to minimize litter from the flowering stalks.
The green petioles are removed in two cuts of a sharp carpet knife
(has a hooked point) leaving V-shaped tips (inset).
The petiole bases will be removed a year or so later.

Multiple-stemmed palms form new shoots below ground. Though there is little need to prune cluster palms, an entire stem is sometimes removed. You may have to remove individual stems if they grow close to a building, if the cluster occupies too much space, or if you want a specimen plant. In some monocarpic clustering palms, the old stems should be pruned out after the final fruiting, because they will die near their base shortly afterwards (Ledin). The new shoots that surround the old will maintain the clump. Stems should be cut off as close to the ground as possible.

Palms that produce hazardously large fruit or frequently drop old fronds may have to be pruned every three to six months. Though it is one of the most commonly planted palms in tropical landscapes, the coconut palm must be pruned every six months to keep it safe. Both fruiting stalks and fronds are removed at each pruning. The fruiting stalks are pulled down and cut as close to the base as possible. On regularly pruned coconut palms, only the bottom one or two rings of fronds are removed. A thin, fibrous sheath surrounds the frond base and makes cutting difficult unless the sheath is first pulled down

FIGURE 14-43
Holes punched in palm trunks
by using climbing spurs
remain for the life of the tree.
In humid regions these holes
could become sites for decay.

and cut off. The exposed fronds are green and easily cut with a pruning knife or machete. An experienced pruner can climb and prune a mature coconut palm in less than ten minutes. Unfortunately, most pruners use spurs, which create holes in the palm trunks that do not callus (Fig. 14-43).

Some palms, including royal palms, readily shed their old leaves. When the fronds begin to turn brown, they can easily be pulled from the trunk. This should be done fairly regularly on tall trees. The fronds of other palms are usually removed less frequently, depending on the species and the amount of effort required. Some of the date palms are pruned with a chain saw every two years. The cut frond bases can be quite acceptable if pruning is done carefully. The chain housing may have to be enlarged to minimize clogging with palm fibers.

In landscapes that receive more intense care, palms are pruned at least annually and with greater attention to detail (Williamson 1972). When yellow fronds are cut, 200 to 300 mm (8–12 in) of the base is left. The ends of the frond stubs may be shaped to a point by two cuts with a sharp knife (Fig. 14-42). The frond stubs left by the previous pruning are cut off close to the trunk.

Take care when cutting fronds so as not to cut into live growth, which could cause unsightly scars. Most fronds should be cut from below to minimize ripping fibers down the trunk.

Palms injured by cold will usually suffer most damage in the lower leaves. Ledin (1961) recommends removing all injured fronds, except the inner two or three, at the terminal. Even when all the fronds are injured, do not remove an entire palm for at least six months to determine if it will survive. If the terminal bud is alive, it will send out new leaves. If the top leaf bases have hardened, the new growth may become stunted as it tries to push up. To prevent this, loosen the dried leaf bases as the new growth begins to show (Ledin).

SHRUBS

Shrubs are woody plants with several stems originating at or near the ground. They are usually smaller than trees, but their distinctive features are form and branching pattern. Most shrubs have abundant latent buds at their base and along their branches; new shoots from the base replace weak and dying branches and, in effect, keep a shrub young. A mature shrub that normally forms few or no new shoots from its base will do so readily if its top is pruned or its base opened to light.

Shrubs range in height, spread, vigor, and flowering characteristics, each of which may influence pruning techniques. Slow-growing evergreens need little or no pruning, even in confined spaces. Many vigorous evergreen shrubs may do well with no pruning if they grow in ample space and will be viewed from a distance or at high speed. Most fast-growing evergreen and deciduous shrubs require moderate or extensive pruning to contain them, maintain their attractiveness, and balance vegetative growth and flowering. Pruning frequency for mature shrubs depends on their vigor, flowering habit, and desired size.

Shrubs are pruned for essentially the same reasons as trees are pruned. Pruning for structural strength, however, may not be as critical except for large shrubs and those exposed to wind or wet snow. Unfortunately, people too often prune shrubs primarily to control their size; in many situations this is a sign that the wrong species was planted. Nurseries like to market fast-growing shrubs because they are easy to propagate and grow, are attractive in the nursery, and are popular. Many landscape architects and horticulturists similarly favor fast-growing plants because they generally do well and look attractive early. It is not until several years after planting that the shrub becomes overgrown and difficult to contain. For small landscapes and moderate-sized beds, shrubs should be chosen according to their mature size.

Pruning should begin at planting and be part of the regular maintenance program for shrubs. Plants need not be pruned at each inspection or even annually, but their growth and form should be assessed. All too often, shrubs receive little or no attention after initial care at planting until they are too large for their allotted space. By then, inside and lower foliage has probably been weakened and shaded out. If the shrub is pruned to a smaller size at that point, large cuts will result in a stiff, unattractive plant with little or no low foliage; regrowth will be vigorous and upright, further detracting from the natural appearance of the shrub. Pinching the tips of vigorous, nonbranching shoots will make the shrub more compact and symmetrical. Shrubs whose growth must be contained should be on a regular pruning program before they reach the size desired.

As many shrubs mature, shaded lower branches weaken, lose leaves, and become unattractive. The solution is not to cut the low branches partially or completely off but to thin the shrub so light can enter it. If plants are thinned in time, existing branches and new ones will usually be stimulated enough to keep the shrub full to the ground. Blue blossom ceanothus and Victoria tea are exceptions because they do not sprout readily from old wood (Chandler and Cornell 1952). If you head branches instead of thinning them, more dense, upright growth in the top will further shade lower growth.

As stated, pruning severity and frequency depend in large measure on shrub vigor and flowering habit. More specific comments can be made about pruning slow- and fast-growing shrubs.

Slow-growing Species

Slow-growing shrubs, usually broadleaved evergreens, make most of their growth from terminal buds or buds near shoot ends. With many shoots forming, the growth of each is reduced, producing a slowly expanding, fairly symmetrical, compact plant. Most plants of this sort develop a dense outer shell of foliage with few leaves in the center. They require little pruning, except when an occasional branch outgrows the general contour of the shrub. Such a branch should be cut clear to its point of attachment or to a lateral arising deep within the plant.

Slow-growing shrubs develop more of a main-branch framework than do many of the fast-growing shrubs that sprout primarily from the base. Slow-growing shrubs seldom require severe pruning; usually only a few shoots are cut back to their attachment or to another lateral. A light thinning will open up a compact shrub so more light and air can reach the center. Such pruning can create pleasing branch and leaf patterns and stimulates inside leaves. Slow-growing plants should seldom be headed back and should never be sheared.

Fast-growing Shrubs

Both deciduous and broadleaved evergreen shrubs occur in fast-growing species. Some species form a main framework from which most new shoots grow. Other species, particularly many deciduous shrubs, sprout vigorously from the base, which in a few years can become a tangle of branches unless pruned. On some species, shoots are so vigorous that they cannot remain upright and bend over instead. New laterals on vigorous shoots are usually stimulated at, and just behind, the top of such a bend. The original shoot bends even more with the added weight of laterals. With proper pruning, the shrub can have a graceful arching form; if left unattended, it will continue to enlarge and become a confusion of branches. Fast-growing shrubs need regular, fairly severe pruning to keep them attractive and contained. You can maintain the natural shape by thinning out branches and cutting others back to a bud or lateral to control the direction of future growth. This pruning should also enhance renewal of the plant.

When pruning a fast-growing shrub, remove unwanted branches first, branches that lie on the ground, have little foliage, are weak, broken, crossing, dead, diseased, or infested with insects. Extremely vigorous unbranching shoots that originate from the base and outgrow the general outline of the shrub should be removed close to the base. Next, cut back some of the largest and oldest branches to their origin. Before each cut, visualize what the plant will look like with the branch in question removed. Pruning should not leave a void in the shrub form. Distribute succeeding cuts to provide a uniform thinning (Fig. 14-44). On vigorous, mature shrubs, remove at least 25 to 30 percent of the older branches annually. Finally, you may need to shorten some of the younger, smaller branches to maintain shrub size or shape. Take care to vary the depth of these cuts in order to leave the shrub with a textured, natural look. A shrub so pruned will produce and maintain vigorous shoots with low foliage and healthy appearance.

FIGURE 14-44
A shrub (center) can be pruned so that it becomes more spreading (left)
or more upright (right) in habit.
Short stubs have been left to indicate where the cuts have been made.

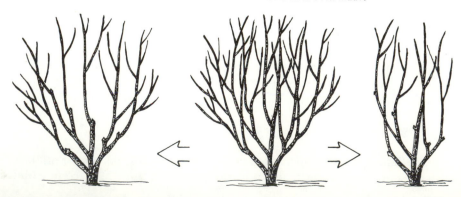

A well-pruned shrub will add more character and interest to a landscape than one that has become a dense tangle of branches. If a mature, overgrown shrub is to be considerably reduced in size, reduce it over a two- to four-year period.

Severe pruning is recommended only for old, overgrown shrubs in certain situations, but it will usually not endanger the plants. In fact, most fast-growing shrubs can be cut close to the ground (50–150 mm or 2–6 in) (Brown 1972). New shoots soon establish a shrub with renewed vigor and foliage clear to the ground (Fig. 14-45). Plants pruned so severely would have less vigorous regrowth if some of their roots were cut, but roots are seldom pruned. Vigorous regrowth will develop more girth if it is headed a few times during the growing season. Some of the new basal shoots should be thinned to allow the greater development of others and to give the shrub a more natural look. Without thinning, the shrub might consist of many spindly upright shoots growing close together.

Shrubs that must be kept from spreading or overgrowing must be pruned with care if they are to maintain an informal, natural look. If shrubs are sheared back, the resulting surface will reveal many dead twigs. The offending shoots should be cut back varying distances to make the surface irregular and the plant more natural looking.

FIGURE 14-45
Many overgrown shrubs can be pruned close to the ground
when there seems to be no other way to reduce their size
without exposing bare branches. The escallonia on the left was pruned
within 150 mm (6 in) of the ground two years before the photograph
and the one on the left one year. Compact symmetrical spheres result;
frequent thinning will be necessary to develop any character to the plants.

Prostrate shrubs can be cut back considerably without appearing to have been pruned. Cuts should be buried back in the shrub center and pruning should be to a lateral growing from the top of its parent branch.

Shrubs into Small Trees

A shrub that has grown too large to be easily reduced can often be trained to become an attractive small tree (Fig. 14-46). The space under the canopy can either be left open or planted with ground cover or small shrubs. Low-growing branches that do not carry into the top of the plant, and usually some of the main branches as well, should be removed; the remaining branches will then become more prominent visually. You will seldom need to create a new branch structure but should thin the existing one. Severe pruning and exposure of lower trunks to more light may stimulate water sprouts. If you make close cuts and treat the wounds with naphthaleneacetic acid, water sprouts will be minimized (see Chapter 15). If several shrubs in the same planting are to be trained into small trees, some of them should probably be removed to provide room for the others to develop individually.

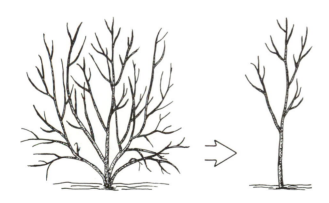

FIGURE 14-46
Occasionally an overgrown shrub can be pruned into an attractive small tree (top). A more attractive plant will usually result if fewer branches are left than on the callistemon (lower left), though they can be more than the single-trunked pyracantha (lower right).

Hedges

A hedge is a row of closely spaced shrubs or trees, usually of a single species, grown either informally with little or no pruning, or formally, trained and sheared. Hedges serve as visual and physical screens and dividers. They are more common now in public and large private landscapes than in small residential plantings. Informal hedges are allowed to grow naturally and usually require considerable space; Wyman (1971) suggests that hedges, particularly conifers, be allowed to grow nearly as wide as they are high. Formal hedges are sheared fairly regularly to retain their form, size, and attractiveness.

Plants—particularly for formal hedges—should have small leaves, short internodes, and dense branches, and should be able to sprout from old wood. Evergreens are preferred for effective year-round screening. In colder climates, conifers may be the only evergreens hardy enough to survive. Plants with larger leaves are often used for informal hedges, but the cut foliage of large-leaved plants is often unattractive when it has been sheared for more formal effects. Clark (1979), Grounds (1973), and Wyman (1971) recommend specific species for particular uses and geographical areas.

Informal hedge plants are spaced 500 to 900 mm (20–36 in) apart, and all growth is headed back at least halfway to encourage low, dense branching. After this, only occasional pruning is needed to maintain moderately uniform growth and symmetry and to keep the hedge in bounds. Pruning individual branches with hand shears instead of shearing many branches at once will maintain an informal appearance.

For a formal hedge, small-growing evergreens may be planted as close together as 450 mm (18 in), while large-growing deciduous plants should be planted at least 900 mm (36 in) apart. Depending on the size of the shrubs or trees at planting, small evergreens should be cut back to within 75 or 100 mm (3 or 4 in) of the ground and large deciduous plants to within 150 or 300 mm (6 or 12 in) of the ground. No further pruning takes place in the first growing season, so plants can become well-established.

Before growth begins in the second year, the shoots of broad-leaved plants should be headed to within 100 or 150 mm (4 or 6 in) of the pruning height at planting. Conifers should be tip-pruned. Until the desired height is reached, head back new shoots one-half to two-thirds of their length each time they grow 150 to 300 mm (6–12 in). Severe heading of new growth encourages plants to spread and develop a dense, low-branched structure. When the developing hedge has filled in fairly well at the base, it should be allowed to grow more in height than width. Hedges, particularly conifer hedges, should be wider at the base than at the top to ensure that the lower foliage will receive

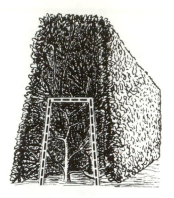

FIGURE 14-47
To maintain low foliage,
shear a hedge so that it is
wider at the base than at the top,
particularly if oriented
in an east-west direction.
When a hedge has become too tall and wide,
prune it back to about the size
indicated by the dark broken line;
after two or three shearings
the hedge will be well clothed in leaves.

enough light (Fig. 14-47). In areas subject to early frosts or cold winters, pruning should not take place after midsummer.

After a hedge has reached the desired height, time shearing according to the amount and cycle of growth. The effects of shearing will last longest if it is done after growth has ended for the season. Growth may be so vigorous, however, that the hedge becomes unattractive and must be sheared during the growing season. With experience, you will be able to compromise between minimizing the number of shearings and maintaining an attractive hedge. Most small formal hedges begin to look unkempt when new growth is 50 to 100 mm (2-4 in) long.

Shear new growth on a mature hedge to within 10 mm (0.4 in) of the previous shearing, leaving only one to three new leaves (and buds). This will clothe the hedge in new foliage and slow its increase in height and spread. After each shearing, the hedge will grow a little larger than it was after the previous shearing. Eventually most hedges become too large for their sites or lose their lower foliage. Light pruning removes most foliage without greatly reducing the size of the hedge. In contrast to many conifers, most broadleaved hedges can be pruned back severely. It is best to cut broadleaved hedges back to about half the desired mature height and width (Fig. 14-47). The pruned plants will have little or no foliage but should soon develop new shoots. After one or two shearings the hedge will return to its original beauty. Such heavy pruning should be done just before growth begins in spring, so the hedge is bare for the shortest time. Conifers must be pruned with greater care; be sure new shoots will grow from older wood (see Table 14-2).

Roses

Roses are usually pruned more severely than other shrubs, not only to keep them in bounds but to balance vegetative growth and flowering. There are three common habits of growth and flowering (Brown 1972):

Rambler roses produce from their bases long, limber canes that bloom profusely the following spring (Fig. 14-48).

Climbing roses produce long canes that remain vigorous and productive for many years. They bloom on current shoots from laterals that grew in the previous season from the main canes or their branches (Fig. 14-49).

Tea and hybrid tea rose bushes, among others, bloom on shoots growing from canes or laterals that developed in the previous season and on shoots growing from other shoots that developed earlier in the current season (Fig. 14-50).

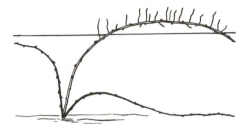

FIGURE 14-48
Rambler rose produces vigorous canes (lower right cane), which can replace those that grew the previous year and have flowered (upper right cane). Train seven to ten vigorous new canes on the trellis in late summer after the old canes have been pruned out (for the sake of simplicity, only three canes are shown).

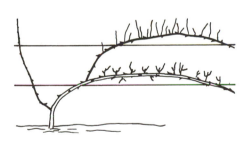

FIGURE 14-49
The canes of climbing roses are productive for several years and should be trained horizontally or in arches on a trellis. Laterals along the cane that bloomed the previous year (upper right cane) should be headed (lower right cane) before spring growth. Should a cane become unproductive (lower right), replace it with a more vigorous basal cane (left) or cut it back to a lower vigorous cane (upper right).

FIGURE 14-50
Prune hybrid tea rose bushes (left) annually, selecting three to six canes (depending on the vigor and purpose of the bush) (right). Replace one or two of the oldest canes each year to maintain a vigorous and productive bush.

You can control the balance between vegetative growth and flowering fairly well by varying the number and length of canes selected at pruning. In deciding how severely to prune, evaluate the plant's vigor and the size of the previous season's roses. If you want more vigorous growth or larger blossoms, leave fewer or shorter canes than previously. On the other hand, if growth has been vigorous with large blooms, you may leave more wood to take advantage of the plant's flower-producing capacity.

You can find details for pruning roses in many books specializing in roses and pruning, particularly those by Brown (1972) and Williamson (1972).

VINES

Vines, also called climbers or wall shrubs, have vigorous, slender, flexible stems that require support in order to grow upright. Vines are versatile and can grow quickly as a ground cover, a screen on a trellis, a cascade of foliage and flowers against a wall, a leafy cover on an arbor, or a climber on a pole or tree. Since they are vigorous growers, most vines must be pruned annually to keep them attractive, healthy, and in bounds. To be effective in the landscape, most climbing vines require as much attention as roses do. Vines can greatly enhance even a small landscape, but the species must be selected with regard to function and space.

Not all vines can climb. While most have some means of holding onto a support, some need help. Vines climb upright in four basic ways (Fig. 14-51): aerial roots, holdfasts, or sucker discs (ivy and Virginia creeper); tendrils (clematis and grape); twining stems (star jasmine and wisteria); or artificial means of support (bougainvillea and rose). Thorns and hook-like projections on stems and leaves help other plants attach themselves to supports. Vines with aerial roots or holdfasts can climb almost any surface, while other vines need a trellis, fence, or wire for support.

Managing vine growth consists primarily of directing the stems on a support (training), pruning out old growth, and containing plants in their allotted space. Particularly when plants are young, you may wish to arrange stems on their support to form a pleasing pattern. If the stems do not branch readily, pinch their tips where branching is wanted. Pinched stems cover a trellis more completely, especially stems that radiate from the base. Many vines overgrow their space quickly and become unattractive tangles unless they are pruned regularly. New shoots that will not be useful replacements must be pruned back or out to keep a vine from becoming too dense and heavy.

FIGURE 14-51
Vines attach themselves for support in several ways:
English ivy by aerial roots (top left); Boston ivy by holdfasts (top right);
grape by tendrils (bottom left); and wisteria by twining stems (bottom right).

Older branching systems that fail to produce vigorous flowering shoots must be cut back to main branches or all the way to the base. Vigorous stems that grow beyond their allotted space or begin to climb where not wanted should be cut back or trained back onto the vine's support.

The growth and flowering habits of vines provide clues for training and pruning the plants to carry out their landscape function. Vigorous vines that cling to a wall or support, such as ivy, can be clipped close to the wall at least once a year. Prune them before growth begins and again as needed during the growing season. This keeps the foliage close to the wall and, in ivy, prevents mature branches from forming. Some vines, such as wisteria, form a heavy structure on which vigorous shoots grow; buds near the bases of shoots flower during spring following shoot growth. The first shoots of a young vine should be trained to become the permanent framework. Later shoots from the framework branches should be pinched or cut back to two or three leaves during the growing season. Succeeding growth flushes, in turn, must be headed (Fig. 14-52). Before spring growth, cut back each headed shoot system to the lowest one or two buds. Short shoots will grow from these basal spur buds and form terminal racemes of flowers.

On vines that flower along vigorous one-year-old canes, such as honeysuckle and rambler rose, the canes should be pruned out or back to new low-growing shoots after they have flowered in the spring. The pruned stems must be removed carefully so that the new shoots are not damaged, particularly if the shoots twine around each other. Other vines, such as weeping forsythia, develop main branches from which pendulous shoots grow readily for several years. Flowers are borne on these one-year-old shoots. The main branches should be trained to provide height from which the long shoots can hang. Although annual pruning is recommended, it need not be as severe as most vines require; remove about one-third of the older wood each year after flowering. Likewise, thin about one-third of the drooping shoots on remaining branches. Vines that flower on vigorous current-season's growth, as some clematis do, should be pruned severely each year, often close to the ground, before growth begins in the spring. This annual pruning keeps the plant vigorous and in bounds.

A number of vines, such as ivy and star jasmine, are also grown as ground covers. These spread quickly, and some, such as ivy, root where shoots contact moist soil. These vines must be trimmed two or three times a season to keep them within their borders. Within a few years, the mass of canes builds up so high that it must be cut back. If ivy canes are not too large, a rotary power mower or hedge shears can be used to remove almost all the canes back to where they have rooted. Do this just before spring growth begins, so that new growth quickly covers the bare canes. Many vigorous vines must be planted with care near buildings. Small shoots can grow between cracks in

FIGURE 14-52
A thirty-year-old wisteria in bloom. A wisteria shoot system
that was headed to two to four buds during the growing season (lower right);
the spur system could be cut back to the second spur before spring growth.
Short flowering shoots grow from the basal spur buds (lower left).

walls, roofs, fences, and other structures and can cause considerable
damage as they increase in diameter. They can clog downspouts and
rain gutters. If not regularly pruned, some vines become so heavy
they collapse their support. Vines can engulf shrubs and small trees
and sometimes kill them by blocking sunlight.

SPECIALIZED PRUNING

Espalier

Some trees and shrubs can be trained in a vertical plane on a wire or wooden trellis, either free-standing or against a wall (Fig. 14-53). An espalier is a trellis or a plant grown on a trellis; in the United States, espalier is also used as a noun or verb to describe a method of training. Fruit trees have been espaliered along walls and stone fences that face the equator in European gardens for centuries. The heat capacity of walls and stone fences usually hastens the bloom and fruit maturity of plants growing against them, and protects the plants from frosts and winter cold. Espaliered plants, sometimes referred to as wall shrubs, also require less space, an important factor in small gardens.

Species that flower on spurs, such as apple and pyracantha, are best suited for the formal espaliers, especially if branches are to be horizontal. Species that flower on vigorous one-year-old shoots, such as peach, or on current growth, such as citrus, are better suited to more informal or upright patterns.

Espaliers are usually trained on tightly stretched wires or wood supports. Horizontal and vertical supports are usually 400 to 500 mm

FIGURE 14-53
A pear tree at Wisley Gardens, England, trained as a palmette espalier (left).
(Photo: H. T. Hartmann, W. J. Flocker, and A. M. Kofranek. *Plant Science: Growth, Development, and Utilization of Cultivated Plants*, © 1981, p. 352. Reprinted by permission of Prentice-Hall, Inc., Englewood Cliffs, New Jersey)
A candelabra or double-U espalier of pear at Cornell University, New York (right). (Photo courtesy of Clarence Lewis, East Lansing, Michigan)

(16–20 in) apart. A trellis 100 to 150 mm (4–6 in) from a wall allows the sun to create interesting shadows on the wall, particularly behind deciduous plants in winter. Fruit injury due to high temperatures can be reduced by encouraging more foliage to shade the wall or espaliers of fruit can be grown in the open. Species that must be sprayed for pests should not be espaliered on wooden walls, painted or unpainted, because pesticides can discolor or stain the surface.

   Palmette and *cordon* are forms of espalier whose distinctions have become unclear, at least in America. A palmette plant has one vertical trunk with parallel pairs of laterals arising at about equal intervals along it (Fig. 14-53). A proper cordon is a plant grown to a single trunk or to two main branches trained in opposite directions (Bailey 1916).

Palmette Training   Palmette training is the most common form of espalier (Fig. 14-53). Laterals should arise from the leader just below the horizontal support on which they are to be tied. Where laterals are available, tie one in each direction from the leader and remove those not tied to horizontal supports. At planting, if no laterals can be satisfactorily spaced, head the leader at or just below the lowest horizontal support.

   During the growing season, when the leader has grown about 100 mm (4 in) above a horizontal support, head it back to the support or just below it. Develop the new topmost shoot as the leader and train two laterals onto the horizontal support in opposite directions. Pinch off other shoots that grow. Tie each shoot down about every 150 mm (6 in) as it grows, but do not tie down the growing tip, however; allow it to grow freely and maintain vigor. Repeat this process until all horizontal supports have laterals; then keep the leader headed at the top laterals.

   While laterals are kept horizontal, vertical branches may be allowed to develop along them to carry the main foliage, as well as flower and fruiting spurs. Frequent pinching of vertical branches checks growth, so they will not subdue the horizontal branch from which they arise. The shoots pinched to hardened wood will not grow back as quickly as they would from a softer pinch.

Cordon Training   Cordon training is best for fruit trees that form or are pruned to spurs. You can develop a cordon by bending an unbranched trunk (whip) about two meters (6 ft) tall onto a single-wire trellis about 500 mm (20 in) above the ground. The trunk should usually be twisted at the bend to slow growth and hasten flower-bud formation.

   Some cordons are trained to one straight trunk, either vertical or inclined at about a 30° angle from the vertical. Many cordons have one vertical trunk terminating in two lateral branches that grow horizontally in opposite directions.

FIGURE 14-54
This pear tree at Wisley Gardens, England, has been trained as a fan espalier.
(Photo: H. T. Hartmann, W. J. Flocker, A. M. Kofranek. *Plant Science: Growth, Development, and Utilization of Cultivated Plants*, © 1981, p. 352. Reprinted by permission of Prentice-Hall, Inc., Englewood Cliffs, New Jersey)

Fan Training To form a fan, head the trunk 200 to 300 mm (8–12 in) above the ground; select about ten shoots and tie them to a trellis in a narrow fan (Bailey 1916). Before the next growing season, cut back the top two branches about halfway to reduce their total growth and to encourage branching. Head back the other branches to mature wood. Tie remaining branches to the trellis in a broad fan (Fig. 14-54). Species that flower on vigorous one-year-old wood or on current growth are more productive when fan-trained than when trained as palmette or cordon espaliers.

Informal Training Many plants are better adapted to informal or more upright espalier training (Fig. 14-55). Informal espaliers are trained

FIGURE 14-55
A young ginkgo informally espaliered against a wall at the Morton Arboretum, Lisle, Illinois. (Photo courtesy of Clarence Lewis, East Lansing, Michigan)

similarly to fan espaliers except that shoots of informal espaliers are tied in a desired pattern. Shoots of species that do not readily form spurs should grow at an angle of at least 30° above the horizontal or should have their terminal growth so inclined. This will help maintain branch vigor and flower production.

All espaliers require persistent attention so growth will be directed where it is wanted and unwanted growth removed before it weakens the rest of the plant and mars the beauty of the espalier.

Pleach

To *pleach* is to weave or intertwine. Occasionally, suitable trees can be grown in one or more rows to a height of 3 to 5 meters (10–15 ft), where they are headed and the more horizontal vigorous shoots are tied to horizontal wires or another support. Eventually, as the shoots from adjacent trees intermingle, they are pleached (woven together) (Fig. 14-56). Upright branches and those from along the trunks are cut off. Species suitable for pleaching have supple branches that can be woven without breaking. Beech, hornbeam, buttonwood, apple, peach, and pear have been successfully pleached (Free 1961).

A double row of closely planted trees can be arched and pleached upon a strong wood or metal frame to form a covered walkway, or allée. Individual trees can be trained as informal espaliers to provide a wall of foliage. Upright branches are arched over, tied to the frame, and eventually tied and pleached with branches from trees in the other row. When these branches graft together, the arch can stand without support. The upright shoots should be cut back so leaves on the inside of the allée will receive enough light.

FIGURE 14-56
This pleached alleé of goldenchain trees at the Royal Botanic Garden, Kew,
has been trained on a pipe and cable trellis (left). The branches are intertwined
and could eventually support the alleé without the trellis.
The London plane trees in a Whittier, California park
have been pleached to form a pergola (right).

Pollard

Pollarding is heading back one- or two-year-old branches to approximately the same place year after year, usually at the terminals of framework branches. Trees can be kept small in this manner and will have a lush, formal appearance. Pollarding is described earlier in this chapter (see Fig. 14-39).

Topiary

Topiary is the practice of training and shearing plants into various formal shapes, geometric or mimetic. Topiary was practiced by the Romans and is widespread even today (Fig. 14-57). Small-leaved evergreen plants that grow readily from latent buds are best adapted to topiary. Creating the more intricate designs requires skill and patience. Wire may be used to hold developing branches in desired positions. Trees may also be sculpted to simple geometric shapes. Plants must be clipped often during the growing season to keep them attractive.

Small-leaved evergreen vines are sometimes trained over wire-mesh frames to create topiary-like forms. This is a quicker way to

FIGURE 14-57
English holly topiary in a home garden near Stratford-upon-Avon, England (left); waxleaf privet topiary at Oki Nursery, Sacramento, California (right).

develop garden sculptures, and few people will be aware that the plants are not actual topiary forms.

Bonsai

This ancient Asian art produces miniature replicas of mature trees (Fig. 14-58). The essentials of bonsai are careful pruning of the top, the occasional use of wire to bend a trunk or branch, maintenance of a small but healthy root system in a shallow container, and sensitive placement in a complementary microlandscape, which includes rocks, moss, and lichen. Species selected for bonsai should have small leaves, flowers, or fruit that will be in scale on a plant less than 600 mm (24 in) tall. Frequent pruning of shoots not only keeps the plant small but increases the taper of the trunk and branches and makes branch growth more tortuous, to simulate old trees. Roots should be pruned back and repotted with new soil every two to four years. Plants are kept in moderate vigor, but growth is pinched back to control size and form. Bonsai plants must be protected from extreme heat and cold and must be watered frequently because of their limited roots. A number of good books are available on the art of bonsai (Wyman 1971).

FIGURE 14-58
Japanese maple (left) and ginkgo (right) grown as bonsai by N. William Stice, Sacramento, California.

FURTHER READING

BRIDGEMAN, P. H. 1976. *Tree Surgery*. London: David & Charles.

BROOKLYN BOTANIC GARDEN. 1981. Handbook on Pruning. *Plants & Gardens* 37(2).

BROWN, G. E. 1972. *The Pruning of Trees, Shrubs, and Conifers*. London: Faber and Faber.

HALLIWELL, B., J. TURPIN, and J. WRIGHT. 1979. *The Complete Handbook of Pruning*. 2nd ed. London: Ward Lock.

HUDSON, R. L. 1972. *The Pruning Handbook*. Englewood Cliffs, N.J., Prentice-Hall.

WILLIAMSON, J. F. 1972. *Sunset Pruning Handbook*. Menlo Park, Calif.: Lane.

CHAPTER FIFTEEN

Chemical Control of Plants

A number of chemicals are used to stimulate, reduce, or stop the growth of shoots and to initiate, delay, or prevent flowering and fruit set. Those who use chemicals for these purposes want to interfere as little as possible with the health and appearance of plants. Other and some of the same chemicals are used to kill unwanted plants. Certain chemicals have been employed for years, some are recent, and others are still experimental. An arborist should understand the effects of these growth-regulating chemicals and the conditions under which they can be used safely and beneficially.

STIMULATION OF GROWTH

Fertilization

Fertilizers, particularly nitrogen, have been used for many years to increase the growth and development of plants. Fertilization has been perfected to a science for many food, fiber, and greenhouse crops and is particularly important for stimulating young plants in a landscape (see Chapter 11).

Irrigation

Irrigation is necessary to maintain the normal, vigorous growth of young plants and to support all plants when moisture in the root zone becomes limiting. Too much or too little soil moisture can seriously reduce growth or even kill plants (see Chapter 12).

Application of Growth Stimulants

Several growth-regulating chemicals increase the height of shoots or the dry weight of leaves and stems. Some substances normally used as growth retardants or herbicides will stimulate growth in very dilute concentrations.

Gibberellins are a group of more than 50 closely related natural compounds that stimulate cell division or cell elongation (Weaver 1972). They stimulate stem elongation in many plants that are genetically dwarf or rosette, particularly herbaceous plants. Gibberellins can also be used to satisfy the cold, light, or dark requirements for seed germination or bud break in some species. When gibberellic acid (Gibberellin A_3) was sprayed at 100 and 1000 ppm in midspring on greenhouse-grown sour cherry trees that had set terminal buds, it stimulated a second flush of growth (Hull and Lewis 1959). Stems elongated, the diameter of trunks enlarged, and tops increased in both fresh and dry weight. The acid did not affect root growth, but a second spray in July caused spindly shoots, many of which died back at the tips.

In southern California, foliage sprays of gibberellin have stimulated the growth of young carob seedlings planted in outdoor containers (Goodin and Stoutemyer 1962). The researchers applied gibberellin at 10 to 50 ppm for 20 weeks beginning in midfall; the intervals between applications ranged from a week to a month. The gibberellin spray apparently overcame the growth-suppressing effects of long nights in fall and winter. By late winter, when applications were stopped, the growth rate of the untreated plants was about the same as that of the treated plants. Weekly spraying, however, produced spindly shoots, which died on three of 25 trees. Although gibberellin is used commercially on a few crop plants (to encourage melon fruit set and grape enlargement, for example), it has not been used for woody landscape plants.

CONTROL OF PLANT HEIGHT

Chemicals can control the height growth of many herbaceous and woody plants. Growth-regulating chemicals should be chosen and used with care, however, because each chemical has specific properties; their use should depend on plant species and environment.

Plant Height Control Mechanisms
and Chemicals

Chemicals restrict height growth in at least two ways (Sachs and Hackett 1972):

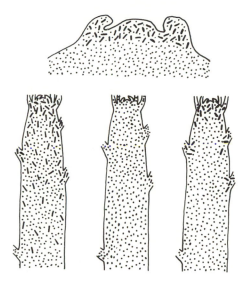

FIGURE 15-1
Diagrammatic sketches of the meristematic tissue of shoot apexes.
The short, heavy, black lines represent dividing cells.
The dividing cells in the apical meristematic dome (top)
constitute the approximate zone for formation of leaf initials,
two of which are shown on either side of the apical dome.
In contrast to the untreated shoot (lower left),
the one treated with a general inhibitor such as maleic hydrazide (lower center)
exhibits almost complete inhibition of cell division.
Subapical inhibitors, such as daminozide, cause relatively restricted inhibition
in the subapical tissue and none at the apex (lower right).
In the lower figures the basal shoot diameter is about 3 mm (0.1 in);
the leaf bases are indicated as broken stumps.
(Adapted from Sachs and others 1970)

General inhibitors may (1) kill terminal buds or severely inhibit apical meristematic activity and (2) reduce apical dominance, allowing lateral buds to
grow, which in turn reduce terminal growth (Fig. 15-1).

Subapical inhibitors retard internode elongation without disrupting apical
meristematic functions (Fig. 15-1).

As would be expected, growth-retarding chemicals differ widely
(Tables 15-1 and 15-2). Their effectiveness varies with

Plant species and clone
Formulation and concentration of the chemical
Time of application in relation to growth cycle of plants
Method of application
Frequency of application
Weather during and just after application

TABLE 15-1   Growth-inhibiting chemicals registered for that purpose (adapted from Sachs and Hackett 1972; Sachs 1975)

	Common Name[a]	Chemical Name	Trade Name	Manufacturer
General Inhibitors	Maleic hydrazide (MH)	6-Hydroxy-3-(2H)-pyridazinon	Royal Slo-Gro®	UniRoyal
	Chlorflurenols	Methyl-2-chloro-9-hydroxy-	Maintain CF 125®	E. Merck
	(Morphactins)	fluorene-9-carboxylate		
	Dikegulac	2,3,4,6-di-o-isopropylidene-2	Maintain A	Maag Agrochemicals
		Keto-L-gulonate	Atrinal®	
	Fluoridamide	[N-4-methyl-3-[[(1,1,1-	Sustar®	3M
		trifluoromethyl) sulfonyl]		
		amino] phenyl] acetamide		
	Fatty acid esters	Methyl nonanoate	Emgard 2077®	Emery Ind.
		Methyl decanoate + other esters	Off-Shoot-O®	Proctor & Gamble
	Naphthaleneacetic acid	Ethylester of naphthalene-	Tree-Hold®	Amchem Products
	(NAA)	acetic acid	Inhibitor-Fortified	
			Tree-Paint®	
Subapical Inhibitors	Daminozide (SADH)	Succinic acid-2,2 dimethyl-	Alar® or B-9®	UniRoyal
		hydrazide		
	Chlormequat (CCC)	(2-chloroethyl) trimethyl	Cycocel®	American Cyanamid
		ammonium chloride		
	CBBP (Phosfon)	2,4-dichlorobenzyl-tributyl	Phosfon-D®	Mobil Chemical
		phosphonium chloride		
	Ancymidol (El-531)	α-cyclopropyl, α-4-methoxy-	A-Rest®	Elanco Products
		propyl, α-5-pyrmidine		

[a]The preferred common name is listed with synonyms in parentheses.

TABLE 15-2   The uses and effects of the more common growth-inhibiting chemicals[a]

	Chemical	Method of Application[b]	Concentration Range-ppm	Responsive Plant Groups[c]	Effect on Flowering and Fruiting[d]
General Inhibitors	Maleic hydrazide (MH)	Foliage Injection	400–4000 70–115	Trees* Shrubs	Inhibits
	Chlorflurenols (Morphactins)	Foliage Bark	125–1250 10,000	Trees* Ivy* Turf Ice plant*	Inhibits
	Dikegulac	Foliage Injection	750–6000 10–75	Shrubs Trees	Generally inhibits (can enhance azaleas)
	Fluoridamide	Foliage	500–5000	Trees & shrubs Turf	Inhibits
	Fatty acid esters	Foliage	25,000	Pot plants*	Inhibits
	Naphthaleneacetic acid (NAA)	Pruning cuts Trunk sprouts	10,000	Trees*	Little or none
Subapical Inhibitors	Daminozide (SADH)	Foliage	1000–10,000	Pot plants* Shrubs No conifers	Enhances
	Chlormequat (CCC)	Drench Foliage	500–3000 500–3000	Pot plants	Enhances
	CBBP (Phosfon)	Drench	1–500	Pot plants*	Enhances
	Ancymidol (El-531)	Foliage Drench	10–200 0.5–5.0	Pot plants* Shrubs & trees Few conifers	Enhances

[a]Compiled primarily from Sachs and Hackett (1972), Sachs (1975), Cathey (1975), and Roberts, Brown, and Wilson 1979).

[b]The method listed first is the primary method of application; chemicals are applied to foliage as a spray; chemicals are applied to the soil as a drench.

[c]An asterisk indicates that the chemical is used commercially on species in that group of plants.

[d]Depending on species and timing, flowering and fruiting may be unaffected by the general inhibitors or even enhanced by the subapical inhibitors.

General inhibitors injure or kill immature shoots and may reduce flowering, depending on the timing of application. Subapical inhibitors, on the other hand, generally improve plant appearance by making the plant more compact and enhancing flowering. Although the subapical inhibitors produce more desirable effects, woody plants respond more reliably to general inhibitors. In certain situations, the advantages of general inhibitors outweigh the disadvantages of terminal shoot death, foliar discoloration or distortion, and abscission of immature leaves. When viewed from more than 10 meters (30 ft) away or from a fast-moving car, most plants seem to be altered very little by such phenomena. The inhibition of flower initiation and subsequent fruiting is desired in many landscapes, so the phytotoxic effects of general inhibitors may be tolerated and even exploited.

Height Control of Trees and Shrubs

Utility companies are particularly interested in methods of inhibiting the growth of trees into overhead lines. Other public and private arborists are also interested in maintaining trees and shrubs in bounds. General inhibitors are therefore attractive for use on woody landscape plants. Inhibitors can be sprayed on foliage, painted or sprayed on bark (banding), or injected into trunks and branches (Table 15-3).

Since 1960, hundreds of thousands of trees have been treated with growth inhibitors each year in Arizona, California, and Nevada. The use of inhibitors in other parts of the United States has been minimal because of erratic and disappointing results; these are probably the effects of growing season, weather during and after application, and time of application. Public concern over the safety of growth-inhibiting chemicals is also an important factor; formulations and techniques that will improve the safety of these chemicals and limit application to targeted plants may do much to allay concern. Economic pressures alone may encourage the wider use of growth-inhibiting chemicals in the landscape.

Since the early 1960s, when growth inhibitors were first applied commercially, the cost of spraying trees in the open has been about 10 to 15 percent the cost of pruning (Sachs and others 1970). Expense varies greatly depending on the number and size of trees, size of crew, precautions taken, and efficiency. Banding is less expensive than spraying in urban areas because one person with a small hand-sprayer can maneuver more easily and efficiently than large crews with large spray equipment (Alvarez 1977). Banding minimizes spray drift, while injecting trunks eliminates it altogether; both are less objectionable to the public than spraying. Banding, however, is usually faster and cheaper than injecting growth inhibitors.

TABLE 15-3   General inhibitors used on trees: application methods and their effectiveness

Chemical	Used since	Primarily Used in	Sprays[a]	Banding[a]	Injections
MH	1960	Northern California	Effective on a wide range of species.*[b]	Not effective.	Experimental.[d] Effective on most species tested.
CF 125®	1971	Central & southern California	Effective on a wide range of species.*[c]	Effective on a wide range of species.*[c]	Experimental.[d] Injury to young silver maple.
Atrinal®	1981		Not widely tested.	Effective on several species.	Effective on most species tested.[d]

[a]Asterisks indicate that the method is used commercially for application to trees.

[b]Sachs and others. 1970.

[c]Hield, Sachs, and Hemstreet. 1978.

[d]Roberts, Brown, and Wilson. 1979.

Maleic hydrazide (MH), used since 1960, and chlorflurenol (CF 125®), used since 1971, are general inhibitors applied to control the top growth of trees (Fig. 15-2). Where traffic and susceptible non-targeted plants cause little or no interference, MH or chlorflurenol can be sprayed on the foliage of large plantings. In populated areas, where spraying may not be acceptable, chlorflurenol is sprayed or painted on the bark of trees. Some arborists prefer chlorflurenol over MH because it is effective on more species, usually causes less injury, can be used on either the foliage or the bark of many species, and is used in more dilute concentrations. The right amount or concentration of chlorflurenol can slow growth directly and indirectly by inhibiting the dominance of shoot terminals, often without seriously injuring them. The resulting growth of laterals further reduces terminal and lateral elongation. Sprays of MH, on the other hand, are effective only when they kill or severely inhibit immature shoots. Injecting MH or dikegulac (Atrinal®) into treetrunks is a promising growth-control practice (Roberts, Brown, and Wilson 1979); in 1981, Atrinal® was registered in the United States for use by trunk injection in several fast-growing trees.

FIGURE 15-2
Sycamore trees on the left were sprayed with maleic hydrazide
in early summer by Sohner Tree Service; those on the right were not sprayed.
All the trees had been pruned the previous winter.
Note the new growth on both rows of trees and the inhibited branch elongation
on the treated trees the following spring. Trees were in Marin County, California.
(Sachs and others 1970)

General Inhibitors in the Distant Landscape

Large trees (regardless of location), small trees, and shrubs, can be considered part of the distant landscape when they are viewed along streets, highways, and utility rights of way, or from more than 10 meters (30 ft). The injury or even death of shoot tips is seldom noticed in these situations.

Spraying Deciduous Trees and Shrubs　Follow label directions to determine the appropriate concentration of growth inhibitor and the species that will be responsive. Recommended concentrations range from 250 to 4000 ppm for MH and 150 to 1200 ppm for chlorflurenol, depending on the species and number of applications. Species that grow in more than one flush may need two sprayings, which should be less concentrated than a single spray application.

The timing of spray applications is important for controlling growth without excessive injury (Fig. 15-3). Unnecessary damage

FIGURE 15-3
Stage of growth, not calendar date, is all important in timing the application of maleic hydrazide. The dashed lines indicate the level from which growth started the year of spraying. Spraying on May 1 was too early and killed the new shoot, so that the following year growth had to come from two-year-old wood. Spraying on June 15 was too late; growth had already taken place. (Adapted from Sachs and others 1970)

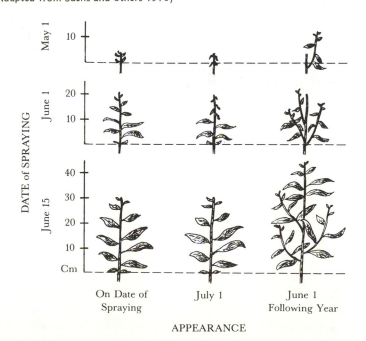

occurs when inhibitors are applied to deciduous species in the spring before leaves have developed sufficiently. On the other hand, if inhibitors are applied when shoots have almost ceased to elongate and leaf formation has nearly stopped, they will retard growth very little if at all. In some cases, growth will be delayed and even deformed the following spring if inhibitors are applied too late. Application should be timed according to the stages of plant growth, not according to the calendar.

For best results, determine application timing for each species. Timing is based on a given density of foliage, usually four or five full-sized leaves per shoot. Before spraying, count and record the number of fully-expanded leaves on ten representative shoots on each of two trees. If density stays at the desired level after growth has been inhibited, then in the future apply the inhibitor when the same number of leaves have formed. If density after spraying is undesirable, base the next application on a different count of fully expanded leaves. When you must treat many trees at once, base timing on the most vigorous trees. One spraying each spring is usually adequate. For extremely vigorous plants or those that normally have more than one flush of vigorous growth a year, make a second application in summer just as the first new leaves of renewed growth reach full size.

In the spring following growth-inhibiting treatments, shoots will grow from axillary buds below the dead terminals (Fig. 15-4). The

FIGURE 15-4
Tip dieback in MH-treated Shamel ash (left) and American elm (right)
as photographed in the spring following application. (Sachs and others 1970)

trees should be treated again when the desired number of leaves expand.   As growth-inhibiting treatments continue, trees and shrubs appear more compact because terminal buds die and axillary branching increases.   Annual treatments will produce approximately the same effects as annual removal of the terminal bud or reduction of apical dominance; they will keep some trees near the same height for up to ten years without pruning (Sachs and others 1970).   Before beginning a chemical growth-control program, prune trees to the desired size and shape.   You will usually spray only the top third of a tree.   Spray when the air is calm and either cool or moist, or both; sprays are less effective under arid conditions.   When applying MH under dry conditions, some arborists increase the amount of surfactant (wetting agent) up to four times the recommended level (R. G. Alvarez 1981, personal communication).   Under humid conditions, the surfactant can be reduced to half the recommended amount.

Spraying Evergreen Trees and Shrubs   To achieve maximum growth inhibition, spray evergreen trees and shrubs with general inhibitors after pruning or in the spring when axillary buds begin to grow (Sachs and others).   There are three primary reasons for this:

> Newly developing leaves absorb the chemical more readily than do fully expanded leaves.
>
> Inhibition occurs in terminal buds, close to the site of absorption in newly developing leaves.
>
> Applications during bud break prevent new shoots from elongating beyond the desired pruning line.

This last factor is particularly important in rapidly growing species, such as cotoneaster, eucalyptus, pyracantha, and xylosma. Earlier or later applications in summer will stop or slow growth less effectively.

In evergreen or deciduous species with no true dormancy (whose shoot growth is intermittent throughout the growing season), the timing of applications is not critical, but multiple applications are usually required.   In eucalyptus, for example, axillary buds will start growing shortly after the terminal is killed if the temperature is warm enough (Sachs and others).   Eucalyptus may sometimes require two or three applications at 60- to 90-day intervals for annual control. Lower dosages should be used whenever multiple applications are necessary to control axillary branch development.   Many vigorously growing species are more sensitive to growth inhibitors than are slow-growing species.

Banding Treetrunks   Chlorflurenol applied on trunks inhibits the growth of many deciduous and evergreen trees (Hield 1979).   It is currently registered for application at 10,000 ppm (1 percent) in a band equal in width to the diameter of the trunk. The most stable carrier for the chemical is a mixture of 30 percent toluene and 70 percent No. 1 diesel oil. Hield reports that effective band widths of chlorflurenol ranged from one-fourth of the trunk diameter for the California pepper tree to eight times the diameter for blue gum (Table 15-4). Container labels will give recommended rates for species on which the chemical is effective. If the tree canopy is considerably smaller than normal in relation to trunk diameter, band width should be reduced proportionally. Young trees (with immature bark) and those that are diseased or under stress should not be treated.

As with spray applications of general inhibitors, the banding of deciduous trees should be done in the spring as soon as four or five full-sized leaves form on each terminal shoot. To minimize needle and candle distortion, treat pines when they are dormant, or immediately after pruning if the topmost terminals and obvious lateral terminals have been removed. Timing is less critical for broadleaved evergreen trees, which can be banded in spring and early summer. Inhibition becomes greater as application time becomes earlier, but there must be adequate leaf surface for tree health and appearance.

TABLE 15-4

Recommended band widths of a 1 percent
chlorflurenol solution (Hield 1979)

Species	Band Width in Trunk Diameters[a]
Acacia longifolia	$\frac{1}{2}$-1
Ceratonia siliqua	1
Cupressus sempervirens	$\frac{1}{2}$-1
Eucalyptus globulus	6-8
E. sideroxylon	<2
Fraxinus uhdei	2
F. velutina	1-2
Ficus nitida	4
F. macrophylla	1-2
F. elastica	1
Pinus radiata	1
P. canariensis	1
P. halipensis	$\frac{1}{2}$
Ulmus parvifolia	1-2
U. pumila	1
Schinus molle	$\frac{1}{4}$-$\frac{1}{2}$

[a]Container labels will provide recommended rates for
species on which the chemical is effective.

Those evergreens, particularly certain pines, that only retain leaves for two or three years should not be treated more than once every other year. Do not band broadleaved evergreen or deciduous trees between midsummer and the time when growth begins in spring; do not band conifers between midsummer and midwinter. Late applications would probably result in disappointing growth reduction and small and distorted leaves the following spring (Hield 1979).

Chlorflurenol can be sprayed or painted on bark. You can minimize spray drift and splatter by using a 12-liter (3-gal) hand sprayer at low pressure, which will form large droplets. Give the first treatment to the base of a tree alone; it is less visible and allows you more control over the area treated. Wet the bark only enough to begin runoff. Do not treat exposed surface roots. Make the next application higher on the trunk or on individual limbs to concentrate more of the chemical on vigorously growing branches (Alvarez 1977). The inhibiting effects of chlorflurenol are more pronounced on vigorous terminal shoots than on less vigorous laterals.

Some leaf curl usually results from chlorflurenol banding, but you can minimize this by decreasing the width of band application and by applying the inhibitor after heavy pruning. From a distance, however, leaf curl, some chlorosis, and immature leaf fall will generally not be noticed. Chlorflurenol will also tend to hasten leaf senescence and the aging of bark (Hield 1979). The immature bark of young trees (five years old or younger) may increase in thickness and become corky (Fig. 15-5) if it is banded with chlorflurenol. This thickening appears to be confined to treated bark; the trunk xylem and roots are unaffected.

FIGURE 15-5
The increased development
of mature bark characteristics
on a young Shamel ash.
Beginning when the tree
was five years old,
it was bark-banded annually
for three years.
(Photo courtesy of Henry Hield,
University of California, Riverside)

General Inhibitors
in the Near Landscape

Shrubs and small trees that may be viewed from less than 10 meters (30 ft) should exhibit little or no injury. Some injury can be tolerated, however, on shrubs that grow rapidly and can recover within a few weeks. Chlorflurenol and dikegulac inhibit the terminal growth of shoots and promote the lateral branching of many shrubs and vines, yet they cause less injury than MH. Most shrubs and vines that require inhibition are vigorous and grow throughout the season or have multiple flushes of growth. More than one spray per season is usually necessary to keep them under control.

Except when chlorflurenol is applied to conifers, foliage sprays should be applied until runoff within a week after pruning or when new growth reaches the desired density or length (usually 25–75 mm or 1–3 in). Apply chlorflurenol to conifers before buds expand, however, to prevent the distortion of new growth. Conifers are more sensitive to these chemicals than are broadleaved evergreens, and broadleaved evergreens are more sensitive than are deciduous hardwoods. Mature

FIGURE 15-6

A *Cotoneaster lacteus* hedge (left photo) treated with 2500 ppm MH-30 at 60-day intervals after pruning; on the right, an unsprayed part of hedge. Excellent growth control was achieved, but terminal buds were damaged or killed. (Sachs and others 1970) This creeping fig was sheared and sprayed by Green Valley Landscaping (San Jose) with growth regulator in order to confine it to the shape of the arches (right photo). Two or three sprays of Maintain CF 125® during the growing season reduced the number of shearings from six down to two or three; each shearing took five days to complete. Embark was used when Maintain CF 125® was not available.

foliage is usually not affected by inhibitors when they are applied at recommended concentrations, but chlorflurenol may cause immature shoots to twist and leaves to curl. Dikegulac can cause chlorosis of new growth; this is usually transient but may persist up to six weeks on certain shrubs. Even though MH will effectively retard the growth of some shrubs without injury, MH commonly causes injury or terminal dieback (Sachs and others 1970).

In California, Sachs and his co-researchers found that evergreen hedge species could be successfully treated with MH in early fall, four to six weeks after pruning (Fig. 15-6). Spraying at 1500 to 4000 ppm stopped growth completely for six to seven months by preventing axillary bud break, and even delayed bud break the following spring, maintaining a hedge's original trim line for several weeks after spring growth began. Subsequent treatments of 750 ppm MH at 45- to 60-day intervals (or whenever growth resumed) maintained a hedge at essentially the desired size and shape. Some time for leaf renewal must be allowed, however; some axillary branches should develop a few new leaves each year. Continuous inhibition with general inhibitors will lead to death of the branch and eventually the plant.

Rejuvenation of Shrubs and Hedges   Chlorflurenol, dikegulac, and MH applications promote axillary bud break even on low older branches (Fig. 15-7). Shrubs and hedges treated with these chemicals will maintain or increase fullness, much as frequently pruned vigorous hedges will. In fact, these sprays are even more effective than pruning, because all vigorous growing points are inhibited and axillary buds are stimulated (Sachs and others 1970).

FIGURE 15-7
Increased branching on the large limbs of old pyracantha shrubs as a result of MH-30 treatment. (Sachs and others 1970)

Flowering and Fruiting of Inhibitor-treated Shrubs   Chlorflurenol or dikegulac applied just before flowering can reduce or eliminate bloom or fruit set on most shrubs. If either of these chemicals or MH are sprayed in early summer on shrubs that bloom on one-year-old wood, they are likely to interfere with flower-bud differentiation. In order to retain flowering and fruiting of these spring-flowering shrubs, apply inhibitors after early summer, even though growth will be reduced less than it would be by an earlier application. MH, however, may be applied to pyracantha soon after growth starts but before bloom begins; flowers initiated the previous summer will develop normally, and fruiting will also be near normal (Sachs and others 1970). Growth-inhibiting chemicals applied during the early summer to most shrubs and vines that bloom on current season's growth will reduce or suppress flowering for three to six weeks. As you gain more experience, you will use these growth-inhibiting chemicals more effectively and will be able to minimize adverse effects on flowering and fruiting when you wish to do so.

Long-term Effects
of Using General Inhibitors

As long as plants retain enough foliage, they should suffer no adverse effects from long-term applications of general inhibitors (Sachs and others). MH has been applied to the same trees annually for more than ten years, and chlorflurenol has been applied as needed for at least five. However, general inhibitors should not be applied to weak, diseased, or stressed plants: Growth-control sprays could further weaken or kill affected parts. Foliar discoloration caused by low winter temperatures (as low as $-4°C$ or $25°F$) has been observed in MH-treated subtropical evergreen shrubs in California (Sachs and others). None of the foliage on affected species, however, was permanently damaged.

Subapical Inhibitors
in the Near Environment

Subapical inhibitors retard cell division and elongation in the region below the apex so that leaf initiation and expansion continue at nearly normal rates, while at the same time stem elongation is severely inhibited. This produces shorter internodes. Plant appearance is usually improved by the denser foliage and closer lateral branches. These compounds, if their effectiveness were broader, would be preferred for general application in the landscape, because they cause no terminal dieback or leaf discoloration. Daminozide is used commercially to reduce the growth of herbaceous plants such as chrysanthemum and poinsettia. Because daminozide is expensive and is effective on only a few woody species, however, it is not generally useful in the landscape. Ancymidol, introduced in 1970, is effective on a wider range of plants, but it too is expensive and has not been found suitable for general use on landscape plants. Ancymidol may be useful, however, for inhibiting the growth of interior landscape plants.

In addition to retarding stem elongation, subapical inhibitors may affect plants in one or more of the following ways (Cathey 1975):

Increase the green color of leaves

Enhance and accelerate flowering

Cause greater formation of yellow pigments in flowers

Reduce visible injury from ozone and sulfur dioxide

Increase resistance to cold and transplanting shock (particularly for ground-cover plants)

A California wholesale nursery that specializes in ground-cover plants regularly treats them with a subapical inhibitor shortly before delivery to a retailer. This slows growth and enables plants to better withstand handling and transplanting.

Safety of Growth Retardants

The growth retardants listed in Table 15-1 are unlikely to be toxic to animals. The chlorflurenols are somewhat more toxic than the other chemicals, but all have a low acute toxicity (of interest to applicators). Growth retardants are much safer than herbicides, which themselves are safer than fungicides and insecticides. Chronic toxicity (prolonged or repeated exposure or ingestion of small amounts of toxin) has not been determined for many of these chemicals. In 1981, the U.S. Environmental Protection Agency (EPA) suspended the pesticide registration of diethanolamine maleic hydrazide (DEA-MH) formulations so that it could no longer be manufactured for sale in the United States. Potassium maleic hydrazide (K-MH, sold as Royal Slo-Gro®) remains registered for controlling growth of trees, sprouting of potatoes and onions, and suckering of tobacco (Federal Register 1981).

Although growth inhibitors are relatively safe, many people are opposed to the application of any chemicals to food and landscape plants. Since most growth inhibitors are used by or on behalf of public utilities and governmental agencies, unfavorable public relations may make the practice impractical even though it is cost effective and is apparently completely safe. In general, growth inhibitors are not applied to edible crops and continued exposure in the landscape is unlikely except when the chemicals are applied to turf. Even so, it is wise to follow all precautions on container labels.

CONTROL OF TRUNK SPROUTS

Sprouts on the trunks of trees and suckers from roots are a nuisance to nurserymen, fruit growers, and arborists. Young trees, however, benefit when shoots of low vigor grow along their trunks to provide shade and speed trunk development. When a trunk is 75 to 125 mm (3–5 in) in diameter, you can begin to remove the shoots along the trunk below the first scaffold branch. On grafted trees, keep the rootstock free of sprouts.

In many species, however, pruning vigorous sprouts from the trunk and trunk roots does not eliminate the problem and may even encourage the development of more sprouts, particularly if pruning has not been close to the trunk. In many trees, the periodic removal of sprouts as the trees mature will overcome this problem, but some trees continue resprouting for years or throughout their entire lives.

Researchers and professional arborists have treated pruning wounds in an effort to control regrowth. Since 1970, the practice has been in widest use in the eastern part of the United States. The chemicals most commonly used for this purpose are naphthaleneacetic acid (NAA) (Fuller, Bell, and Kazmaier 1965) and chlorflurenol (Maintain A) (Domir 1978), both carried in an asphalt paint. A one percent solution of an ethyl ester of NAA reduces the number of sprouts growing near pruning cuts by about 50 percent. The sprouts that do develop usu-

FIGURE 15-8
The bases of these coast redwoods were excavated about 150 mm (6 in)
to expose basal sprouts. The sprouts to the left of the pencil
in each photograph were cut close; those to the right were headed
near the original ground level. The left tree was not sprayed;
the base of the right one was sprayed with NAA in mid-spring.
The photographs were taken eight weeks later.
The close-cut bases produced only 20 percent as many sprouts
as the stubbed bases produced; NAA reduced the sprouting in both areas
to 5 percent. (Harris, Sachs, and Fissell 1971)

ally achieve about 50 percent of their normal size (Fuller, Bell, and Kazmaier). A one percent NAA solution sprayed on the trunks of northern California black walnut, olive, and crape myrtle sharply reduced the number of new sprouts for four to five months after treatment (Harris, Sachs, and Fissell 1971). This solution has produced successful results on nearly all species tested: The length as well as the number of sprouts was reduced; sprouts present at the time of spraying were killed; there appeared to be little translocation, even though NAA was applied at high concentrations (Fig. 15-8). Even when treated areas were only 25 to 50 mm (1–2 in) away from untreated areas, the former exhibited almost complete suppression of sprouts, while adjacent bark areas whose sprouts had been removed but not treated resprouted. In most cases, annual treatment is necessary.

KILLING WOODY PLANTS

Techniques for clearing land of unwanted brush and trees for agriculture and forestry can be used to kill woody plants in the landscape. Resprouting of stumps or large roots left in the ground after tree removal frequently is a problem.

Low-volatile esters of 2,4-D (2,4-dichlorophenoxyacetic acid) and ammate are commonly recommended for killing woody plants. These are applied as basal trunk sprays to shrubs or poured into cuts in the bark or on the stumps of trees (Leonard and Harvey 1965). Translocatable herbicides should not be applied to plants that may be root grafted to plants that are to remain.

Nonchemical Control of Sprouting In the midwestern and eastern United States, trees are sometimes girdled a year or two before being cut down to prevent stump sprouting. Some arborists remove a ring of bark at least 75 mm (3 in) wide; others strip the bark from the lower trunk. After a growing season, the stump will be less likely to sprout than if it had not been girdled. This process is slow and usually not acceptable once a decision has been made to remove a tree from a landscape. Grinding a stump and basal roots with a stump cutter reduces sprouting from the remaining roots except on species that sucker readily from roots.

CONTROL OF ROOTS IN SEWERS

Tree roots constantly break and clog sewer lines, a perennial problem that wastes millions of dollars each year. Roots cannot enter intact sewer lines, but as they enlarge they may break lines and later enter the cracks. A sewer district in Washington reports that two distinct types of roots create problems. The roots of trees such as poplars and willows enter sewers through the lower quadrant of joints and grow longitudinally along the bottom of the sewer. The roots of trees like honey locust, on the other hand, enter any part of a joint and form a ring in the pipe but seldom extend far from their point of entry (EPA 1977). Once a root enters a sewer, the conditions of aeration, moisture, and nutrients are so favorable that it inevitably grows until it clogs the sewer (see Fig. 1-5).

The standard procedure for unplugging a sewer line is to use a mechanical router (a powered rotary blade attached to a flexible steel cable) to physically cut and remove the roots. A crack in a sewer line is usually not repaired unless the same line must be routed out more than once a year; roots are cut back only to their place of entry. Unless some other action is taken, the sewer clogs again in one to three years.

Metham and Dichlobenil Treatment

To reduce root growth in sewers, workers may block a section 75 to 125 m (250–400 ft) long with an air plug and fill it for an hour with a solution of 1000 ppm metham (Vapam®) and 100 ppm dichlobenil (Carsoron®) (Leonard and Townley 1971). By starting at the top of a sewer line, they may use the same solution several times. In early trials in Sacramento County, California, root stoppages in 23 km (14 mi) of sewer lines were reduced 50 to 100 percent for six to 18 months after treatment. Even though results have varied from excellent to only moderate, Sacramento County has used this mixture (marketed as Vaporooter Plus®) since 1970. In problem lines, mechanical routing must usually take place annually, whereas chemical treatments last for two years and sometimes longer.

Leonard, Bayer, and Glenn (1974) report that metham alone or with dichlobenil was at least 20 times more effective when applied in an air-aqueous (19 to 1) foam than as an aqueous mixture. Foam uses less of the chemicals and is easier to apply, particularly in uneven terrain, but special equipment is required to create it. The foam is introduced through a sewer line clean-out under pressure of less than 2 kg/cm^2 (30 psi) and fills the sewer so that roots in even the upper portion are killed. Care must be taken to prevent backup into household fixtures and sewer vents. Metham treatment kills roots a short distance beyond the area treated; no injury to other roots or to the tops of plants is apparent. Dichlobenil inhibits regrowth. The foam mixture is available as Sanafoam Vaporooter® and can be applied with foaming equipment such as the Foamaker®.

The city of St. Petersburg, Florida, has used foam since 1975 and has reduced root-control work from 300 lines in 1973-1974 to only 69 in the first six months of 1975-1976 (Monck 1976). The cost of chemical treatment has been $95/100 m ($290/1000 ft) of sewer line, and the control period will probably be at least two years. The cost of mechanical treatment ran $70/100 m ($210/1000 ft) of line for a control period of only one year. Sewer lines damaged by roots cost about $1300/100 m ($4000/1000 ft) to repair. Monck estimates that chemical treatment reduced sewer maintenance costs by about $100,000 per year.

Sprays of Vaporooter Plus® are also used in hilly terrain or in large drains that would require large amounts of chemicals. Spraying is not as effective as soaking or foaming: It may kill roots in the drains but not in the cracks or joints. It also requires special equipment. Spraying must be done more frequently than soaking or foaming.

The results of metham and dichlobenil treatment vary with tree species, soil conditions, and thoroughness of application. The chemicals apparently create no problems in sewage treatment plants.

Copper Sulfate Treatment

The town of Ridgewood, New Jersey, has been treating sewers with copper sulfate since 1937 (Davis 1975). One-liter (1-qt) of 6-mm (0.25-in) diameter crystals are poured into porcelain toilet bowls. They are never poured in sinks, which have thin-walled traps. The toilet is flushed and the water in the bowl agitated with a plunger to make sure all crystals are carried down the drain. Semiannual treatments have controlled roots in tile sewer connections without discernible injury to the plants whose roots are killed. When the copper sulfate program was discontinued for one year, in 1961, the cost of maintaining sewer lines rose about 40 percent. If an entire community were to use as much copper sulfate as Ridgewood does, sewage sludge contamination might result.

A sewer district in Washington has placed copper sulfate crystals in permeable plastic bags, which are then hung in manholes in contact with flowing wastewater. The crystals are replaced, as needed, over a period of several months. Root growth is impeded for three years with no detrimental effects reported at the treatment plant (EPA 1977).

CONTROL OF FLOWERING
AND FRUIT SET

People have long been interested in inhibiting the flowering or fruit set of certain trees and shrubs to minimize allergies, unsightliness, safety hazards, or odor caused by falling flowers and fruit. NAA (naphthaleneacetic acid) has been the standard spray for preventing or reducing fruit set on olive; it has met with only minimal success on other troublesome trees. Even on olive, NAA must be sprayed more than once, especially along the coast or in cool springs when bloom is prolonged. Under normal conditions, one spray of 150 ppm NAA should be applied two to three days before full bloom and a second spray one to two weeks later to prevent fruit set of late-opening flowers. If bloom is short, one spray at full bloom can eliminate most of the olive set (Opitz 1970).

Dikegulac (Atrinal®) appears to be more effective and versatile than NAA (Hamilton 1979); it can be used on more species and can minimize flowering and/or fruit set, depending on time of application. Dikegulac is registered for use on Japanese holly, olive, and glossy privet. Allergy sufferers will especially welcome the elimination of olive bloom. Label directions must be followed for appropriate species, concentration, and timing. Recommended concentrations range from 1000 to 8000 ppm; leaves may yellow at the higher concentrations, but only for a few weeks.

Tender ornamental plants growing under or near those to be treated should be covered with a plastic tarp. If you accidentally spray non-targeted plants with NAA or dikegulac, hose them immediately and thoroughly with water. To reduce spray drift, use moderate pressure, apply a coarse spray (avoid fogging), spray only on calm days, and direct the spray into the tree or shrub.

In San Jose, California, two applications of MH two weeks apart during bloom were found to greatly reduce the fruit set of ginkgo (Hamilton 1977). Arborists applied the potassium salt of MH (Royal Slo-Gro®) at 5000 ppm (50 ml/100 liters or 2 qt/100 gal of water) to both male and female trees for the best control of fruiting. Surrounding vegetation showed no ill effects.

FURTHER READING

SACHS, R. M., and others. 1970. *Chemical Control of Plant Growth in Landscapes.* Calif. Agr. Exp. Station Bull. 844.

CHAPTER SIXTEEN

Preventive Maintenance and Repair

No matter how well plants are maintained, some will be injured, develop weak structure, cause damage to buildings, or require removal. These problems usually become serious only as plants grow and age; most can be avoided or at least postponed by proper selection, planting, and care. You must take care of problems when they occur in order to safeguard the plants, people, and property. In this chapter, I will discuss some of the more important plant maintenance problems, how they might be avoided or minimized, and how to treat them should they occur.

CABLING AND BRACING

All too often trees develop weak or poor structure and require special care to preserve them or to prevent further injury. Even trees that are routinely maintained may develop weaknesses that affect their own safety and that of people and property. The most common problem is the weak attachment that may occur when two or more branches of about equal size arise at approximately the same level on the trunk. Horizontal branches can become heavy and dangerous, particularly those of conifers in snowy regions. Branches can also be weakened by decay and storms. Proper pruning can shorten, lighten, and thin hazardous branches, but it may be inadequate to keep certain limbs or trees safe. In such cases, cabling and bracing may be required to reduce stress.

Cabling involves the attachment of a flexible steel cable between branches to limit excessive limb motion and to reduce stress on a crotch or branches (N.A.A. 1979). Sometimes trees are cabled together for support. Bracing uses bolts or threaded rods to rigidly secure weak or split crotches, unite split trunks or branches, and hold rubbing limbs together or apart (Thompson 1959). Before you undertake cabling or bracing, however, assess the internal condition of trunk and main branches and the value of the tree as compared with the cost of labor, supplies, and continued maintenance. Is the tree worth the effort?

Almost all cabling and bracing recommendations are based on those given by Thompson (1959), including the standards of the National Arborist Association. Shigo (1981, personal communication) recommends that cabling, bracing, and other practices that wound a tree should not be done during spring growth flush or leaf fall. Use sharp tools to make a clean-edged wound, which will close faster than a ragged wound.

Cabling

One of three cabling systems is usually used to safeguard main branches and stems. If a single cable (*simple direct system*) is used, two limbs of about the same size arising from a weak or split crotch can directly support each other, and a large limb or the trunk can support a weak or heavy horizontal branch (Fig. 16-1). Three cables attached to three limbs in a *triangular* arrangement provide direct support to weak crotches and lateral support to minimize branch twisting, which can put considerable strain on branch attachments. A *box* or *rotary* system of cables connects the main limbs of a multi-branched tree and permits movement of individual branches within safe limits. A box cable system can be made more rigid if interior cables are placed to create a series of triangles. The system chosen will depend on the tree and the branches that need support.

FIGURE 16-1
Cabling systems: Simple direct cabling involves a single cable
between two branches of approximately equal size (left); box cabling
connects the large branches of a tree and allows movement within limits (center);
triangular cabling gives direct support and minimizes twisting.
(Adapted from Thompson 1959)

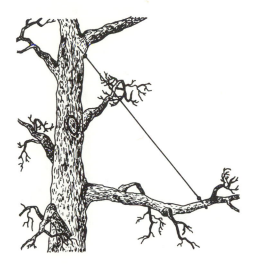

FIGURE 16-2
When supporting a weak horizontal branch,
attach the cable as far out on the branch
and as high on the trunk or support branch
as possible without unduly distorting the
branch; the cable and branch should form
an angle of at least 45°.
(Adapted from artwork by M. J. Whitehead
in Bridgeman, P. H. 1976. *Tree Surgery*.
Devon, England: David & Charles)

It is important to position cable attachments so as to combine
adequate support and flexibility. Cables should be attached to limbs
at about two-thirds of their length from the crotch, but each cable
attachment should be approximately the same distance from a crotch,
so some compromise will be required when limbs vary substantially in
length. To support a horizontal branch, however, Bridgeman (1976)
recommends that the cable be attached as high as possible on the supporting branch, so that its angle from the weak branch is not less than
45° (Fig. 16-2). If the supporting branch leans toward the one supported, it may be wise to install a second cable from the support branch
to a third branch for counterbalance. Place cables so that they do not
rub against limbs or one another.

Thompson (1959) found that lag screws (Fig. 16-3) are as satisfactory as bolts for cabling sound hardwood branches, but bolts are
safer for softwoods or decayed hardwoods. Shigo and Felix (1980)
confirm that lag screws should not be put into decayed wood: When
a lag or rod touches internal decayed wood, decay spreads to surrounding tissues. Where there is decayed wood, insert a bolt through a branch

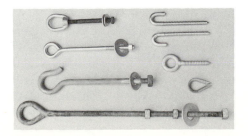

FIGURE 16-3
Cabling hardware includes
(clockwise from bottom):
forged eyebolt, hook bolt, bent eyebolt,
bolt with amon nut (left end),
right-hand threaded lag hook,
left-hand threaded lag hook,
forged lag eye, and thimble;
the bolts have washers and nuts.

TABLE 16-1

Relative strengths of cable anchor hardware 15 mm ($\frac{5}{8}$ in)[a] in diameter (Thompson 1959)

Cable Anchor Hardware	Safe Loads[b]	
	Kilograms	Pounds
Bent hook eye bolts	540	1200
Drop-forged eye bolts	1350	3000
Ring nuts (Amon nuts)	1485	3300

[a]This represents the smallest size available in all three types of hardware.

[b]The weights listed are 25 percent of the maximum static load each type was able to support.

and secure it with a large round or oval washer at the threaded end. Trace or scribe the bark so the washers are set against the wood and can be callused over more easily. The ends of the traced bark should be rounded not pointed (Shigo and Felix).

Lag screws are cheaper and easier to install than bolts, but are not as secure in softwood and do not allow for the adjustment of cable tension unless right- and left-hand lags are used (Bridgeman 1976). Lag screw hooks or bolt hooks are commonly used in the United States even though they do not have the tensile strength of drop-forged eye bolts or ring nuts (Amon nuts) (Table 16-1). Lag screw hooks ("J" lags) should have at least a 12-mm ($\frac{1}{2}$-in) unthreaded shank above the threads and should be inserted the full length of the unthreaded shank, because the threaded portion is more vulnerable to sway damage and crystalization (Mayne 1975). Bernatzky (1978) advises against the use of lag screw hooks and hook bolts for attaching cables because they will not hold a slackened cable as securely as screw eyes or eye bolts will. Cables seldom come loose, however, if they are properly tensioned, if the hooks are tightened until they almost touch the bark, and if the wood is sound. Mayne (1975) found that a cable without a thimble will secure as much weight as a cable with a thimble and it takes less time to install and splice a cable without thimbles, but the cable without a thimble may stretch more.

Lag screws should be screwed into holes drilled 1 to 2 mm ($\frac{1}{16}$ in) smaller in diameter than the lag and slightly deeper than the length of the threaded shank. Mayne (1975) recommends that the hole in softwood should be 3 mm ($\frac{3}{32}$ in) smaller than the lag. Drill the hole so that the cable and lag will form a straight line when attached (Fig. 16-4). The closer to 90° you can make the angle between each lag and its branch, the more mechanical advantage the cable will have. Lag screws

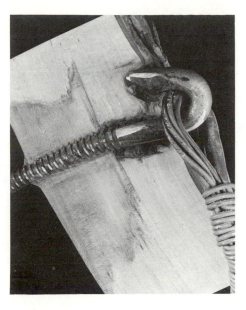

FIGURE 16-4
Install the lag screws and cable
in a straight line; otherwise the lag
will be pulled laterally and/or vertically,
further injuring the wood and reducing
the holding strength of the lag.
(Photo courtesy Alex Shigo, U.S. Forest
Service)

should be screwed in far enough to just allow the cable to be slipped into place. A lag screw should penetrate at least two-thirds the diameter of the smaller branch. Tighten each screw the full distance without stopping, or the wood may expand and hold the screw so tightly that it cannot be moved. Align each hook or eye with the grain of the wood for smoother coverage as the branch grows. Do not align a lag screw or bolt with the crotch junction or place it within 300 mm (12 in) of a weak crotch. Attach only one cable to each lag screw or bolt. Place lag screws or bolts at least 250 mm (10 in) apart on the same branch to avoid weakening the branch.

If hook or eye bolts are to anchor cables, install them as you would install lag screws, but drill holes that are the same diameter as the bolts, and countersink the nuts and washers to the cambium. If a bolt is 50 mm (2 in) longer than the diameter of the branch in which it is placed, the extra threads can be used to adjust cable tension. Place a bolt through a weak or decayed branch for greater support; use a lag screw in a support branch or trunk so large that it would require a bolt much longer than needed to support the weak branch.

The required strength of cable and other hardware will depend on the species, size, and general condition of the tree; the size and weight of branches to be supported; the presence of decay and cavities; and exposure to wind and snow. These factors have not been integrated into a formula for determining cable size. Thompson (1959), however, examined several hundred cabled trees and concluded that a 7-wire galvanized cable 6 mm ($\frac{1}{4}$ in) in diameter is safe for limbs up to 150 mm (6 in) in diameter at the point of cable attachment. A 7.5-mm ($\frac{5}{16}$-in)

TABLE 16-2

Sample substitutions of extra high strength (EHS) cables
for common grade cables when Preformed Tree-Grip dead ends are used
(Jeffers and Abbott 1979)[a]

7-Wire Cable Diameter Sizes (mm)	Minimum Cable Strength (Kilograms)		Weight of Cable (Kg/100 meters)	
	EHS	Common	EHS	Common
48 (3/16)[b]	1795 (3990)[c] ┌	515 (1150)[c]	11 (73)[d] ┌	11 (73)[d]
64 (1/4)	2945 (6550) ┐	855 (1900)	18 (121) ┐	18 (121)
80 (5/16)	└	1440 (3200)	└	30 (205)
96 (3/8)	┌	1910 (4250)	┌	40 (273)
112 (7/16)	└	2565 (5700)	└	59 (399)

[a]Figures in parentheses are English equivalents.
[b]Inches.
[c]Pounds.
[d]Pounds per 1000 feet.

cable is satisfactory for limbs up to 250 mm (10 in) in diameter at the point of attachment. Although common grade galvanized cable is not as strong as steel cable of higher tensile strength (Table 16-2) and is more subject to stretching, the common grade is regularly used in the United States because it is easy to splice. Improved methods of joining cables, however, should increase the use of extra high strength (EHS) cable.

Before cabling, prune the tree to correct any hazardous structure, reduce weight at the ends of long branches, and balance the tree. The effectiveness of a given cable tension is determined by the size of the limbs, the weight of the limbs and foliage, and distance to the crotch, but guidelines to proper cable tension are not very specific. Thompson (1959) advises that a slack cable may have to bear sudden, heavy loads during gusty winds but that tree shape may be distorted if the upper foliage is drawn too tightly together. A cable should be snug but not tight when the branches are least strained (during winter for a deciduous tree). Bernatzky (1978), on the other hand, recommends that a cable be slackened so that in an unloaded condition it sags 2 to 3 mm for each meter (0.1 in/3 ft) of cable length. During a gusty wind, such a practice might create too much slack, followed by extreme stress as the cable is pulled taut. Since very few cables or cabled limbs have failed regardless of which of these approaches were used, most cables and their attachments and the cable tensions must be adequate.

A tree that is rigidly cabled may lever the trunk and roots unduly, possibly causing wind-throw. Compression spring inserts (see Fig. 9-10) in cables would reduce this stress on the tree, the tension necessary, and the maximum load under wind. Cables with compression springs would allow more movement of the branches and would foster strong branch growth. In a triangular or a box cable system, a compression spring in one cable would provide flexibility to all cabled limbs (Mayne 1975).

After the lag screws or bolts are in place, pull the limbs together to install the cable on the hooks or in the eyes. You can pull limbs together by tying a rope loop around them and twisting it tight or by using right- and left-hand lag screws, bolts with extra long threads, a block and tackle, a lineman's come-along, or a running bowline (Thompson 1959). Cable tension can be adjusted more precisely if at least one of the anchors is a bolt or if right- and left-hand lag screws are used.

Common grade cable can be attached to an anchor lag screw or bolt by means of an eye splice, cable clamps (bulldog or U), or Tree-Grip® dead ends (Fig. 16-5). Extra-high-strength steel cable (EHS) can be held with two or more cable clamps or preformed Tree-Grip dead ends. Tree-Grips, introduced in 1977, speed installation, particularly for less experienced workers, by simplifying cable attachment (Jeffers and Abbott 1979).

Thompson (1959) reports that a properly constructed eye splice is as strong as common grade cable, whereas a single cable clamp will slip when the strain exceeds one-seventh of cable strength. That is the reason for using two or three cable clamps. Tree-Grips will hold EHS

FIGURE 16-5
Cables are joined together by (top to bottom): splicing; at least two clamps for each cable union; a twisted cable loop (Tree-Grip) whose ends are wrapped around the end of the cable; the Tree-Grip is held by an Amon ring nut but without a thimble.

cable securely well beyond the rated breaking strength of the cable (Jeffers and Abbott). Cable clamps or Tree-Grips allow for the use of EHS cable and a consequent reduction of cable weight for a given cable strength. With these fasteners, you need only keep two cable sizes on hand, rather than five.

An eye splice in a seven-wire strand is made as follows (Fig. 16-6) (Thompson): Make a loop by bending the cable about 250 to 300 mm (10–12 in) from the end, then insert a thimble in the loop, pass the cable through the eye, or loop it on the hook. Unwrap the wires of the 250-mm section and lay them parallel to the main piece of cable. Using pliers, wrap one strand tightly two times around both the cable and the remaining six strands and then cut it off. Wrap the rest of the strands, one at a time, in the same fashion. The resulting tapered eye splice is not only effective but inconspicuous. Though one wrap of each strand has been shown to be adequate, two wraps are practically as easy to make and will provide a margin of safety and improve appearance. Tighten cable clamps (placed 50 to 75 mm or 2–3 in apart) securely to the cable and cable end, but not so tightly as to flatten the cable, which could cause the galvanizing to break and allow rust to enter (Bridgeman 1976).

Tree-Grips come in 4.5- and 6-mm ($\frac{3}{16}$- and $\frac{1}{4}$-in) sizes for the corresponding sizes of cable. Since the Tree-Grip dead ends are left-hand lay, the cable must be the same for proper holding. Tape EHS cable before cutting it to prevent unraveling. Lay one end of the cable in the short leg of the dead end and wrap it with the short leg (Fig. 16-6). Insert a thimble in the Tree-Grip loop, or insert the loop in the anchor eye. Then wrap the longer leg around the cable between the wraps of the short leg. Additional instructions are provided with the Tree-Grips.

Inspect cabling at least annually to check cable tension and the stability of the anchor lags and bolts. If a tree is vigorous, you may need to raise the cabling in eight to ten years to provide adequate support.

FIGURE 16-6 (See opposite page)
Steps in making a seven-wire splice that will provide a strong
and attractive union for a looped cable end (top). (Thompson 1959)
Photos, upper left to lower right: Taped end of the cable is laid in the
Tree-Grip dead-end's short leg (painted black for visibility),
slightly above the cross-over paint mark; the short leg of the dead-end
is wrapped around the cable; a thimble is inserted and the longer leg
of the dead-end (painted white) is wrapped around the cable;
the strands of the dead-ends can be separated to finish wrapping the last two
turns of each subgroup individually or each dead-end can be wrapped separately;
the finished Tree-Grip dead-end.

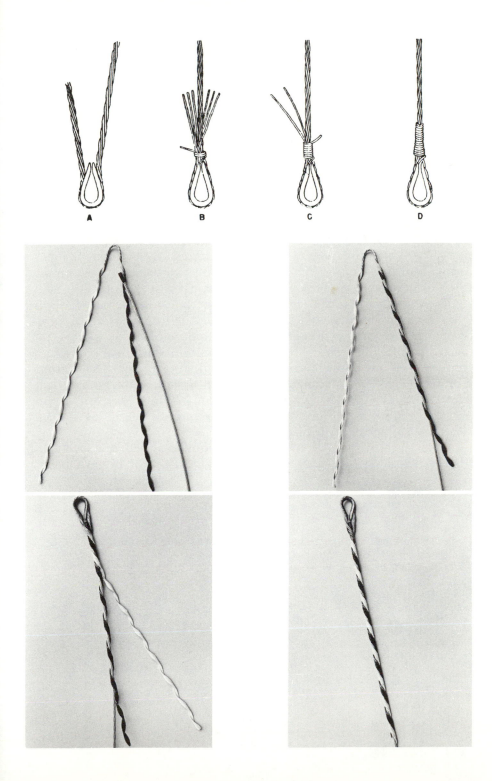

Bolting or Rod Bracing

Rod bracing provides rigid support for weak branch attachments and split branches and trunks. It also provides cavity bracing and holds limbs together or apart. Machine threaded steel rods of various diameters can be used alone or with nuts and round or oval washers. For maximum holding power, a rod is screwed through a hole that is 1 to 2 mm ($\frac{1}{16}$ in) smaller in diameter; a washer and nut are then tightened at each end. When two branches are to be brought closer together, the rod must be able to move in one of the holes and washers and nuts used.

Weak Crotches Split crotches are usually rigidly braced with a screw rod to supplement cable bracing higher in the tree. Weak but unsplit crotches of large trees are often braced in a similar manner. A single rod is usually sufficient to hold the two sections of an unsplit crotch together securely (Fig. 16-7). Large split or weak crotches usually require two or more rods to hold the two sections together and to minimize twisting. Holes should be centered on the trunk. When two rods are used to brace a single crotch, they should be placed just below the crotch, parallel to each other in the same horizontal plane and at a distance from one another approximately equal to a radius of the trunk or branch (Fig. 16-8). Thompson (1959), however, cautions that parallel rods should rarely be placed closer than 120 mm (5 in) or farther apart than 500 mm (18 in). If the rods are more than 500 mm apart, use a third rod in the same plane.

If at least 150 mm (6 in) of sound wood is available at each end, self-threaded holes and rods should provide sufficient holding power

FIGURE 16-7
If only one rod is to be inserted, it should be just below the crotch (left).
If nuts are used, round or oval washers should be countersunk
flush with the cambium (center and right).

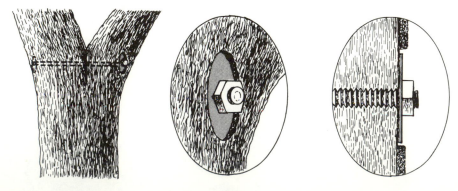

490

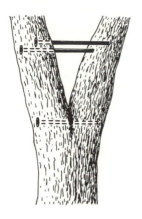

FIGURE 16-8
For large, weak or split crotches,
insert a rod through the crotch
and two more above it, separated by a
distance equal to twice the diameter of the
limbs. The two rods give direct support
and reduce twisting strain on the crotch.

(Thompson), but if any of the wood in the crotch region has rotted, use round or oval washers and nuts on the rods (Shigo and Felix 1980). In a large tree where the branches to be braced are close together for a considerable distance above the crotch, use an additional safety bolt 0.3 to 1 m (1–3 ft) above the crotch to provide extra rigidity.

When a screw rod is to be inserted at only one level, Thompson (1959) recommends that it be inserted directly through the crotch or just above it. Pirone (1978b) recommends that a single rod be inserted somewhere above the crotch or that two rods be placed at a distance from one another equal to twice the average diameter of the limbs. Bridgeman (1976), on the other hand, recommends that one or more rods be inserted at or just below the crotch. From an engineering standpoint, insertion at or just below a crotch would provide the greatest stability in gusty winds. Cables will not prevent the branches from being blown together, and self-threaded rods located above a crotch can act as a fulcrum; if the tops of the bolted limbs are blown together, the trunk or branch below the crotch might split. Rods above the crotch will be effective if they are situated at more than one level or are combined with rods inserted through the crotch.

Split Trunk or Branch Bolting   A long split in a branch or trunk can be held together by one or more bolts or rods (lip bolts) (Fig. 16-9). You may also be able to close or reduce the split in width before bolting. You may partially close a split in a stressed branch by moving the branch into a favorable position. Frost cracks in the trunk (see Fig. 4-1) usually close when temperatures rise above freezing.

Rods are usually installed with washers and nuts and are not self-threaded through the wood. "Lip bolts" should be placed about 300 to 400 mm (12–16 in) apart; take care so that consecutive bolts are not located in the same wood grain but are staggered to minimize the chance of weakening the wood.

FIGURE 16-9
Bolt a frost crack closed during
warm weather when the crack is closed.
The washer and nuts have been painted
with asphalt on this London plane tree.
(Photo courtesy of E. B. Himelick,
Illinois Natural History Survey)

Cavity Bracing   A cavity opening may need mechanical strength-
ening without closure. Bracing will partially replace the lost strength of
the decayed or removed wood and will hold the cavity walls in position.
When you use lip bolts in thin-walled wood, washers and nuts used on
each side of the wood wall will provide rigidity and keep the walls from
being pulled together (Thompson 1959). Cross-bracing is used to keep
the cavity walls from spreading and to minimize twisting strains that
can damage fill material. Screw rods are placed in alternating diagonals
across a trunk cavity: If seen from above the rods form an *X*. Cross-
bracing has limited benefits, however, because the stresses set up in a
large cavity and the leverage exerted by a tree in strong winds are so
great (Thompson 1959). In many cases, it will be better to replace
the tree.

Bracing Rubbing Limbs   Occasionally two limbs grow so that
they rub together, to the detriment of both. Usually one can be removed to the advantage of both the tree and the surviving limb. If both
crossing limbs must be preserved, however, you can protect them by
bolting them tightly together or bracing them a short distance apart.

When branches are to be held together to foster grafting, remove
the bark so that the two cambiums will be in contact. Drill a hole
through the two limbs for a bolt. After the two branches have been
bolted together, seal the slit around their union and around the washers

and nuts with asphalt emulsion or grafting wax. The dressing will protect the cambium from drying out. This approach-grafting should be done in the spring just as growth begins.

When branches are to be held apart, block them at the desired distance while you drill a hole in a direct line through both. Washers and two nuts or a short length of pipe with washers at either end can constitute the permanent spreader on the rod. Washers and nuts tightened on the outside of either branch will hold them as desired.

Rodding Standards   Thompson (1959) gives extensive instructions for the selection and use of rods. The most important are listed below:

Drill holes 1 to 2 mm ($\frac{1}{16}$ in) smaller in diameter than the rods, except when limbs or split trunks must be drawn together; in this case drill at least one of the holes the same diameter as the rods.

Countersink round or oval washers to the cambium (Shigo and Felix 1980); cut the bottom of the holes so that the washers will lie flat and at right angles to the rod.

Except for countersinks, ream all rod holes below the cambium so as to avoid loosening the bark when the rod is inserted.

Cover exposed rods with tightly fitting pipe that extends the full length of the exposed area to provide extra strength and protection.

When rods and nuts are used to draw split crotches together, use two or more washers under each nut for additional strength. Prevent the washer against the wood from twisting by separating it from the nut with a second washer. Shigo and Felix (1980) recommend round or oval washers instead of diamond-shaped washers, whose sharp points can lead to cambium dieback and trunk splitting.

When a rod is to be broken off below the cambium, cut it about halfway through before you insert it into the hole. Make this cut at least 40 mm (1.5 in) from the rod end so there will be space to clamp a pipe wrench. Take care when inserting a rod so that the wrench bites do not damage the threads that have to engage wood or nuts. When it is screwed in far enough, break the rod by bending the exposed portion to and fro.

When you drill for screw rods, line up the holes in the opposite limbs exactly. A long extension bit is useful for this purpose.

Take care when you tighten nuts on screw rods so that you do not injure the bark and cambium. A socket wrench is good for this purpose. Expose some of the threads after you tighten each nut so that the rod end can be spread with a ball peen hammer; this will prevent the nut from being worked loose or easily taken off.

Thompson (1959) recommends that all rods be covered with a mastic or tree wound dressing as they are being inserted and that all wounds be dressed after exposed metal has been covered with a metal preservative paint. He suggests that exposed rods, pipe, nuts, and washers also be painted. Except in especially wet regions or near the seacoast where rust and corrosion are prevalent, painting the metal is of doubtful value. Wound dressings are probably not worth the effort either, except for cosmetic purposes.

Propping

If pruning alone does not make low branches and leaning trees safe, these trees may have to be propped after pruning (Fig. 16-10). The main reasons for propping are (Bridgeman 1976)

> The upper branches and trunk are missing, too weak, or wrongly positioned to support a cable.
>
> The entire tree leans or exerts leverage against the root system.
>
> Supporting cables would be unsightly.

Metal props are usually of pipe, angle iron, or I, U, or box beams. Wooden props, particularly straight branches with open forks selected from felled trees, are also common. If wood is used, it should have a

FIGURE 16-10
A Japanese pagoda tree supported with five wooden props
at the Royal Botanic Garden, Kew. Some of the props have "V" crotches
to more securely cradle the limb or trunk they are supporting.

straight grain. The top end of the prop should form a curved fork to cradle the branch or trunk, but propping can injure or girdle the bark at the point of support. A bolt inserted horizontally through a branch and supported by a U end on the prop will carry a load without pressure on the bark (Bernatzky 1978). Metal props are more permanent and usually less obvious than those made of wood.

Place props so that they provide the maximum support and are least noticeable. If the load is heavy, place the prop on a firm base of concrete, stone, or other durable material. Do not embed the props in concrete because they should allow some branch movement (Fig. 16-10). A prop should be vertical to support a low branch but angled to support a leaning tree. Raise the branch or tree enough so that when the prop is in place it need not be tied to the tree. A hydraulic jack or winch can be used to raise the branch or tree until the prop is in place. Take care not to lift the branch or tree so much as to damage it.

Occasionally a tree may be partially uprooted from wet soil during strong winds or a flood. Often a small or medium-size tree can be rehabilitated if it is straightened soon after the mishap and if the root system was fairly well formed beforehand. If most of the roots on the upwind (upstream) side are broken, dig out soil from that side to make room for the root portions still attached to the tree. Take care not to injure unbroken roots. Make sure the soil is very muddy; then attach a cable or stout line to a strong branch or the upper trunk at a good angle for leverage and use a winch or block and tackle to pull the tree upright while you work the exposed roots into the muddy soil. When the tree is in a reasonable position, guy it and possibly prop it upright. Roughly level the soil surface and thin the tree to compensate for root loss. Irrigate carefully during the following growing season. At the end of that season, determine the stability of the tree without its guys and props; it will probably need some support for two to three years.

Improper braces are all too common. Many have been used as temporary expedients and subsequently forgotten; some are used with no reference to the way trees grow. Whatever its cause, improper bracing can jeopardize the tree it is supposed to help. The most common mistake is to wrap supported branches or encircle them with cable. The weight of the branches or the constriction of the cable begins to cause girdling as branches and trunk increase in girth. Growth beyond the attachments may die back or break off.

Take care when removing cable or bands that constrict branches; not only the tree's well-being but your own safety is at stake. The constricting collar may be under such pressure that fragments of bark, wood, and metal will be hurled at dangerous speeds when it is cut. If the collar is deeply embedded, it will be difficult to remove without injuring the embedding tissue. Occasionally callus will overgrow and enclose the collar. Even though new continuous wood is laid down outside the collar, that area will always be a point of weakness. In such cases, the portion beyond the girdling collar should be thinned and shortened to reduce the stress at the collar.

After removing a constricting collar, you may hasten recovery by cutting longitudinally through the bark in several places on either side of the girdle. Extend cuts to the cambium at the bottom of the girdle (Bernatzky 1978). It may also be wise to thin the branch to reduce its weight.

WOUNDS, DECAY, AND WOUND TREATMENT

Foresters, fruit growers, arborists, and builders—practically all people who grow trees and use tree products—are concerned about tree wounds and decay. Bark wounds not only interfere with movement of organic compounds between the top and roots but also open wood to microorganisms that can cause decay. Bark can be wounded by animals, by wind, by lightning, sun, ice, cold, and fire, by cars and other machinery, by vandals, and even by arborists who expose internal tissues when they prune, cable, and inject trees. Unless a wound encircles a trunk or branch, the main concern is decay.

Decay in trees is the breakdown of wood cell walls. Although many pathologists have studied this process and ways to prevent it, Robert Hartig's 1878 publication on the subject was supplemented by only a few significant discoveries in the following 50 years. Hartig's concept was that wood deterioration followed fungus infection of a fresh wound into heartwood (Merrill, Lambert, and Liese 1975). Welch (1949) sums up what was learned about decay during the next 50 years:

> Some tree species decay faster than others, and individuals within a species differ in their susceptibility to decay.
>
> Fungi vary greatly in their ability to attack and disintegrate wood.
>
> Decay is influenced by the amount of available water and air. Moisture in the wood of living plants is probably always present in quantities large enough to support the growth of fungi. If water can enter a plant through wounds, fungus spores and other microorganisms can also enter. The amount of air in the wood is probably more significant than the amount of water that enters from outside. Superficial wound rot and sapwood rot require a great deal of air, whereas deep rot is able to get along with very little; some rot does quite well when completely sealed up inside a tree. It appears, therefore, that sealing a cavity may stop some types of decay but not others.
>
> The activity of secondary agents—insects, rodents, and birds—may introduce air and water so as to influence the development of decay.

Compartmentalization

George Hepting (1935) is credited with stimulating research into a plant's role in decay: He made the first definitive observations on compartmentalization of decay in trees (Shigo and Marx 1977). To this has been added more information about the succession of different microorganisms in the decay process. Shigo and coworkers have dissected thousands of trees, both hardwoods and conifers, to study tree responses to wounding and decay. The following discussion summarizes some of their observations and postulations.

Wounds begin the processes that can lead to decay. When tissue is wounded, its metabolism and that of adjacent tissue changes. Certain ions concentrate in wounded tissue, phenolic compounds in angiosperms and terpenes in gymnosperms oxidize and polymerize, enzymes are activated, tissue pH rises, and affected wood discolors, usually darkening (Shigo 1979b). These metabolic changes form a chemical protective zone around the wound.

In addition, materials from surrounding live cells plug the vascular system to retard the vertical movement of microorganisms and plant substances (including newly formed chemicals) into, out of, and through wounded wood (Wall 1). On the interior side of the wound, chemical changes and the small, thick-walled cells of summer wood deter the progression of decay (Wall 2). The radial sheets of ray cells act in a similar way to prevent the spread of decay around a trunk or branch (Wall 3) (see Fig. 2-7).

The chemical and physical changes that follow wounding, the hard rings of summer wood, and the side walls of ray cells act to inhibit decay or contain it in the wound. Usually wood-inhabiting fungi and bacteria will be inhibited, but when a wound is severe, the tree weak, environmental conditions favorable, or the fungi virulent, some microorganisms may penetrate the protective zone and infect surrounding wood. Metabolic changes then extend to adjacent tissue and the next line of physical constraints is called upon to stop or deter decay. The pioneer infecting organisms usually do not cause wood decay directly but begin a succession of microorganisms that can lead to decay and complete rotting of wood.

The vertical vascular system (Wall 1) is the weakest restraint on infection, even when its elements become plugged. This accounts for the greater development of decay above and below a wound than in other directions (Fig. 16-11). The vessels and tracheids of the inner walls (Wall 2) are not as decay-resistant as the xylem rays (Wall 3).

After a tree is wounded, the xylem cells formed by the cambium differentiate to form a barrier zone (often darker than surrounding tissue) separating the wounded tissue from the cambium and the new

FIGURE 16-11
Compartmentalization of discoloration
and decay in sugar maple. Clockwise from
upper left: longitudinal section through
an injection hole, vertical spread of
discoloration confined by walls 1, spread
from the bottom of the hole slowed by
wall 2 but contained by wall 4 formed by
the cambium after the injection; wall 4 has
confined decay to the wood that was present
at the time of wounding—all the other walls
failed; strong compartmentalization by walls
2, 3, and 4 have confined wood discoloration
close to the original wound; zones of
discoloration from old injection wounds
have coalesced primarily due to failure of
walls 2—walls 3 were fairly effective in the
younger wood; well developed wall 4
(dark arcing line) and wall 3 confined
discoloration to the gray wedge upon failure
of wall 2 to retard discoloration toward the
center—the radial crack is from drying after
cutting. (Photo courtesy Alex Shigo,
U.S. Forest Service)

xylem that develops afterward (Shigo 1979b). The barrier zone (Wall 4) confines the decay and, more importantly, it protects the cambium from infecting microorganisms. Unless further injury occurs, the greatest diameter of decay will be the diameter of the tree at the time of wounding. The barrier zone is highly protective but lacks strength and may become the site for ring shakes and frost cracks.

A healthy tree is usually able to contain or repel decay by means of this internal compartmentalization (Shigo and Marx 1977). Infection does not necessarily lead to decomposition, even in wood that is near a wound. How far the succession of organisms progresses depends on the tree's genetic makeup and vigor; the type, size, and position of the wound; the virulence of microorganisms and their antagonists; and environmental conditions. Decay need not be completed. Callusing of wounds, antagonism among organisms, environments unfavorable to decay organisms, and other forces may stop the process in any stage.

Discoloration Preceding Decay

After wounding, wood will discolor before decay begins (Shortle 1979). Discoloration involves the alteration of living cell contents in response to metabolic changes regardless of the presence of micro-organisms. Discoloration itself does not weaken wood but is a warning that decay may not be far behind.

Not all discolored wood is predisposed to decay, however; heart-wood darkens in the normal process of aging. Some species (oak, pine, and black walnut) have heartwood, and some (birch and maple) do not. Darkened heartwood no longer transports water and nutrients but retains its strength and ability for changing chemically; in fact, heartwood is more resistant to decay than active sapwood (Shigo and others 1979). Normal heartwood and wood that has discolored and become predisposed to decay are quite different. Heartwood, however, can discolor and decay if exposed to infection; active sapwood (formed before the wounding) adjacent to discolored heartwood will usually discolor also. Heartwood reacts to infection by the same compartmentalization process that sapwood does (Shigo 1981, personal communication).

Detecting Decay

Decay may be indicated by open or callused wounds, cavities, frost cracks and other trunk splits, broken or dead branches, or fruiting bodies of decay fungi (conks) on exposed wood. Even when the surface of a wound appears sound, it may be decayed underneath. Check the trunk base closely for signs of root injury. It is possible for a dangerously decayed tree to appear healthy: Its bark may be intact and top growth fairly vigorous. Striking the bark of a hollow trunk with a mallet will often reveal the defect with a hollow, drum-like sound.

You can check either living or dead wood for internal soundness (Shortle 1979). Use an electric drill with a bit 200 mm (8 in) long or longer and 2.4 mm ($\frac{3}{32}$ in) in diameter to penetrate the wood in question. During drilling, a sudden increase in motor speed or bit penetration indicates a void or pocket of decayed wood. You can estimate the diameter of the void by noting the differences in ease of bit penetration through the specimen. Determine the vertical extent of decay by probing along the specimen. If you find no indication of decay in the first probe, make a second probe at the trunk base and another on the opposite side 1.5 meters (5 ft) above ground (Shortle).

As wood discolors and decays, its electrical resistance decreases (Skutt, Shigo, and Lessard 1972). You can locate discolored wood—the precursor of decay—by measuring the electrical resistance of wood to a pulsed electric current. A pulsed-current resistance meter, like the Shigometer®, will distinguish among sound, discolored, and decayed areas in both live and dead wood (Shigo and Shigo 1974).

Factors Affecting Wounds and Wound Closure

Plant wounds do not "heal" in the sense that animal wounds heal. An animal wound usually heals from the inside out, replacing dead and damaged tissue; a surface scar may be the only sign of an earlier wound.

In woody plants, however, wounding breaks or destroys the cambium so that the original wound remains, and though it may eventually callus over, wood may decay in what appears to be a sound trunk. It is more appropriate to speak of wound closure or callusing than wound healing.

Wounds from improperly pruned, broken, and dying branches not only occur in greater numbers than surface wounds but also have the potential for more serious decay. Because branch stubs originate near the center of a trunk or branch, all of the wood present at the time of breakage or pruning can become infected (Fig. 16-12). Most wounds are naturally or accidentally caused, although some are deliberately made in the course of normal tree care, particularly when pruning, cabling, injecting, or diagnosing is improperly performed. Some of these practices can be performed at a certain time or in certain ways so as to minimize the amount or seriousness of wounds. Early pruning, for example, will minimize the size of pruning wounds and hasten their closure.

FIGURE 16-12
Pockets of decay have begun in the discolored sapwood in a black walnut stub (large arrows).
The barrier zone will confine decay to the inner side of the barrier (small arrows).
(U.S. Forest Service photo, Shigo and others 1979)

Characteristics of the Tree   The rate of wound closure is most closely correlated with tree vigor as measured by radial stem growth at the wound site (Neely 1979b). Vigorous trees callus more quickly than those of low vigor. Practices that encourage growth will not only speed wound closure but will also reduce the possibility and the progress of decay. Decay will spread more easily in some species, such as mulberry and willow, than in others. Individuals within a species also differ: The clones of poplar hybrids (Shigo, Shortle, and Garrett 1977) and red and silver maple (Santamour 1979a) exhibit quite varied resistance to the spread of decay. You may wish to consider this when you select trees for landscape planting.

Wound Characteristics   The wider a wound, the greater is the danger of decay. Small holes in sound wood are usually of little consequence. In a bark wound, width is the important dimension. Marshall (1931), McQuilkin (1950), and Neely (1970) each report that callus develops primarily from the sides of a wound, much less from the top and least of all from the bottom. When Shigo (1981, personal communication) examined elliptical wounds that were parallel to the wood grain, he found three times as much cambial dieback (and hence less callusing) above and below those with pointed apexes as above and below those with rounded apexes. Neely (1970), however, made square wounds and reported that callus formed first at the corners and less along the sides. He also found that the shape of a wound has little influence on rate of closure: After the first year, elliptical, circular, and square bark wounds of the same width callused at essentially the same rate.

Because loose and torn bark can harbor insects and disease organisms, cut it back cleanly to the edge of a wound, taking care not to increase unnecessarily the width of the wound. Peninsulas of live bark in a wound will speed callusing; therefore leave as much live bark as possible that is firmly attached (Fig. 16-13).

A pruning wound calluses most quickly when the cut is made at the outer edge of the shoulder rings or the collar at a branch attachment. This is true even though such a wound is 50 to 100 percent larger than a wound that leaves a short stub. Avoid, however, making a wound too close (flush) and large, which will be open to infection and relatively slower closure.

Callus growth is most rapid when xylem growth is most rapid. Trunk wounds directly above large, vigorous roots callus more rapidly than those located in depressions between roots. On vigorous trees, wounds near the base will usually callus more rapidly; on trees of low vigor, wounds higher on the trunk will probably callus more quickly than will lower wounds. Neely (1970) found that trunk wounds between 0.6 and 1.8 m (2-6 ft) above the ground callused at about the same rate.

FIGURE 16-13
Even though the peninsulas of bark do not
callus as rapidly as the edges of the wound,
they protect wood and help to close this
200-mm (8-in) wide wound on the trunk
of a mature lime tree in England.

Season   Bark is easily damaged or torn in the spring and will slip when the cambium is actively dividing. Bark can easily be knocked loose from the trunk or torn when holes are drilled or pruning is done in the spring (Felix and Shigo 1977). Torn bark, however, can often be replaced before the cambium dries.

Neely (1970) examined ash, honey locust, and pin oak during the growing season after wounding: Wounds made in the spring, summer, and winter suffered little dieback, but those inflicted in the fall callused about 20 percent more slowly. Three to six times more area callused during the first growing season on wounds inflicted in spring than those made in summer. The latter had less time to callus. The bark died back around wounds made in summer and fall, especially on honey locust. It would appear that most of the wounds were not exposed to direct sunlight, because callusing did not vary with the side of the trunk on which a wound was made. Wounds made late in the growing season, when many species of wood-invading fungi are sporulating, are often infected more rapidly than are wounds made at other times (Felix and Shigo 1977).

Other Factors   Microorganisms antagonistic to wood-decaying fungi may prevent or at least delay decay. In red maples heavy inoculation of the fungus *Trichoderma harzianum* may delay the invasion of wood-decaying Hymenomycetes for up to two years (Shortle 1979). Attempts to sterilize a wound may do as much harm as good because antagonistic organisms will also be affected. Decay usually develops more rapidly when a wound is alternately exposed to wet and dry conditions.

Injection Wounds

Trees may be injected with insecticides, fungicides, fertilizers, and growth regulators. In urban areas and in plantings of mixed species, the injection of chemicals that would otherwise be sprayed or dusted minimizes toxic hazards and usually improves effectiveness. Injection holes, particularly those that are large and deep, can lead to decay and weakening of the tree, particularly if annual treatments are necessary. The possible damage from injection wounds must be weighed against the possible benefits of the chemical treatment.

Injection wounds alone seldom create problems in vigorous trees, but the chemicals injected are not without their hazards. When Shigo, Money, and Dodds (1977) examined injection holes in red maple, white oak, and shagbark hickory in West Virginia, they found little discolored wood and cambial dieback associated with one-year-old injection holes (no chemicals added) 5 mm ($\frac{3}{16}$ in) in diameter and 18 mm ($\frac{3}{4}$ in) deep. Columns of discolored wood and some cambial dieback, however, were associated with injection sites of Bidrin® or Meta-Systox-R®. Less injury resulted when Fungisol®, Stemix®, or micronutrients were injected singly.

Shigo and Campana (1977) found discolored wood associated with every injection wound made for the control of Dutch elm disease in 80 large American elm trees in the northeastern United States (see Fig. 2-7). Injection wounds made in several successive years caused severe internal injuries. More than half of the trees exhibited decay near the injection sites. In one case, obvious decay began within one year of injection. This study indicates that

> After an injection hole is made, discolored wood develops, microorganisms infect the wood, and decay may set in.

> If holes are made in the same trees in successive years, the individual columns of discolored wood associated with each wound begin to coalesce to form large columns of dead discolored wood; these may contain some early decay.

> The rate and extent of infection, decay, and coalescence of injured tissue vary greatly among trees, even those of the same species.

Shigo and Campana recommend that, until better methods of injection are developed, holes be made shallow, clean-edged, small in diameter, and few in number. Injection holes should be drilled infrequently (once every three to five years) on a given tree. Make first injections on the ridges of roots, not in the depressions between the ridges, and as low as possible on the trunk. Establish subsequent holes at least 500 mm (20 in) above and at a diagonal to the grain from old holes. Shigo and Campana (1977) conclude that the injection of chemicals might be a reasonable treatment for a tree infected with Dutch elm disease, but that the use of injections for preventing the disease poses some serious problems.

Treating Wounds

Decay problems can be largely prevented if you plant species or clones known to compartmentalize effectively, to close wounds rapidly, and to resist decay. Wounds are inevitable, but their number, size, and severity can often be minimized. Felix and Shigo (1977) recommend that needed pruning, cabling, scribing, and injection be done at times other than spring growth flush and leaf fall, when plants seem to be most vulnerable. If you choose safe locations for planting and use protective stakes or guard rails, you can prevent or reduce plant injury from cars, machinery, and people. Early training of trees can prevent sunburn and reduce the number and size of later pruning cuts; well-trained trees should not have to be cabled or braced. When possible, use methods other than injection for applying chemicals to trees.

Maintaining Vigor   Tree vigor is the best antidote to lingering wounds and decay. Vigor, particularly in conifers, greatly reduces the possibility of borer infestation. Vigor can be maintained and enhanced by wise fertilization, irrigation, and pest management.

Cleaning and Shaping Wounds   Remove loose bark from in and around a wound; the wound will look better and insects will not have protective cover. Arborists commonly recommend that the bark around wounds and even around pruning cuts be shaped, but experiments have brought the value of this practice into question, particularly in cases where it widens the wound. Shaping a wound is of little or no value in hastening wound closure (Neely 1970).

If you do shape a wound, take care to increase its width as little as possible and to create an ellipse with no sharp apexes. Cambium at the apexes of pointed ellipses usually dies back, and the new xylem often develops cracks (Felix and Shigo 1977; Mercer 1979). Arborists usually shape bark wounds so that the long axis is parallel to the grain of the wood immediately under the cambium; this may or may not be parallel to the length of the trunk or branch. Shaping a wound otherwise will increase its width and the time of closure. An elliptical wound supposedly favors the flow of organic compounds around and along its edge so as to enhance callusing. A circular wound, however, will callus

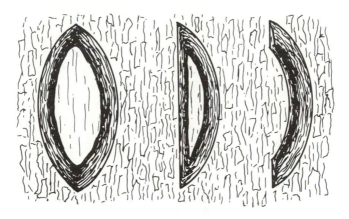

FIGURE 16-14
The value of tracing wounds in the shape of an ellipse has been refuted
by experiments performed by Neely (1970). Neely cut a series of wounds
in the bark of several tree species: The right edge of each wound
was half an ellipse, the left side of the wounds was half an ellipse or concave
(left), a straight line (center), or a convex curve that reduced
the width of the wound to one-half that formed by half an ellipse (right).
Even though the convex edge callused (dark-lined areas)
only 25 percent as quickly as the concave edge (standard ellipse),
the wound it surrounded (right) closed in one growing season
while the others did not close until the next.
(Adapted from Neely 1970)

as quickly as an elliptical one (Neely 1970). Neely found that the rates of callus growth from a straight edge (parallel to the grain) and a convex edge were about 50 and 25 percent respectively of the growth rate from a concave edge (the shape usually recommended) (Fig. 16-14). In no case did it prove worthwhile to widen a wound.

Leave peninsulas of live bark in a wound to speed callusing. Even though callus formation at the edge of a peninsula of bark is only 10 percent of that on the inside of an ellipse, the peninsula callus has to fill less area than if the entire wound had been "properly" shaped. When cleaning a wound, use a sharp wood chisel, gouge, or pruning knife to cut the loose bark at right angles to the wound surface, usually the cambium layer. Cut to firm bark. If the wood is damaged, smooth it so it will not trap water and debris, hamper callusing, or present an unpleasant appearance. Do not unnecessarily enlarge a wound.

Dressing Wounds Ideally a wound dressing should prevent decay, stimulate callus growth, and improve tree appearance. To date, however, no dressing commercially available in the United States has been shown to prevent decay. Shigo and Wilson (1977) applied asphalt, polyurethane, and shellac to separate wounds on red maple and American elm and found that none affected vertical extensions of discolored and decayed wood or the presence of decay fungi. In fact, the electrical resistance of discolored wood indicated that wounds dressed with asphalt were most serious and untreated wounds the least serious. Of more than 40 wound paints tested in the laboratory by Mercer (1979), all those containing a fungicide, except one, were phytotoxic. The one exception was Santar® ; because it contains mercury, it is not registered for use in the United States and probably never will be. Neely (1970) and Shigo and Wilson (1977) concluded after testing six wound dressings available in the United States that the materials had no effect on the rate of wound closure. Lac Balsam®, a latex emulsion available in Europe, was the only material tested by Mercer (1979) that stimulated callus formation. Shigo (1980, personal communication) maintains, however, that unless a wound calluses over within three years, the rate of callusing has little or no effect on decay. Once decay begins, callusing will not reduce its development.

The main function of wound dressings, therefore, is cosmetic (Shigo 1981). Many wounds are less noticeable when painted with a dark-colored covering. Many people regard wound dressings as the sign of professionalism. The most common dressings are asphalt-based; they should be applied in thin coats to minimize cracking. If wound coverings crack, moisture can accumulate and foster decay.

In some cases, pruning and bark wounds should be treated with fungicides to reduce the chances of a serious vascular infection or other disease. Some diseases are more serious and faster-acting than wood decay. Cankerstain of plane trees (*Ceratocystis fimbriata* var. *platani*), for example, is serious in the eastern United States and can be fatal. Where this disease is known to be a problem, pruning tools should be disinfected and wounds treated with a mild disinfectant, such as Benlate® or thiabendazole (Pirone 1978b). Crown gall, a bacterial disease that afflicts both young and established trees, will be less of a problem if wounding is kept to a minimum and wounds that do occur are treated with Galltrol A® (see Chapter 19). There may be diseases other than those directly involved in decay infection that can be reduced in severity by proper treatment of wounds.

Replacing Loose Bark Many bark wounds occur in the spring and early summer, when cambium is active and bark easily separates from wood. Loosened bark can often be replaced with little or no damage to the injured area, but must be replaced soon after it has been loosened. Keep the moist cambial regions of both the wood and the bark from drying out. Carefully remove all shredded bark and debris from the wound so that the bark will fit tightly against the film of cambium on the xylem. Press the patch of loose bark into place and hold it with a few small lacquered nails or duct tape. Cover the wounded area with a moist pad of cloth, paper towel, or peat; wrap a sheet of polyethylene film around the pad and the trunk or branch. Hold the film in place with a tight tie of tape or line above and below the wound; you may have to smooth rough bark to ensure a snug fit. Shade the wound from the sun.

If you perform this operation carefully and soon after an injury occurs, much or all of the loosened bark may reunite with the wood. Reunion will be most likely during the period of rapid cambial growth. Within a week or two, inspect the wound to see whether the bark patch is still alive and has begun to unite with the wood. You can remove the plastic film and the moist pad if callus has formed around the cracks, but shade the wound. If the bark patch does not reunite with the trunk or if the wound is an old one, free the wound area of dead bark and leave it exposed to the air.

Grafting Bark Girdling wounds can be bridged with bark implants or bridge grafts. Bridge grafts have been used by fruit tree growers to reunite the top and roots of trees whose trunks have been damaged by rabbits or other rodents (Hartmann and Kester 1983), but the technique is seldom used in landscapes. Bark implants, however, have been used to bridge girdling bark wounds, sometimes caused by winch cables or vandals. If the cambium is still active, you can use bark from other parts of the tree to bridge the break (Fig. 16-15).

Begin by cleaning the girdled wound of shredded bark and debris. If the wound is fresh and the cambium moist, remove rings of bark about 20 mm ($\frac{3}{4}$ in) wide from around the trunk, one each from the top and the bottom of the girdle. Cut bark patches the same height as the enlarged girdle and about 50 mm (2 in) wide from other places on the trunk or branches. Cut the patches from widely separated areas on the trunk and large branches of the same tree or other trees of the same species. Implant these patches right end up and close together in the enlarged girdled wound. The patches can be held in place with small lacquered nails; both wound and patches should be moist. Cover the patches and about 25 mm (1 in) of bark above and below the girdle with duct tape to hold the patches snugly and to reduce the loss of moisture. Also place duct tape on the wounds made when patches were cut. Keep all wounded areas moist during the operation and shade them afterward. Within a week or two, the bark implants should

FIGURE 16-15

A girdle injury can often be repaired by bark implants. A girdle through the bark and into the wood is shown at the left. When the bark will slip easily (during the early growing season), cut bands of bark about 20-mm ($\frac{3}{4}$-in) wide from both sides of the girdle (center). Cut a bark patch the width of the enlarged girdle from another area of the trunk or a branch and tack it into place, covering the girdle and newly exposed cambium (right). Keep the exposed cambium, new bark patch, and the new patch wound from drying out.

be callused in place and the patch wounds callused over from the film of cambium that remained on the xylem.

By using bark implants, Herbert Warren, the former park director of Victoria, British Columbia, once saved two 400-mm (15-in) diameter trees that had been girdled with a chain saw.

CAVITIES AND CAVITY TREATMENT

In the advanced stages of decay, wood is consumed by fungi and insects or falls from the wound. Cavities develop from bark wounds, breakage, and pruning wounds. Older trees occasionally have large cavities in their trunks and main branches. A considerable portion of the trunk can be cavitated without greatly affecting the strength or vigor of a tree. A cylinder is almost as strong as a rod of the same diameter and material; so it is with a treetrunk. Sapwood and bark active in transporting water, nutrients, and organic material are unaffected except for the cavity opening. As a treetrunk grows in girth, most of the new wood is continuous with the spreading surface roots that provide tree stability. Some mature trees have stood for years with less than 25 percent of their trunk caliper intact.

If a tree is vigorous, new wood usually more than compensates for wood lost to decay; due to compartmentalization, a cavity will become no larger than the diameter of the tree at the time of the injury (Shortle 1979). Decay and cavities are slow to develop except in softwood trees of low vigor. Wood-decaying fungi are not the pioneer invaders of wounds and often do not infect a wound for several years (Shigo 1979). A tree of low vigor, however, may be doomed unless decay is arrested and tree vigor improved.

When deciding whether and how to treat a cavity, consider: the importance of the tree; its species, age, and condition; and size and location of the cavity. A cavity may be so large or close to large branches that the tree is irremediably unsafe. On the other hand, a seriously cavitated tree may be of such historic or landscape value that considerable work is justified. Short-lived softwood trees are usually of questionable value compared with longer-lived hardwoods and conifers.

Decaying wood in cavities is sometimes infested with termites and ants. The most destructive species in the United States are the subterranean termites (*Reticulitermes* spp.) and carpenter ants (*Camponotus* spp.) (Pirone 1978a). Termites can feed in sound wood as well as decaying wood. Ants and termites, however, are usually contained within the chemical and physical barrier zone formed by the cambium after wounding. Carpenter ant larvae can create large tunnels in decayed and sound heartwood. Ants can be distinguished from termites by the pronounced constriction of the body between the thorax (where legs and wings are attached) and the abdomen; termites have no such constriction. Ants can be controlled by dusts or sprays of Diazinon® in the infested cavity (Pirone 1978a); termites are more difficult to control. Chlordane has been the standard spray material used to control termites, but in some states it can be applied only by licensed applicators. The solvents in some materials used to protect buildings against termites are toxic to woody plants and should not be used on them. Consult a knowledgeable specialist (see Chapter 17) about the chemicals registered for control of ants and termites in your particular area.

Cavities usually follow the development pattern of wood decay in that they are long and narrow. Cavities almost always open to the outside: Some openings will callus over and others may be concealed in branch crotches, but the cavities to which they connect may still be enlarging. Most cavities extend farther above and below their openings through the bark; water, debris, and insects collect in the lower portions. The cavity opening often restricts access so that it is difficult to assess the extent of decay or take remedial steps without enlarging the opening and damaging sound wood.

The tissue around a cavity wound will callus, particularly along the vertical edges. Callus extends in a smooth, thin layer over a firm wound or fill surface, but it rolls inward on itself in a cavity and can become quite thick (Fig. 16-16). A callus roll will not close a wound as quickly as will a thin callus, however. For this reason some arborists recommend filling or covering to obtain faster cavity closure. If a cavity is large, though, a callus roll may provide needed mechanical strength to the trunk (Armstrong 1925, Bernatzky 1978, Pirone 1978a).

You can usually determine the extent of a cavity by sight if the cavity is open or by sound if you tap on the bark with a mallet—a hollow sound indicates a possible cavity. The amount of decay and discolored wood associated with a cavity and sometimes the size of the cavity itself can be determined only by probing with a drill and a resistance meter as described earlier.

FIGURE 16-16
At the edge of a cavity
in this eucalyptus cross-section,
the callus rolls inward on itself.

Treating Cavities

Arborists have recommended that cavities be cleaned, sterilized, braced, sealed, and filled or covered so that decay would be stopped, the tree strengthened, and a hard surface provided for callusing to close the wound (Collins 1920; Bernatzky 1978; Pirone 1978a). In the early 1900s, sterilization of a cavity after cleaning was added to the list of treatment procedures, though the value of filling cavities has been questioned. Shigo and coworkers (Shortle 1979) have found that these procedures will do little to help the tree and can in fact be harmful if they cause the protective barrier to be broken. Though some may be of questionable value, the various procedures will be described and discussed in sequence:

Cleaning The essence of almost all recommendations is to clean all decayed wood out of a cavity. Many people also advise removing discolored and water-soaked wood. Because decay fungi have been found 0.3 to 1.4 m (1–4.5 ft) in advance of decayed wood (Welch 1949), however, removing infected wood from a cavity is impractical if not impossible. Any attempt to remove even all of the rotten wood will probably breech the compartmentalized barriers to decay and will open sound wood to infection (Shortle 1979).

Most cavities collect water, but providing drainage probably causes more harm than would water, particularly if sound wood is exposed (Shigo and Felix 1980). Alternating wet and dry conditions usually promote wood rotting more than continual saturation; air is more often a limiting factor than is water (Welch).

If cavities are to be cleaned, take care not to cut into sound wood even if it is discolored (Wilson and Shigo 1973). Even more

FIGURE 16-17
Small cavities can be partially filled (left) or drained (right)
to keep water from accumulating. (Adapted from artwork by M. J. Whitehead
in Bridgeman, P. H. 1976. *Tree Surgery.* Devon, England: David & Charles)

 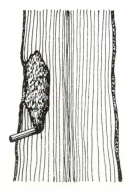

importantly, do not cut into healthy wood formed since the wound originally occurred. A pocket fill would probably be most satisfactory to minimize cavity water (Fig. 16-17).

Sterilizing   None of the recommended sterilizing chemicals, such as creosote or copper sulfate solution, penetrates wood more than a few millimeters (a fraction of an inch). As already noted, decay infection can extend more than a meter (3 ft) into sound wood. In addition, most sterilizing agents are poisonous or corrosive and must be handled with care.

Sealing   No wound dressing available in the United States has been shown to inhibit decay; in some trials, asphalt dressings actually increased the extent of infection (Shigo and Wilson 1977). The main reason for painting the inside of a cavity is to improve appearance if the cavity is to be left open.

Filling   Most arborists agree that filling a cavity is of little or no value in promoting the health and longevity of a tree.

Fill material seldom strengthens tree structure as much as would the callus roll that develops around an open cavity. Both the movement of a tree and differing expansion characteristics loosen a fill in its cavity and decrease its efficacy as a support and air, moisture, and spores will diffuse or seep into the openings. Further decay is favored. Tree appearance remains the main reason for filling cavities.

Cavity fill material should be durable, nontoxic to trees, flexible, plastic, and waterproof (Pirone 1978a). Many substances, including patented fill materials are used to fill cavities. *Concrete* has been the standard fill material, primarily because of its availability, low cost, and durability (Fig. 16-18). However, it is not very flexible or waterproof; it is heavy, requires skill to install, and is difficult to remove. Concrete is most satisfactory in small cavities and trunk cavities subject to little or no movement in storms. Place in a clean cavity (decayed wood removed), a relatively dry concrete mix in layers 75 to 125 mm (3–5 in) thick separated by heavy tar paper to allow for expansion and some movement. Pack the concrete firmly and form a smooth surface even with the cambium.

A mixture of *asphalt and sand* is also commonly used, particularly for basal trunk and pocket cavities (Bridgeman 1976). Asphalt mixtures are better fillers than concrete but are difficult to prepare and apply (Collins 1920). The mixture must be stiff so that it will not slump at the base when it is tamped into place. Asphalt mixtures are most satisfactory for pocket cavities (Fig. 16-17) or trunk cavities that will not be warmed excessively by the summer sun, which could cause the mixture to soften and slump. Since asphalt mixtures are malleable, no expansion joints are necessary.

FIGURE 16-18
Although seldom done any more, trunk cavities are sometimes filled with cement
to improve the appearance of trees, such as these two elms:
The basal cavity was filled and the main branches rodded on this elm
on the University of Illinois campus shortly before being photographed.
(Photo courtesy of E. B. Himelick, Illinois Natural History Survey);
the elm on the State Capital grounds in Sacramento in the right photograph
is actively callusing over the filling installed 35 years earlier.

Urethane foam has several advantages over other fill materials: It is easy to use, is relatively inexpensive, and requires little time to install (Fig. 16-19) (King, Beatly, and McKenzie 1970). When installed, the foam is lightweight, nontoxic, and somewhat flexible. It can be used to fill any size cavity. Hamilton and Marling (1981), however, report that foam fills have not held up. Bridgeman (1976) reports that polyurethane foams tend to shrink and draw away from the cavity wall, but that urethane foams do not do so. King (1978, personal communication) found that cavities filled with polyurethane and covered with auto body putty were in good shape after eight years.

In the past, two liquids had to be mixed and quickly poured into the cavity where their interaction produced foam. Workers had to use a cover to contain the expanding foam until the cavity was filled and the foam hardened. You can now obtain a pressurized container kit that releases the two liquids simultaneously and allows you to build up

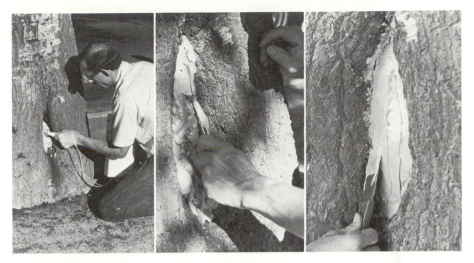

FIGURE 16-19
A small cavity (about 4 liters; 1 gal) in the trunk of a live oak is cleaned
and filled with urethane foam from a pressurized kit (Insta-Foam®).
After the cavity is cleaned of loose decayed wood and the liquids are mixed
in the small chamber in the applicator's hand, the cavity is partially filled (left);
the fill surface is shaped with a pocket knife (center); auto body filler and hardener
is applied with a putty knife to give a hard surface (right).
After the cavity was cleaned, it took about thirty minutes to complete
the fill, with fifteen minutes of the time spent waiting for the foam
to harden so it could be shaped.

the foam fill from the bottom and the back of the cavity. This method allows for improved speed and control in filling small cavities; however, large cavities can be filled faster at less cost by mixing the two liquids. After the cavity is filled, trim the foam down to the wood at the edge of the cavity with a knife. Because the foam is soft and easily compressed, it should be covered with auto body filler and hardener for protection. The hardened surface can then be painted. The urethane ingredients foam properly at temperatures above 20°C (65°F) (King, Beatly, and McKensie 1970); Bridgeman (1976) indicates that temperatures above 10°C (50°F) are adequate.

A form can be used to hold expanding foam in a cavity until it is firm. Line the form with a sheet of aluminum foil sprayed on the inside surface with silicone. Hold the foil in place with a form of a sheet of aluminum, linoleum, or other flexible material tied tightly to the trunk. You may have to smooth the bark at the cavity edge to provide a seal. Leave an opening at the top of the cavity into which you will pour the foam mixture. Several minutes after the two liquids are poured, the foam will expand to fill the cavity; if you added insufficient material, you can add more ten minutes later. Ten minutes after the foaming is complete, the cavity cover can be removed. What little heat is generated by the foaming will not affect the cambium.

Covering   You can cover a cavity with sheet metal when you want neither to leave the cavity open nor to fill it in (Collins 1920). However, there is little to recommend this practice. A sheet metal covering can be collapsed inward by the pressure of the growing callus (Keith Davey 1975, personal communication). Such a covering may not serve its purpose long.

Recommendations

A tree with a cavity will be best served if you do everything possible to improve tree vigor so that new wood and bark grow faster than decay advances within the compartmentalized barrier. A cavitated tree may have to be thinned to reduce the weight and wind resistance of the top. If a cavity is not conspicuous or endangering the tree, you need only remove branch stubs or decayed wood that might interfere with callus development. If a cavity is so extensive that it weakens tree structure, you may also have to brace the tree. Decayed wood that can be easily removed should be extracted from the cavity before bracing rods are installed.

If cavity appearance is important, remove enough decayed wood to give a smooth, firm surface, taking care not to cut into sound wood. Clean the portion of the cavity that is below the opening, dry it (with a blow torch if necessary), and fill it with asphalt (if the cavity is small), an asphalt-sand mixture, or urethane foam so that water will be shed (Fig. 16-17). Paint the entire cavity and any fill you have installed.

Trees with large cavities may need to be protected from people who climb on bracing rods, build fires in the cavities, or otherwise deface them. Protect the entire trunk with a free-standing screen fence or cover the cavity with a screen, always making allowance for trunk expansion.

MANAGING ROOT GROWTH

Root growth varies with tree species, soil conditions, surface covering, rainfall, and irrigation practices. Root growth and function are essential for healthy, sturdy plants, but roots can also cause problems. Roots crack and plug sewer lines; they lift and break curbs, sidewalks, pavement, and building foundations (Fig. 16-20). When roots dry the soil under pavement and buildings, certain soils will shrink and

FIGURE 16-20
Even though the curb and sidewalk have been replaced at least once,
this eucalyptus tree again has broken and raised the curb, sidewalk,
and driveway.

settle so that structural damage occurs. Shallow roots in lawns can be
unsightly and difficult to mow around. Root suckers in lawns and
flower beds are a nuisance.

Roots in Sewers

People rarely give thought to possible root problems when they
plant a young tree or shrub; not until a sewer has been blocked do most
people become aware of the problem. Roots can crack a sewer line
and invade it if there is an opening. When the sewer is plugged, roots
need to be removed or chemically treated (see Chapter 15).

The problem is often difficult to address until failure occurs, but
there are measures that will prevent or at least delay the invasion of
sewer lines. One of the most important measures is to avoid planting
near a sewer or a septic field any fast-growing species that are known
to have invasive roots, such as willow, poplar, and silver maple. Unfortunately, homeowners often find that utility trenches are the easiest
places to plant. Less than four years after a fruitless mulberry was
planted in a newly filled sewer trench in Sacramento, California, two
of its roots crushed an orangeburg sewer line 100-mm (4-in) in diameter

(see Fig. 1-5). Only small trees and modest-size shrubs should be planted near sewer lines, particularly if the soil is shallow or poorly drained and the sewer trench is deeper than the surrounding soil.

In new building sites, root problems can be minimized if watertight flexible telescopic sewer joints are installed (Morling 1963). Some contractors wrap sewer joints with copper wire screen, which is toxic to small roots. The copper wire may delay sewer invasion, but it will do little to prevent cracks caused by enlarging roots. Copper will probably remain effective longer in acid than in alkaline soils but will eventually be eroded away. In alkaline soils, the surface of copper will oxidize: This will decrease its effectiveness but will also protect the remaining copper so that it will last longer than the more effective copper in acid soils.

Surface Rooting

In a 1975 survey of street tree root problems in 22 northern California cities, the city arborists cited most often the following four causes of pavement and sidewalk damage (Hamilton 1976):

Planting in narrow parkways

Selection of shallow-rooted, large-growing trees

Planting in compact, poorly drained soil

Frequent, shallow, or excessive watering

The first three causes of damage could be largely avoided if plantings were properly planned. Each of the last three practices favors surface rooting, which can be discouraged by several methods described in Chapter 10. Infrequent or deep irrigation or none at all will favor deeper rooting, but because, in most situations, homeowners water the street trees, it is difficult to establish uniform, proper irrigation. This is true despite the fact that almost 75 percent of the cities surveyed provide written instructions for the care of street trees. Hamilton (1976) observed that trees planted in deep, non-compacted soil ten years or more before the installation of curbs, sidewalks, and landscape irrigation created fewer concrete breakage problems than the same species planted in a compacted soil after curb and sidewalks were in place.

Damage caused by tree roots is expensive to repair. Root problems of the estimated 970,000 street trees in the 22 cities surveyed (Hamilton 1976) necessitated repairs costing $589,000 in 1975; this cost may have included sewer as well as curb and sidewalk repair. Sidewalk damage began when the trees were between seven and 20 years old; the median age was 15 years. Twelve of the cities used root-cutting machines as needed; three used them on a regular schedule (Fig. 16-21). Surface roots are cut when a trench is dug between the trunk and the pavement; the trench should be 50 to 100 mm (2–4 in) wide and 200 to 300 mm (8–12 in) deep 150 to 300 mm (6–12 in) from the curb or sidewalk. Roots should be cut on only one side of a tree in any one year. To minimize wind throw, the tops of the trees are usually thinned before roots are pruned. Care must be taken not to damage underground utility and irrigation lines.

Few arborists are satisfied with present root-pruning practices (Hamilton); regrowth is extremely rapid from most cut root stubs. Sometimes sidewalks are lifted within one year after they are repaired and the trees root-pruned. A barrier of heavy plastic or fiberglass placed in the trench at a slight angle downward and away from the tree should deflect roots underneath pavement and structures. Such barriers, however, have not been adequately tested.

The long-term solution to root damage will have to combine selection of better species, more thorough preparation of compacted soils,

FIGURE 16-21
A Vermeer root-cutting machine speeds the cutting of large surface roots.
(Photo courtesy Vermeer Manufacturing Company, Pella, Iowa)

root barriers or deflectors, wise irrigation practices, and root pruning when necessary.

Soil Contraction and Expansion Caused by Plants

When tree roots grow under buildings and pavement in "shrinkable clay soils," the alternating shrinkage and expansion (Morling 1963) may pose a serious threat to structures. In England, the shrinking and swelling of clay soils due to moisture extraction by roots and subsequent rewetting causes pavement, foundations, and walls to crack and doors and windows to stick. Surface cracking of a soil indicates its susceptibility to shrinking and swelling.

Soil Subsidence   Soil shrinkage is particularly severe during droughts. Following the drought of 1975–1976, English insurance companies paid about $70 million to cover building damage caused by the shrinkage of clay soils (Reece 1979). Even though soil subsidence rarely makes a building unsafe, it may seriously affect its appearance, value, and marketability (Pryke 1979). Uneven subsidence is the most disruptive. After a dry summer a clay soil may subside 500 to 750 mm (2–3 in) (Biddle 1979; Pryke). Soils under buildings may also dry and shrink after repair of leaking drains or installation of underground central heating pipes.

Fast-growing, large trees, such as poplar, willow, elm, and ash, can cause problems in clay soil over a horizontal radius equal to one-and-a-half times their mature height (Morling 1963). Young trees, with rapidly enlarging roots, usually cause more damage than mature trees. Shrinking and swelling are greatest near the surface, where most roots are located, and decrease with depth to about 1.5 or 2 meters (5 or 6 ft) below the surface, where soil movement is slight. During a dry year, shrinkage near large trees may extend to 3 m (10 ft) below the surface (Morling; Pryke).

Soil Expansion   Most cracks in buildings caused by soil subsidence will nearly close during the next rainy season or irrigation period. However, the swelling of a dry clay soil after a building has been constructed may cause even more serious problems. Morling (1963) reports that the surface of a clay soil increased 100 mm (4 in) in elevation in the six years after two large elm trees were removed and still rose 10 mm (0.4 in) in the seventh year. Soil moisture in the root zone must have been near the wilting point when the trees were cut down. Even with ample rain or irrigation, many clay soils under buildings are slow to rewet and will continue expansion over long periods of time.

Minimizing Problems Caused by Shrinking and Swelling of Clay Soils   A number of things can be done to reduce the problems caused by trees growing in clay soils:

Provide buildings on expansive clay soils with particularly deep and strong foundation footings. Short-bored pile foundations (augered holes filled with cement so they are integral with the foundation) 3 m (10 ft) long can provide needed stability at reasonable cost (Pryke 1979).

Near buildings and pavement, plant moderately vigorous trees that will grow to a small or medium size. Avoid fast-growing trees that become large when mature, such as poplar, willow, elm, and ash, particularly if building foundations are inadequate.

If shrubs, vines, or young mature trees begin to cause damage around existing buildings, consider removing them. Do not repair settling cracks in buildings or pavement until the soil has had a chance to rewet and the cracks close, at least partially.

While a tree is young, or before planting, improve the tilth, fertility, and moisture conditions of the soil under what will be the mature canopy. Between the building and the treetrunk, treat only that half of the soil area that is nearer the trunk. This should increase root branching in the improved soil and minimize root growth toward the building.

Biddle (1979) suggests that heavy pruning (reducing trunk growth 75 percent) will reduce water use (transpiration), root growth, and soil subsidence. If the pruning were done during a drought year only, transpiration would be proportionately reduced only early in the growing season. If the heavy pruning were done every year or two, it might be a way to prevent severe subsidence. More information on this strategy is needed.

You may create a barrier by killing invading roots with a soil fumigant (Biddle 1979) or may prune roots regularly. These solutions are acceptable primarily where no other plants are involved.

Where subsidence causes only minor cracking, you can reduce it by slowly soaking the soil along the foundation or pavement. This would be especially important during a drought. I am reminded to irrigate an ivy groundcover planting next to my house when the kitchen cupboard doors begin to stick.

You may be tempted to cut down a large tree that has been creating subsidence problems. However, this may cause more damage due to heaving as the soil rewets.

Keep trees growing in shrinkable clay soil at a construction site adequately watered so that the soil does not subside or expand unduly.

If a shrinkable clay soil has been severely dried by plants, irrigate the soil thoroughly and allow it to drain before grading for construction.

TREE REMOVAL

A tree may be removed for a number of reasons:

It may be unsafe because of old age, storm damage, or death.

It may be diseased or be host to a disease that can spread to more valuable trees.

It may be located in a future building site or roadway or may obstruct a view.

It may crowd other trees.

It may be wanted for timber or firewood.

If a tree is in the open, it may be pushed down with a large track-laying bulldozer, pulled down with a power winch, or cut down with a chain saw. In more restricted areas, a tree may have to be felled in a certain direction or removed in sections. Bridgeman (1976) gives directions for felling trees.

Sometimes large trees are cut close to the ground and the stumps allowed to remain. The stumps of trees and large shrubs susceptible to *Armillaria mellea*, however, (see Chapter 19) should be removed even though they are not infected. As Rishbeth (1970) has shown, airborne spores of the fungus can infect stumps. You can control or minimize stump sprouting and root suckers by treating the stump chemically or girdling the trunk a year before the tree is cut down (see Chapter 15).

Stump removal is not easy. The stumps of small trees are usually dug out by hand with a shovel and ax; a winch makes the job easier. When the equipment is available and the stump accessible, a stump-cutting or grinding machine is the quickest and most satisfactory means of removal (Fig. 16-22). It can chip out a stump to 200 or 300 mm (8 or 12 in) below ground level in minutes. A stump-cutter can remove stumps in confined spaces, like those between curb and sidewalk. You may lift the stump out with a hydraulic lift—a "Stumpmaster" can lift stumps up to 375 mm (15 in) in diameter (Bridgeman 1976). Blasting the stump out of the ground with explosives should be done by a specialist and only when the stump is in open country. Finally, you may burn the stump in place by pouring saltpeter (KNO_3) and paraffin in holes in the stump and igniting them—this method is slow and can be a fire hazard in organic soils.

FIGURE 16-22
A Vermeer stump cutter will chip a stump away to a depth of about 250 mm (10 in) below the soil surface. (Photo courtesy Vermeer Manufacturing Company, Pella, Iowa)

CHAPTER SEVENTEEN

Diagnosing Plant Problems

Most landscape plantings get off to a good start, particularly in the first season, and begin their growth under nearly ideal conditions of spring-like weather and moist soil. Both new and existing plants may also retain vigor and food reserves from earlier care and conditions. Their new surroundings, however, may be less favorable than their previous environment. This is particularly the case in inner cities and other intensively developed sites. Plants can be harmed as their roots grow out into surrounding soil and as new shoots encounter the increasingly hostile environment of summer and following seasons. It is remarkable that many plants grow as well as they do.

Problems encountered by plants in new or redeveloped sites may include

Stratified and compacted soil

Infestation by unknown pests and disease-producing organisms

Reflection and reradiation that increase air temperature

Light intensity that varies from deep shade to full sun

The extension of daylength by night lighting

The reduction of humidity by arid expanses of pavement

A sparse or excessive water supply

Insufficient nutrients

Air pollution

Accidents, vandalism, and neglect

You can minimize or avoid most such hazards by choosing a favorable site, preparing the site for planting, selecting plants that are tolerant to the site and relatively pest-resistant, and caring for plants to ensure vigorous growth. Be alert to environmental changes that may jeopardize existing plantings.

Even in what appear to be the best growing situations, plants may not perform as desired: They may become unattractive, grow slowly, or become so seriously diseased that they must be removed. Plant problems are diagnosed so that they may be corrected or minimized. Diagnosis may also establish responsibility for the purposes of litigation or for insurance or tax purposes.

THE DIAGNOSIS

Efforts at treatment may be unavailing or may even compound a problem if plants are treated for the wrong malady. Accurate diagnosis of a plant problem depends on

The observation of subtle differences from the normal in plant appearance or the surrounding environment

A knowledge of plants, soil, climate, cultural practices, pests, diseases, and their interactions

Accurate information about the recent history of affected plants, the site, the climate, and cultural practices

The availability of a few simple tools

An analytical approach to difficult problems

Careful observation is paramount and can be developed with practice and interest. Familiarity with normal plant appearance and growth and the effects of environment and treatment will increase with experience.

Normal Plants

Inexperienced observers often think they have encountered a serious problem when, in fact, the "symptom" that alarms them is only a normal feature of a plant, a seasonal change in its appearance, or the sign of a nonparasitic agent. Thus, recognizing the appearance of healthy plants at different stages of growth and at different seasons of the year can be as important as recognizing the symptoms of particular disorders.

Leaf Color   The leaves of a number of plants may vary from the green, gray-green, or reddish color that is normal for most of the growing season. The early spring growth of some plants, such as apricot and Chinese photinia, may be copper or reddish in color, giving way to green as the leaves and shoots mature. Other plants may have yellowish foliage early in the season, particularly in springs that are cool and wet.

In some plants, the older leaves lose their green earlier than younger leaves; the oldest year's leaves of some evergreens will lose their color just before they drop. Half of the needles of Eastern white pine, which holds its needles for only two years, will turn yellow in the fall before they turn brown and drop (Tattar 1978). Trees may look quite unhealthy during this normal transition. Many species develop yellow, orange, or red foliage in the fall as chlorophyll fades. The onset and intensity of leaf color change will vary for a given plant or clone from year to year, as summer growing conditions and fall weather vary.

Yellow foliage is usually thought to indicate a nutrient deficiency, some other soil problem, or root abnormality. Frisia locust, sunburst honey locust, and several junipers, however, are often selected for landscape use because of the normal yellow in some of their foliage. A number of shrubs, trees, and vines are vegetatively propagated for their characteristic variegated patterns of yellow or white on otherwise green leaves. Gold spot euonymus has dark-green leaves with golden yellow blotches and edges; tricolor dogwood has variegated leaves of white, pink, rose, and green in spring and summer.

On a single plant, some branches will have green foliage, others yellow, and still others variegated yellow and green (Fig. 17-1). In such

FIGURE 17-1
Most of the foliage on this Gold Spot euonymus is variegated yellow and green, but one portion has reverted to leaves that are entirely green and another portion has all yellow foliage.

cases, the variegated combination of green and yellow is the result of a chimera, a genetic vegetative mutation that involves only a sector of the tissue. Chimeras, when propagated vegetatively to preserve the variegated leaf characteristic, may not be stable. That is, leaves may form from tissue that has all (yellow), none (green), or only some (variegated) of the mutation. This mixture of differently colored branches may be more likely after a vigorous variegated plant has been severely pruned. If you wish to preserve the variegated characteristic in a specimen, remove shoots with pure yellow or green leaves as soon as you notice them; cut as close to the parent branch as possible. If you wait until severe pruning is necessary, many more or all of the new shoots will have leaves that are all green.

Occasionally the leaves on one or more branches of a tree or shrub will differ in color, size, or some other characteristic from leaves on other branches. This could be the result of a mutation. More likely, one set of the branches derives from the rootstock and the other from the top grafted onto the rootstock. You can remove the branches with the less desirable characteristics or can allow both kinds of branches to remain, depending on your preference.

Juvenile and Mature Leaves Many plants, such as ivy, acacia, and eucalyptus, have distinct juvenile and mature phases, which are reflected in the characteristics of their leaves (see Fig. 2-16). These differences are normal. If desired, juvenile types of foliage can sometimes be maintained by certain practices.

Leaf Abscission Some deciduous plants drop their leaves at the first indication of water stress. California buckeye often drops its leaves in early summer unless it is well supplied with water. These plants essentially go dormant until the following spring, with no apparent ill effects. Broadleaved evergreen plants vary in the season of year when the oldest leaves are shed. Southern magnolia, depending on the plant and the climate, may drop its oldest leaves at almost any season over a period of only a few to many weeks. Cork oak usually drops its oldest leaves just before or as new growth begins in the spring. Some evergreens may drop a few leaves all year. Leaf drop will be earlier and heavier than normal if the plant is stressed.

Most deciduous plants shed their leaves in the fall. Leaf fall may be hastened by an early frost. Some plants, however, such as beech, chestnut, and certain species of oak, retain many of their dead leaves through the winter (Tattar 1978). A few leaves drop with each wind and storm until the spring, when the rest are shed and new growth begins. Some plants, however, both deciduous and evergreen, retain dead brown or black leaves into the winter because branches and leaves were killed quickly by a disease such as fire blight (*Erwinia amylovora*).

Inspect buds and bark on such plants to ascertain whether they too are dead.

Bark Characteristics   Most young plants have fairly smooth bark; as trunks and branches grow, most bark eventually peels or cracks. The bark of canoe birch, plane tree, and Russian olive will flake or peel. The bark of black walnut, black locust, and mulberry will crack and furrow as trees grow. Although it is entirely normal for such bark to crack or peel as a tree increases in size, inexperienced observers may become concerned when they first notice this phenomenon. In fact, the deeply furrowed bark of some liquidambar trees is especially prized by some growers (Fig. 17-2).

FIGURE 17-2
The peeling bark of red gum (top left), the dark rough areas
of birch bark (top right), and the rough cracked bark of cork oak (bottom left)
are normal as these trees mature. The bark of young liquidambar branches
can be deeply furrowed (bottom right) and is particularly prized by arborists.

Flowers, Fruit, and Seeds   Most woody plants produce flowers that set fruit with viable seeds; this is the primary way by which plants have reproduced themselves through the ages. Many landscape plants, particularly shrubs, are selected and grown for their attractive flowers and fruit. On the other hand, plants have been selected and vegetatively propagated to produce showy blooms but set very little or no fruit. Some of the flowering fruit trees, such as the double and semi-double flowering cherry, peach, and plum, are good examples. Some trees that have objectionable (messy or smelly) fruit are dioecious; that is, staminate and pistillate flowers occur on separate plants. The male trees of dioecious species are fruitless. When such trees are wind-pollinated, as is the Chinese pistache, they can aggravate allergies. Fruitless mulberry, ginkgo, and honey locust are dioecious trees often selected for their fruitlessness.

On the other hand, the absence of fruit may be a concern when plants are prized for their berries, as are holly plants. This absence may occur when male plants are not close enough to fertilize the female flowers. Even when perfect flowers or both male and female flowers occur on a single plant, the flowers may be incapable of setting fruit without cross-pollination. The female flower or flower part may not be receptive when the pollen is shed. These forms of unfruitfulness are normal for some plants.

Young plants may not flower for several years. This is most common on plants whose flowers grow on spurs. When young plants with a spur flowering habit respond to severe pruning and heavy nitrogen fertilization with extreme vigor, spur formation and flowering can be delayed for many years. Plants under low light conditions can also be slow to set fruit. Trees and shrubs may also develop an alternate bearing cycle, so that a year of heavy fruiting is followed by a year in which very few flowers and fruit are produced. Pyracantha, crab apple, and apricot may develop alternate bearing cycles. The poor crop in the "off year" can worry an inexperienced grower. Of course, flowering and fruiting can be reduced or eliminated by unfavorable weather, non-parasitic disorders, diseases, insects, birds, rodents, and raccoons. It is important to distinguish between inherent and invasive causes.

A number of plants normally set fruit with no seeds or no viable seeds. Most edible figs, some grapes, and navel oranges may mature without seeds. Perfectly good seeds may fail to germinate because they have impervious seed coats or contain inhibitors that must be changed chemically or leached out.

Pollen shed by conifers, particularly pines and junipers, may be so abundant that it coats foliage with what may appear to be spores of disease organisms. Many people are bothered by the pollen, but it is harmless to plants and will soon disappear with wind or rain.

Galls   Insects, nematodes, and disease organisms can cause swellings or galls on the leaves, fruit, twigs, branches, trunks, or roots of plants. These may or may not impair the growth and appearance of a plant (see Chapters 19 and 20). On the other hand, swellings may be an integral part of certain plants or may be caused by beneficial fungi and bacteria.

Certain mycorrhizal fungi cause small roots to enlarge until they are lobed, beaded, or brainlike in shape. Mycorrhizae help plants to absorb water and nutrients, particularly in infertile soil (see Chapter 6). The small roots of legumes, such as black locust and silk tree, may have nodules containing bacteria that fix nitrogen from the air into nutrient forms. This ability can be especially beneficial in infertile soils. A number of species, such as redwood, manzanita, and eucalyptus, may develop *burls* or *lignotubers* on their branches or trunks (see Fig. 3-9). These swollen masses of woody tissue are reservoirs of latent buds. A plant with well-developed lignotubers is able to sprout quickly if its top is injured. In contrast, no sprouts have been reported to grow from crown galls (caused by *Agrobacterium tumefaciens*) or the knots caused by nematodes.

Lichens, Mosses, Algae   Certain lower order plants grow harmlessly on the bark of woody species. Algae and mosses, both green plants, are often found in moist, shady areas on the bark of trees. Some mosses, such as Spanish moss, hang from branches in wooded areas with mild winters.

Lichens are composed of algae and fungi growing in symbiotic relation for their mutual benefit. They grow in a variety of colors on soil, rocks, bark, and other moist surfaces. Lichens are common on woody plants throughout the southern and southwestern United States but are less numerous in the northern states (Pirone 1978a). Since lichens are sensitive to air pollution, they are seldom found in or downwind from large cities or industrial complexes. The presence of lichens usually indicates clean air (Tattar 1978).

An objection to their appearance is the only reason to remove algae, moss, or lichens from plant surfaces. For this purpose, a copper fungicide should be effective (Pirone 1978a).

APPROACH TO DIAGNOSIS

Diagnosing the problems of landscape plants can be very complex: The characteristics and requirements of these plants are much more diverse than those associated with agricultural crops, and landscape gardeners often attempt to grow plants at the limit of their tolerance. In a given area, however, you can become familiar with the plants commonly grown, their problems, and their responses to environment and cultural practices. Talking with experienced growers in the area can be invaluable.

Not all disorders require an analytical approach. Many are easily identified and easily remedied by routine corrective measures. Plants that wilt during a prolonged drought very likely need watering. Shoot dieback of pyracantha during a warm, wet spring is probably fire blight, caused by *Erwinia amylovora*, and infected shoots should be promptly removed. A preliminary diagnosis is often possible even without inspection of the plants involved, on the basis of plant species, season, and recent weather alone. It may be more difficult to diagnose other problems or to predict how a malady will affect a particular plant. In such cases, you will need a systematic approach for quick and accurate identification.

Plant problems can be classed or grouped according to several criteria: symptoms, causes of the problem, kinds of plants affected, or plant parts involved. Neely (1972) proposes that plant disorders be placed in one of three categories: (1) injury, (2) noninfectious (physiogenic) disorder, or (3) infectious (pathogenic) diseases. Diseases and noninfectious disorders result from continued irritation or association with a causal agent and involve normal plant responses and observable symptoms (Table 17-1). The causal agents of infectious diseases are living organisms that become intimately associated with the tissues of the host plant. Noninfectious disorders are caused by nonliving agents. An injury is the result of a single or discontinuous event that is detrimental to plant health. There are, of course, exceptions to this scheme. If, for example, the amount of a chemical involved or the speed of plant response is great enough, a problem may be considered an injury

TABLE 17-1

Stress factors that may cause plant injury or disease[a] (Adapted from Smith, W. H. 1970. *Tree Pathology: A Short Introduction*. New York: Academic Press)

Causes of Injury		Causes of Disease	
Abiotic	*Biotic*	*Abiotic*	*Biotic*
Temperature extremes	Insects		Bacteria
Wind	Mammals		Fungi
Snow	Birds		Mycoplasmas
Ice		Mineral deficiencies	Viruses
Lightning		and excesses	Higher plants
Salt		Salt	Nematodes
Air pollutants		Air pollutants	
Moisture extremes		Moisture extremes	
Pesticides		Pesticides	
Radiation		Radiation	

[a]Certain abiotic factors may cause either injury or disease (noninfectious disorder), depending on the severity of the stress factor.

instead of a noninfectious disorder, analogous to acute or chronic toxicity in animals. A girdling wire might be classed as an injury but a girdling root a noninfectious disorder, since the former is associated with a specific act while the latter occurs naturally. Both girdling problems might exist for a long period before symptoms appear. Similarly, moderate air pollution may cause a general weakening of plants, whereas a single exposure to higher levels of the same pollutants will cause serious leaf symptoms and dieback. Despite ambiguities, these three categories should assist in identifying specific problems.

Plant Injury

Mechanical, chemical, or thermal injury may result from specific actions, chemical excesses, or temperature extremes.

Mechanical Injury Mechanical injury is usually the easiest to diagnose. Leaves are tattered by wind or chewed by insects or animals; limbs are broken by wind, ice, snow, or high temperatures; bark is lost, pulverized, cut, or separated from the wood by rodents, insects, accidents, or vandals. Insects can injure plants by feeding or ovipositing on leaves, stems, trunks, or roots. They can chew, suck, mine, bore, or cause galls and dehydration of tissue. Live insects or their remains may often be present on the affected parts of the plant. Snails and slugs feed on leaves and immature shoots. They hide during the day but may betray their presence by a trail of dried slime.

Chemical Injury A wide range of chemicals, including certain insecticides and fungicides, can injure or kill leaves, particularly if temperatures are above 25°C (77°F). The most obvious and frequently observed chemical injury of landscape plants is caused by careless or inadvertent application of herbicides (see Fig. 18-7). Leaf and shoot distortion, discoloration, or death may result from spray drift, spray vaporization, or root uptake of chemicals.

Ocean spray and salt used to reduce ice on pavement can cause leaves to develop necrotic spots and margins. Since brown leaf margins may develop quite gradually from increasing soil salinity, the problem would properly be classed as a noninfectious disorder.

Thermal Injury Low temperatures injure plants more often than high temperatures. The season of the year and the stage of growth greatly influence a plant's ability to withstand the cold. Low temperatures may produce a wide range of symptoms.

Some plants are injured when air temperatures are high. If the soil is dry, the root system inadequate, or the trunk restrictive to water movement, transpiration demand may exceed absorption and translocation of water from the soil to the leaves. In this situation, the plant becomes desiccated and the leaves may wilt, become dry and brittle, or fall off. Fire is a frequent cause of high temperature injury. Plant parts exposed to fire may shrivel, blacken, and become misshapen.

Noninfectious Disorders

Most noninfectious disorders involve the entire plant and result from air, soil, and root problems. Soil and root problems usually affect leaf color and size and rate of growth uniformly. Proper assessment of shoot and trunk growth is discussed in Chapter 2. Deficiencies of essential elements often result in reduced growth and leaf chlorosis.

Mineral excesses and chemical pollutants in the soil or air, like other noninfectious problems, generally have uniform effects on a plant, but iron deficiency symptoms may be confined to particular limbs (see Chapter 11).

Excess water, natural gas, compacted soil, and fill soil reduce the level of oxygen and in some cases increase the carbon dioxide in the root zone. This reduces root activity and, if conditions are severe, can injure or even kill the roots. Lack of water can result in leaf fall or necrosis of leaf tips or margins. Circling roots or wire can girdle the trunk of a plant, causing swelling above the constriction, limiting growth (particularly of roots), and causing a physical weakness that can eventually result in death. Plants grown in relatively small soil volumes may become "pot bound." The limited capacity of the rooting medium may combine with infrequent irrigation to deny the plant moisture and nutrients sufficient for normal growth.

If temperatures in winter are below 7°C (45°F) for an insufficient period, the growth and flowering of some temperate plants may be delayed the following spring (see Chapter 4).

Infectious Diseases

Bacteria, fungi, viruses, mycoplasmas, nematodes, and parasitic flowering plants can result in unsightly appearance; poor growth; necrotic tissue of leaves, branches, and fruit; abnormal growth; and even death of an entire plant. If any part of the causal organism can be seen on the outside of the infected plant, this visible portion is called a *sign*: Rust, powdery mildew, and mistletoe are examples. Other disease organisms may cause distinctive *symptoms* or abnormalities of the host plant: necrotic leaf spots, shoot dieback, cankers on branches, mottling or mosaic patterns on leaves, and wilting (Fig. 17-3). With study and experience, you can readily identify distinctive symptoms of the more common diseases on plants in your area. When infectious diseases attack a mixed planting of woody plants: (1) Only one

FIGURE 17-3
Leaves are susceptible to many diseases:
powdery mildew (*Oidium euonymijaponici*) on euonymus (upper left),
anthracnose (*Gnomonia plantani*) on sycamore (upper right),
and peach leaf curl (*Taphrina deformans*) (lower left).
Insects, however, such as aphids (*Periphyllus*), can similarly deform leaves
like these of green ash (lower right). (Sycamore photo: McCain 1979a)

genus or species is likely to be affected; (2) only one plant in a group may be affected; (3) symptoms should be of the same type on corresponding plant parts; and (4) diseased plants will not succumb overnight (Neely 1972).

Several causal factors may be involved in a plant problem. Each may mask the typical symptoms of the others, and some may interact. There is mounting evidence that nonparasitic stress and injury can increase the disease susceptibility of plants, and diseased plants are certainly more vulnerable to environmental stress (Schoeneweiss 1973). Trees and shrubs subjected to drought or freezing stress are often prone to stem diseases, which are usually not a problem in similar unstressed plants. Plants weakened from any cause may be more vulnerable to insects, environmental stresses, and disease-producing organisms of low virulence.

A Diagnostic Approach

Most arborists use similar approaches for diagnosing the more difficult problems (Neely 1972, Pirone 1978a). The following might be used as a diagnostic checklist:

Depending on the situation, obtain information about:
  The history of the plant(s) and the area;
  Cultural practices used on affected and nearby plants;
  Any unusual weather conditions.
You will want to look at the plant(s) and surroundings for yourself. Even though an owner or caretaker may unconsciously or deliberately withhold certain facts or give incorrect information, careful questioning and observation should reduce the likelihood of your being misled.

If you are familiar with the area, you should be aware of:
  Problem soil areas;
  Previous weather conditions;
  Recent air quality levels;
  The more serious pest and disease problems.
Survey the general site and its plants:
  How many plants are affected? If more than one, are they of the same genus and species?
  Do they form any pattern? Are they clustered together or in a line? Are they grouped in relation to the topography or wind?
  Does the site appear to be unfavorable for the plants? Are there low areas where frost or drainage could be problems? Has there been recent construction, paving, excavation, or soil filling?
  Where are the gas, water, and sewer lines and the septic field? You may have to inquire about these features.
  Are plants affected uniformly or on only one side or branch?
Examine the leaves and shoots or the dormant twigs:
  Are the leaves and twigs normal in appearance and size?
  Has annual growth been normal for the last few years?
  Are there dead leaves on some branches, live foliage or none at all on others? Such dead leaves may indicate that the condition set in too quickly for leaf abscission to occur. The healthy plants of some deciduous species, however, may retain leaves long into the winter even though they are brown and dry.
Check the trunk and main branches:
  Does the bark appear normal?
  Is the trunk sound?
  Are there abnormal swellings, sunken areas, or signs of girdling?
  Does the trunk flair uniformly around the base as it enters the soil? If not, a girdling root may be a problem on the flat or indented side.

FIGURE 17-4
A London plane tree stressed and severely weakened from lack of water has many weak spindly epicormic shoots on large branches and the trunk. Low auxin levels, caused by weak growth and increased light on the bark, release latent buds.

Do vigorous suckers emerge from the trunk base? Do water sprouts emerge below the affected portion of the plant? The portion of the plant below such vigorous growth is probably healthy, and the cause is probably above ground.

Is top growth generally weak, characterized by many short twigs, 250 to 300 mm (10–12 in) long and small leaves growing along the upper surface of one or more main branches (Fig. 17-4)? Such symptoms indicate a severely weakened plant that has been in poor health for some time. The cause is probably a nonparasitic agent, a slow-acting disease, or old age.

Determine root stability and health:

If the plant is not too large, try to move the trunk to and fro. If the trunk is not securely anchored, suspect a defective root system or a poorly prepared planting hole.

If necessary, expose and inspect two or more main roots. Are roots girdling the trunk or each other? Are the roots solid? Is the bark firmly attached?

Cut the bark to expose a small area of the trunk cambium below the soil level. Do the xylem and cambium appear normal? A fanlike white fungal growth between the bark and wood indicates shoestring root rot or oak root fungus (*Armillaria mellea*).

Probe the soil under the plant with a soil tube or shovel:

Is the soil soggy, moist, or dry?

Is moisture uniform with depth?

Does the soil have a normal color and odor, or is it dark-gray with a smell like rotten eggs? The latter symptoms indicate extremely wet soil, a natural-gas leak, or landfill gas.

Take a closer look at nearby plants, particularly those of the same species:

Can you find similar symptoms on those plants? Could the condition be normal?

Could these plants be in the beginning stages of the same difficulty?

You may want to check or obtain additional information from the owner or caretaker of the plants.

For adequate examination of a plant, you may require one or more tools (Table 17-2). A pocket-knife is probably the most useful and most easily available. Even better is a folding linoleum knife, which is sturdy and versatile for a variety of examining needs. You can slice shoots and small branches to examine young wood and bark; you can cut into bark to check the condition of larger branches and the trunk; you can explore cankers, decayed areas, and insect holes to determine the extent of injury; when no other implement is available, you can even use a sturdy knife to remove soil next to the trunk for a look at the trunk root.

You can use a shovel or spade and a sturdy trowel to uncover the trunk base and roots and to examine the texture, compaction, moisture, color, and odor of soil. A soil tube is particularly useful for sampling soil below 250 mm (10 in). You can use an auger instead, but soil changes with depth are easier to assess from a soil core than from a pulverized auger sample.

Use pruning shears to collect twig and branch samples and to cut them for internal inspection. A folding tree saw will conveniently remove limbs that are too thick for shears. You may need a pole pruner to remove high twigs and branches for closer observation. A pruner with a sectional pole can be easily carried and stored.

A folding 10-power hand lens may be helpful for detecting mites or inspecting plant tissue or symptoms. With binoculars, you will be able to examine specimens that are out of reach.

TABLE 17-2

Certain tools and equipment are recommended by the Council of Tree and Landscape Appraisers for diagnosis (Neely 1979a)

Binoculars	Mallet
Chisel, gouge	pH meter
Clipboard	Photographic equipment
Compass	Pole pruner
Diameter (caliper) tape	Pruning knife
Engineer's pocket scale	Pruning shears
Entrenching tool	Shigometer
Gas detector	Shovel, spade
Glassine or plastic bags	Small saw
Hand lens	Soil auger, profile tube
Height and distance meters	Soil moisture meter
Ice pick	Vials for insects
Increment borer	

You may need a wood chisel and hammer or mallet to cut the thick bark of large trees. You can use an ice pick to probe bark and wood for decay. Use any of these three tools to tap the bark if you wish to determine whether it is sound.

With a camera, you can record plant symptoms and surroundings for future reference or evidence. Plastic bags of various sizes will keep specimens separate and prevent them from drying out. Use labels, string tags, and rubber bands to identify specimens.

You may need measuring tapes to check shoot growth, trunk caliper, or other sizes and distances. Optic meters will measure heights and distances fairly accurately, even through or around obstructions and voids. A compass may be useful for determining direction and location when you prepare a plot plan of a planting.

In some cases an increment borer is the only satisfactory way of sampling the xylem of a large tree or shrub to check its condition or to determine its growth rate for a period of years (Fig. 17-5). The

FIGURE 17-5
An increment borer is useful for measuring tree growth over a period of years.
The width of annual rings of the upper core indicates it is from a more vigorous
pine than the lower one (the beginning of each growth ring has been darkened;
the diagonal marks are near the trunk centers).

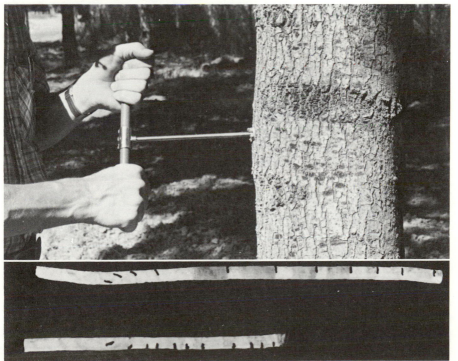

width of annual rings in conifers and broadleaved plants having ring porous xylem are easy to see; the spring wood cells are larger than summer wood cells. Annual rings in diffuse porous xylem are more difficult to see. An increment borer removes a small core that represents the cross-section of bark and wood. Take such a sample only when necessary, since the process is time consuming and injures the trunk. You will usually make such a wound in a tree of low vigor, and it will consequently be slow to close.

Instruments that measure the resistance of wood to a pulsed electrical current, as does the Shigometer®, may be used to determine the extent of wood decay, internal injury, or even the vigor of plants (Skutt, Shigo, and Lessard 1972). Minerals leak from weak and injured cells and concentrate in the intracellular spaces, thereby reducing the resistance to an electric current. Instruments that measure combustible gas, oxygen, and carbon dioxide levels in soils may be useful where you suspect natural-gas leaks, landfill gases, or exceedingly poor drainage (Leone and others 1977).

Collecting Samples

You may have to take samples of affected and healthy plant parts when you cannot diagnose the problem in the field or must verify a field observation. Choose leaf, shoot, branch, or root samples that include the transition from diseased to healthy tissue, and collect them as soon as you notice symptoms, particularly when an insect or disease is involved. In any case, gather both diseased and healthy specimens.

Keep leaf samples fresh and do not allow them to dry or to become too wet; otherwise, they may decay or foster secondary organisms that mask causes or symptoms. You can keep leaves in good condition if you keep them cool and seal them in plastic bags from which excess air has been pressed out. A dry paper towel will absorb excess moisture and reduce mold growth. Immature fruit left on leafy twigs will provide water to the twig and leaves and will thereby delay wilting.

Woody specimens of branches, trunks, and roots should include both bark and wood. Large plastic bags are convenient for transporting specimens. If samples are to be mailed for analysis, however, pack them so that they are not damaged in transit. Keep small bark and wood samples from drying by placing them in plastic bags before packing.

Take soil samples from representative sites of diseased and healthy plants, or from problem and favorable soils. You may need samples from depth increments of 150 to 300 mm (6 or 12 in).

Label all specimens with the name of the plant species, location of the plant(s), and the part of the plant from which the specimen was collected. Also include the date of the sample and the name of the collector.

Sources of Help

You can employ many sources of information to help solve difficult problems or to confirm a diagnosis. University and governmental agriculture, forestry, and conservation extension services publish many useful bulletins on the diagnosis and treatment of unhealthy landscape plants. You can phone or visit most extension services to discuss specific problems. Local agricultural commissioners, who must make sure that only healthy plants are sold and transported, are particularly knowledgeable about insect and disease problems. Retail nursery personnel are usually familiar with problems common in the area. Some park and urban forestry departments will provide assistance. The staff members of botanical gardens and arboreta can be quite helpful, particularly in identifying plants and many of their pests.

Sales representatives for distributors of landscape plant supplies will often provide advice on plant problems, particularly if some of their products can alleviate the malady. You may have to employ commercial arborists and arboricultural consultants for difficult or extensive problems; many can provide treatment as well as diagnosis. Be certain, however, that the people you consult are qualified to diagnose landscape problems. A good indication of competence is membership in a recognized professional organization, such as the International Society of Arboriculture (primarily in the United States and Canada), the National Arborist Association (United States), the American Society of Consulting Arborists (United States and Canada), and the Arboricultural Association (Great Britain).

FURTHER READING

JOHNSON, W. T., and H. H. LYON. 1976. *Insects That Feed on Trees and Shrubs.* Ithaca, N.Y., Comstock.

NEELY, D. 1972. Hints on Diagnosis of Tree Problems. *Proc. Intl. Shade Tree Conf.* 48:33–37.

PIRONE, P. P. 1978. *Tree Maintenance.* 5th ed. New York: Oxford University Press.

SMITH, M. D. 1982. *The Ortho Problem Solver.* San Francisco: Ortho Books.

STREETS, R. B. 1969. *The Diagnosis of Plant Diseases.* University of Arizona Agric. Exp. Sta. Extension Service, Tucson, Arizona.

TATTAR, T. A. 1978. *Diseases of Shade Trees.* New York: Academic Press.

CHAPTER EIGHTEEN

Noninfectious Disorders

Noninfectious disorders often lead to poor growth, damaged appearance, or even the death of plants. Such problems are more common in landscape situations than in commercial nurseries or fruit orchards, where plants are usually given better care and where sites are particularly suited to the species grown. Landscape plants may be subject to unfavorable soil, climate, cultivation practices, traffic, or a number of other hazards.

PROBLEMS RELATED TO SOIL OR ROOTS

Most noninfectious disorders are related to plant roots or the soil. You can usually determine the ability of a soil to support the adequate growth of woody plants by observing plants that grow at the site or nearby on the same type of soil. An absence of plants or signs of poor growth may indicate poor soil conditions. You may need to provide drainage, mound the soil to increase depth, amend the soil, or replace it with a more favorable rooting medium.

An established site may feature mature trees and shrubs that have begun to decline. Soil-borne diseases or pests may be responsible, or physical changes may have affected soil aeration and moisture. Changes in grade, new paving, construction, trenching, changes in drainage or moisture supply, soil compaction, and gas leaks can all affect the ability of a soil to support plant life. Other soil or below-ground problems include: mineral deficiencies and excesses, toxic chemicals, girdling roots, and rodent damage.

Poor physical structure of the soil will usually restrict root systems and reduce their activity. As a consequence, plants may receive inadequate water, air, and nutrients and may generally decline, may show symptoms of nitrogen deficiency, and may temporarily wilt on hot days. Short twigs with small leaves may grow on the upper surfaces of branches (see Fig. 17-4). It should be emphasized that such symptoms are not necessarily caused by the physical condition of the soil. The same symptoms may be induced when roots encounter disease, nematodes, or nutrient deficiencies.

Adverse changes in the soil may decrease aeration, thus reducing oxygen and increasing carbon dioxide and other gases, some of which may be toxic. Poor aeration reduces the ability of roots to absorb water and nutrients. Certain practices may increase or decrease water in the root zone. Excess water impairs aeration and can encourage root-rotting organisms to infect susceptible plants. Lack of soil moisture reduces growth, wilts leaves and young shoots, and in arid areas can produce a more saline soil. In a seeming paradox, plants may also wilt because of excess water in the soil; damaged roots may be unable to supply the top of the plant with adequate water.

Impervious Soil Coverings

Aeration and moisture in the rooting zone can be seriously decreased by pavement, fill soil, buildings, soil compaction, and frequent, shallow waterings of heavy soil. Pavement and other construction near plants can accentuate the effects of drought and high temperatures. Depending on the species and health of a plant, the texture and depth of the soil, the thickness or permeability of soil covering, and subsequent care, some portion of the soil surface under a tree or large shrub can be covered or compacted without adversely affecting the tree.

Mature trees of many species survive fairly well even when asphalt or cement covers the soil around the trunk to more than 50 meters (150 ft) in all directions (Fig. 18-1). On the other hand, a mature oak

FIGURE 18-1
This mature New Zealand Christmas tree does well in Wellington, N.Z., city center even though the soil has been covered for at least 50 m (150 ft) in all directions.

growing with moderate vigor in a clay soil may be weakened if less than one-third of the soil surface within the drip line of the tree is paved or if the entire area is covered with a thin layer of fill soil. In general, healthy plants withstand adverse soil changes better than plants of low vigor.

Symptoms   When you suspect poor physical condition of the soil, examine the tree or shrub trunk at ground level. The trunks of most woody plants flare out just above or below the original ground level, and the soil around many trunks will be 50 to 100 mm (2-4 in) higher than the surrounding soil surface. If a trunk enters the ground with little or no flare and no raised soil surface, it is very likely that soil

FIGURE 18-2
The soil level has been maintained near these two mature live oak trees
but the soil has been mounded on either side (upper left).
The turf was sprinkler irrigated. Within two years the tree
in the foreground was doing poorly (upper right) and soon died.
Two years later, the remaining tree began to weaken (lower left);
however, when irrigation was severely reduced during the droughts of 1975–1977,
the tree made a spectacular recovery (lower right).
(Photos courtesy Robert Raabe, University of California, Berkeley)

has been added to the planting area. If fill soil is coarser than the original soil, it should cause no problems unless it is compacted, unusually deep, or placed over an uneven or compacted surface. Fill soil of finer texture than the original not only will restrict air and water movement but also will form a perched water table above the soil interface, further impeding air movement into the root zone.

Impervious coverings (pavement, buildings, compacted surface soil) can greatly impede water and air movement into the soil below. Lateral movement through the soil is extremely slow without a positive flow of water from a higher elevation. Unless they are graded to drain in another direction, impervious surfaces can concentrate excess water around plants (Fig. 18-2).

Although compaction can seriously impair water infiltration and aeration in most soils, several studies have shown that compaction of some forest soils may be beneficial. The compaction of shallow, friable soil in forest campgrounds provides greater root firmness and improves the stability of trees and shrubs (Lull 1959; Magill 1970). It may also improve growth by allowing the soil to retain greater supplies of moisture and nutrients because microflora are reduced in compacted soil.

Recommendations You can protect plants from impervious soil coverings by installing drainage (see Chapter 7). You can also take less expensive steps to increase aeration and water movement: You may fracture tight soil with compressed air or bore holes into tight soil with an auger or a water jet. To remain effective, these holes must be open to the surface. Boring new holes every two or three years may be more effective and less troublesome than filling them with coarse gravel to slow the accumulation of soil and debris. Even when gravel is present, soil and roots can soon fill holes. You may help a plant adjust to improved soil conditions by thinning out the top, applying a moderate amount of nitrogen (1 kg/100 m^2 or 2 lbs/1000 ft^2), and taking care to water in proper amounts. On the other hand, young or newly planted trees should adjust to confined and paved areas with little harm if paving and buildings do not funnel water to the planting basins or increase reflected heat unduly. Such trees may of course grow more slowly than they would in less confined circumstances.

If possible, keep fill soil away from the trunks of trees and shrubs. Soil covering the bark may keep it moist and make the plant vulnerable to various crown-rot fungi. It may also seriously reduce aeration. Some tolerant species will put out new roots from portions of the trunk that are covered by soil (see Chapter 7). You can place a tree well around the trunk to keep fill soil away; see that the soil surface in the well slopes away from the trunk, that drainage is adequate, and that the soil next to the trunk is not kept unduly moist (Harris and Davis 1976).

Impaired Drainage

Symptoms   Impaired soil drainage can produce symptoms similar to those caused by impervious soil coverings. As plants mature or lose vigor, they become more sensitive to excess soil moisture. Construction on slopes or in canyons can reduce the flow of water in the soil so that, at least at certain times, water accumulates closer to the surface on the uphill side of the construction site. Soil moisture may increase more quickly than it can drain away, especially if rainfall is higher than normal or unaccustomed sources (a leaking water or sewer pipe, a newly irrigated landscape, drainage from buildings and pavement) contribute. Without proper drainage and aeration, plants may suffer root suffocation or root rot.

Recommendations   To improve plant health, identify the cause of the impaired drainage and correct it if this is economically feasible. Check soil moisture with an auger or probe. If the soil is too wet, block incoming water or improve surface drainage. It may be necessary to provide internal drainage (see Chapter 7).

Gas Injury

Plants can be injured and killed by gases in the soil; such gases may be directly toxic or may reduce oxygen to critically low levels. Manufactured gas derived primarily from coal and oil and gas produced by the decomposition of organic matter in landfills are directly toxic to plants at certain concentrations.   Although natural gas may not be directly toxic to tree roots (Pirone 1978a), a wide range of plant species have been killed or severely weakened by leaks in natural gas lines.

Manufactured gas contains a number of toxic compounds, including carbon monoxide, hydrogen cyanide, and unsaturated hydrocarbons (such as ethylene and acetylene) (Hitchcock, Crocker, and Zimmerman 1932).   Manufactured gas has been less commonly used in the United States since the 1940s, when natural gas was piped to many northern regions from the southern and southwestern states. As supplies of natural gas decrease, however, manufactured gas may become more common again.

The decomposition of organic matter, primarily wood and paper, in landfills produces methane and carbon dioxide (Leone and others 1977); the latter can reach levels that are toxic to roots. A leak of natural gas (saturated hydrocarbons, including methane) can displace much of the soil atmosphere, reducing the oxygen level. Bacteria in the soil consume methane and oxygen, producing carbon dioxide and water (Hocks 1972). This further reduces the oxygen and increases carbon dioxide levels. Garner (1974) has also found that, under the anaerobic conditions caused by natural gas leaks, bacteria will transform sulfates in the soil to hydrogen sulfide. Nitrous oxides also form under these conditions. Both substances are toxic to plants.

Since most gas leaks or landfill gases build up slowly, symptoms usually develop over several months or years and may be similar to those caused by poor soil aeration, excess moisture in the soil, and certain soil-borne fungus diseases. Therefore, diagnosis may be difficult. The timing and severity of symptoms will depend on soil permeability, plant species, and the rate at which gas has evolved. If all plants in the area are affected, this is a good indication that an infectious disease is not involved. There have been cases (Davis 1971), however, in which trees were killed at some distance from a gas leak while intervening plants were apparently unaffected. In such a case, a compacted soil layer may prevent the gas from escaping near its source and cause it to diffuse beneath the impervious layer until it escapes to the surface or dissipates in the soil. Under pavement, frozen soil, or a compacted soil layer, gas can travel 50 m (150 ft) or more. Gas has been found to damage plants up to 30 m (90 ft) from the edge of landfills (Leone and others 1977).

Symptoms Gas-damaged plants grow poorly, with sparse shoots and small leaves. Leaves and shoots may wilt and turn brown, and the most seriously affected branches may die. Roots may develop a bluish, water-soaked appearance and may also be killed. Many gaseous soils become blue-gray or black, similar in color to poorly drained swamp mud; these soils often have the "rotten egg" smell of hydrogen sulfide. The soil atmosphere can be tested for combustible gases by trained diagnosticians or employees of a gas utility. The preferred instrument for gas analysis uses the catalytic combustion principle (Leone and others). Instruments used by many gas companies work on a thermal conductivity principle that is calibrated for methane alone. A landfill gas mixture contains primarily air, methane, and carbon dioxide. Meters that work on the thermal conductivity principle may give inaccurate results for methane from landfills because high concentrations of carbon dioxide also exist. Catalytic combustion gas meters are available from the Mine Safety Appliance Company and the Bacharach Instrument Company, both of Pittsburgh, Pennsylvania.

Other methods of detection might involve the measurement of the oxygen and carbon dioxide in a problem area (Leone and others). Uneven settling and surface cracks in the soil may be clues that organic material is below the surface (Table 18-1). Gas injury is likely to become an increasingly common problem as gas lines age and corrode, manufactured gas is used in greater amounts, and more landfills are landscaped.

TABLE 18-1

Guide to evaluating gas problems in landfill soil (Leone and others 1977)

Characteristic	Anaerobic Soil	Aerobic (Healthy) Soil
Odor	Septic	Pleasant
Color	Dark	Light
Moisture content	High	Low
Friability	Poor	Good
Temperature	High	Low
Combustible gas levels	High	Very low or zero
Oxygen levels	Low	High
Carbon dioxide levels	High	Low

Recommendations   In a gaseous planting site, have the gas leak repaired, remove the weak and dead plants, and replace all discolored or seriously affected soil with uncontaminated soil.   Then replant. On a landfill site, you may need an underground system to collect toxic gases (see Chapter 10).   If large plants appear to have a chance of survival, remove the most heavily contaminated soil; the roots in that soil are probably dead anyway.   Aerate the remaining root zone with an auger, water jet, or aerator and replace the contaminated soil with a good planting medium.   Fertilize moderately with nitrogen (10 g/m^2 or 2 lbs/1000 ft^2).   To compensate for root loss and damage, thin the tops of the plants.

Loss of Roots

Roots may be lost directly when the soil surface near trees and large shrubs is lowered, when trenches are dug for utility pipes, or when storms cause breakage.

Symptoms   Plants so affected will be more subject to drought, will show symptoms of nitrogen deficiency (poor growth and light green foliage), or will be unable to stand in a strong wind.   Unless the grade has been lowered too drastically, plants will adapt to new moisture and nutrition regimes after initial shock.   A storm-damaged root system, however, may require considerable care while new roots become established.

In an area exposed to the sun and sky, removal of surface soil will force the tree to root into deeper soil, which is usually less fertile and well-aerated than topsoil.   If a mulch or low-growing plants are removed, the remaining unprotected soil will be subjected to drying and possibly extreme temperatures.   The plant may be blown down in a wind if many large surface roots are cut. Lowering the soil level will have less severe effects in a deep, well-drained soil of a sandy texture than in a shallow, poorly drained clay. Vigorous, moderate-size young trees of species that normally root deeply should be able to withstand soil removal better than shallow-rooted large trees of low vigor. The less soil removed, the smaller will be the effect on a tree. Several cultural practices will materially enhance the survival and growth of plants from which soil has been removed.

Recommendations   Remove as little soil as possible and avoid injury to large roots. A raised planter on a level area or a terrace on a slope can retain the original soil surface around plants (see Figs. 7-4 and 7-9). In some cases, large roots can be exposed for a distance of 1 to 3 m (3-10 ft) from the trunk provided that they then continue below the lowered ground level. Exposed roots can add visual interest to the landscape. Newly exposed roots must be protected from direct sunlight and freezing, however, until they have had time to acclimate. Protect the bark of exposed roots from foot and vehicular traffic.

You can usually remove one or more large surface roots without serious damage if vertical or other horizontal roots grow below them on the same side of the plant and if the tree is not generally shallow rooted. Each situation, however, must be appraised individually. Cut the roots smoothly and thin the top of the plant to compensate for the loss in water and nutrient uptake and to reduce the weight and wind resistance. No treatment of the wound has been shown to be beneficial. When a plant has had more than 20 percent of its root system disturbed, water it more frequently and apply nitrogen moderately (1 kg/100 m^2 or 2 lbs/1000 ft^2) on undisturbed or fill soil areas. To minimize injury to shallow feeder roots, apply only half as much nitrogen to excavated areas. After the soil level is lowered, you can protect plants further by placing a 50 to 100 mm (2-4 in) organic mulch over the newly created surface. The temperatures of the surface and roots will thus be moderated, and moisture conserved.

Drought

Plants or portions of plants can become deficient in moisture for a variety of reasons: lack of soil moisture, poorly aerated or frozen soil, injured or diseased roots, which reduce water uptake, and injured or diseased trunk and branches, which restrict water movement. Sometimes an excessively large leaf area will develop in cool spring weather; if the longer days of later spring are accompanied by high temperatures and wind, the root system may not be able to supply the top with enough water to prevent wilting or leaf scorch.

Symptoms   The first symptom of inadequate moisture in the top
of a plant is a wilting of the leaves and the tips of young shoots. These
symptoms will commonly appear on healthy plants during warm afternoons in late spring and early summer. If the leaves and young shoots
are not turgid again by the following morning, however, lack of moisture could seriously affect the plant. Should wilting persist or recur
frequently, the areas between veins and on the margins of leaves may
fade and turn brown; entire leaves may turn brown and fall. Pirone
(1978a) indicates that the species most susceptible to leaf scorch are
maple, horse chestnut, ash, elm, and beech. Lack of moisture usually
affects all the leaves on one or more branches. The leaves affected first
and most severely are those exposed to the afternoon sun and prevailing
winds. Older leaves, leaves that are small, thick, and rigid, and most
conifer leaves may not wilt visibly but may turn brown entirely or just
at the tips or margins.

A number of broadleaved evergreen shrubs and conifers can
withstand considerable periods of moisture stress without becoming
unattractive (Sachs, Kretchun, and Mock 1975). In two trials in
California coastal valleys, little or no leaf injury occurred even when
transpiration was sufficiently reduced to produce leaf temperatures
6° to 15°C (11°–27°F) above ambient. The exposed leaves of plants
with adequate soil moisture are usually at or slightly below ambient
air temperature. In the California experiments, tree and shrub growth
was markedly reduced; this can be an advantage when plants near their
desired landscape size.

Relatively few trees and large shrubs were lost in California during
the 1975–1977 drought, even though annual rainfall was below 250 mm
(10 in). Some surviving plants had been irrigated and some unirrigated
before the drought; most could not be irrigated when water was scarce.
Most trees that were lost had shallow root systems or were growing on
shallow soil. Whether irrigated infrequently or not at all, trees and large
shrubs growing in Mediterranean climates (with essentially dry growing
seasons) can survive without rain or irrigation for relatively long periods,
as contrasted to similar plants that grow in areas where summer rainfall
occurs. The latter plants may have less extensive root systems than
plants accustomed to arid seasons. It may be that some plants can grow
in soils of limited moisture-holding capacity if they receive summer
rainfall but are seriously affected when moderate drought occurs.

Recommendations   If plants cannot survive in good health and
appearance on their own during dry periods, one or two good waterings
during the growing season will usually keep them attractive. The
number of irrigations and the amount of water required will depend on
the water-holding capacity of the soil and the rooting depth of the
plants. One good irrigation that brings most of the rooting zone to
field capacity would be better than the same amount of water applied

more frequently in smaller doses. Observations (Harris and Coppock 1977) indicate that, in areas where dry summers are normal, most woody plants with extensive root systems can survive on only half the water they would otherwise transpire.

Girdling Roots

Trees and shrubs may be weakened or killed by roots that girdle the trunk or main roots (Fig. 18-3). Circling roots usually start in the nursery, particularly when plants are container grown or balled and burlapped. Plants that are started in seed flats and liner pots may have kinked and circling roots, not readily visible, that are difficult to correct at planting.

The outer circling roots of container-grown plants can usually be corrected at planting. Make the planting hole large enough or shorten the roots so that they can radiate out from the trunk or root ball without undue bending. The soil beyond the planting hole should be friable enough that roots will not circle in the planting hole before growing out. If the plant is healthy, roots grow wherever conditions are favorable. The branches of some roots may grow back toward the trunk or large basal roots, particularly if they encounter impervious or poor soil farther out from the trunk. Such roots may create a girdling problem as they enlarge (Fig. 18-3).

FIGURE 18-3
Not all circling roots cause serious problems, as is evidenced by the oak tree at the left. The Monterey pine trunk and roots below, however, exemplify the fate of more than 75 trees on one northern California golf course; the trunk diameter at the girdling root is less than half that immediately above. Circling roots were not corrected when the trees were planted.

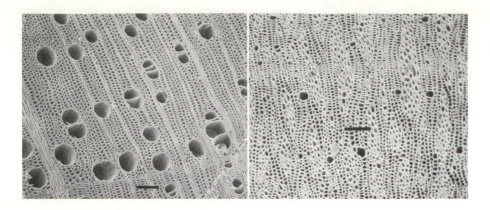

FIGURE 18-4
Transverse sections of Norway maple stem: normal xylem (left);
malformed xylem from stem portion compressed by a girdling root (right).
Scale of each bar is 100 μm. (Electron photomicrographs courtesy
George W. Hudler and M. A. Beale, Cornell University, 1981)

When roots touch one another and continue to enlarge, each will deform at the point of contact, and each will slow the caliper growth of the other (Fig. 18-4). Initially, the movement of food from the top to the roots is restricted; this in turn reduces root growth and function below the point of deformation. Less water and fewer nutrients will reach the top part supplied by the affected roots.

Symptoms   The symptoms of a weakened, ineffective root system have been described earlier. Except in newly planted plants, the symptoms of girdling roots, which are identical to those described, may take years to develop. A gradual decline in growth and appearance occurs as the girdling becomes more intense. It is most apparent in autumn, when leaves color and fall earlier than is normal.

If you suspect girdling roots, inspect the trunk where it enters the ground. A normal flare at the ground will rule out girdling roots. If the trunk enters the soil vertically or you notice an indentation just below the ground around part or almost all of the trunk circumference, a girdling root or roots are likely. If the entire trunk circumference enters the soil vertically, the problem is more likely due to soil fill. If a trunk is completely encircled by a girdling root or roots, it will be larger in caliper just above the girdle than below. Do not be misled by a swelling of the trunk above a graft union, for that occurs often. Graft swelling may indicate a mild incompatibility between the rootstock and top but is seldom a problem.

Girdling roots are often visible at the soil surface. If they are not, remove soil carefully from around the trunk until you find girdling roots or encounter the main horizontal roots. If unencumbered roots radiate out from the trunk, girdling roots are not a problem.

Recommendations   Girdling roots can be removed most easily with a chisel and mallet. Cut each at its attachment to the trunk or to a large root and again beyond the girdled area. Remove the cut portion. Painting the cut root surface is of little or no value. When all of the girdling roots have been cut, use the soil removed to fill the excavation back to the original soil level. If the tree has been weakened by the girdling, thin out the top of the tree accordingly, apply about 10 g of nitrogen per m^2 (2 lbs/1000 ft^2), and water wisely. The girdling may be so severe that the girdled trunk cannot support the tree in a storm (Fig. 18-3). Besides thinning the top to reduce wind resistance, you may have to guy the tree (see Chapter 9). You may even have to remove it if the tree is so severely weakened by the girdling that it constitutes a hazard.

Leaf Scorch

Symptoms   A number of conditions can cause the death of scattered areas between the veins or at the tip and along the margins of leaves (Fig. 18-5). Affected areas become light or dark brown and appear scorched. In severe situations, the entire leaf dries up. The primary causes are drought, a marked increase in light intensity, high temperatures, frozen soil, or drying winds, which either decrease the amount of water available or increase transpiration. Potassium deficiency or excesses of boron, chloride, or sodium can also result in leaf scorch.

Drought and increased transpiration were discussed earlier in this chapter. Light intensity or temperatures can increase markedly when a plant is moved from a cool, shady location to one of more intense sunlight, when the top of a plant is pruned, when a structure or another plant is removed, or when the sun nears the summer solstice.

Recommendations   Nothing can be done to correct the scorched leaves. You may keep the condition from becoming worse by ensuring that the plant has sufficient moisture and is not subjected to a hot, drying environment. To minimize future leaf scorch, encourage root growth by fertilizing and wise watering. If the plants are irrigated, be sure water reaches most of the roots. If the problem recurs, you might thin the top to reduce transpiration.

Marginal or tip burn on leaves may be due to a mineral imbalance. Potassium deficiency can cause yellowing of leaves, followed by scorching of the margins or entire leaves. See recommendations for potassium application in Chapter 11. Particularly in areas of low rainfall, excesses

FIGURE 18-5
Conditions that cause leaf scorch cannot be easily distinguished by symptoms alone: desiccation (top) of Japanese maple (left) and of holly; salinity (center) on tulip tree (left) and hibiscus; and excess boron (bottom) on southern magnolia (left) and coast redwood. (Salinity photos courtesy Leland François, U.S. Salinity Laboratory, Riverside, California)

of boron, sodium, and chloride can cause leaf scorch and reduced growth. Soil salinity and its correction are primarily irrigation problems (see Chapter 12). Excess boron is somewhat similar to excess salinity, in that the most feasible solution is leaching with good quality water. Successful leaching depends on adequate drainage.

Leaf Chlorosis

The general yellowing of a leaf, often referred to as chlorosis, can be caused by a variety of factors and is often associated with poor growth. Chlorosis may be caused by disease or insects, excess soil moisture, cold weather, air or soil pollutants, high levels of certain minerals in the soil, or nutrient deficiencies (Pirone 1978a). Camphor trees and pin oak are particularly sensitive to iron deficiency. Even if the soil contains iron, its availability decreases rapidly at a soil pH above 7.0 (see Chapter 6).

Animal Damage

In snowy areas, the roots of young plants and the trunks of large plants may be injured by pocket gophers, moles, and mice. Gophers can also feed on tree roots from their underground burrows. For most effective control, place Macabee traps in the main burrow runs. For large areas infested with gophers or for landscape plantings that border upon open fields and farmlands, underground tunnels similar to gopher burrows may be formed with special equipment and poison bait placed at intervals (Marsh and Cummings 1976). This effective control is commonly used in California in soil that is free of tree and shrub roots.

Moles are less serious threats than gophers but they disfigure lawns and ground surfaces with their burrows, heave up small seedlings, and eat or break small roots. Moles are also more difficult to control than gophers. Traps (choker, harpoon, or scissor-jaw) and poison baits are used in control attempts.

When a coarse organic mulch gives mice an opportunity to tunnel to the trunks of plants, they may girdle the bark around trees and shrubs, particularly when snow restricts their access to other sources of food. Therefore, mulch must be kept away from trunks of plants. You may place a ring of gravel about 200 mm (8 in) wide and thick around each trunk, or use a sleeve of hardware cloth or fine wire mesh and bury the lower portion of the sleeve around the base of each plant. Either barrier should keep mice and other rodents from burrowing to trunks.

Squirrels, porcupine, beaver, cattle, deer, elk, horses, and other grazing animals can also severely injure bark of trees and shrubs. You can fence trees individually or in groups to keep larger animals away. A hardware cloth sleeve around the trunk will protect a plant from small and large animals. Dogs can injure the bark of young trees by urinating on it. A plastic sleeve around the lower trunk will protect a tree from such injury, though if the problem is serious tree roots may be injured; a repellent may be needed. The trunk should be protected from the sun when the sleeve is removed.

Sapsuckers and woodpeckers disfigure the trunks of many trees when they search for insects. Woodpeckers normally work only on decayed and dead wood; the only harm they cause is visual. Sapsuckers may damage live bark, even girdling small treetrunks. The most practical control appears to be the removal of nesting places in trees that have cavities or internal decay (Tattar 1978).

WEATHER-INDUCED PROBLEMS

Temperature and moisture are the environmental factors most influential in determining the distribution and well-being of plants. Irrigation can modify the influence of rainfall and evapotranspiration. Plants grown near their climatic limits are usually more subject to the vagaries of weather, particularly if temperatures fluctuate markedly near critical levels. Chapter 4 discusses the influence of climate on plants, the types of temperature-related plant injuries, and the ways many weather-induced problems can be avoided or minimized.

AIR QUALITY

Pollution

Symptoms Leaves are the plant parts most likely to show symptoms of air pollution injury. On broadleaved plants, leaves may develop interveinal necrotic areas, marginal or tip necrosis, stippling of the upper surface, or silvering of the lower surface (Table 18-2). Conifer needles affected by pollution may exhibit patterns of necrosis and chlorosis. These are symptoms of acute toxicity, since they usually result from a short exposure to high concentrations of a gaseous pollutant.

Symptoms of acute injury can be highly variable, depending on the plant species and stage of growth, the type and concentration of pollutants, the length of exposure, the amount of moisture in the leaves, humidity, light, temperature, wind, and other factors (Scott 1973) (Fig. 18-6). Different leaf tissues respond with different sensitivities to the gaseous pollutants that enter the leaves through the stomates (Table 18-2).

Long periods of exposure to low levels of pollution result in chronic injury, which may be hard to distinguish from other forms of poor growth. Affected plants will be low in vigor and fruit yield, leaves will be pale green, and leaves and fruit will color early and more brightly and will drop earlier. Affected plants are usually more easily subject to other disorders. Chronic injury from gaseous or particulate

TABLE 18-2

Four major air pollutants: common sources, concentrations toxic to sensitive plants, and plant symptoms (after Scott 1973 and USDA 1973a)

	Sulfur Dioxide	*Fluoride*	*Ozone*	*PAN*
Source	*Point*	*Point*	*Atmospheric*	*Atmospheric*
	Atmospheric	*Atmospheric*	Combustion of coal and oil; auto exhausts are the primary source.	Similar to ozone.
	Power plants and industries that burn coal and high sulfur oil or use compounds in processing.	Industries that use catalysts and fluxes containing fluorides.		
Toxic Levels	0.7 ppm for 1 hour 0.18 ppm for 8 hours	1 ppb for 24 hours	0.05 ppm for 4 hours	10 ppb for 4 hours
Symptoms Broadleaved plants	Light brown interveinal necrotic areas on both leaf surfaces.	Marginal or tip chlorosis or necrosis.	Flecking or pigmented stippling on upper leaf surface.	Silvering or glazing of lower leaf surface.
Conifers	Needle tip necrosis and chlorotic banding.	Nonspecific needle tip necrosis and chlorosis.	Needle tip necrosis, tip chlorosis, chlorotic banding and mottling.	Whitish necrotic areas on the needle portion most sensitive at the time of exposure.

FIGURE 18-6
Right birch leaf exhibits typical interveinal loss of color
followed by necrosis after exposure to sulfur dioxide (upper left).
(Photo courtesy Edward Lee, USDA Beltsville Agriculture Research Center)
Ozone causes mottle of ponderosa pine needles (upper right).
Susceptibility to natural smog (primarily PAN) in southern California
of petunia leaves of different ages from the same plant
(upper left is youngest leaf formed after critical smog level;
oldest leaf at lower right). (Pine and petunia photos courtesy
Clifton Taylor, University of California, Riverside)

air pollutants occurs when pollutant concentrations are high enough to
interfere with or change certain plant processes.

Airborne chemicals have been known for some 150 years to injure
and kill plants. The earliest reports were of sulfur dioxide injury from
coal-burning factories in Europe (Scott 1973). The major phytotoxic
air pollutants are the gaseous compounds: sulfur dioxide, the fluorides
(HF, SiF_4), ozone, and peroxyacetyl nitrates (PAN). Ammonia, the
chlorides (HCl, Cl_2), ethylene, hydrogen sulfide, and nitrogen oxides

are also toxic to plants but are not as widespread or as frequently encountered at toxic levels. Carbon monoxide, a major product of gasoline combustion, does not cause plant injury in ambient concentrations (USDA 1973a).

Sulfur dioxide and the fluorides are generally considered point-source pollutants, since injury usually occurs within 15 km (10 mi) of the source (Scott). Sulfur dioxide is produced by the burning of coal, refining of petroleum and natural gas, and smelting and refining of ores. Fluorides are released when fluoride-containing materials are heated or chemically treated in the production of steel, aluminum, ceramics, glass, and phosphate fertilizers. Concentrations of these pollutants have been reduced or at least contained by the use of specialized filters (which reduce toxic emissions) and tall chimneys (which allow greater dilution before susceptible plants are reached).

Ozone and PAN are oxidants formed in the atmosphere by photochemical reactions between hydrocarbons and nitrogen oxides in the presence of sunlight. Photochemical oxidants can cause plant injury as far as 120 to 200 km (75–120 mi) from the origin of the primary pollutants. The hydrocarbons and nitrogen oxides come from the combustion of coal and petroleum; vehicular exhausts constitute the major source.

Although the sources of the major air pollutants are the activities of people, natural phenomena release considerable quantities of these chemicals into the atmosphere. Volcanoes release sulfur dioxide. Ozone and other pollutants are generated photochemically by hydrocarbons volatilized from vegetation. Rasmussen estimated in 1972 that the world's forests emitted 192×10^6 metric tons (175×16^6 tons) of reactive hydrocarbons annually, six times the emission produced by human activities (Rich 1975). In rural areas of upstate New York, Stasiuk and Coffey (1974) found ozone concentrations as high as those in New York City. Ozone levels remained high at night in the rural areas, but dropped to low concentrations after sunset in cities. Emergence tip burn of eastern white pine, now associated with ozone injury, was known before 1900 (Hepting 1968). Since the hydrocarbons produced by human activities at that time were only a fraction of what they are today, the toxic oxidants must have had natural origins. In some situations, natural sources may even account for toxic pollutant levels.

Particulates in the air may have chronic or acutely toxic effects on plants. Particulates are seldom acutely toxic when they are dry, but dew or fine mist may increase their toxicity. Rain may wash the particulates from leaf surfaces before injury occurs. Even from a single source, particulate matter is not homogeneous but is a mixture of different chemicals, so generalizations are difficult. Sulfuric acid mist may form in the air when sulfur dioxide oxidizes to sulfur trioxide, which quickly hydrates to form aerosols. Necrotic spots on the upper surfaces of leaves have been observed in the Los Angeles area after periods of heavy air pollution accompanied by fog (Middleton, Darley, and Brewer 1958).

When particles settle on leaves, shoots, and flowers, surfaces become coated and reduce the amount of sunlight reaching the plant. In some cases stomates may be clogged with dust (Lerman and Darley 1975). Photosynthesis can be greatly reduced, leading to poor growth, leaf drop, and death of twigs or, in severe cases, the entire plant.

The source of particulate pollution is usually local industrial activity, refuse burning, or transportation. Specific particulates include: kiln dust from cement plants; fluoride dusts; soot; sulfuric acid mist; lead, rubber, and oil particles from vehicles; and particulates from various types of metal processing (Lerman and Darley). Cement-kiln dust primarily coats the upper surfaces of plant parts. If it remains dry, the dust will be removed by wind or hard rain. On the other hand, fog, dew, or light rain will form a crust that may continue to build up. Along heavily traveled streets and highways, plants can become coated with particles of oil, rubber, lead, soot, and soil. Along the Pasadena freeway in southern California, roadside plantings are close to the heavily-used traffic lanes. Plant leaves and stems become grimy, unattractive, and sickly if they are not cleaned. Pirone (1978a) reports that trees planted along busy streets were healthier in the middle of the block than trees near intersections with traffic lights, where autos and buses have to stop or slow down and then accelerate.

Recommendations   The most obvious solution is to reduce pollutants at their sources. Filters, modified processing, and strategic industrial location can accomplish much. Much more is needed, however, to protect human health as well as the health of plants. In areas that are likely to be highly polluted, plant trees and shrubs that are most tolerant of the expected pollutants (Appendix 7). You can reduce particulates on shrubs and small trees by hosing them with water. If the leaf coating is oily or adheres to the leaves, use a detergent solution.

In regions with arid summers, plant sensitivity to gaseous air pollutants can be influenced by irrigation practices (Rich 1975). In almost all reported observations, moisture stress before and during exposure to toxic pollution levels has reduced injury. In a similar manner, low relative humidity has increased the resistance of plants to air pollutants. In both situations, stomata close more quickly and completely, thereby reducing pollutant uptake. Experiments have shown, however, that moisture stress following toxic exposures increases subsequent injury (Rich). Rich also cites an observation that dew on tobacco leaves increases resistance to ozone by impeding the movement of gases into the leaves. You may, therefore, be able to reduce pollution injury to sensitive plants by spraying them with water during and after toxic exposures. Apply enough water to create some runoff but not enough to wet more than the surface soil, at least if the pollution level is likely to continue or recur soon.

Certain fungicides and growth retardants increase the resistance of plants to air pollutants. Several of the dithiocarbamate fungicides have been reported to protect plants against ozone injury (Rich 1975). These materials, however, gave protection only when the lower surface of the leaves was treated (where stomata are located), and protection was restricted to the portion of the leaf covered. Benomyl and closely related compounds act systemically to protect plants against pollutants. Rich reports that naturally occurring oxidant injury on azaleas was suppressed by soil drenches and foliar sprays of benzimidazole and oxathiin compounds. To date, these fungicides are still used primarily for disease control, and the protection they afford against pollution damage can only be considered an added benefit.

The subapical growth inhibitors, such as daminozide (SADH, sold as Alar® or B-9®) and CBBP (Phosfon®), increase the hardiness of responsive plants in the face of a number of environmental hazards, including ozone and sulfur dioxide (Cathey 1975). Only the growth that develops after treatment has a special resistance to ozone. One large ground-cover nursery in California routinely sprays most of its plants with those chemicals as much to increase plant hardiness as to retard growth. These responses raise the possibility of a practical treatment to minimize pollutant injury to plants.

The relation of nutrition and susceptibility to air pollution is unclear. Most observations have involved annual herbaceous plants, primarily tobacco and vegetables. Plants grown at moderate fertility levels are usually more tolerant than plants grown at either lower or higher fertility levels (Rich 1975).

Some landscape developers have proposed tree and shrub plantings to reduce air pollution within a "sanitary protective area" for human activity (Bennett and Hill 1975). An absorptive canopy of vegetation would have to be dense enough that exchange between polluted air above and less polluted air within the canopy would be no greater than about 150 m/sec (0.1 mi/hr). Exterior plants would displace the wind and oncoming polluted air upward. These plants should have dense canopies and tolerance to the pollutants. Plants within the protected area should be selected for their ability to remove pollutants from relatively calm air. They might be somewhat less tolerant of air pollution than exterior plants. More information is needed before we can judge the feasibility of such a proposal.

CHEMICAL INJURIES

A number of chemicals used for various purposes in and around landscape plantings can cause poor growth, injury, and sometimes plant death. These may act by direct application, through toxic levels in the soil, or as air pollutants.

Herbicides

Herbicides are usually the most phytotoxic of the pesticides used around landscape plants. Herbicides are designed to inhibit or kill undesirable plants but not others. Non-targeted plants, however, can be injured by too concentrated a mixture, too heavy an application, careless application, or subsequent movement in soil or water. In urban landscapes, damage is likely to result from the misapplication of products that combine fertilizers and herbicides or from the drift of sprays. In rural areas, damage may be caused by farm chemical applications. The chemical 2,4-D (2,4-dichlorophenoxyacetic acid) has injured more non-targeted plants than almost any other herbicide. There are three reasons: (1) 2,4-D is systemic, so that it may affect the whole plant even though it comes in contact with only a part of the plant; (2) early formulations (primarily esters) were quite volatile; and (3) since 2,4-D is used to eliminate broadleaved weeds in lawns, it can easily affect adjacent trees and shrubs.

Symptoms   Pirone (1978a) noted that pin oak produces distorted leaves the spring after a fall application of 2,4-D to nearby lawn. The leaves and shoots of affected broadleaved plants may exhibit epinasty, depending on concentrations of the herbicide, plant species, temperatures, and stage of growth (Fig. 18-7). The leaves become cup shaped, with margins curling up or down. Leaf petioles bend down, giving the plants a wilted appearance. Shoot tips become twisted. The leaves of seriously affected plants lose their bright green color and may die. Plants will recover, however, unless they have received an exceptionally heavy dose of the herbicide.

Although most of the preemergent herbicides can be sprayed on the foliage of landscape plants without injury, toxic amounts can accumulate in the root zone. That is particularly true in sandy soils when rainfall or irrigation is heavy or where the watering basin for a newly planted tree or shrub has been treated with a preemergent herbicide. Subsequent watering or rain can erode the surface soil on the berm into the basin, leaching toxic concentrations into the root ball. Plant symptoms vary with the herbicide, its concentration, and the plant species, but in most cases affected plants will exhibit veinal or interveinal and marginal chlorosis in the leaves, followed by necrosis and leaf fall.

FIGURE 18-7
Herbicides can injure nontargeted plants if care is not taken.
Carob shoot deformed by 2,4-D drift (left). Simazine injury to pyracantha
(upper right). Green ash leaves injured by excessive application of Amitrol to the
soil (lower right). (Carob and pyracantha photos courtesy Clyde Elmore,
University of California, Davis)

Soil sterilants used to prevent weeds in and around pavement, around buildings, and along railroad rights-of-way, ditch banks, fences, and roads can seriously affect plants that grow nearby or are subsequently planted in treated soil. Water runoff and erosion can carry the herbicide a considerable distance, injuring plants at some distance from the treated area.

Recommendations You can minimize the chances of 2,4-D injury by using low-volatile acid or amine formulations and applying a coarse spray at low pressure on calm, cool days or during the cool part of the day. Spray equipment used for herbicide application should not be used to apply other materials to landscape plants. If leaf symptoms appear within a few weeks of preemergent herbicide application, reduce further injury by removing the top 50 mm (2 in) of soil around the plants. Stir a 25-mm (1-in) layer of activated charcoal or 50 mm (2 in) of finely divided compost or fortified sawdust into the 25 mm (1 in) of remaining surface soil so that it may adsorb some of the remaining herbicide. Sprinkle water lightly to moisten the mixture of soil and organic matter. Bring the soil up to its former level by adding clean soil. Do not irrigate for several days and take precautions to minimize surface erosion during rains or irrigations; such erosion might accumulate the herbicide close to plants. Later, you may heavily irrigate sandy soils to leach unadsorbed herbicide from the root zone or to dilute it. Remove severely affected young plants and plant replacements after you have substituted clean soil for the contaminated soil.

Chapter 10 contains information about planting in soil with toxic levels of residual herbicide.

Salt

Symptoms Sensitive plants exposed to saline soils or sprays will usually show marginal and tip leaf scorch somewhat similar to boron excess (see Fig. 18-5). If the salt comes from the soil, injury tends to be more severe on the portion of the leaf that has the greatest concentration of stomata or vein endings. Spray-borne salt usually accumulates on the windward side of the plants. As severity increases, entire leaves are affected and may drop; buds fail to grow, and dieback occurs. In extreme cases the plant dies. The first and most common symptom of salt injury, however, is simply a slowing of plant growth that often goes unnoticed or is thought to be caused by something other than salinity. Reduced growth is usually accompanied by early fall color and defoliation. The extent of injury will depend on the species of plant, the concentration and kind of salt(s), the medium through which salt reaches the plant, soil moisture and rainfall, and many other factors. Salts that affect plants may be present in the soil initially or may be added by irrigation water, fertilizers, salt water intrusion, or salts applied to icy or dusty roads, or ocean spray. Soil salinity can become an especial problem in low rainfall areas subject to irrigation and fertilization (see Chapter 12).

Millions of metric tons of sodium chloride are used each winter in the United States to de-ice roadways and sidewalks. The salt is mixed with sand, volcanic cinders, or other gritty carriers to promote uniform distribution and improve traction. Even though the salt is applied on roadways, most of it eventually ends up in the soil or on plants. Melting snow and rain runoff carry the salt to low areas along the roadway. Salt is thrown from the road during snow removal and may drift considerable distances in the turbulent mists created along high-speed roads. As would be expected, the amount of salt in the soil and on plants and the severity of plant injury decrease with distance from the road, lower traffic speed, and lower traffic volume (Lumis, Hofstra, and Hall 1973). In New Hampshire, Rich (1971) found that most trees affected by salt were within 9 m (30 ft) of the road.

The effects of repeated annual applications of salt depend on the amount of salt applied, the amount blown on plants or concentrated by runoff, the soil texture and drainage, and the rainfall. In Maine, Hutchinson (1968) analyzed soil to a depth of 150 to 200 mm (6–8 in) at different distances from the road. In the first five years of salting, most of the increased sodium and chloride was within 5 m (15 ft) of the road. After 18 years, however, the levels were much higher and extended at least 15 m (45 ft) from the road; salts were about half as concentrated at 15 m as at 2 m (6 ft).

Plants seem to be more seriously injured by de-icing salt that is thrown, splashed, or allowed to drift onto them than by salt that accumulates in the soil (Lumis, Hofstra, and Hall 1973). Injury will be more severe on the downwind side of a road and on the side of a plant that faces the road. Lumis and coworkers found that plants covered with snow or in sheltered locations were not injured. Damage became evident in late winter on coniferous species but not until spring growth began on deciduous species.

On conifers, the injury appears as tip necrosis of the needles and progresses toward the base as injury increases. Conifers with more cuticular wax on the needles are less injured by salt spray. Lumis, Hofstra, and Hall also found that deciduous plants with resinous buds, such as horse chestnut and cottonwood, were resistant to injury, as were those trees whose buds are submerged in the twig, such as black locust and honey locust. Within a given genus, plants with naked buds were injured more than plants with scaly buds.

Since the usual symptoms of salt injury are poor growth and scorched leaves, it is easy to confuse this problem with a host of other ailments. In the California Sierra, poor growth and needle browning of conifers along salted highways have been caused as much by borers or fill soil around the trunks as by salt (Leiser and others 1980). In Massachusetts, symptoms resembling those caused by salt injury have been associated with a lack of moisture (Pirone 1978b). Careful observation and leaf analysis may be needed to ascertain the cause of poor growth and leaf scorch.

Not only are sodium and chloride ions toxic to plants but high concentrations of sodium disperse colloidal particles and produce tight, poorly drained silt and clay soils. Poor drainage aggravates the salinity problem by reducing leaching. High sodium levels will have little effect on the physical properties of sandy soil and decomposed granite, however.

Calcium chloride is sometimes used on dirt and gravel roads to keep the dust down. Like de-icing salt, calcium chloride may be carried to nearby vegetation by rain runoff, splashing from puddles, or air turbulence caused by passing vehicles. Calcium chloride is not nearly as toxic as sodium chloride but can cause similar injury to plants.

Plants growing near the sea or other bodies of salty water can be injured by salt spray and salt blown by the wind. Again, plants nearest the water are affected most severely. Hurricanes and typhoons can carry salt considerable distances inland. Injury from salt spray will vary with the seasonal development of leaves, particularly on deciduous plants. Deciduous plants will be little affected by wind-borne salt just before leaf fall. On the other hand, evergreen plants may be more seriously injured during this period.

Recommendations   Salt deposits on leaves and twigs can be washed off with a garden hose or high-pressure sprayer if heavy rain does not follow the salt buildup. Severely injured plant parts should be pruned off. In soil that is moderately or well drained, you can leach salt by applying large amounts of water (see Chapter 12). Frequent irrigation in summer will keep salt at lower concentrations in the soil.

Long-term measures to minimize salt injury may be more effective. Select species that are relatively salt tolerant, but be aware that plants tolerant of salt in soil may not tolerate salt spray, and vice versa. Grade salt-treated paved areas and adjacent planting areas and install barriers so that surface drainage water does not accumulate near plants. Apply de-icing salt sparingly and only when needed. To reduce slipping on pavement near plantings, use sand, decomposed granite, or sawdust whenever possible. Plant in the spring, so that plants may grow for a full season before they are subjected to winter de-icing salt. If the soil structure deteriorates because of high sodium, apply gypsum (calcium sulfate) to improve structure and drainage and to facilitate leaching.

Boron

Symptoms   In some areas, primarily those regions with low rainfall, excess boron in the soil can cause serious plant injury and even death. The leaves of affected plants will have dark or necrotic margins and tips (see Fig. 18-5); they may also have dark interveinal necrotic spots. Leaves will progressively worsen from the tip to the base of affected shoots. On some species, shoots will swell and crack below the buds. Boron builds up in soil primarily near septic tank leach fields or when irrigation water contains more than 1 ppm boron. Areas with excess boron may be located within a few kilometers (a few miles) of boron-deficient areas. By wise irrigation of soils with good drainage, you can grow quite a variety of plants with water containing as much as 2 ppm boron.

Recommendations   The only way to overcome excess boron is through leaching with good-quality water. This treatment is only moderately effective in sandy soil. A few woody landscape plants are known to be relatively tolerant of the high boron levels found in some soils.

Other Chemicals

Pesticides can sometimes injure plants. Since many species may be grown in a single landscape, take care that treating one species for pests does not injure sensitive plants nearby. Read pesticide labels carefully to learn which species are sensitive and what concentrations and timing are recommended. Do not assume that if a little is good, more will be better. Plants of low vigor are often especially susceptible to spray injury. Lime sulfur and sulfur sprays and dusts are more likely to cause injury if temperature and humidity are high.

Cinders of burnt coal should not be used as a soil covering near plants, since they contain several toxic chemicals, particularly if they have not been leached by rain or sprinklers. Rain or sprinkler irrigation could leach these chemicals into the root zone.

MECHANICAL DAMAGE

The bark of trunks is the part of trees most often subject to mechanical damage, especially in spring and early summer, when the cambium is active and the bark "slipping." A slight blow at that time can cause the bark to loosen or tear off. Autos and lawnmowers, the most common cause of bark damage, can debark a tree at any season of the year (Fig. 18-8). Particularly vulnerable are trees planted in parking lots and near street curbs and driveways. Trees in lawn areas

FIGURE 18-8
The trunk of a two-year-old London plane tree is 75 percent girdled by continually being hit with a lawn mower (left).
The bark of a mature London plane growing near a driveway has been repeatedly injured by vehicles (right).

are often partially girdled by lawnmowers. Such damage can severely weaken or stunt young trees. Bark wounds close to the ground are particularly susceptible to decay organisms; they may also force latent buds to grow.

Trees are often vandalized: Their tops may be damaged or destroyed; their bark may be cut, beaten, or torn from the trunk. Treetrunks have been girdled by saws and axes. Trees subject to bark injury should be protected by two or three guard stakes (see Chapter 8). Plant young trees so that they are least likely to be struck by vehicles. A turf-free area around the trunks of young trees will reduce the likelihood of lawnmower injury.

You can often minimize bark injury if you take action before the cambium has had a chance to dry. Loosened bark can often be replaced, and even trees with girdled trunks can be saved (see Chapter 16). If a young tree or shrub is in good health when all or part of its top is broken off, a new top will usually grow. Prune with a slanting cut so that the cut surface faces away from the mid-afternoon sun. Keep the trunk and cut surface cool by shading them or painting them with white latex.

Improper staking or guying can result in damaged bark and wood. Young treetrunks increase in girth quickly, and unless the tree ties are inspected regularly and adjusted when needed, they can girdle the trunks. This is a particular problem when low branches or leaves along the trunks hide the ties. A tie may occasionally break loose from the stake but continue to circle the trunk, becoming imbedded and causing poor top growth and a ring of weak wood that is subject to breaking (see Fig. 8-10). When tree ties break, are fastened too low, or become loose, the trunk bark can be seriously damaged by rubbing against the stakes or tree guard (see Chapter 8).

Ice and wind storms can cause branches to split and break. Pirone (1978a) notes that certain deciduous trees are more liable than others to such damage. These include:

Ash	Red maple
Catalpa	Siberian elm
Empress tree	Silk tree
Hickory	Tulip tree
Horse chestnut	Yellowwood

Although the phenomenon is not widely reported in the literature, apparently sound branches up to 600 mm (24 in) in diameter can break out of trees without warning, most commonly on hot, still summer afternoons (Harris 1972). This form of breakage was originally observed in the southwestern United States. More recently, similar observations have been confirmed in Australia, England, and South Africa (see Chapter 4).

Snow can accumulate on the branches of evergreen trees and shrubs to the point where they break or split out. You can minimize damage by carefully knocking the snow off with a broom, flexible rake, or pole. If the branches are heavily loaded, touching them from below will remove the snow with less strain on the plant. The species most often damaged by snow are juniper, yew, and hemlock (Pirone 1978a). If transplanted to a heavy snow area, moderate to large conifers from areas of little or no snow may be more subject to limb breakage than trees native to the area. Many conifers growing in snow areas can shed and withstand considerable snow loads because their branches became more pendulous from snow loads while the trees were young.

If temperatures rise above freezing, ice-covered branches may be spared the breakage that often occurs when winds follow an ice storm. Hose the most heavily coated branches with a coarse spray of water until most but not all of the ice has been melted. The amount of ice will only be increased by a fine spray applied when temperatures are at or below freezing.

FURTHER READING

Pirone, P. P. 1978. *Tree Maintenance*. 5th ed. New York: Oxford University Press.

Smith, W. H. 1970. *Tree Pathology: A Short Introduction*. New York: Academic Press.

Streets, R. B. 1969. *The Diagnosis of Plant Diseases*. Tucson: Univ. of Ariz. Agr. Exp. Sta. Ext. Serv.

Tattar, T. A. 1978. *Diseases of Shade Trees*. New York: Academic Press.

CHAPTER NINETEEN

Diseases

Fungi, bacteria, mycoplasmas, and viruses are able to invade and infect plants, causing poor growth and appearance, disruption of plant processes, distortion of certain plant parts, and even death. Some higher (green) plants can also adversely affect the function and appearance of woody plants. Nematodes are considered by some plant pathologists to be disease-causing agents, since some invade plants and have adverse effects. Nematodes are animals and are often studied by entomologists. They are considered separately by nematologists and are discussed in the next chapter.

Although some new disease agents may be introduced into a given area, most disease damage to landscape plants is associated with pathogenic agents that are endemic or have been present in the area for some time (Schoeneweiss 1978). Many examples are familiar: leaf spots and blights, rusts, mildews, and vascular wilts. Healthy plants are likely to become diseased if

> The disease organisms are aggressive
> The plants constitute susceptible hosts
> Climate and soil conditions are favorable for disease development

Although a few of the pathogens that cause stem cankers, dieback, decline, and some root rots aggressively attack healthy plants, many attack only plants that are weakened or low in vigor.

Reductions in plant vigor usually arise from exposure to stress: drought, poor soil aeration, freezing or extreme temperature fluctuation, defoliation, nutrient deficiency, chemical injury, mechanical damage, or transplant shock (Schoeneweiss 1978). One or more of

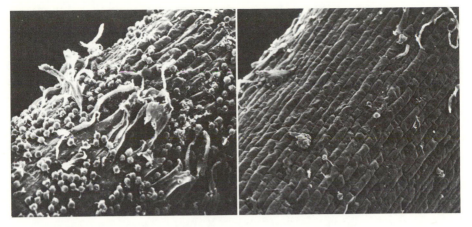

FIGURE 19-1
Root tip of salt-stressed chrysanthemum (left) attracts many oospores
(spherical objects on root surface) of Phytophthora root rot fungus
(*Phytophthora cinnamomi*) while a root tip of a nonstressed plant (right)
growing in a half-strength Hoagland's solution with the same concentration
of *Phytopthora* oospores attracts few oospores. Similar responses have
been observed with roots of woody plants and under different types of stress.
(Photo courtesy James MacDonald. 1982. *Phytopathology* 72:214–19)

these stresses may cause visible injury, even when no disease organisms
are present. Stress can also have a more subtle effect; it may weaken
a plant and increase its susceptibility to pathogens (Fig. 19-1) and
attack by boring insects. Exposure to severe or prolonged stress may
increase disease susceptibility at almost any time. In general, however,
plants are most readily predisposed to infection when they encounter
stress just after planting or when they age and begin to decline in
vitality.

Most organisms associated with stress-related disease normally
grow as saprophytes on dead plant tissue or other organic matter in
the soil. Thus, a pathogen may be present on healthy plants without
causing damage until the plants become stressed. Even though unaggressive pathogens may not penetrate healthy plant surfaces directly,
nearly all plants have dead branch stubs or twigs and at least minor
wounds that can provide access. Fungus spores and bacteria can be
spread by pruning tools and splashing rain during wet weather.

Some of the widespread diseases are discussed below to illustrate
means of infection, symptoms, and measures used to control the different pathogens. Federal and state publications and several commercially printed books give more detailed information on these diseases
and on many diseases not included here. Some of these publications
are listed at the end of this chapter.

FUNGI

Fungus parasites are the most common cause of infectious plant disease. Fungi are filamentous nongreen plants that are widely distributed on plants, in soil, and in the air. The filamentous strands (hyphae) are microscopic; mushrooms, conks, and mold are their visible reproductive stages. These structures produce millions of spores, which are dispersed by wind or rain to new hosts. Under favorable conditions of warmth and moisture, spores germinate and give rise to hyphae, which can infect susceptible plants. Thus the cycle continues.

Verticillium Wilt

Verticillium wilt is a soil-borne fungus disease known to infect several hundred woody and herbaceous species throughout the world. The causal fungus, *Verticillium dahliae* (*V. albo-atrum*), invades and plugs the water-conducting tissues. It can enter the xylem through root and trunk wounds, root grafts, and possibly intact young roots.

Symptoms The name of the disease describes its most obvious symptom: The leaves on one or more limbs wilt. Infection often takes place in the season before symptoms appear. If the following spring is mild and soil moisture is adequate, a relatively large succulent leaf canopy can develop only to wilt during the first hot days because the large transpiring surface cannot be adequately supplied by the fungus-impaired water-conducting system. If the disease spreads slowly and weather is not severe, small cup-shaped leaves may develop before wilting occurs. Depending on the severity of wilting, leaves may turn yellow, first at the margins and between the veins, then turn brown, and later die from the base of the plant or branch upward. If the symptoms develop slowly, the leaves will probably drop; if the leaves wilt and die quickly, they may hang on the tree for some time.

The xylem of infected plants will often turn olive green, dark brown, or black, depending on the species. If you cut diagonally through the base of a shoot or small branch, you will expose the discolored xylem (Fig. 19-2). In some infected plants, including olive, ash, and rose, xylem discoloration will be slight or nonexistent (McCain 1979b). To complicate diagnosis further, discolored xylem may indicate a number of other diseases, including Dutch elm disease. A positive diagnosis can be made only if the fungus from affected wood is cultured in a laboratory. Even when verticillium has caused xylem discoloration, it may be impossible to isolate the fungus on wilted shoots. The fungus may be lower in the branch or may be no longer viable.

FIGURE 19-2
Stripping the bark from a branch infected
with verticillium wilt or cutting diagonally
through the branch will usually expose dark,
discolored streaks in the young xylem.
(Photo courtesy Robert Raabe,
University of California, Berkeley)

One or more branches and occasionally the entire plant may wilt,
depending on how far and how quickly the disease progresses. Young
plants are more susceptible than old and may be more severely injured.
Mildly affected plants usually will recover from a verticillium attack.

Treatment   You may select plants that are resistant or immune
to verticillium wilt. A number of state Cooperative Extension publications list both susceptible and resistant plants (Himelick 1969; McCain
1979). As far as is known, conifers and broadleaved evergreen plants
are resistant to verticillium (Pirone 1978b). In California, McCain
discovered that olive trees under sod culture are less severely affected
by verticillium wilt than olive trees grown in weed-free groves. Sod or
grass culture might reduce the severity of this disease in other species
as well.

No cure is known for verticillium wilt once a plant is infected.
You can aid the recovery of mildly affected plants by fertilizing with
nitrogen (10 to 20 g/m^2; 2 to 4 lbs/1000 ft^2) as early in the season as
possible and by assuring that the plant has sufficient water. This will
stimulate the growth of additional xylem to replace some of the wood
that has become plugged. Some plants may recover from initial infec-

tions and escape further difficulties; others may continue to lose branches in succeeding years but will remain essentially viable; yet others will be so seriously affected they must be removed. When trees are to be saved, delay the removal of affected branches until you can determine the amount and quality of regrowth. Removing dead or wilted branches will not remove the fungus from the plant, although it will improve appearance.

After removing seriously affected and dead plants, replant with resistant species. If you must replant susceptible species, remove as much of the root system of diseased plants as possible. Fumigating the soil with chloropicrin will be partially effective against verticillium (Sinclair and Johnson 1975); do not get too close to adjacent plants. Soil solarization may have promise, even in existing plantings (see Chapter 20). The fungus, however, forms resting black structures (microsclerotia) that are capable of surviving in soil for many years; some apparently survive fumigation. When a newly infected tree grows among a group of healthy trees of the same species, Tattar (1978) recommends cutting any possible root grafts to reduce the spread of disease. Cutting roots, however, could open healthy roots to infection.

Armillaria Root Rot

The common names of this soil-borne fungus disease are shoe-string root rot, oak root fungus, mushroom root rot, and armillaria root rot. The organism, *Armillaria mellea*, attacks a wide range of land-scape, orchard, and forest plants throughout the world. Although it infects healthy plants of susceptible species, the disease is often associated with plants of low vigor (Wargo 1980).

Symptoms Affected trees and shrubs usually exhibit a general loss of vigor. Symptoms are similar to those caused by poorly aerated soil or an impaired root or trunk system. Plant decline is often accompanied by yellowing of foliage and leaf drop. Infected plants may die back slowly, a branch at a time, or may wilt and die quickly.

The most reliable signs of armillaria are fan-shaped plaques (mycelium) of white or cream-colored fungus tissue (Fig. 19-3) (McCain and Raabe 1972). These appear between the bark and wood in roots and trunk just at or near the soil surface. Although the fungus can infect when spores germinate on a favorable plant surface, it most commonly enters a susceptible plant when a rhizomorph (a dark, root-like fungus structure) contacts a root. Rhizomorphs (which resemble shoestrings, hence the name "shoestring root rot") grow on the surface of affected roots and for short distances into the soil (Fig. 19-3).

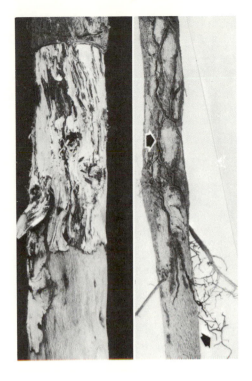

FIGURE 19-3
Armillaria mellea may weaken and kill only part of a tree or cause a general weakening; death usually follows.
Diagnostic signs of the fungus disease are the white or cream colored fan-shaped mycelium between the bark and wood on the main roots and trunk base (left) and the string-like rhizomorphs along the roots or in the soil (right).
(Photos courtesy Robert Raabe, University of California, Berkeley)

Other wood-rotting fungi are sometimes mistaken for *Armillaria mellea*. These other fungi are usually secondary invaders, however, that spread throughout the bark and sometimes into the wood, causing it to crumble easily. Armillaria-affected trees retain firm wood except in very advanced stages of the disease, when wood becomes wet and soggy (McCain and Raabe). In the late summer or early fall (as late as early winter in milder climates) honey-colored mushrooms may appear at the base of infected plants, though they are not common in arid regions. Many other fungi, not necessarily pathogenic, also form mushrooms at the base of plants.

Treatment   You can select plants for resistance to armillaria root rot. Consult state or regional publications for recommendations (Raabe 1979). The disease is often associated with plants of low vigor or plants under stress. The health and vigor of uninfected plants can be improved by fertilization with nitrogen (10 to 20 g/m^2; 2 to 4 lbs/1000 ft^2), irrigation (if necessary), and soil aeration. When you aerate the soil, take care not to spread infected soil and roots to areas that are clean. The best way to improve aeration with minimal root disturbance is to scarify the surface soil, apply a light organic mulch to reduce further compaction, minimize foot and vehicular traffic under the trees, and prevent excess water from accumulating on or in the soil.

There is no control for armillaria root rot once a plant is infected. You can slow or stop the progress of the disease, however, if you act before much of the root system is infected. In the spring or summer, expose the root collar and the larger roots to about 600 mm (2 ft) from the trunk. All soil removed should be carted away or placed on other infected soil. Remove roots with rotted bark, remove affected firm bark up to healthy wood and bark that shows no fungal mats, and cut away any rotten wood. You need not sterilize cutting tools (Robert Raabe 1981, personal communication). All wood, bark, and rhizomorphs removed, however, should be collected and thoroughly dried or burned. The bark and wood surfaces exposed to the air will dry, decreasing the chances of further infection. Keep the roots dry and exposed to the air as long as is practical. In areas where exposed roots may freeze, however, place clean soil around the roots in late fall. To keep the root collar surface dry, place about 250 mm (10 in) of coarse gravel around the trunk from the bottom of the hole to 50 mm (2 in) above the original ground level. Fill the hole up to the gravel surface and slope away to the original soil level at the edge of the hole so that water will not accumulate at the treetrunk.

Only valuable trees are worth the efforts to save them. The roots of severely infected trees will rot, making the trees hazardous. Remove severely infected and dead plants and as much of their roots as possible, at least those that are 20 mm (1 in) in diameter and larger. Soil disturbed when the plant and roots are removed should be kept in the infected area and fumigated or dried to kill the fungus. Alternatively, remove soil from the infected area to a depth of at least 600 mm (2 ft). Take care not to drop any soil en route to the disposal site, lest you spread the disease further. Choose the disposal site carefully so that spread of the fungus will not cause problems.

When only one tree in a group has been killed or severely infected, remove the tree and infected roots as described above, even if infected roots are found under the canopies of seemingly healthy adjacent trees. Soil can be fumigated with methyl bromide to within two-thirds of the canopy radius on one side of a living tree. The soil should be dry and the surface friable or lightly cultivated before fumigation, but, even at best, control will probably not be complete.

Dutch Elm Disease

Dutch elm disease (DED) is one of the most widely known and one of the most destructive plant diseases of the twentieth century (Sinclair and Campana 1978). The disease appeared in 1918 in Europe, and the causal fungus, *Ceratocystis ulmi*, was identified in 1922 in the Netherlands. The fungus, introduced into North America on veneer logs imported from Europe, caused disease that was discovered in Ohio in 1930 (Schreiber and Peacock 1974). By 1980, infected elms were found in almost all of the contiguous United States, although the disease appears to be less destructive in the Southwest, where no indigenous elms are found, where Asiatic species constitute an unusually high proportion of elms grown, and where summer temperatures and humidities are unfavorable to the fungus (Campana 1978).

The DED fungus is transmitted in North America by two insects: the smaller European elm bark beetle, *Scolytus multistriatus*, which preceded the fungus from Europe, and the elm bark beetle, *Hylurgopinus rufipes*, which is native to much of the midwestern and northeastern United States and southeastern Canada (Johnson and Lyon 1976). The bark beetles breed and lay eggs in the inner bark of weakened, dying, and dead elm trees. After the larval and pupal stages, adult beetles emerge and fly to healthy elms (or healthy parts of the same elm), on which they feed briefly, or to weakened elms, in which they breed (see Chapter 20). If the elm in which the beetles breed is infected with DED, the beetles will carry the fungus to healthy trees. Since DED weakens and kills elms, an epidemic may easily be started. The fungus can also spread from one elm tree to another through root grafts.

Strains of *C. ulmi* vary in virulence. England has had two DED epidemics: The first followed the introduction of the fungus in 1927, and a second occurred in the 1970s, caused by a more virulent import from North America. There are 20 to 30 species of elm; the American elm and the European white elm are highly susceptible to DED, whereas a high proportion of the Asiatic species, particularly the Siberian and Chinese elms, are fairly resistant (Sinclair 1978). No elm species has been found to be immune. Plants in related genera, *Zelkova* and *Planera*, have become diseased when artificially inoculated with the fungus (Schreiber and Peacock 1974). *Zelkova carpinifola* has been severely damaged by natural infection in Iran (Sinclair), but *Zelkova serrata* has generally escaped infection when growing close to diseased elms (McCain 1979).

Symptoms   The symptoms of DED, their sequence, and the rate of development depend on the species of elm, tree nutrition, soil moisture, temperature, and site of inoculation (VanAlfen and MacHardy 1978). Healthy elms are most commonly infected in spring and early summer, when elm bark beetles feed in the crotches of one- and two-year-old twigs (*S. multistriatus*) (see Fig. 20-4) or in branches (*H. rufipes*). These beetles make wounds in the outermost sapwood as they feed and introduce spores of the disease fungus into these wounds.

The first observable symptom is usually the sudden discoloration or drooping of leaves at the tip of one or more branches (Fig. 19-4). The leaves may then yellow, roll upward at the edges, and turn brown,

FIGURE 19-4
An early symptom of Dutch elm disease is "flagging" (wilting) of leaves
on one or more branches; the rest of the foliage may appear normal.
(Photos courtesy USDA)

or dry rapidly, turn dull green, and then brown. When the leaves yellow
or curl, they usually drop quite soon, but when leaves shrivel and die
rapidly, they may remain attached for several weeks. Shoot tips may
also wilt and droop; this is called flagging. Small flags develop after
elm twigs are inoculated by European elm beetles, but large flags develop after inoculation by native elm bark beetles (VanAlfen and
McHardy 1978). Trees infected in early summer may be killed the
first year, others may live for several years, and some may recover.

Infected elm twigs may have brown discoloration just beneath
the bark (Fig. 19-5). Cut a twig diagonally with a knife: If the tree is
infected, dark spots or a continuous brown layer in the current season's
growth ring will usually be exposed. Other vascular wilt diseases,
however, cause many of the same symptoms, so only laboratory identification will be reliable (Schreiber and Peacock 1974).

Treatment   No single method has successfully controlled DED,
so the primary emphasis is on prevention. To combat Dutch elm
disease

Eliminate beetle breeding sites
Protect healthy elms from feeding beetles
Prevent underground transmission of the fungus
Plant trees that are resistant to the disease (Schreiber and Peacock)

An integrated approach to DED control is presented in Chapter 21.

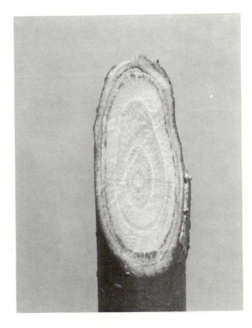

FIGURE 19-5
If you peel the bark from branches infected with Dutch elm disease (left)
or cut diagonally through them (right), you will expose brown streaks.
This symptom can be easily confused with that of verticillium wilt.
(Photos courtesy USDA)

Leaf Diseases

Leaves not only reflect the health of plants but can be directly injured or disfigured by a variety of causes, including fungi. Most leaf diseases of deciduous woody plants, however, are not serious. Plants may be weakened by leaf diseases, but the effects usually end with leaf fall, unless premature defoliation depletes food reserves.

In contrast, the leaf diseases of evergreens often severely affect plant health and can even kill trees and shrubs. Because evergreens normally hold individual leaves for several years, infection and loss of leaves have great physiological and visual effects. Evergreens are slow to recover from leaf drop, since several years of foliage may be lost at a single time. Evergreens with severe leaf infections are also more susceptible to insects, environmental stress, and other pathogens.

Symptoms The leaf diseases of deciduous plants have been grouped into six categories, according to symptoms (Tattar 1978):

Leaf spot: Dead area on leaf, well distinguished from healthy tissue

Leaf blotch: Dead area on leaf, often diffusing into healthy tissue

Anthracnose: Irregular dead areas on leaf margins, sometimes killing entire leaves; shoots and small twigs often injured

Powdery mildew: White or gray-white fungus on leaves and shoots

Leaf blister: Leaf spot or blotch that is swollen or raised, producing a blister-like area on upper surface

Shot hole: Holes caused when dead areas inside leaf spots drop out

The leaf diseases of evergreens can also be grouped (Tattar):

Leaf spot: Broadleaved evergreens may show spots of infection similar to those on deciduous plants.

Needle spot, needle cast, needle blight: Fungus infections of conifer needles may progress through a series of increasingly severe stages. *Needle spots* can occur at random along a needle. The spots may enlarge, coalesce, and eventually kill the entire needle, which may then drop; this last stage is called *needle cast*. When all or most of the plant is severely affected, the symptom may be called *needle blight*. *Needle cast* and *needle blight* are also used interchangeably.

Needle rust: Swollen white or orange blisters may form on the needles, sometimes causing needles to drop.

Treatment   Obviously the best long-term control involves selecting species or cultivars that are resistant to known leaf diseases. Resistant cultivars have been selected from a number of susceptible species. Many existing landscape plants, however, are susceptible to one or more leaf diseases. Control measures can be recommended only when the health of the plants is poor or the appearance of an important landscape planting is jeopardized. Since evergreens are usually more seriously affected, they should be given priority if control measures must be limited.

As is the case for most diseases, protection against infection is the only effective control. Infection for most leaf diseases occurs during warm wet weather in spring, when the pathogens become active and developing leaves and shoots are most tender. If fungicides are to be effective, they must be applied during or before bud break and about every two weeks thereafter; this will keep newly developing foliage covered until the temperatures become hot or the danger of wet weather is past, usually at the beginning of summer. After symptoms occur, control measures may reduce subsequent symptoms but will not improve the leaves already infected. As temperatures rise and if wet weather subsides in late spring, infection will decrease and normal growth will prevail. Leaves infected earlier, however, may drop throughout the summer.

You can reduce conditions favorable to leaf infection by planting and pruning to permit greater sun penetration and air movement through susceptible plants. This is especially important for evergreens, since they may have a relatively dense canopy of leaves in the early spring, when chances of infection are greatest. If foliage dries quickly and humidity within the leaf canopy is lowered, the possibility of leaf infection will be decreased. It is not clear, however, that you must avoid sprinkler irrigation in all situations. When irrigation is necessary, leaf surfaces during the day will be dry, humidity will probably be low, the temperatures high, and most of the foliage mature and less susceptible to infection. Under these conditions, foliage will dry quickly. It may be unwise to irrigate with sprinklers during inclement weather, since drying conditions will be slow and temperatures cool. Sprinkler irrigation is a common practice in California vineyards even though powdery mildew (*Uncinula necator*) must be restrained by preventive dusts on foliage and grapes (Hewitt and Jensen 1979). Following irrigation, high temperatures dry the leaves quickly.

It is of doubtful value to prune off infected twigs and burn them together with infected leaves because mycelia (the masses of fungus hyphae) and spores of leaf pathogens will remain on plants and their immediate environment, some infected twigs and leaves will escape burning, and spores can be blown great distances. Pruning off infected small branches can improve the appearance of most affected plants, however. If only the leaves on a few twigs have been seriously infected, their removal could retard further infection, particularly if it is combined with other control measures. Keep pruning to a minimum so that you do not stimulate new growth, which is particularly susceptible to mildew.

BACTERIA

Bacteria are microscopic organisms, devoid of chlorophyll, and usually single-celled; they obtain their food from living or dead organic matter. Like fungi, most bacteria are beneficial. They decompose organic matter, convert complex chemical compounds to more simple forms, synthesize gaseous nitrogen into a usable nutrient, and can be used in many industrial processes. Some species, however, cause diseases in plants and animals.

Bacteria are smaller than fungi and can penetrate healthy plants through natural openings, such as stomata, lenticels, the terminal openings of leaf veins, the glands in flowers, and wounds of all sorts. Some bacteria are spread by contaminated pruning and grafting tools. Insects commonly transmit certain bacterial pathogens.

Fire Blight

Fire blight frequently attacks plants in the rose family. Certain pears and quince are extremely susceptible; apple, crab apple, and pyracantha species are also often damaged. Fire blight is caused by *Erwinia amylovora*, which survives the summer and winter in blighted twigs and cankers. During warm, moist spring weather, the bacteria ooze from the holdover cankers in small sticky drops. They are carried to blossoms and tender shoots by insects and splashing rain and from blossom to blossom by bees and flies. Temperatures between 20° and 30°C (70° and 85°F) and high humidity caused by dew, rain, fog, or irrigation favor development of the disease (McCain 1975).

Symptoms Fire blight infection causes a sudden wilting, shriveling, and blackening of blossoms, tender shoots, and small fruit (Fig. 19-6). Affected parts look as though they have been scorched by fire—hence the name of the disease. Leaves are killed so quickly that they remain on dead shoots throughout the summer. Infection may progress down a shoot and into the bark of larger limbs, where it produces sunken dark cankers. Only a few shoots on a plant may become infected, but if cankers enlarge they can completely girdle branches or the main trunk, killing large portions of the plant.

Treatment Prune out affected shoots or branches as soon as possible in the spring; this should stop the spread of the infection in

FIGURE 19-6
When blossoms and young shoots of plants in the rose family
are infected with fire blight bacteria, they wilt, shrivel, and turn black
in the spring (left). Infection may be carried systemically to older branches,
causing cankers on dead bark and wood (rough bark area) (right).

that area and reduce the inoculum available to infect healthy blossoms or shoots. Cut well below (300–450 mm or 12–18 in) the infected area, whose edge will usually be easily seen. It is commonly recommended that you dip the tool into a disinfectant after each pruning cut. However, Dr. Robert Raabe, plant pathologist at the University of California, Berkeley, has found that contaminated pruning shears, whether they have cut through a diseased shoot or been dipped in a bacterial slurry, will not infect healthy woody pear shoots (1981, personal communication). Prunings should be burned or removed from the vicinity of susceptible plants.

If at all possible, remove fire blight cankers before the next spring. Prune off branches on which cankers occur. Shrubs with trunk cankers may not be worth saving. When cankers occur on the trunk or scaffold branches of large trees, remove the bark from an area that includes healthy bark 75 mm (3 in) on either side and 300 mm (12 in) above and below the canker (Tattar 1978). The exposed wood can be left untreated. Inspect the wound later to see whether any further infection occurs; if so, it should be treated as described above.

Fire blight occurs most frequently on succulent growth. Where the disease has been a problem, cultural practices (particularly nitrogen fertilization) should be designed to avoid vigorous growth, especially early in the growing season, when conditions most favor the infection. Blossoms are the most susceptible part of the plant. They can be protected from the infection by sprays that contain copper or streptomycin. Make the first application before more than ten percent of the plant has bloomed; repeat at 4- to 5-day intervals until bloom is over.

In areas where fire blight is known to be a problem, select species or clones that are known to be resistant or nearly so. The callery pear is quite resistant.

Crown Gall

The bacterium *Agrobacterium tumefaciens* causes galls on many woody and herbaceous plants throughout the world. Many species of the rose family are susceptible, as are a number of other landscape plants.

Symptoms   Crown galls start out as small swellings on the bark of the lower trunk and roots, although they can occur higher on the plant, particularly on poplar and willow. These swellings continue to enlarge, engulfing more of the stem or root (Fig. 19-7). At first, the galls resemble callus tissue similar to a potato tuber in density but with a tougher texture. As the galls enlarge, their surfaces become rough and fissured. The outer portions of older galls may slough off.

FIGURE 19-7
Crown galls were on roots of the left young almond when dug in a nursery;
the infected plant is doomed to poor growth and a possible early death;
roots to the right are not infected. (Photo courtesy Milton Schroth,
University of California, Berkeley) A gall on the trunk has seriously weakened
this young flowering cherry tree (lower left), while the large gall
on a major limb of an elm has not jeopardized more than the limb
on which it occurs (lower right).

Burls and similar growths more closely resemble normal wood tissue than do crown galls. Galls on black walnut, however, may be difficult to distinguish from burls, which often form at the crown of old trees. The internal cavities of old galls are sometimes lined with dead bark and decayed tissue, whereas burls are solid (Ross and others 1970).

Crown gall bacteria commonly present in the soil enter plants through wounds. Young plants are particularly susceptible during handling in the nursery, planting, and early care in the landscape. Fumigating soil before planting can decrease the incidence of crown gall (Moore 1979). The bacteria or the plasmids that they form stimulate large numbers of undifferentiated cells, which form the galls. These galls can interfere with the movement of water and nutrients in the xylem. Crown gall is particularly serious on young plants, since one infection can engulf a main trunk, branch, or root within a year or two, seriously weakening or killing the plant. Crown gall infections on large trees seldom cause serious reduction of growth.

Treatment Crown gall can be largely prevented if you select plants free of galls and avoid injury to the roots and trunk during planting. Most nurseries produce clean stock.

In Australia, a root dip solution of a non-disease-producing strain of the crown gall bacterium (*Agrobacterium radiobacter* 'Strain 84') has proven extremely effective in protecting young fruit and landscape plants from crown gall (New and Kerr 1972). The *A. radiobacter* 'Strain 84' bacterium, now marketed in the United States as Galltrol A®, produces an antibiotic. Crown gall bacteria resistant to Galltrol A have been reported, however (Moore 1979).

Young plants that do poorly in the landscape should be inspected for crown gall. You may need to expose the trunk and main roots for a complete inspection. A 25-mm diameter (1-inch) stream of water at a pressure of about 4 kg/cm^2 (60 psi) will remove soil from the base of a tree with little or no injury. Any galls exposed should be allowed to dry for several days. Painting the galls and 10 to 15 mm (0.5 in) of surrounding healthy bark with Gallex® (formerly called Bacticin) has eradicated over 95 percent of treated galls on several fruit species (Ross and others 1970). Large galls may require two applications. Leave the treated areas exposed for several weeks. You can treat galls at any time, but if there is little danger of winter injury to exposed roots, treatment during the dormant season will cause the least disturbance to maintenance practices and use of the landscape.

Wetwood (Slime Flux)

The most common evidence of wetwood is bleeding or fluxing from wounds in the trunk or large branches. Wetwood is a chronic disease that contributes to the general decline of many trees, especially old trees growing under adverse conditions. Although it is common in elm, wetwood also occurs in apple, birch, fir, hemlock, hickory, linden, maple, mesquite, mulberry, oak, pine, poplar, redbud, sycamore, tulip tree, and willow (Carter 1975). In elm, wetwood is caused by the bacterium *Erwinia nimipressuralis*. Almost all elms show wetwood to some degree, but the Siberian elm is unusually susceptible.

Symptoms Wetwood occurs when bacteria invade and grow in the spring wood of trunks and large branches. Affected wood becomes dark brown and waterlogged; when the wood is cut, sap oozes out. Discolored spring wood occurs in the annual rings, primarily in the heartwood and older sapwood. Fluxing commonly occurs from the spring through the fall, with the most active period in the heat of summer, when the bacteria are most active. In mild climates, some fluxing may occur in winter.

The bacteria ferment organic constituents in the sap, producing gas, primarily methane and nitrogen, that can develop considerable pressure. Carter (1975) reports pressures up to 4 kg/cm^2 (60 lbs/in^2). More common pressures, however, are 0.4 to 0.8 kg/cm^2 (5–10 lbs/in^2). These pressures force fermented sap and gas out of wounds in the trunk and branches. The sap is colorless or tan but darkens upon exposure to air and wets the bark below wounds. Growth of bacteria and fungi in the sap often produces foamy gray or slimy brown masses at and below wounds (Fig. 19-8). This is commonly called slime flux. Pirone (1978b), however, does not associate slime flux with wetwood.

Fermented sap is more alkaline than sap from healthy wood but is toxic enough to retard or prevent callus formation at the base of wounds. Slime flux can kill leaves, shoots, and grass on which it may drip. When dry, slime flux leaves a gray or white encrustation on the bark. In addition to fluxing, if current season wood is infected, leaves may yellow, wilt, and drop, branches may die, and the entire tree may decline in health and vigor. Brown streaking in the wood of young branches can easily be confused with similar symptoms of other wilt diseases.

Carter (1975) cautions that not all trees with flux are affected with wetwood. Healthy trees of a number of species, including some susceptible to wetwood, may bleed profusely from pruning wounds, particularly if large limbs are removed just before growth begins in the spring (see Chapter 14).

Wounds limited to the bark may also produce slime flux, but that should not be confused with the flux that results from wetwood. Wounds associated with wetwood extend through the bark and into the old sapwood. Carter (1975) recommends that affected bark around bark wounds be removed and the exposed wood disinfected.

FIGURE 19-8
Sap oozes out of an elm trunk crotch.
The lighter colored bark beyond the dark
wet area has been discolored from earlier
heavy flows of sap.

Treatment Since the heartwood and old sapwood are infected with the bacteria, no chemical or biological treatments will eliminate the disease from infected trees. Since the bacteria are usually widespread in the trunk and large branches of infected trees, removing branches that exude slime flux or show other symptoms will not eliminate the disease. Remove only dead wood, so that the tree will retain as much landscape appeal as possible. Nitrogen fertilization will stimulate growth and may help overcome the adverse effects of the wetwood infection.

A drain hole may be bored below a fluxing wound to reduce or stop sap flow, though this will not cure the disease. Carter (1975) recommends that the hole be 10 to 15 mm (0.5 in) in diameter and 150 to 350 mm (6–14 in) below the fluxing region. Slant the drainage hole so the sap can drain out. Bore it through the heartwood to within 50 or 100 mm (2 or 4 in) of the bark on the opposite side. Insert semi-rigid polyethylene tubing that is large enough in diameter to give a tight fit and long enough that the toxic sap will drip on the soil, not on the trunk below. Coat the inner surface of the hole to a depth of 25 mm (1 in) with asphalt-emulsion grafting wax. Insert the tubing only far enough so that it is held securely and seal the tubing and bark with more grafting wax. The winter after the fluxing stops, replace the tubing with a short plug to seal the hole. The plugged wound should heal quickly if the tree is vigorous. Shigo (1979b) states that the benefits of boring a hole to drain slime flux is balanced by the possibility of spreading the wetwood to tissues that surround the hole.

In proper care of young trees, all pruning cuts should be small. Small wounds on healthy trees callus quickly, reducing the possibility that slime flux will develop. Prune in late summer or fall to further minimize the bleeding of either normal wood or wetwood.

VIRUSES

Viruses are submicroscopic in size and consist of a nucleic acid core with a protein covering. Living as parasites in plants and animals, they can reproduce themselves only inside living tissue. A virus stimulates the host cell to create more viruses identical to itself. The virus and its multiplication interfere with some of the normal functions of plant cells. Viruses are usually spread from infected to healthy plants by sucking insects (aphids and leaf hoppers) and possibly by pruning tools. Budding or grafting infected scions onto healthy rootstocks will transmit viruses. Few viruses are transmitted through seed.

Symptoms Symptoms of viral diseases commonly fall into one or more of the following categories (Tattar 1978):

Lack of chlorophyll formation in normally green tissue

Inhibition of vegetative growth

Distorted growth of all or part of the plant

Necrotic areas

Even though viruses may dwarf or seriously weaken a plant, they are seldom fatal. Plants weakened by viruses, however, may be more susceptible to other diseases and to physiological stress.

The viral diseases of commercial fruit species have been studied more intensively than those of landscape plants. Some viral diseases have become serious in certain fruit species, primarily in cultivars that have been propagated vegetatively for so long. Many of the flowering fruit tree cultivars used in the landscape are infected with the same viruses.

Viruses are named or described by their symptoms and the host infected. Some infect more than one species of plant and may cause either similar or different symptoms in the various species. You may determine the identity of a virus by grafting infected buds onto cultivars that produce characteristic symptoms in response to known viruses. Some of the more common viral diseases of trees have been listed by Tattar (1978) (Table 19-1).

Treatment The best protection against viral diseases is to select virus-free or virus-indexed plants. Rather complex procedures have

TABLE 19-1

Some of the more common viral diseases of trees
(after Agrios 1975 and Tattar, T. A. 1978.
Diseases of Shade Trees. New York: Academic
Press)

Host	Disease
Apple	Apple mosaic
	Chlorotic leaf spot
	Stem pitting
	Flat limb
Ash	Ash ringspot (tobacco ringspot virus)
	Ash line pattern (tobacco mosaic virus)
Birch	Birch line pattern (apple mosaic virus)
Black locust	Black locust witches'-broom
	(tomato spotted wilt virus)
Cherry	Stem pitting of *Prunus* (tomato ringspot virus)
	Line pattern
	Prunus necrotic ringspot
	Raspleaf
Elder	Tobacco ringspot virus on elder
Elm	Elm mosaic virus
	Elm ringspot virus
	Elm scorch virus
	Elm zonate canker virus
Lilac	Lilac ringspot virus
Maple	Maple mosaic (maple mottle)
	Peach rosette virus on maple
	Tobacco ringspot virus on maple
Oak	Oak ringspot
Peach	Ringspot
	Yellowbud mosaic (stem pitting of *Prunus*,
	tomato ringspot virus)
	Mosaic
Pear	Stony pit
	Vein yellows and red mottle
	Ring pattern mosaic
Plum	Stem pitting of *Prunus* (tomato ringspot virus)
	Sour cherry yellows (prune dwarf virus,
	Prunus necrotic ringspot virus)
Poplar	Poplar mosaic virus

been established by some nurseries and governmental agencies to ensure the propagation of material that is free of the more serious viruses. Heat treatment of plants will inactivate viruses and allow disease-free plants to be propagated for nursery "mother" blocks. Healthy "mother" plants are maintained in isolated blocks and are periodically indexed to ensure that they are still free from disease. Plants that are grown to be sold are propagated with care to prevent infection.

Experience with flowering cherries indexed for the more serious viruses illustrates the importance of virus-free plants: The percentage of successfully budded Yoshino cherry trees increased from 40 to 100 when virus-indexed trees were used (Leiser and Nyland 1972). Shoot growth was 2.5 times greater on the virus-indexed trees than on trees infected with necrotic ringspot (Fig. 19-9).

Mature trees infected with a virus will usually live for many years and will be satisfactory landscape features. Optimum levels of moisture and nitrogen nutrition should increase plant vigor and tolerance to other diseases and should prolong the life of the tree.

FIGURE 19-9
This Yoshino cherry tree (left), which is free
of the recurrent necrotic ring spot virus, is more vigorous than the same clone
infected with the virus. The trees have grown three years from being budded
on young established rootstocks; the marker stick is 1800 mm (6 ft) tall.
(Photos courtesy Andrew Leiser and George Nyland,
University of California, Davis)

MYCOPLASMAS

In 1967 some plant diseases thought to be viral were found to be caused by mycoplasmas. No fungi or bacteria were associated with diseased plants, but the disease-causing agents were transmissible by grafts and by insect vectors, characteristics of virus diseases. In contrast to viruses, which resemble organic chemicals, mycoplasmas are similar to bacteria but have a plasma membrane instead of a cell wall. Although mycoplasmas have limited mobility, they can move throughout the plant in the phloem. Disease-causing mycoplasmas are parasites of plants and animals and multiply in living cells by division, like bacteria. The most common vectors of mycoplasmas are leaf hoppers. Mycoplasmas must incubate 10 to 20 days within the insect's body before they can be transmitted to a new host (Tattar 1978). Once an insect is infected it can transmit mycoplasmas the rest of its life.

The multiplication and spread of mycoplasmas block the movement of essential organic materials in the vascular system and can cause proliferation of shoots and roots, commonly known as witches'-broom or hairy root. Roots are affected first because they are soon weakened by lack of food from the tops. Active phloem is killed in the small roots and then in progressively larger roots. Later, the upper parts of plants undergo a progressive decline that usually results in death. Plant decline may be accompanied by any or all of the following: leaf chlorosis, weak growth of many shoots (witches'-broom), branch dieback, lack of flowering and fruiting, phloem necrosis, and proliferation of roots. Agrios (1975) lists the plant diseases known or believed to be caused by mycoplasmas (Table 19-2).

TABLE 19-2

Some of the more common mycoplasma diseases of trees
(after Agrios 1975 and Tattar, T. A. 1978. *Diseases of Shade Trees.*
New York: Academic Press)

Host	Disease	Host	Disease
Apple	Proliferation disease	Pear	Pear decline
Green ash
White ash	Ash witches'-broom	Pecan
Hickory (native species)	Pecan bunch disease
Cherry	Rusty mottle
X-disease	Black walnut
Butternut	Walnut bunch disease
Black locust	Black locust witches'-broom	English walnut
Japanese walnut	
Peach	Little peach
Peach rosette
X-disease
Yellows	Willow	Willow witches'-broom

MISTLETOES

Mistletoes are perennial evergreen parasites that grow on the stems and branches of trees and shrubs (Fig. 19-10). Most of the 1300 species or forms of mistletoe are tropical, but the common mistletoes of Europe (*Viscum album*) and the United States (*Phoradendron flavescens*) and a number of other species are found up to latitude 40° and at high elevations closer to the equator (Gill and Hawksworth 1961). *V. album* damages trees from the Mediterranean to Germany but is not a significant pest farther north in Europe. *Phoradendron* does not occur in Canada, although several species infect both hardwoods and conifers in the United States. When winter temperatures sink to -18°C (0°F) or below for several days, they have been observed to kill the aerial parts of *P. flavescens* in Illinois. The northern limit of *Phoradendron*, from east to west, is considered to be: southern New Jersey, Illinois, Colorado, northern Oregon. *P. flavescens* is most serious as a pest in the western states but is also a problem on Florida pecans and Texas citrus (Gill and Hawksworth).

Of greater economic importance are the dwarf mistletoes (primarily *Arceuthobium vaginatum* and *A. campylopodum*), which destroy many conifer species. Trees are stunted and heart-rotting fungi easily become established.

The seeds of leafy mistletoes, such as *Viscum*, are dispersed primarily by birds. Dwarf mistletoes, such as *Arceuthobium*, discharge their seeds with considerable force, throwing them up to 9 m (30 ft), and strong winds have been known to carry these seeds 400 m (0.25 mile) (Gill and Hawksworth). The seeds of mistletoe are covered with a mucilaginous layer that attaches the seed to the bark of the host. When the seed is firmly cemented, it germinates, and its hypocotyl

FIGURE 19-10
These branches of green ash are infected with *Phoradendron flavescens* var. *macrophyllum*; note the swelling of the branches at the infection sites and the differences in stem diameter above and below the mistletoe (right).

(that portion of the seedling between the embryo and the seed leaves) grows along the surface of the bark. When it reaches a suitable point for invasion, it forms a holdfast. The holdfast of *Viscum album* is a dome-like structure that attaches to the bark. The holdfast lifts the bark surface until a crack forms, through which the primary haustorium can grow into the host tissue. Although haustoria may differ morphologically from normal plant roots, they perform for mistletoe the same functions: absorption of water and nutrients. Haustoria may invade the cortical region of the bark and form a close union with the host xylem, or a strand from the original holdfast may grow along the bark to form a new plant at a second location. Berry-producing trees and shrubs near susceptible trees have been observed by the author to increase the number of infections. As might be expected, proximity to trees with mistletoe also increases the likelihood of infestation.

Signs and Symptoms   Leaves vary from the minute scales of the dwarf mistletoes to those more than 25 mm (1 in) long, which we associate with the Christmas season. As is typical of xerophytes, the leaves are thick and leathery, with sunken stomata that reduce transpiration. Some mistletoes are rather specific and grow on only a single genus; others occur on a wide range of genera. Even though they are completely parasitic, mistletoes manufacture much of their own food and in general require only water and nutrients from the host plant.

Although *Viscum album* has some use as a medicine and, along with *Phoradendron flavescens*, is prized for winter holiday decoration and for animal feed when normal feed sources are short, mistletoe damages forest, fruit, and landscape trees and shrubs. The amount of damage varies with the species of mistletoe, its longevity, and the intensity of its parasitism. The effects of mistletoes may include any of the following: reduced vigor, poor fruit or seed crops, malfunction of woody tissues, sparse foliage, top death, predisposition to insect and disease attack, and premature death (Gill and Hawksworth 1961).

Treatment   You can keep mistletoe establishment in new plantings to a minimum by selecting trees that are immune or highly resistant to mistletoe infestation. It is particularly important to remove mistletoe before it produces seed. Removal is easiest during the dormant season, when the green mistletoe is easily seen growing on the bare branches of the trees. You can remove mistletoe most completely by pruning out infected tree limbs. Cut off an affected limb at least 300 mm (1 ft) below the point of mistletoe attachment, preferably at the next lower crotch.

If mistletoe grows on main branches or the trunk, so that removal would markedly damage tree structure or appearance, cut off the aerial part of the mistletoe. Make the cut close or treat the stub with herbicide to kill the haustoria imbedded in the tree. Cutting out the aerial portions without applying an inhibitor will prevent fruiting for at least two seasons.

More than 9000 of the approximately 85,000 street trees in Sacramento, California, were infected with mistletoe in 1977; 95 percent of affected trees were Modesto or Arizona ash (Torngren, Perry, and Elmore 1980). In 1980, the city began a major removal and pruning project. Seriously infected trees, with permission of the adjacent property owner, were removed; if not removed, the trees were pruned severely and infected branches were removed or cut back. Mistletoe on the trunk and main scaffolds was cut off. Less severely infected trees were pruned with similar severity. In addition, the city lent pole pruning equipment to citizens who pruned mistletoe out of private and public trees. The purpose was to improve tree health and to minimize further infection.

Dormant application of a 2,4-D (2,4-dichlorophenoayacetic acid) amine and dicamba mixture as either a paint or a foam (Super D Weedone Foam Weed Control®) is recommended to control *P. flavescens* on limbs too large to be removed (Torngren, Perry, and Elmore). Mistletoe clumps should be broken or headed back to 45- to 90-mm (1.5- to 3-in) stubs. Cover the freshly cut stub thoroughly with the 2,4-D herbicide, taking care to confine the material to the area immediately around each stub. Treat mistletoe growths less than 90 mm (3 in) long with herbicide, but do not cut them back or remove them. Herbicidal action is quite slow: Mistletoe growth will initially stop, and later the treated stub will turn brown. The stub should be dead in six to ten months. Monitor treated trees annually to remove new infections and retreat persistent infections.

Although Torngren and Chan (1978) report that in California tests 90 percent or more of *P. flavescens* was controlled by dormant applications of 2,4-D and dicamba mixture to flush cuts and short stubs, mistletoe now appears to be emerging from haustoria that were not killed by earlier treatment. Whether retreatment will improve control has yet to be determined.

Earlier, arborists were advised to cut the mistletoe flush with the bark, wrap the area of attachment with a band of black polyethylene wide enough to exclude light, and tie with twine or flexible tape (Johnson 1977). In the absence of light, most haustoria systems should die within a year or two. Another species (*P. libocedri*), however, has been reported to live without aerial shoots for well over 100 years (Gill and Hawksworth 1961). Even if effective, wrapping mistletoe cuts is time-consuming and leaves an unsightly record in the landscape.

The removal of infected limbs and trees has proved to be more effective with dwarf mistletoes than with *P. flavescens*. Dwarf mistletoe is spread and intensified primarily by explosive fruits rather than by animal vectors. Once the parasite is eradicated or materially reduced in an area, there is little chance for reinfection from the outside (Gill and Hawksworth).

FURTHER READING

AGRIOS, G. N. 1969. *Plant Pathology*. New York: Academic Press.

MANION, P. D. 1981. *Tree Disease Concepts*. Englewood Cliffs, N.J.: Prentice-Hall.

PIRONE, P. P. 1978. *Diseases and Pests of Ornamental Plants*. 5th ed. New York: John Wiley.

SMITH, W. H. 1970. *Tree Pathology: A Short Introduction*. New York: Academic Press.

STREETS, R. B. 1969. *Diseases of the Cultivated Plants of the Southwest*. Tucson: University of Arizona Press.

TATTAR, T. A. 1978, *Diseases of Shade Trees*. New York: Academic Press.

CHAPTER TWENTY

Insects
and
Related Pests

Some insects pollinate crop plants or serve as sources of food, medicine, and dyes. Others may seriously damage crop, forest, and landscape plants and transmit serious plant and animal disease pathogens. The life cycle of most insects and mites is short, and many species produce several generations a year and exceptionally large populations when conditions are optimal. Several factors account for these large numbers and persistent survival: a high rate of reproduction, small size, hard external skeletons, and, in many insects, an ability to fly, and an inactive or resting stage in the life cycle. Of the more than one million species of insects and mites in the world, about 2500 species damage ornamental plants in the United States alone (Johnson and Lyon 1976).

Insects (*Insecta*) and mites (*Arachnida*) are arthropods, which have an external skeleton, a segmented body, and jointed legs. Insects have three pairs of legs; a body composed of head, thorax, and abdomen; and usually wings (see Fig. 20-4). Mites (spiders) have four pairs of legs; bodies with abdomen, a fused head, and thorax (cephalothorax); and no wings (see Fig. 20-7). Many insects go through a complete metamorphosis, or four stages of development: egg, larva, pupa, and adult. Plant injury may be caused by the larva or by the adult (Fig. 20-1). Mites and other insects go through incomplete metamorphosis, involving only egg, nymph, and adult stages. Control measures depend on the number of individuals present at the most vulnerable stage in a pest's life cycle.

Insects can be conveniently separated into those with chewing mouthparts and those with sucking mouthparts; they differ in their

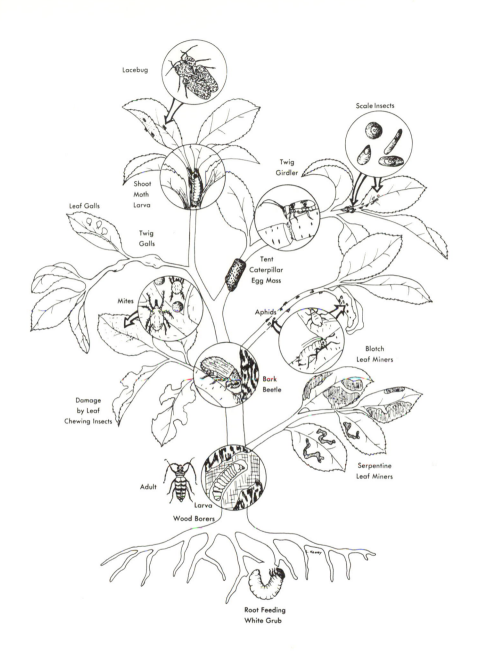

FIGURE 20-1

Insects and mites may infest all parts of most plants.

This sketch depicts some of the types of insects and the injury they can cause.

(Reprinted from Warren T. Johnson and Howard H. Lyon.

Insects that Feed on Trees and Shrubs. Copyright © 1976 by Cornell University.

Used by permission of the publisher, Cornell University Press) manner of feeding, the type of food they eat, and the type of plant injury they cause. They also respond to different kinds of control measures. Most larvae and some adults have chewing mouthparts and can be particularly destructive. Usually larger than insects with sucking mouthparts, they eat holes or bore in the plant. (Snails and slugs, although they are not insects, have rasping habits and can cause a loss of tissue on landscape plants.) Sucking insects insert their mouthparts into plant tissue and suck the juice. Other sucking pests include mites and nematodes. Insects commonly transmit the causal agents of plant and animal diseases. They can also seriously damage or weaken plants, making them more subject to further insect or disease attack.

A few insects are discussed in this chapter to illustrate their life cycles, the damage they do, and the control measures recommended. Federal and state publications and several commercially published books discuss these insects and many more in greater detail.

CHEWING INSECTS

Insect larvae and adults with chewing mouthparts feed on various plant parts. Caterpillars, cutworms, hornworms, slugs, and worms eat holes in leaves or along the margins. Leaf miners and skeletonizers are larvae that feed primarily between leaf veins, leaving only the epidermis. Leaf miners feed between the two leaf surfaces; skeletonizers may leave only the upper epidermis or none at all. Larvae that feed on roots are usually referred to as grubs, maggots, weevils, and wireworms. Beetles and grasshoppers are adult insects that eat above-ground parts, principally the leaves. The larvae of some species live in the soil, feeding on plant roots and later emerging as adults to eat above-ground parts. Borers are the larval stage of any insect that feeds inside roots, trunks, branches, or shoots of plants. Some adult beetles may also bore into plants.

The Japanese Beetle

Popillia japonica, the Japanese beetle, causes extensive damage to plants in eastern North America. A native of Japan, it was first found in this country in New Jersey in 1916. It continues to spread and by 1980 could be found from southern Nova Scotia and Ontario to Georgia, Tennessee, Missouri, and Iowa. The beetle's northern spread will be limited by winter cold. Its western movement and activity will probably be limited by low summer rainfall (Johnson and Lyon 1976). Because of irrigation, however, favorable conditions exist in many parts of the Pacific coast states. Several infestations have been discovered and eradicated in California since 1961.

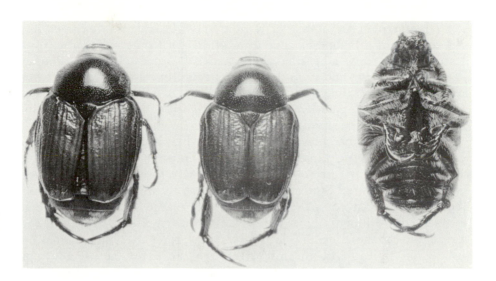

FIGURE 20-2
Adult Japanese beetles are about 10 mm (0.4 in) long with shiny,
metallic green bodies, coppery-brown front wings, and six small patches
of white hair along the sides and back of the body, under the wing edges.
(Photo courtesy Leland Brown, University of California, Riverside)

The Japanese beetle spends about ten months of the year in the
ground as a white grub that measures about 25 mm (1 in) long at
maturity. The grubs feed on plant roots, particularly grasses, in late
summer, early fall, and again in the spring. They cause the most serious
root damage in late summer. Adult beetles emerge from the soil in late
spring and early summer, depending on the climate (Fig. 20-2). Their
period of greatest activity lasts four to six weeks, after which they
gradually disappear. In the summer the females lay eggs in small groups
25 to 50 mm (1–2 in) deep in the soil, usually in turf sod. The eggs
hatch within two weeks, and the new grubs begin feeding on roots.

Symptoms Larval injury to the roots manifests itself as poor
plant growth, sometimes leading to death. Grubs severely injure turf
and shallow-rooted annuals, but not deeper-rooted healthy plants.
Adult Japanese beetles will feed on more than 275 different plant
species (USDA 1973b) but seldom bother conifers (Becker 1938).
They often congregate and feed on flowers, leaves, and fruit of plants
exposed to the sun. They skeletonize the foliage, leaving only a lace-
work of veins. Beetles are most active during the warmer parts of the
day. They usually begin by feeding on the upper and outer parts of
plants and work downward and inward (Becker). A severely attacked
tree or shrub may lose most of its leaf surface in a short time.

Treatment   A number of natural phenomena govern the extent and severity of a Japanese beetle infestation. Dry summer weather may destroy many eggs and kill young grubs. Conversely, wet summer weather favors egg and grub development and usually results in greater numbers of adult beetles. Beetles and grubs are subject to several diseases, to insect parasites, and to several other natural enemies, including birds, moles, skunks, and shrews.

Many naturally isolated areas protected by quarantine and inspection are free of the Japanese beetle. When adult beetles are active, commercial aircraft are routinely sprayed with insecticide before they leave eastern United States airports for west coast destinations. Dead beetles have been found on a number of treated airplanes. Federal and state governments have initiated cooperative efforts to retard the spread of Japanese beetles into new areas; they obtain information on the distribution of the insect through the extensive use of traps along the margins of infested areas, at airports, and at other key areas (USDA 1973b).

Japanese beetle grubs can be brought under long-term control if milky disease (*Bacillus popilliae*) spore dust is applied to infested soil areas. This bacterial spore dust, available commercially, can be applied any time the ground is not frozen. Large areas must be treated, since beetles from adjacent untreated areas can fly in to attack the tops of plants. Adult beetles are not affected by milky disease. Control may take several years, since the bacteria must increase and spread in the soil so as to infect newly hatching grubs. During this period, chemical insecticide should not be applied to the soil; though it will reduce the grub population to a low level, it will also delay or prevent the buildup of milky disease bacteria (USDA 1973b).

Chemical treatment is needed if there is more than one grub per 100 cm^2 (ten grubs/ft^2) present and if turf is being damaged. Diazinon, Dursban®, or Dylox® applied in late summer will protect turf for one year. Adult Japanese beetles seldom become a problem when regular spray schedules are used to control other insects and where grub control is practiced over a large, contiguous area. If beetle control is needed, the foliage and fruit of most plants will be protected if they are sprayed with carbaryl or malathion. Spray thoroughly as soon as beetles appear and before damage is done. Treat only plants that need protection, but treat them repeatedly throughout the period that beetles are active. Follow all label directions on any pesticide.

Insecticides will not fully protect rose blossoms, which open quickly and are especially attractive to beetles. When beetles are most abundant, nip the rose buds and spray to protect the leaves. When the beetles become scarce, let the bushes bloom.

When only a few small plants need protection, you can physically remove beetles by shaking the plant or individual branches early in the morning, when it is cool and the beetles are quiet. Collect the beetles on a sheet placed under the plant and put them in a bucket of water containing a little kerosene. Do this every day because more beetles will fly in.

Plants in poor health are particularly susceptible to attack by beetles (USDA 1973b). Keep plants in vigorous condition with proper fertilization and other necessary cultural practices. The odor of ripe or diseased fruit attracts beetles, which then attack sound fruit. Therefore, remove ripe fruit from the plants and ground.

The Smaller European Elm Bark Beetle

The smaller European elm bark beetle (*Scolytus multistriatus*), introduced into the eastern United States about 1909, had been found by 1974 in all but four of the contiguous United States (Arizona, Florida, Montana, and North Dakota) (Johnson and Lyon 1976). It is also found in Ontario. Although this boring insect can seriously weaken trees directly, its function as the principal vector of the fungus (*Ceratocystis ulmi*) that causes Dutch elm disease constitutes its main threat. The bark beetle attacks all species of elm and zelkova (Brown and Eads 1966).

The adult beetle is shiny, reddish brown or black, and about 3 mm (0.12 in) long. It feeds primarily on the 2- to 4-year-old twig crotches of living elms (Schreiber and Peacock 1974). Beetles breed in weakened, dying, or dead elm wood whose bark is intact. Although healthy elms are readily attacked for feeding, they are rarely exploited for breeding purposes.

The adult beetle and larvae make a distinctive gallery of tunnels between the bark and the wood. The tunnel made by the adult female follows the grain of the wood and is usually 25 to 50 mm (1–2 in) long (Fig. 20-3). Eggs are laid individually in cavities created along the tunnel and packed with some of the frass, the refuse left behind by boring insects, produced by the tunneling. The hatching larvae tunnel between the bark and wood away from the original tunnel and bore into the bark to pupate. As the adult emerges from the pupa, it chews its way through the bark to the surface. Overwintering larvae complete their development during the spring and emerge from the wood as adults at about the time elms begin to leaf out. In summer, the development from egg to adult is complete in about six weeks. Two to three generations occur each year.

Symptoms The smaller European elm bark beetle confines its activities to species of elm and zelkova. It can seriously damage and even kill weakened trees. A heavy infestation can girdle a treetrunk with its system of galleries, disrupting movement of water and food in the tree. The first symptoms that may be noticed are a general weakening of the tree and a lack of foliage on some of the smaller

FIGURE 20-3
Smaller European elm bark beetle
engraving patterns. The bark has been
removed to show adult egg-laying tunnels
parallel to the wood grain and fanning out
from these the many larval tunnels of
increasing diameter.
(Photo courtesy Leland Brown,
University of California, Riverside)

limbs (Brown and Eads 1966). The bark of affected trees will exhibit many "shot holes" on the more severely injured branches or on the trunk. Much paprika-like frass will appear in the limb crotches, clinging to the roughened bark, or on the soil around the trunk. Peeling back bark with shot holes (Fig. 20-3) will reveal single tunnels, parallel to the wood grain, and many smaller debris-filled tunnels running per-

FIGURE 20-4
A smaller European elm bark beetle adult
and a feeding site in a twig crotch.
(Photos courtesy Wayne Sinclair, Cornell
University, and Illinois Natural History
Survey, respectively)

pendicular to the single tunnels. In summer, adult beetles will be found on the bark surface or in their tunnels. They may also feed in the crotches of small healthy elm twigs (Fig. 20-4).

The wilting, yellowing, and loss of foliage caused by Dutch elm disease may be the first prominent evidence of attack by the elm bark beetle. In such cases, beetle damage is insignificant compared with the Dutch elm disease, which is usually fatal. In arid regions, however, the smaller European bark beetle can be directly fatal to elms suffering from drought.

Treatment   You can reduce populations of elm bark beetles and keep them at low levels by keeping trees vigorous and free of weak and dead branches. Beetles are reluctant to bore into the bark of healthy elm trees. To reduce breeding sites, burn or bury branches removed from trees, or remove the bark and burn or bury it. Destroy or debark elm stumps and wood piles. In order to minimize beetle infestation and infection by the Dutch elm disease fungus in areas where the disease is absent, elm firewood should be sprayed with Sevin®.

Do not prune elms during beetle emergence in the spring and summer, since fresh pruning cuts attract beetles. If wounds must be made a short time before or during beetle emergence, paint them with a thin coat of asphalt, which may reduce the attractiveness to beetles. Urea or nitrate fertilization and appropriate irrigation will help revitalize a tree weakened by elm bark beetles and will reduce the likelihood of further infestations. Methoxychlor may be applied as a dormant spray in the spring before beetle emergence (Schreiber and Peacock 1974). Carbaryl (Sevin®) has also been registered for use in certain areas. Thorough coverage of the bark is essential, even in the tops of the trees. These pesticides upset the nervous systems of insects and thus lead to their death.

The Gypsy Moth

Gypsy moths (*Porthetria dispar*) defoliate millions of trees annually in the northeastern United States, eastern Canada, and Europe. Probably no tree pest has received more publicity or provoked more expensive control efforts than the gypsy moth (Johnson and Lyon 1976). The insects were brought into Massachusetts in the late 1860s in an attempt to establish silk moths in New England. Within 12 years after they had escaped, gypsy moths had become a serious nuisance: Trees were defoliated by mid-spring; streets and walks under trees were slippery with crushed caterpillars. The gypsy moth became such a problem that cities established positions for arborists who in many communities are still called moth wardens. Because its natural enemies were absent, the gypsy moth spread rapidly and in 1976 infested more than 50 million hectares (200,000 square miles) of forest land in the northeastern United States (Johnson and Lyon).

Adult female gypsy moths have whitish wings with a 50 mm (2 in) span and wavy, dark bands across the forewings. The males are dark brown. The females deposit eggs on branches, posts, buildings, automobiles, camping trailers, or almost any surface. Masses of 100 to 600 eggs are covered with buff-colored hairs (Johnson and Lyon). The larvae hatch in early to mid-spring and within a few days climb into trees to feed on foliage. As they mature, the caterpillars feed primarily at night and descend to spend the day in protected places. Mature caterpillars attain lengths of up to 65 mm (2.5 in). The larval stage lasts about seven weeks, after which the caterpillars find protected places and pupate. Moths emerge in the early summer, though the female does not fly. Adults die without feeding after they mate and deposit eggs.

Symptoms   Defoliated trees with severely chewed leaves are the obvious mark of the gypsy moth. The decline and death of many oaks in New England have been associated with defoliation by the gypsy moth (Houston 1980). Most deciduous trees can withstand one or two years of infestation before serious decline takes place, but conifers will die after one complete defoliation (Johnson and Lyon 1976). Trees are weakened most severely if they put out one or more flushes of regrowth after defoliation early in the growing season.

TABLE 20-1

The hosts of gypsy moth larvae, in order of preference (leaves) (Adapted from Johnson, W. T., and H. H. Lyon. 1976. *Insects That Feed on Trees and Shrubs.* Ithaca, N.Y.: Cornell University Press)

Preferred Hosts	Less Preferred Hosts	Occasional Hosts	Least Preferred Hosts
alder	elm	beech	ash
apple	black gum	white cedar	balsam fir
basswood	hickory	hemlock	butternut
hawthorn	maple	pine	black walnut
oak	sassafras	spruce	catalpa
some poplars			red cedar
willow			dogwood
			holly
			locust
			sycamore
			yellow poplar

Treatment   A number of plants listed in Table 20-1 should remain relatively free of damage by gypsy moths. Despite early attempts at quarantine, chemical eradication, and control programs, the gypsy moth has continued to spread, particularly in the South.

The bacterial pathogen *Bacillus thuringiensis* is sometimes sprayed on large forested areas to reduce the gypsy moth population. The larvae of many caterpillar species will become infected by the bacteria and die. Sprays of Imidan®, methoxychlor, carbaryl (Sevin®), and Sevimol® are also effective against the gypsy moth (Pirone 1978b), especially on young larvae. DDT, because of its persistence, was more effective and was widely used until it was banned. Parasites and predators of the gypsy moth have been introduced from Europe and Asia and seem to be having some effect on moth populations (Johnson and Lyon 1976).

SUCKING INSECTS AND MITES

Like insects with chewing mouthparts, sucking insects attack all parts of plants, though they usually prefer leaves and developing shoots. They may stunt and deform new growth, curl leaves, form galls, and seriously weaken plants. They may cause mature leaves to yellow or glaze and, if infestation is heavy, to dry up.

Aphids

Aphids are soft-bodied small insects with sucking mouthparts. Some have wings; others do not. Species may be black, brown, green, pink, purple, red, or yellow. Most aphids have naked bodies, although many secrete a white, cottony, wax-like material for protection. In most areas, overwintering eggs hatch into wingless females that, without fertilization, give birth to living females. The ability of females to reproduce without mating is termed *parthenogenesis*. Young are brought forth throughout the summer, accounting for rapid increases in aphid populations. Winged females are produced later in the summer or when crowding occurs. In late fall, sexual males and females are born. After mating, the female deposits one or more overwintering eggs (Essig 1958), and the cycle continues the following year.

Symptoms   Aphids live in colonies on the bark, leaves, blossoms, fruit, and roots of most plants at one time or another (Fig. 20-5). Most species excrete large amounts of honeydew, which coats the host plants and often drips onto objects below. The honeydew attracts ants, flies, and bees and serves as a medium for the growth of sooty mold fungus. In most cases, injury from aphids is inconspicuous, but the

FIGURE 20-5
Rose aphids (*Macrosiphum rosae*)
on a young stem include wingless stem
mothers and young. Note the spherical
black parasitized aphids.
(Photo courtesy Leland Brown,
University of California, Riverside)

plant gradually weakens as its food is sucked. Young growth may be stunted and deformed.

Treatment A number of natural enemies prey on aphids. Important predators are the larvae of green lacewings (Neuroptera), ladybird beetles (*Coleoptera*), and syrphid flies (*Diptera*). Lacewing eggs are available commercially for biological control of aphids.

Knock aphids off infested plants with a spray of water; use a hose for a few small plants or a power sprayer for more numerous, larger plants. A mild soap solution is more effective than water alone and will kill some of the aphids that may remain on plants.

Control ants in the landscape. Ants often "domesticate" aphids and use their honeydew as a source of food. Keep ants out of trees and shrubs by spraying the ground and the base of plants with diazinon. Alternatively, band the trunks of trees and shrubs with a sticky material, such as Stik-Em® or Tanglefoot®, that traps crawling insects.

Aphids can be controlled chemically with any one of a number of contact insecticides or systemics. Diazinon and malathion are broad-spectrum contact insecticides. Disyston®, Meta-Systox-R®, and Orthene® move within plants and poison sucking insects. Systemics can be watered into the soil for root uptake; this has proved successful with small shrubs such as roses.

The San Jose Scale

The San Jose scale (*Quadraspidiotus perniciosus*) afflicts a wide range of broadleaved woody plants in many parts of the world. In about 1870, this pest was introduced from the Orient into California, whence it quickly spread and began to afflict deciduous fruit trees throughout the United States (Essig 1958). Until the newer synthetic organic insecticides became available in the late 1940s, the San Jose scale was one of the most common, destructive, and widely distributed pests to infest woody plants (Johnson and Lyon 1976). If not controlled, it can kill plants.

Most scale insects are very small. The San Jose scale is an armored insect, characterized by a protective shell or scale that covers its body (Fig. 20-6). The young, called *crawlers*, are born alive or from eggs and are able to crawl considerable distances to find a feeding site. During the first molt, they lose legs, antennae, and anal filaments and assume a stationary form, consisting primarily of mouthparts (Essig). The scale covering is black at first and gray when more developed and has a tiny, yellowish, nipple-like formation on the insect's back. Armored scales, in contrast to unarmored or soft scales, produce almost no honeydew. As many as five overlapping generations of scales develop in a single season (Johnson and Lyon 1976). The bark of shoots, limbs, and trunk is often completely encrusted with scales, which suck sap from the plant. Small scales, gray to black, can be seen on bark and on fruit surfaces. The bodies under the scale shell are bright yellow.

Symptoms Plants seriously attacked by San Jose scale will weaken and die unless the pests are controlled. Leaves on weakened branches may drop prematurely, and seriously affected branches may be killed back. Dark discolored areas may appear on infested young bark and the tissue below. The scale has an apparent toxic effect on plant tissue (Essig 1958).

FIGURE 20-6
An old heavy infestation of San Jose scale on an apple twig. Several females (large) and many males (small, lighter gray) are present. (Photo courtesy Leland Brown, University of California, Riverside)

Treatment   Even though several beneficial insects prey on the San Jose scale, they can seldom reduce scale numbers during periods of rapid population increase.   Susceptible plants should be fertilized sparingly with nitrogen or not at all; scale insect population can be double on vigorous plants what it is on plants of low vigor (Johnson 1982).   Dormant oil sprays may be used on deciduous plants as late in winter as possible but before the buds begin to swell (Moore, Davis, and Koehler 1979).   Spray broadleaved evergreen plants, tender, dormant deciduous plants, and plants in leaf with a refined summer oil to control scale insects with little or no injury to the host plant.   To increase their effectiveness, fortify either the dormant or the summer oil sprays with malathion or diazinon.   During the spring growing season, apply the first spray when you first see the tiny crawlers, or immature scales, on the plants.

Spider Mites

Mites have eight legs when mature and have the other distinguishing characteristics of spiders (Fig. 20-7), rather than those of insects. Of the plant-feeding mites, spider mites are the most injurious to agricultural crops and landscape plants throughout the world (Essig 1958). Many species of spider mites attack woody plants.

As their name implies, mites are extremely small.   Adult spider mites range in length from 0.20 to 1.0 mm (0.01–0.04 in).   The various species may be yellow, green, red, or brown; the "red spider" is familiar to most horticulturists.   Many of these mites also spin extensive webs, which protect them and their eggs.   As adults, the mites hibernate in the soil in cold regions, and in trash and plant crevices in warmer areas. Adults appear in late spring.   Each female lays 50 to 60 eggs, primarily on leaves.   The eggs usually hatch in three days, the mites reach maturity in another 10 to 12 days, and they live but one to two weeks in the summer.   Needless to say, mite populations can increase rapidly under favorable conditions.

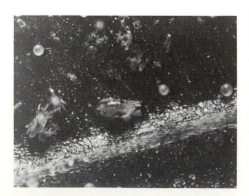

FIGURE 20-7
Pacific spider mites, *Tetranychus pacificus*, shown near vein of a bean leaf used in rearing. To left: slender, adult male near small, fat nymph; adult female in center; several shiny, translucent, spherical eggs are seen, without webbing; near top: whitish surface cells, lacking chlorophyll, from mite feeding.
(Photo courtesy Leland Brown, University of California, Riverside)

Symptoms   With a pair of needle-like teeth (*chelicerae*), the spider mite punctures and drains the epidermal cells of the host plant. Injured leaf, fruit, or shoot surfaces have a flecked or stippled appearance. Leaves become yellow or bronze and drop; fruit may be scarred; weakened plants can be killed. Hot, dry weather and a dusty environment favor mite buildup and injury to plants. Some mite species, however, are most destructive during the cooler weather of spring or fall.

Treatment   To minimize spider mite buildup, keep plants vigorous and well watered, and the landscape free of dust. Sprinkler irrigating or spraying susceptible plants with water will help. In arid regions, mites usually disappear with the first fall rains.

Several insects and mites prey on spider mites but effectively reduce their numbers only after leaves are severely damaged (Johnson and Lyon 1976). Unfortunately, sprays often do more harm to mite predators than to mites. In any case, biological control of spider mites has not been particularly effective.

Galls

Tiny wasps and flies commonly cause a wide range of swollen abnormalities, or *galls*, on plants. In North America alone, more than 800 different kinds of galls are found on oak trees and more than 400 on beech (Schuder 1978). Galls are found on almost all species of landscape plants but are particularly numerous on rose, willow, walnut, pine, elm, and birch. Practically all plant parts may be attacked by insects that stimulate galls. Slightly more than half of the different galls are found on leaves; about 14 percent are found on twigs and branches; 7 to 9 percent each are found on buds, roots, flowers, and fruits.

Each species of insect causes a particular type of deformity on a particular plant part on a single or closely related plant species. Each type of gall has a distinct shape, size, structure, and color (Fig. 20-8). The insect involved can usually be identified entirely on the basis of the gall produced.

When a wasp lays eggs, plant tissue is stimulated into abnormal growth that usually surrounds or engulfs the eggs. The gall protects and feeds the larvae developing within. After wasp adults emerge from galls, they usually lay eggs in the early spring that cause gall formation and larval growth to occur when plant growth is most active. One generation per year is normal. Some wasp species produce alternating generations, so the adults that emerge from a gall are different from the adults that laid the eggs. The new adults usually lay eggs on a different plant part; the resulting larvae and adults are similar to those that existed two generations before.

FIGURE 20-8
Willow apple galls caused by the sawfly (*Pontania pacifica*)
on leaves of willow (left). "Oak apples" up to 50 mm (2 in) in diameter
induced on a branch of valley oak by the gallwasp *Andricus californicus* (upper
right). Walnut purse galls are caused by an eriophyid mite (lower right).
(Photos courtesy Carlton Koehler, University of California, Berkeley, 1980)

Symptoms   Galls seldom cause serious injury, although twig galls
produced by two species of *Callirhytis* in eastern North America can
seriously injure or kill oak trees.   The horned oak gall, caused by
C. cornigera, can injure black, blackjack, pin, scrub, and water oak.
C. punctata produces gouty oak galls on the twigs of black, pin, red,
and scarlet oak (Johnson and Lyon 1976); these galls, up to 50 mm
(2 in) in diameter, often grow together and engulf more than 300 mm
(12 in) of a twig.   Felt (1940) saw one red oak that bore an estimated
20,000 gouty oak galls. Beyond the galls, branches may die back.

Treatment   Control measures are seldom taken against gall-causing
insects.   Plants are sometimes sprayed with a dormant lime sulfur or a
dormant miscible oil to reduce the number of overwintering insects.
Valuable trees and those heavily infested with gall insects should also
be sprayed with carbaryl (Sevin®) or methoxychlor-Kelthane just after
mid-spring and again a month later (Pirone 1978a). Dead and heavily

infested branches should be removed and destroyed before the adults emerge in early spring.

NEMATODES

Plant-parasitic nematodes are microscopic worms 0.5 to 5 mm (0.02–0.2 in) long. Unlike most other nematodes, they have a stylet (which resembles a hypodermic needle) in their mouth which they use to pierce cell walls. They can then feed and move into the penetrated tissue. Mature nematodes may be cylindrical or pear-like in shape, while most of the young are slender and worm-like. They may live in the soil and feed on roots, or may spend most of their adult lives inside roots. Once nematodes have entered roots, they spread easily and are difficult to control.

Symptoms Nematodes attack almost any plant part but frequently attack woody plants only in the roots. The pinewood nematode (*Bursaphelenchus lignicolus*), however, invades the vascular system of twigs, limbs, and trunks in pine trees. Foliar nematodes have been reported on woody plants. Nematodes feed most commonly in or on the tips of small roots. The injury they cause reduces absorption and movement of water and nutrients. The most common symptoms are therefore typical of many other soil and root disorders: poor growth, leaf chlorosis, early leaf drop, and so forth. Diagnosis in woody plants is difficult without an examination of the roots and soil.

Root-knot nematodes are by far the most common and attack the greatest number of plants; *Meloidogyne incognita* alone infests more than 1700 species (Streets 1969). Root-knot nematodes enter roots, causing the cells and tissue to enlarge. The smaller infested roots will have many small to medium-sized swellings or galls (Fig. 20-9). Large swellings differ from crown gall in that they extend along the root instead of adopting a spherical shape, and are associated with many smaller knots. *M. incognita* is most common in areas with mild winters, while *M. hapla* infests crops and landscape plants in the northern United States and Canada. Nematodes are not particularly active below soil temperatures of 18°C (65°F) (Streets). Porous sandy and loam soils favor their spread and activity. When they enter the soil, nematodes essentially live in the film of water surrounding soil particles; thus wet soils favor nematode infestation (Tattar 1978).

Meadow or root-lesion nematodes (*Pratylenchus* spp.) also attack a wide range of plants, entering the small roots and feeding in the cortical tissue of the root bark. The injured tissue dies, forming a dark-colored lesion visible on the surface of the root. Young roots can be

FIGURE 20-9
Beadlike galls of root-knot nematode,
Meloidogyne sp., on the small roots of
mulberry. (Photo courtesy William Moller,
University of California, Davis)

girdled or seriously injured and the upper plant condemned to poor growth.

Stubby root nematodes (*Trichodorus* sp.) feed externally on root tips, causing the tips to thicken or become corky. As secondary roots begin growth, they too are attacked. The root systems of affected plants, including the feeder roots, are usually sparse. The curtailed root system results in a weakened top.

Treatment You can check a planting site for root-knot nematodes by examining the roots of the most susceptible plants. If no susceptible plants grow on the site, test the soil in containers in which you plant seeds or seedlings of fast-growing species like beans, squash, or tomatoes. After the above-ground portion has grown for at least three or four weeks, carefully remove the roots, wash them, and examine them for root-knot galls. The number and size of galls will give a rough estimate of nematode infestation (Streets 1969).

If the soil is free of nematodes, be sure that planting stock is free of them also. Most nurseries use nematode-free soil if government regulations require pest-free plants. Even so, before accepting plants, check the roots for nematodes. If you find any root-knot galls, reject all plants from that source.

Although many landscape plants are susceptible to root-knot nematodes, a number are resistant or tolerant. If nematodes are a problem in your area, check with local government publications or extension personnel for lists of susceptible and resistant or tolerant plants. Such lists are provided by Streets (1969) and Lear and Johnson (1975). You may use resistant rootstocks for plants that would otherwise be susceptible, such as Nemaguard or S-37 rootstocks for flowering and fruiting peaches.

Isolated planting areas (containers, interior planters, roof gardens, and the like) can be freed of nematodes, by heat sterilization or chemical fumigation before planting. Place pest-free soil in self-contained planters to ensure healthy plants and to minimize future maintenance. Steam or electric heat will pasteurize the soil. Temperatures above 45°C (110°F) will kill nematodes; those above 60°C (140°F) will kill most soil disease and insect organisms but will not kill many of the beneficial microorganisms. Solar heat may be used when space and time allow. When moist soil is covered with clear polyethylene film and exposed to the summer sun for three to four weeks, nematodes, verticillium wilt fungus, and some annual weeds may be killed to a depth of 450 mm (18 in). In open landscapes, nematode populations can be effectively reduced for a year or two to help new plantings get off to a vigorous start. Nematodes from adjacent untreated soil move back into the treated areas, and nematode kill will be uncertain if infested large roots (more than 25 mm or 1 in. in diameter) remain in the fumigated soil.

No presently registered chemical nematicide is selective; any one of them must therefore be used before planting. Soil for self-contained planters can be fumigated in airtight containers, in the planters themselves, or on an outside work area. You can fumigate small landscape areas by injecting the nematicide into the soil with a handgun or applying water-soluble materials in solution on the soil surface. Follow label instructions carefully. After application, wet the soil to a depth of 20 or 30 mm (1 in) to help keep the fumigant in the soil. Allow most fumigants a two-week period to dissipate before you plant.

DBCP fumigants have proved extremely effective against nematodes and may be applied without plant injury to certain plantings (Lear and Johnson 1975). Its use, however, is no longer permitted in the United States because underground water supplies have become contaminated in some areas and it has been linked to sterility of male employees at manufacturing plants of the chemical.

FURTHER READING

ANDERSON, R. F. 1960. *Forest and Shade Tree Entomology*. New York: John Wiley.

JOHNSON, W. T. and H. H. LYON. 1976. *Insects That Feed on Trees and Shrubs*. Ithaca, N.Y.: Comstock.

PIRONE, P. P. 1978. *Diseases and Pests of Ornamental Plants*. 5th ed. New York: John Wiley.

CHAPTER TWENTY-ONE

Pest Management

Key goals for those who grow plants in the landscape are plant health and attractiveness. Much time and effort may be needed to protect plants from pathogens, insects, nematodes, and other pests. The main approach to pest control has been chemical since the 1940s, when the chlorinated hydrocarbons, the organophosphates, the carbamates, and a host of fungicides, herbicides, and other pesticides were first developed. The effectiveness of many chemicals has led to the reduction or discontinuance of some earlier control procedures—plant sanitation, crop rotation, encouragement of natural enemies, and other cultural strategies.

Unfortunately, heavy dependence on chemical control has caused problems in certain situations. Some major pests have developed resistances to previously effective insecticides. As of 1975, 75 percent of the most serious agricultural insect pests in California had developed resistance to one or more major insecticides. In some cases, the most effective chemicals caused resistant individuals to predominate all the sooner, resulting in a resistant strain. Although the most noteworthy examples involve insects, other pests develop resistances also.

Other problems have emerged from primary reliance on chemicals. After treatment with broad-spectrum pesticides, pest populations, particularly insects, will sometimes drop drastically and then surge to levels higher than before. Resurgence occurs because the pesticides kill not only the pest but its natural predators as well. If predators are not killed outright, many starve or emigrate out of the immediate area in search of food. The surviving pests can then multiply rapidly, since their food supply, unlike that of their predators, has not been interrupted. Some organisms that have not previously been problems will increase to damaging levels when their predators are destroyed.

For example, mites may increase dramatically on plants that have been sprayed with certain insecticides for the control of other pests.

In the face of pest resistance, pest resurgence, and secondary pest outbreaks caused by very effective pesticides, growers have often increased pesticide application in efforts to obtain more complete control. Increases in both concentration and frequency not only intensify the problems but hasten environmental contamination by the more persistent chemicals. DDT is the notorious example. It has been used effectively against many insects that transmit serious animal and plant diseases, damage many crop, forest, and landscape plants, and annoy people. One of its advantages as an insecticide—persistence—has proved to be a great disadvantage. DDT residue is everywhere, carried by wind and water to places remote from the sites of application. In addition, DDT and related chlorinated hydrocarbons accumulate in the food chain and may be concentrated in this fashion to hazardous levels.

The United States banned DDT for most uses in 1972 and has severely restricted the use of aldrin, dieldrin, endrin, heptachlor, and chlordane. Restrictions affect the sale and application of many pesticides. In the United States, chemicals are to be used only on the plants or sites listed on the label, only against the listed pests, and only in the specified concentrations and at specified times. Even though pesticide residues do not cause the problems on landscape plants that they cause on edible crops, the use of hazardous chemicals in the landscape is severely restricted because of the proximity of people, pets, and non-targeted plants.

For a number of reasons, primary reliance on pesticides or any other single pest control method has usually not been effective. Using a combination of techniques to control or reduce pests, while minimizing undesirable effects, is far from a new idea but is now receiving renewed interest. This approach has been greatly strengthened by a broader understanding of the interdependence among living organisms. A landscape planting, an urban forest, an agricultural area, or an entire region must be considered as a unique ecosystem. It may not be possible to reduce the population of one species without changing other components of the system. This concept suggests that even pests may be accepted within certain economic or aesthetic bounds. This approach to pest control is called integrated pest management.

INTEGRATED PEST MANAGEMENT

Flint and van den Bosch (1981) have outlined the essentials of an integrated pest management (IPM) approach. It is assumed that the pests are known to cause serious injury or effects on the host plant or its surroundings. Adapted to managed landscape plantings, IPM includes the following concepts:

A plant or landscape planting is a component of a functioning ecosystem, even though the planting is man-made. Hence, actions should be designed to develop, restore, preserve, or augment natural checks and balances, not necessarily to eliminate pest species.

The mere presence of a pest organism does not necessarily constitute a pest problem. Acceptable population and damage levels must be determined.

All possible pest control options should be considered before action is taken. Techniques employed should be as compatible as possible.

Any IPM program should be based on certain guidelines (Flint and van den Bosch 1981):

Understand the biology of the plants involved, especially the manner in which they are influenced by the surrounding ecosystem.

Identify the key pests, know their biology, recognize the kind of damage they inflict, and study the economic and aesthetic consequences of control measures.

Identify the key environmental factors that impinge (favorably or unfavorably) upon pest species and potential pest species in the ecosystem.

Consider concepts, methods, and materials that, individually and in concert, will facilitate permanent suppression or restraint of pest and potential pest species.

Structure the program so that it can be adjusted to meet change or varying situations.

Seek the weak links in the life cycle and population structure of important pest species, and direct control practices as narrowly as possible at those weak links. Avoid broad impact on the plant ecosystem.

Whenever possible, employ methods that preserve, complement, and augment the biological dynamics that characterize the ecosystem.

Whenever possible, diversify the ecosystem.

Monitor pests, natural enemies, and tree health regularly.

Anticipate unforeseen developments, move with caution, and be aware of the complexity of the landscape ecosystem and the changes that can occur within it.

Aesthetic Injury Threshold

One of the tenets of IPM is that certain levels of injury or pest populations are to be expected. The aesthetic injury threshold is the highest level of pest habitation or damage that would be acceptable to most of the people who use the affected area (Olkowski 1973). Such levels may be difficult to determine because of the variability of plants, climates, and human values or tolerances. Landscapes are used and viewed from both close and distant ranges, similar to the "near" and "far" landscape concept used by Sachs and coworkers (1970) to guide the chemical control of growth. Higher levels of plant injury will be more acceptable in a "far" landscape.

Some pests can cause death or serious injury, but many only change the appearance of plants or create products that annoy people or cause damage to nearby vegetation. Plants can become quite unattractive when leaves are damaged by defoliating insects and diseases. A number of aphids, scales, and leaf hoppers excrete "honeydew," which can drip on sidewalks and cars. Caterpillars may leave the plants on which they originate to injure adjacent plants. Occasionally, a large insect population may disturb viewers even though it causes little or no harm to host plants. Acceptable levels of injury or pest presence usually can be increased if the public is educated about pest management programs and their consequences to plant and human health.

Control Threshold

Control measures must usually be taken before the injury threshold is reached. This requires information about the pest's life cycle and monitoring of the plants, the pest, and its predators. Monitoring and sampling techniques have been designed for many agricultural insect pests, but have not been well worked out or incorporated into control programs for pests of landscape plants. The few exceptions include monitoring systems for fire blight (caused by *Erwinia amylovora*) and the bark beetles that transmit Dutch elm disease (*Ceratocystis ulmi*). These systems help arborists differentiate between the mere presence of a pest and densities high enough to cause unacceptable damage. Needless to say, such information is likely to be developed only for pests that cause serious damage to valuable plants that are cultivated in large enough numbers to make the effort cost-effective.

Integrated Pest Management Options

Not all measures for controlling a pest or minimizing damage need be delayed until the control threshold approaches. Some of the most economical and effective management tactics are preventive and should be employed even before pests are observed. A regular field-monitoring program must be used to determine the continuing efficacy of preventive tactics; if pests rise above control thresholds, more immediately effective measures, such as pesticide application, must be adopted. Integrated pest management programs should, however, employ a variety of practices in concert to obtain the least disruptive, most effective, long-lasting pest control at an acceptable cost. Five major types of control measures—regulatory, genetic, biological, cultural, and chemical—contribute to an integrated pest management program. Much of the following discussion is based on *Introduction to Integrated Pest Management* (Flint and van den Bosch 1981).

Regulatory Control   An individual arborist has little or no choice in the initiation or implementation of regulatory controls, which are usually carried out by governmental agencies. Such controls can exert great influence on pest populations in a given area. Pests may be kept out of an area through quarantine and inspection; outbreaks of new pests in small areas may be eradicated; large-scale cooperative efforts (such as removal of trees infected with Dutch elm disease) may be properly supervised and carried out. Such programs should usually be supported to achieve the greatest good for the greatest number of people.

Genetic Control   Control of pests by genetic means takes two forms: (1) host resistance—selecting or developing a plant that is resistant to particular pests; and (2) autocidal control—modifying the genetic makeup of the pest population so that it becomes self-destructive and cannot survive.

Developing *host resistance* is the most successful and ecologically sound way of avoiding problems caused by insect, disease, and nematode pests. Select plant species and cultivars with proven resistance to noxious pests associated with the growing area. Various regional publications list plant species and their relative susceptibility or resistance to common pests, particularly nematodes and vectors for disease. Grafting may protect certain susceptible plants from soil-borne diseases or nematodes. A flowering peach, for example, can be grown on a rootstock resistant to nematodes. You can only take advantage of pest resistance in perennial plants, of course, by proper selection before planting. This control cannot be applied to existing plants.

Within a susceptible species, individuals may be resistant or relatively resistant to a given pest. Such plants can be propagated vegetatively to preserve that resistance. They can also be used in breeding programs that will convey this resistance to other plants with more desirable characteristics or, vice versa, will improve certain characteristics in the resistant selection.

Breeding for pest resistance is a long-term program, particularly when perennial plants are involved. What is more disheartening, a number of pests have evolved strains able to overcome the host plant's mechanism of resistance—much as pests have developed their own resistance to pesticides. A plant naturally resistant to one serious pest may be susceptible to others, so that careful testing must preceed the release of a new selection for widespread planting. Oriental pear, for example, is used as a rootstock to impart fire blight resistance to pear fruit trees but predisposes the tops to "hard end" fruit and pear decline.

Host plant resistance may be due to physiological factors (such as toxic compounds within the plant that inhibit the pest), mechanical factors (such as cuticle too thick or tough for the pest to penetrate), or the host's ability to survive despite damage from pest populations. Host resistance is usually most effective and long-lasting if it relies on more than one gene and preferably more than one character. Pests are less likely to develop strains able to overcome such "polygenic" resistance. Host resistance and the enhancement of other character-istics may be more likely in the future as gene transfer becomes better understood and techniques improve.

Autocidal control causes the pest to contribute to its own demise. The best known example is control of the screwworm, a serious live-stock pest in the southern and southwestern United States: Releasing large numbers of laboratory-propagated sterile males into the environ-ment has eradicated the screwworm in Florida. This technique has also eradicated two species of fruit flies on the island of Rota in the Marianas (Flint and van den Bosch 1977) and has been used against invasions of Mediterranean fruit flies in California.

Several factors are essential for successful eradication by the sterile male technique: (1) The females should mate only once; (2) the area of release must be geographically isolated so that migrants of the species will not come in from untreated areas; (3) the sterilized males must be sexually competitive with normal males; (4) the pest must be amenable to laboratory rearing, and such rearing must not have a debilitating effect on its field performance or survival; and (5) the budget for the program must be extremely large (Flint and van den Bosch 1977). A successful autocidal control program must eradicate the pest. If a few females continue to produce offspring, or if migrants come from untreated areas, the expensive program must be continued or abandoned.

Biological Control   Control by natural enemies can be effective, long-lasting, economical, and minimally disruptive of the ecosystem. Natural enemies include organisms that feed on pests (predators), para-sitize pests (parasiters), cause diseases among pests (pathogens), or dis-place pests (competitors). When managing any landscape, you should assess present and potential plant pests and their natural enemies. Assessment is complicated by the fact that most landscapes have been greatly changed from their original natural state. Exotic plants may be accompanied by their native pests but not the pests' natural enemies or may be susceptible to organisms that do not damage native plants.

In essentially undisturbed native plantings and in wisely managed landscapes, many potential pests are kept at safe levels by other organ-isms, the food supply, or other environmental factors. The importance of these natural controls is seldom appreciated until the balance is upset by a catastrophe (flood, fire, major land development), some ill-advised cultural practice, or a toxic, broad-spectrum pest control program. In such situations, pests that are seemingly new rise to injurious levels or major pests become even more damaging.

Classic biological control involves the deliberate introduction and establishment of natural enemies where they did not previously exist. Such programs are used mainly to control exotic pests that have been introduced without their key natural enemies. The exotic species may be relatively unknown as pests in their native habitats, where they are naturally kept within bounds. Because of the absence of natural enemies, exotic species comprise a high percentage of major pests in the United States. Seventeen of the 28 most serious pest arthropods (insects, mites, and so forth) in the United States are of exotic origin (Glass 1975).

The introduction of natural enemies has been used successfully against about 100 insect and weed pests throughout the world (van den Bosch and Messenger 1973). The oriental moth (*Cnidocampa flavescens*), which attacks shade trees, has been substantially controlled in Massachusetts by an introduced parasite. Initial results in Australia and the United States indicate that crown gall, caused by *Agrobacterium tumefaciens* (see Chapter 19), can be prevented if the seed or plant is treated with a closely related bacterium, *A. radiobacter* 'Strain 84' (New and Kerr 1972). Under favorable conditions, ladybird beetles (Coccinellidae) and lacewings (*Chrysopa* spp.) can be effective predators greatly reducing aphid infestation.

Certain practices preserve and augment biological control activity. Parasites of the red-humped caterpillar larva (*Schizura concinna*), a serious defoliator of California highway landscapes, were strengthened when they were given access to the nectar of nearby flowering shrubs. State workers changed aphid control on those shrubs from a broad-spectrum insecticide to a dilute soap spray (Pinnock and others 1978) so that the parasitic insects were not killed when they sought food in the shrubs. Flowering plants can also be deliberately established as food sources for insect predators and parasites. Certain cultural operations can favor desirable organisms over pests.

Insect pathogens can greatly reduce pest populations. *Bacillus thuringiensis*, a bacterial pathogen, is particularly effective against a number of plant-feeding caterpillars (Lepidoptera) (Flint and van den Bosch 1977). This bacterium is specific to certain insects and has little or no effect on other organisms. *B. thuringiensis* is available as a registered pesticide and is widely used.

Cultural Control A number of maintenance practices or modifications of them can make the environment unfavorable for pest reproduction, movement, or survival. Other mechanical or physical practices may specifically combat plant pests. Such controls, including some of the oldest cultural practices known, are usually more effective for preventing pest buildup than for correcting an existing pest problem. Pest control or reduction is often the by-product of a practice employed for another primary purpose, as irrigation and fertilization, used to maintain or improve plant growth and appearance, may enhance the plants' ability to withstand infection by unaggressive pathogens. In other cases, pest buildup may be prevented more effectively if the timing or specific method of a cultural practice is altered; selective pruning, used in the spring to open up a tree or shrub to sunlight and air movement, can reduce the severity of some leaf diseases by reducing humidity within the plant canopy. Some of the cultural techniques devised for agriculture (crop rotation, harvest procedures, and so forth) are not readily adapted to diverse perennial landscape plantings. Others are broadly effective for minimizing pest buildup and damage. Timing is critical to the success of much cultural pest control. Certain practices will be described here to indicate the range of possibilities in this realm.

Maintaining plant vigor, as already mentioned, not only enhances plant appearance but can increase tolerance to damage and infestation by borers. The goal is moderate plant vigor through careful management of nutrition, aeration, and moisture. In certain species, extreme vigor can increase susceptibility to diseases (fire blight by *Erwinia amylovora*) or certain leaf-feeding and scale insects. The optimum vigor will depend on the plants, the environment, and the potential pests.

Pest-free plants should be selected for planting so pests are not introduced at the planting site. Plants and the soil in which they are moved should be free of pathogens, insects, nematodes, and weed pests. For some species (flowering cherry, for example), virus-indexed plants are available, and their superior performance and appearance in the landscape justify the additional cost (see Fig. 19-9).

Pruning can remove weak and dead wood, the potential brooding site of boring insects. You can thin out the dense head of a plant to reduce certain leaf diseases (anthracnose, powdery mildew, and scab) and increase spray penetration. Prune off infected or infested branches and shoots to remove pests and any continued buildup. When you transplant or when plant nutrition is low, prune away part of the top to improve plant vigor and resistance to certain pests. Minimize the possibilities of wood decay by training and pruning young trees so that large pruning cuts will not be needed later to correct poor tree structure. Make pruning cuts properly to promote rapid callusing and further reduce the likelihood of decay.

Sanitation in and around plants can effectively remove or destroy the breeding, refuge, and overwintering sites of pests. Sanitation which usually involves removing infected, dead, and fallen twigs, leaves, and fruit is effective against various insects, plant pathogens, rodents, and nematodes.

Cultivation can serve as a form of sanitation in that it can mutilate and bury plant residues that would otherwise be habitats for pests. Cultivation for pest control is usually most feasible in extensive landscapes with deep-rooted plants, where root injury from cultivation can be minimized, and mechanized equipment can be used more easily. At the same time, cultivated soil can increase certain pest problems. Rain or sprinkler drops can splash mud and the spores of fungus pathogens from bare soil onto the canes and foliage of shrubs, thereby increasing the likelihood of disease.

Mulching can often reduce splashing and the drops of mud that would protect spores deposited on plant surfaces. Mulches can effectively eliminate or reduce weeds in landscape plantings (see Chapter 13). If mulch is composed of infected or infested plant material, however, it can be a source of disease inoculum or insect buildup. Fortunately, reports do not indicate that mulch is a significant contributor to pest problems on plants. Mulch can also be sprayed to inactivate or minimize pests. Each planting complex must be analyzed to determine whether cultivation or mulching would favor an increase or decrease in the pests that are present.

Burning of grain and rice stubble both eliminates the straw, which uses nitrogen if it is left to decompose, and controls serious pests that would otherwise overwinter in the stubble. Similar burning in the landscape is usually not desirable, although more extensive areas can be "control burned" to kill undesirable brush, small trees, and weeds so that range species of plants can be seeded or hazard from wildfire reduced. When permitted, prunings, leaves, and even entire plants can be burned to destroy tissue infected with diseases such as fire blight, Dutch elm disease, and Armillaria root rot (see Chapter 19).

Species diversification in the landscape can prevent the kind of devastation that occurred in many communities when Dutch elm disease struck (see Chapter 19); elms accounted for more than 75 percent of the landscape trees in many cities. Should a serious malady strike, single-species plantings are particularly vulnerable; a high density of one susceptible species favors insect and pathogen spread. Some cities and agencies ensure diversification by limiting the trees of any one species that may be planted in a given population. A cultivar chosen for the genetic variation it will introduce may of course be susceptible to unsuspected problems in the future, even if it has superior characteristics. For adaptation and survival of a species or planting in an extensive urban area, genetic variation is extremely important; it is less critical in single, small landscape plantings.

Species diversification can provide alternate food sources and refuge for the natural enemies of pests. As already noted, the biological control of caterpillars was enhanced when flowering shrubs were introduced to provide nectar for parasites of the caterpillars (Pinnock and others 1978). Leius (1967) found that unsprayed apple and pear orchards in Canada encouraged more parasitism on tent caterpillar (*Malacosoma americana*) eggs and codling moth (*Laspeyresia pomonella*) larvae when they had ground cover high in nectar-producing flowers. These kinds of interactions need more study so that landscapes can be planned with such built-in protection.

Timing of planting can be adjusted to avoid periods of insect infestation; this technique has been used successfully with some field and vegetable crops but not with woody perennial plants. Timing can be important, however, in protecting susceptible plants from spring frosts and frozen or waterlogged soil.

Eradication, the death or permanent removal of all individuals of a pest species, may focus on a plant, a landscape, or a region. Eradication of a pest from a region normally requires governmental action in order to be effective. Depending on the size of the area, the number of plants involved, the pest, and the expected speed of infestation, eradication can involve intensive spraying or fumigation, removal of all susceptible plants, the release of sterile male insects, or a combination of these methods.

Several things can be done to eliminate pests at the plant or planting level. An infected plant part can be pruned off and burned so that the pathogen does not spread; this constitutes eradication only if all of the pathogen is eliminated. Soil can be sterilized by heat or steam as well as by chemicals. Although it may not be feasible in existing plantings or large areas, soil sterilization can effectively eliminate a number of soil-borne problems in small outdoor areas surrounded by paving or buildings, individual planters, or nursery or greenhouse soil, where young plants may be quite susceptible. It must be remembered that sterilization kills not only pests but their natural enemies and other beneficial species. A variation on conventional heat treatment is solar treatment of soil (see Chapter 20).

Trapping insects can be used to monitor population levels and determine the need for control measures. Collection jars or sticky surfaces are used, usually in combination with bait or some other attractant. In small areas, flypaper may be used to trap cockroaches (*Periplaneta* spp.) and whiteflies (*Trialeurodes* spp.). Pheromones, chemical olfactory stimulants secreted by insects to communicate and to attract the opposite sex, have not yet been incorporated into an effective control procedure. Pheromone trapping is being used to monitor pest populations, however. The number of insects trapped in a given period can be correlated with the general population in the area to determine control needs. Trapping is, on the other hand, effectively used to control rodents.

Chemical Control As previously stated, the application of chemicals is the pest control method most commonly used in developed countries. Because so many of the common pesticides are broad-spectrum, some people believe they should not be used at all. Many chemicals, however, are the only means of controlling some pests and are safe if used in the proper manner. The proper uses and limitations of chemicals must be known by those who administer any effective pest management program.

Pesticide selection is the key to safe, effective chemical pest control. Pesticides may be applied in various formulations and in various ways for many purposes. Choosing the appropriate formulation and method of application is extremely important for effective pest control, for the safety and well-being of people and protected plants, and for the stability of the ecosystem.

Toxicity of pesticides to animals is expressed as the LD (lethal dose). The LD_{50} of a chemical indicates the dose necessary to kill 50 percent of the test animals under prescribed experimental conditions. Toxicity tests are carried out on a variety of laboratory animals, most commonly on rats. LD_{50} is expressed as a ratio of milligrams (mg) of the chemical per kilogram (kg) of animal body weight. Such figures cannot be precisely extrapolated to a lethal dose for humans, but they do provide a relative measure of chemical hazard (Table 21-1). The lower the LD_{50}, the more toxic the chemical. Chemicals with an LD_{50}

TABLE 21-1

Animal toxicity of some common pesticides and other chemicals in relation to dose (Doull, Klaassen, and Amdor 1980; Meister 1982)

Compound	Acute Toxicity		Chronic Toxicity.	Soil Persistence
	Oral LD_{50}[a]	Dermal LD_{50}	No Effect at:	
	mg/kg	mg/kg	mg/kg/day	
Chlordane	457	840	1.0	6 month half-life
DDT	113	2500	0.05	Persistent
Diazinon	300	200	0.1[b]	4–6 week half-life
Lindane	88	1000	1.25	Persistent
Malathion	1375	4100	0.2	2 weeks
Nicotine	50			
Parathion	13	7	0.05[b]	Not persistent
Sevin	500	4000		
Simazine	5000	5000	50.0	Persistent
2,4-D	300	1500		1–4 weeks
Aspirin	300[c]			
Caffeine	78[d]			

[a]LD_{50} indicates the dose (mg/kg) necessary to kill 50 percent of the test animals under prescribed experimental conditions. The lower the LD_{50}, the more toxic the compound. The lowest concentration cited is listed.

[b]Amounts greater than this will reduce cholinesterase.

[c]Causes severe toxicity (Temple 1981).

[d]Caused death of child (Mace 1978).

of 500 and above are considered quite safe; the lethal oral dose of these formulations would range from 30 to 480 ml (1 oz-1 pt) or 450 mg (1 lb) (Stimmann 1977). A person is very unlikely to accidently ingest even 30 ml of a chemical.

LD_{50} values usually refer to ingestion of the chemical (*oral*). Similar values may be determined for *dermal* toxicity (absorption through the skin) and *inhalation* toxicity. For safe handling and application of pesticides, these latter two toxicities are more important than the ingestion toxicity. LD_{50} values are for *acute* toxicity, symptoms that occur within 24 hours of exposure. *Chronic* toxicity produces symptoms that appear more than 24 hours after exposure or, more commonly, symptoms that appear after repeated exposure to concentrations below acute toxicity. Chronic symptoms might include respiratory ailments, nervous disorders, cancer, and teratogenicity and might take years to develop.

Certain phytotoxic pesticides (those that injure plants) should not be used on or near susceptible plants, so read labels carefully. Volatile herbicides have caused serious damage to susceptible plants several kilometers (miles) from the site of application. Ester formulations of 2,4-D applied to control weeds in California grain fields have caused serious damage to grapevines more than 20 km (12 mi) away. Amine formulations now available are much less volatile and therefore safer to use. The procedures for determining a pesticide's effectiveness and short- and long-term safety have become so costly and time-consuming that few new chemicals may be registered for use on landscape plants. The potential market for such chemicals may not warrant the expense of testing and the manufacturer's potential liability for future, unforeseen problems.

Labels on pesticide containers present valuable information about use, handling, and toxicity. The Federal Environmental Pesticide Control Act requires that every registered label contain specific information. For example, the label must state whether the pesticide is classified for general or restricted use. *General use* pesticides are the least hazardous to persons and the environment when used as directed; these pesticides are available for use by anyone. *Restricted use* pesticides are those that may be hazardous to the environment, the applicator, or other persons, even when label directions are followed. Restricted use pesticides are to be used only by certified applicators.

The toxicity hazard to human beings must be stated on a pesticide label. Each pesticide formulation is assigned to a toxicity category, which designates the level of hazard to human health. Categories were determined by considering the effects of the pesticides when they are ingested, inhaled, spilled on skin, or splashed in eyes (Table 21-2). The most hazardous materials are in toxicity category I; the least hazardous in category IV. The directions for use tell how to mix and apply the pesticide, where and when to apply it, how much to apply,

TABLE 21-2

Acute toxicity categories of pesticides established by the Federal Environmental Pesticide Control Act (Stimmann 1977)

Category	Signal Word Required on the Label	LD_{50}[a] Oral mg/kg	LD_{50}[a] Dermal mg/kg	LC_{50}[b] Inhalation µg/l	Probable Oral Lethal Dose Less than	
I Highly toxic	DANGER–POISON Skull and crossbones	0–50	0–200	0–2000	5 ml	1 tsp
II Moderately toxic	WARNING	50–500	200–2000	2000–20,000	30 ml	1 oz
III Slightly toxic	CAUTION	500–5000	2000–20,000		480 ml (0.45 kg)	1 pt 1 lb
IV Relatively toxic	CAUTION	>5000	>20,000		>480 ml >(0.45 kg)	>1 pt >1 lb

[a]LD_{50} indicates the dose (amount of chemical/body weight) necessary to kill 50 percent of the test animals under prescribed conditions.

[b]LC_{50} is the dose (amount of chemical/liter of air) necessary when inhaled to kill 50 percent of the test animals under prescribed conditions.

and how often it can be applied. The label also specifies against what pests the chemical has been registered for use. It is illegal to use a pesticide on a plant or a pest for which it has not been registered. Read all information on a pesticide label before purchase or application.

Pesticide toxicity is an extremely important property and must be considered when you choose the controls to be used in a specific situation. If pesticides are part of the control program, they must be handled and applied so as to ensure minimal risks to humans and other animals.

Highly selective pesticides are of course the most desirable. 2,4-D can eliminate many broadleaved weeds without harming turf. Pre-emergence herbicides (such as Surflan® and Treflan®) will control most emerging seedlings but do not injure established plants because little or none of the chemical leaches to their root zone. Where possible, choose insecticides that have narrow ranges of effectiveness or minimal residual properties; this will cause as little harm as possible to natural pest enemies, other beneficial insects, and non-targeted species. Until insecticides with greater selectivity are available, care must be taken to focus insecticide use on targeted pests. Pesticide selection practices in the past have usually ignored this crucial consideration.

Timing and methods of application can maximize pesticide effectiveness and minimize adverse reactions. Chemicals used to control disease are usually most effective when applied as protectants before the disease organism invades the plant or begins to increase. Contact and systemic herbicides are most effective and most easily applied while weeds are small. Obviously, preemergent herbicides should be applied before weeds germinate. Time insecticide applications to coincide with the most vulnerable part of the pest's life cycle. Many insecticides are effective only during the early active stages of the cycle. Proper timing of pesticide application requires population sampling to determine when the most vulnerable stage of the next generation will occur and whether pest numbers even warrant control measures. Avoid spray drift by spraying on calm days and in the early morning or evening, and avoid harm to beneficial insects by spraying while they are outside the area.

Pesticides can be confined to a localized area by various methods of application. Seed treatment for disease, insect, and nematode control requires only a fraction of the pesticide that would be needed for soil application. Dormant applications usually require less spray than those performed during the growing season. Certain insects can be attracted to baited traps and poisoned there. Pests often concentrate in certain parts of a landscape, so that treatment can be limited to areas of heavy infestation. Again, periodic observations are most helpful for locating the areas of highest population. Shallow-rooted plants can be treated with soil drenches against certain pests. Trunk and root injections eliminate spraying close to homes and places of activity; such injections are effective only against certain insects and diseases, however, and repeated applications, even on an annual basis, inflict wounds that can lead to trunk decay (see Chapter 16).

You can minimize the drift of spray material by spraying on calm days with equipment adjusted to discharge large spray droplets. Large droplets settle more quickly, reducing the area affected. Droplet size can be increased if the pressure on hydraulic sprayers is reduced; this of course reduces the distance the spray can be projected.

In summary, for effective chemical pest control you must apply the right material at the correct concentration to the right place at the right time. Von Rumker and coworkers (1975) estimate that less than 1 percent of a foliar insecticide application may reach the insects to be controlled. On tall trees, very little spray reaches the tops; most falls on lower branches or the trunk. For more thorough and uniform coverage, direct most sprays, particularly high-pressure sprays, to the treetops or to the highest portion of the target area.

Certification of commercial pesticide applicators and advisers is required by federal and state laws in the United States. Only licensed agricultural pest control operators can apply *restricted use* pesticides for hire, and any person who makes recommendations about pesticide use must be licensed as an agricultural pest control adviser. Agricultural use includes uses in parks, golf courses, roadside landscapes, cemeteries, schoolyards, private landscapes, and other similar areas. State agencies administer the registration and safe use of pesticides and the certification of pest control operators and advisers. To be certified, an individual must demonstrate knowledge about pesticide labeling, safety, use, application, handling, pests, environmental effects, equipment, and laws and regulations. Specific regulations govern requests for permission to apply a pesticide, ways of handling and using a pesticide, and records that must be kept.

Example of Integrated
Pest Management Approaches

The goal of integrated pest management should be to reduce pests or damage to an acceptable level while guarding the safety of workers, other people, and desirable animals and plants, controlling effectiveness and cost, and preserving the stability of the landscape ecosystem. Such an approach may involve only one relatively simple measure, such as selecting pest-resistant plants or applying a thick mulch around shrubs to control weeds. On the other hand, acceptable control may require complex coordinated measures.

Dutch Elm Disease (DED)   Although DED remains a devastating disease, particularly to the American elm, a variety of measures now hold elm losses to acceptable levels, about two percent in some cities.

Two species of elm bark beetle carry the DED fungus, *Ceratocystis ulmi*. The beetles become infective in galleries under the bark of diseased elm trees. When the beetles move elsewhere to feed, they then infect healthy trees. In summer, the beetle can complete its life cycle, from egg to adult, within six weeks.

The long-term solution would be to mix tree species in a landscape or community so that disease may threaten one species without undermining a large percentage of the total tree population. If you plant elms, select clones known to have a high degree of resistance to DED; Christine Buisman, Urban, and Sapporo Autumn Gold clones are the result of breeding programs in Europe and the United States. Trees resistant to DED, however, do not directly protect susceptible trees except by minimizing the potential sources of infection.

You may use several other established practices to minimize the spread of DED (Schreiber and Peacock 1974): (1) Reduce the population of beetles by eliminating diseased, weak, and dead elm trees and by pruning out weak and dead wood in trees that are to be saved. To keep beetles from colonization, debark, burn, or bury any elm branches and trees within 30 days of symptom appearance during summer and fall. (2) Maintain tree vigor through fertilization, irrigation (if necessary), pest control, and pruning. Do not prune during the period of beetle emergence, however—freshly cut surfaces attract beetles. (3) Protect healthy elms from feeding beetles by spraying the trees with Sevin® before substantial beetle emergence in the spring and again before second brood emergence. (4) Prevent transmission of the fungus from diseased to adjacent healthy trees through underground root grafts. Do this by creating a chemical (soil fumigant) or physical (trenching) barrier between the diseased tree and each adjacent healthy tree as soon as DED symptoms appear. Vigorous community-wide deployment of these procedures have reduced elm tree losses in the eastern United States to 1 or 2 percent a year. Such programs have proved to be less expensive than removal of dead trees in areas where no control programs are carried out (Fig. 21-1).

Other approaches, some in developmental stages, are also being used to combat the spread of DED: (1) No elm wood can be legally transported from areas infected with DED to areas still thought to be free of the disease. (2) When infected branches are pruned out soon after they are detected, infection may be arrested (Himelick and Ceplecha (1976). Make the cuts deep into healthy wood to be sure that the parts removed contain all of the disease. (3) Two systemic fungicides (Lignasan BLP® and Arbotect®) and other chemicals have been tested as preventive and therapeutic treatments. Chemicals injected into the trunk or roots have been reported to suppress symptoms of DED (Smalley 1978), an effect that is not entirely desirable. The fungus is apparently not eliminated from an infected tree, but the chemicals inactivate the disease and improve tree appearance. Since treatment must

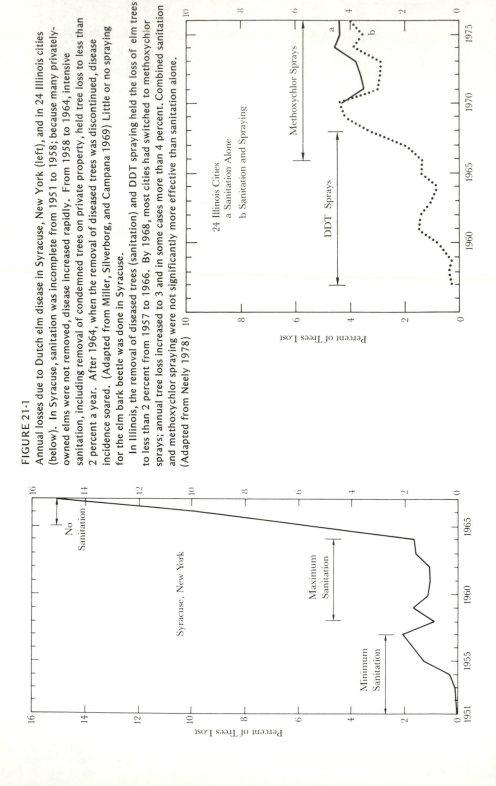

FIGURE 21-1

Annual losses due to Dutch elm disease in Syracuse, New York (left), and in 24 Illinois cities (below). In Syracuse, sanitation was incomplete from 1951 to 1958; because many privately-owned elms were not removed, disease increased rapidly. From 1958 to 1964, intensive sanitation, including removal of condemned trees on private property, held tree loss to less than 2 percent a year. After 1964, when the removal of diseased trees was discontinued, disease incidence soared. (Adapted from Miller, Silverborg, and Campana 1969) Little or no spraying for the elm bark beetle was done in Syracuse.

In Illinois, the removal of diseased trees (sanitation) and DDT spraying held the loss of elm trees to less than 2 percent from 1957 to 1966. By 1968, most cities had switched to methoxychlor sprays; annual tree loss increased to 3 and in some cases more than 4 percent. Combined sanitation and methoxychlor spraying were not significantly more effective than sanitation alone. (Adapted from Neely 1978) be repeated annually or at least every other year, injection wounds could become a problem. Elm populations must be monitored so that infected trees are quickly identified for pruning, removal, or other treatment; this should include trees on private as well as public property.

In the late 1960s, a wasp (*Dendrosoter protuberans*) was introduced in the United States as a parasite of the European elm bark beetle. To date, the wasp has not become established in large enough numbers to be particularly effective. A number of researchers are testing sex-attractant pheromone compounds (chemical olfactory stimulants) in attempts to lure bark beetles away from elm trees or to insect traps, or to confuse their mating pattern. So far, these compounds have been effective only for baiting traps used to monitor the beetle population. Still other workers are testing the herbicides cacodylic acid (Silvisar 150) and potassium iodide (Silvicide), which reduce the number of beetles emerging from treated trees and thus the need to remove diseased trees quickly.

Summary

An integrated approach to pest management holds promise for safer, more effective, and less expensive pest control programs than have been used in the past. Pest management for landscape plants is much more complex than for a single agricultural crop because landscapes are usually small and contain a variety of plant species intimately associated with homes and human activity. Proximate plantings may involve several different individuals or agencies, both public and private, in any efforts to control pest populations. Under such circumstances, an integrated pest management program must be a community effort, particularly if it emphasizes biological control measures. The resolution of many difficult pest problems awaits further study and implementation. On the other hand, much can be done in individual landscapes to improve the health and appearance of landscape plants.

FURTHER READING

COUNCIL ON ENVIRONMENTAL QUALITY (CEQ). 1972. *Integrated Pest Management.* Washington, D.C., U.S. Government Printing Office.

FLINT, M. L., and R. VAN DEN BOSCH. 1981. *Introduction to Integrated Pest Management.* New York: Plenum Press.

GLASS, E. H., ed. 1975. *Integrated Pest Management: Rationale, Potential, Needs, and Implementation.* Entomological Society of America.

HUFFAKER, C. B., ed. 1973. *Biological Control.* New York: Plenum.

Measurement Conversion Factors

Both metric and English units of measurement are given in the text. However, there may be times when you may want to convert a measurement in one system to the other. Most of the needed conversions are given in the following tables.

Column 1[a]	To convert Column 1 units to Column 2 units multiply by	Column 2	To convert Column 2 units to Column 3 units multiply by	Column 3
Length				
millimeter, mm	0.04	inch, in	25.4	millimeter, mm
millimeter, mm	0.0033	feet, ft	303	millimeter, mm
centimeter, cm	0.0328	feet, ft	30	centimeter, cm
meter, m	3.3	feet, ft	0.303	meter, m
meter, m	1.094	yard, yd	0.91	meter, m
kilometer, km	0.62	mile, mi	1.6	kilometer, km
feet, ft	0.00019	mile, mi	5260	feet, ft
micron, μ	0.001	millimeter, mm	1000	micron, μ
millimicron, mμ	0.000001	millimeter, mm	1,000,000	millimicron, mμ
Angstrom, Å	0.1	millimicron, mμ	10	Angstrom, Å
Area				
millimeter^2, mm^2	0.0016	inch^2, in^2	645	millimeter^2, mm^2
centimeter^2, cm^2	0.16	inch^2, in^2	6.45	centimeter^2, cm^2
centimeter^2, cm^2	0.0011	feet^2, ft^2	929	centimeter^2, cm^2
meter^2, m^2	10.76	feet^2, ft^2	0.09	meter^2, m^2
meter^2, m^2	1.2	yard^2, yd^2	0.8	meter^2, m^2
kilometer^2, km^2	0.4	mile^2, mi^2	2.6	kilometer^2, km^2
hectare, ha	2.47	acre, A	0.4	hectare, ha
kilometer^2, km^2	0.1	hectare, ha	100,000	meter^2, m^2
acre, A	43,560	feet^2, ft^2		
mile^2, mi^2	640	acre, A		

Column 1[a]	To convert Column 1 units to Column 2 units multiply by	Column 2	To convert Column 2 units to Column 3 units multiply by	Column 3
Weight or Mass				
gram, g	0.035	ounce, oz (avdp)	28	gram, g
kilogram, kg	2.2	pound, lb	0.454	kilogram, kg
ton (metric)	1.1	ton (U.S. short)	0.9	ton (metric)
ton (metric)	1000	kilogram, kg		
ton (U.S. short)	2000	pound, lb		
Volume				
milliliter, ml	0.22	teaspoon, tps	5	milliliter, ml
milliliter, ml	0.67	tablespoon, tbs	15	milliliter, ml
milliliter, ml	0.03	ounce, oz (fluid)	30	milliliter, ml
liter, l	4.2	cup	0.24	liter, l
liter, l	2.1	pint, pt	0.47	liter, l
liter, l	1.06	quart, qt	0.95	liter, l
liter, l	0.26	gallon, gal (U.S.)	3.8	liter, l
meter^3, m^3	35.3	feet^3, ft^3	0.03	meter^3, m^3
meter^3, m^3	1.3	yard^3, yd^3	0.76	meter^3, m^3
meter^3, m^3	10,000	liter, l		
feet^3, ft^3	7.48	gallon, gal (U.S.)		
gallon, gal (U.S.)	0.83	gallon, gal (imperial)	1.2	gallon, gal (U.S.)

Pressure				
kg/cm^2	14.22	*lbs/in^2, psi*	0.07	kg/cm^2
kg/cm^2	0.968	*atmosphere, atm*	1.033	kg/cm^2
bar	0.987	*atmosphere, atm*	1.013	bar
bar	14.5	*lbs/in^2, psi*	0.069	bar
Rate or Yield				
kg/100 m^2	2.037	*lbs/1000 ft^2*	0.491	kg/100 m^2
kg/ha	0.892	*lbs/A*	1.12	kg/ha
ton (metric)/ha	0.445	*ton (U.S)/A*	2.24	ton (metric)/ha
kg/ha	1000	*kg/100 m^2*		
lbs/A	43.56	*lbs/1000 ft^2*		
Speed				
meters/second, m/sec	3.28	*feet/second, ft/sec*	0.305	meters/second, m/sec
meters/second, m/sec	2.237	*miles/hour, mph*	0.447	meters/second, m/sec
kilometer/hour, km/hr	0.62	*miles/hour, mph*	1.61	kilometers/hour, km/hr
miles/hour, mph	1.47	*feet/second, ft/sec*	0.68	*miles/hour, mph*

[a]English units are in *italics*. Metric units are in standard type.

Temperature Conversion Table

To convert *Fahrenheit to Celsius*, subtract 32
from the Fahrenheit figure, multiply by 5 and
divide by 9.

To convert *Celsius to Fahrenheit*, multiply the
Celsius figure by 9, divide by 5 and add 32.

$F^b\rightarrow$	$\leftarrow C$	$F\rightarrow$	$\leftarrow C$	$F\rightarrow$	$\leftarrow C$
−26	−32	19	−7	64	18
−24	−31	21	−6	66	19
−22	−30	23	−5	68	20
−20	−29	25	−4	70	21
−18	−28	27	−3	72	22
−17	−27	28	−2	73	23
−15	−26	30	−1	75	24
−13	−25	32	0	77	25
−11	−24	34	1	79	26
−9	−23	36	2	81	27
−8	−22	37	3	82	28
−6	−21	39	4	84	29
−4	−20	41	5	86	30
−2	−19	43	6	88	31
0	−18	45	7	90	32
1	−17	46	8	91	33
3	−16	48	9	93	34
5	−15	50	10	95	35
7	−14	52	11	97	36
9	−13	54	12	99	37
10	−12	55	13	100	38
12	−11	57	14	102	39
14	−10	59	15	104	40
16	−9	61	16	106	41
18	−8	63	17	108	42
				212	100

[b]Fahrenheit temperatures have been rounded to the nearest whole number.

APPENDIX TWO

Common and Botanical Names of Plants

Common Name	Botanical Name
Abelia, Glossy	*Abelia* × *grandiflora*
Acacia	*Acacia*
Ailanthus	*Ailanthus altissima*
Alder	*Alnus*
Alfalfa	*Medicago sativa*
Almond	*Prunus dulcis*
Apple	*Malus sylvestris*
Apple, Crab	*Malus*
Apricot	*Prunus armeniaca*
Arborvitae	*Thuja*
Arborvitae, Oriental	*Platycladus orientalis*
Ash	*Fraxinus*
Ash, Arizona	*Fraxinus velutina*
Ash, Green	*Fraxinus pennsylvanica* var. *lanceolata*
Ash, Modesto	*Fraxinus velutina* var. *glabra* 'Modesto'
Ash, Shamel	*Fraxinus uhdei*
Ash, White	*Fraxinus americana*
Aspen	*Populus*
Azalea	*Rhododendron*
Basswood	*Tilia*
Beach Grass	*Ammophila*
Bean	*Phaseolus vulgaris*
Beech	*Fagus*
Bentgrass, Creeping	*Agrostis stolonifera*
Bindweed	*Convolvulus*
Birch, Canoe	*Betula papyrifera*
Birch, European White	*Betula pendula*
Bougainvillea	*Bougainvillea*
Broom	*Cytisus*
Buckeye, California	*Aesculus californica*
Buddleia	*Buddleia*

Common Name	Botanical Name
Buffalograss	*Buchloe dactyloides*
Butternut	*Juglans cinerea*
Buttonwood	*Platanus* or *Conocarpus*
Callistemon	*Callistemon*
Camellia	*Camellia*
Camphor Tree	*Cinnamomum camphora*
Carambola	*Averrhoa carambola*
Carob	*Ceratonia siliqua*
Catalpa	*Catalpa speciosa*
Ceanothus	*Ceanothus*
Ceanothus, Blueblossom	*Ceanothus thyrsiflorus*
Cedar	*Thuja* or *Cedrus*
Cedar, Deodar	*Cedrus deodara*
Cedar, Incense	*Calocedrus decurrens*
Cedar, Red	*Juniperus virginiana*
Cedar, Western Red	*Thuja plicata*
Cedar, White	*Chamaecyparis thyoides* or *Thuja occidentalis*
Cherry	*Prunus*
Cherry, Barbados	*Malpighia glabra*
Cherry, Flowering	*Prunus serrulata*
Cherry, Sour	*Prunus cerasus*
Cherry, Surinam	*Eugenia uniflora*
Cherry, Sweet	*Prunus avium*
Cherry, Yoshino	*Prunus* X *yedoensis*
Chestnut	*Castanea*
Chestnut, Horse	*Aesculus hippocastanum*
Chestnut, Sweet	*Castanea sativa*
Chrysanthemum	*Chrysanthemum*
Citrus	*Citrus*
Clematis	*Clematis*
Corn	*Zea mays*
Cotoneaster	*Cotoneaster*
Cottonwood	*Populus*
Cryptomeria	*Cryptomeria japonica*
Cycad	*Cycas*
Cypress	*Cupressus*
Cypress, Bald	*Taxodium distichum*
Cypress, False	*Chamaecyparis*
Dogwood	*Cornus*
Dogwood, Eastern	*Cornus florida*
Elder	*Sambucus*
Elm	*Ulmus*
Elm, American	*Ulmus americana*
Elm, Asiatic	*Ulmus pumila*
Elm, Chinese	*Ulmus parvifolia*
Elm, English	*Ulmus procera*
Elm, European White	*Ulmus laevis*
Elm, Red	*Ulmus rubra*
Elm, Siberian	*Ulmus pumila*
Empress Tree	*Paulownia tomentosa*
Escallonia	*Escallonia*
Eucalyptus	*Eucalyptus*

Common Name	Botanical Name
Euonymus	*Euonymus*
Euonymus, Gold Spot	*Euonymus japonica* 'Gold Spot'
Ferns	Many genera and species
Fescue	*Festuca*
Fescue, Alta	*Festuca arundinacea*
Fig	*Ficus*
Fig, Creeping	*Ficus pumila*
Fig, Indian Laurel	*Ficus microcarpa*
Fir	*Abies*
Fir, Balsam	*Abies balsamea*
Fir, Douglas	*Pseudotsuga menziesii*
Fir, White	*Abies concolor*
Flannel Bush	*Fremontodendron californicum*
Forsythia	*Forsythia*
Forsythia, Weeping	*Forsythia suspensa*
Ginkgo	*Ginkgo biloba*
Golden Rain Tree	*Koelreuteria paniculata*
Golden-chain Tree	*Laburnum anagyroides*
Grape	*Vitis vinifera*
Grapefruit	*Citrus* × *paradisi*
Guava	*Psidium*
Gum, Black	*Nyssa sylvatica*
Gum, Blue	*Eucalyptus globulus*
Gum, Red	*Eucalyptus camaldulensis*
Gum, Silver-dollar	*Eucalyptus polyanthemos*
Hackberry	*Celtis*
Hawthorn	*Crataegus*
Hazel, Witch	*Hamamelis*
Hemlock	*Tsuga*
Hibiscus	*Hibiscus*
Hibiscus, Chinese	*Hibiscus rosa-sinensis*
Hickory	*Carya*
Hickory, Shagbark	*Carya ovata*
Holly	*Ilex*
Holly, Dwarf Burford	*Ilex cornuta* 'Burfordii Nana'
Holly, English	*Ilex aquifolium*
Holly, Japanese	*Ilex crenata*
Honeysuckle	*Lonicera*
Hornbeam	*Carpinus*
Horsechestnut	*Aesculus hippocastanum*
Ice Plant	*Mesembryanthemum*
Ivy	*Hedera*
Ivy, English	*Hedera helix*
Jacaranda	*Jacaranda*
Jasmine	*Jasminum*
Jasmine, Crape	*Ervatamia coronaria*
Jasmine, Star	*Trachelospermum jasminoides*
Jessamine, Orange	*Murraya paniculata*
Juniper	*Juniperus*
Juniper, Pfitzer	*Juniperus chinensis* 'Pfitzerana'
Larch	*Larix*
Laurel, English	*Prunus laurocerasus*

Common Name	Botanical Name
Lemon	*Citrus limon*
Lilac	*Syringa*
Lime Tree	*Tilia* × *vulgaris*
Linden	*Tilia*
Liquidambar	*Liquidambar styraciflua*
Locust	*Robinia*
Locust, Black	*Robinia pseudoacacia*
Locust, Frisia	*Robinia pseudoacacia* 'Frisia'
Locust, Honey	*Gleditsia triacanthos*
Locust, Sunburst Honey	*Gleditsia triacanthos* f. *inermis* 'Sunburst'
Loquat	*Eriobotrya japonica*
Magnolia	*Magnolia*
Magnolia, Southern	*Magnolia grandiflora*
Mango	*Mangifera indica*
Manzanita	*Arctostaphylos*
Maple	*Acer*
Maple, Japanese	*Acer palmatum* or *Acer japonicum*
Maple, Norway	*Acer platanoides*
Maple, Red	*Acer rubrum*
Maple, Silver	*Acer saccharinum*
Maple, Sugar	*Acer saccharum*
Mayten, Chile	*Maytenus boaria*
Mesquite	*Prosopis*
Mirror Plant	*Coprosma repens*
Mistletoe	*Phoradendron* or *Viscum*
Mulberry	*Morus alba*
Mulberry, Fruitless	*Morus alba* 'Stribling'
Myrtle, Crape	*Lagerstroemia indica*
New Zealand Christmas Tree	*Metrosideros excelsus*
Oak	*Quercus*
Oak, Black	*Quercus velutina*
Oak, Blackjack	*Quercus marilandica*
Oak, Blue	*Quercus douglasii*
Oak, Cork	*Quercus suber*
Oak, Interior Live	*Quercus wislizenii*
Oak, Pin	*Quercus palustris*
Oak, Red	*Quercus rubra*
Oak, Sawtooth	*Quercus acutissima*
Oak, Scarlet	*Quercus coccinea*
Oak, Scrub	*Quercus ilicifolia*
Oak, Silk	*Grevillea robusta*
Oak, Valley	*Quercus lobata*
Oak, Water	*Quercus nigra*
Oak, White	*Quercus alba*
Oleander	*Nerium oleander*
Olive	*Olea europaea*
Olive, Swan Hill Fruitless	*Olea europaea* 'Swan Hill'
Olive, Russian	*Elaeagnus angustifolia*
Onion	*Allium cepa*
Orange	*Citrus sinensis*
Orange, Navel	*Citrus sinensis*
Pagoda Tree, Japanese	*Sophora japonica*
Palms	numerous genera

Common Name	Botanical Name
Palm, Coconut	*Cocos nucifera*
Palm, Date	*Phoenix dactylifera*
Palm, Mexican Fan	*Washingtonia robusta*
Palm, Royal	*Roystonea*
Palo Verde	*Cercidium*
Peach	*Prunus persica*
Pear	*Pyrus communis*
Pear, Bradford	*Pyrus calleryana* 'Bradford'
Pear, Callery	*Pyrus calleryana*
Pear, Evergreen	*Pyrus kawakamii*
Pear, Oriental	*Pyrus pyrifolia*
Pecan	*Carya illinoensis*
Pepper Tree, California	*Schinus molle*
Petunia	*Petunia* × *hybrida*
Photinia, Chinese	*Photinia serrulata*
Pine	*Pinus*
Pine, Eastern White	*Pinus strobus*
Pine, Lodgepole	*Pinus murrayana*
	Syn.: *P. contorta* var. *latifolia*
Pine, Monterey	*Pinus radiata*
Pine, Ponderosa	*Pinus ponderosa*
Pine, Red	*Pinus resinosa*
Pine, Scotch	*Pinus sylvestris*
Pine, Slash	*Pinus caribaea*
Pine, Sugar	*Pinus lambertiana*
Pine, White	*Pinus strobus*
Pineapple	*Ananas comosus*
Pistache	*Pistacia*
Pistache, Chinese	*Pistacia chinensis*
Pittosporum	*Pittosporum*
Pittosporum, Japanese	*Pittosporum tobira*
Plane, London	*Platanus* × *acerifolia*
Plane Tree	*Platanus*
Plum	*Prunus*
Plum, Flowering	*Prunus*
Podocarpus	*Podocarpus*
Poinsettia	*Euphorbia pulcherrima*
Poplar	*Populus*
Poplar, Lombardy	*Populus nigra* 'Italica'
Poplar, Yellow	*Liriodendron tulipifera*
Potato	*Solanum tuberosum*
Privet, Glossy	*Ligustrum lucidum*
Privet, Wax-leaf	*Ligustrum japonicum*
Pyracantha	*Pyracantha*
Quack Grass	*Agropyron repens*
Quince	*Cydonia oblonga*
Raspberry	*Rubus idaeus*
Redbud	*Cercis*
Redwood	*Sequoia sempervirens*
	or *Sequoiadendron giganteum*
Redwood, Coast	*Sequoia sempervirens*
Redwood, Dawn	*Metasequoia glyptostroboides*
Rice	*Oryza sativa*

Common Name	Botanical Name
Rose	*Rosa*
Rose, Rambler	*Rosa*
Rye	*Secale cereale*
Sassafras	*Sassafras albidum*
Sequoia	*Sequoia sempervirens* or *Sequoiadendron giganteum*
Sequoia, Giant	*Sequoiadendron giganteum*
Silk Tree	*Albizia julibrissin*
Sorghum	*Sorghum vulgare*
Spruce	*Picea*
Spruce, Black	*Picea mariana*
Spruce, Norway	*Picea abies*
Spruce, Sitka	*Picea sitchensis*
Spruce, White	*Picea glauca*
Squash	*Cucurbita maxima*
Sweetgum	*Liquidambar styraciflua*
Sycamore	*Platanus*
Sycamore, American	*Platanus occidentalis*
Sycamore, California	*Platanus racemosa*
Tea, Victoria	*Leptospermum laevigatum*
Tobacco	*Nicotiana tabacum*
Tomato	*Lycopersicon esculentum*
Tree-of-heaven	*Ailanthus altissima*
Tulip Tree	*Liriodendron tulipifera*
Tung	*Aleurites montana*
Tupelo	*Nyssa sylvatica*
Virginia Creeper	*Parthenocissus quinquefolia*
Walnut	*Juglans*
Walnut, Black	*Juglans nigra*
Walnut, English	*Juglans regia*
Walnut, Japanese	*Juglans ailanthifolia*
Walnut, Northern California Black	*Juglans hindsii*
Willow	*Salix*
Willow, Weeping	*Salix babylonica*
Wisteria	*Wisteria*
Xylosma, Shiny	*Xylosma congestum*
Yellowwood (American)	*Cladrastis lotea*
Yew	*Taxus*
Yew, Japanese	*Taxus cuspidata*
Zelkova	*Zelkova*
Zelkova, Japanese	*Zelkova serrata*
Zelkova, Sawleaf	*Zelkova serrata*

Plants Recommended for Specific Purposes

Lists of plants recommended for specific purposes or situations must be used with caution. In most cases, compilations are based on field observations and individual judgments rather than actual screening tests. Frequently a compilation is based entirely on other lists whose constraints and caveats have been excluded. Plants of one species can vary greatly in their susceptibility or resistance to environmental, disease, and insect problems. Even plants of the same clone may respond differently to a particular situation depending on prior events; a plant that has been stressed is more susceptible to certain diseases than identical plants that have not been stressed. Plant lists are helpful primarily as a starting point, after which you should obtain local information about promising plants.

These appendices include plant lists that are not widely available in the literature. More extensive lists appear in other books; Appendix 7 constitutes a guide to some of these resources, which the reader is urged to consult.

The botanical names used in this book generally conform to those listed in *Hortus Third* (Bailey and others 1976) and the *Annotated Checklist of Woody Ornamental Plants of California, Oregon, and Washington* (McClintock and Leiser 1979).

Times of Tree Foliation and Defoliation[a]

Dates of Foliation[b]					Dates of Defoliation[c]		
Early $2/26$	*Mid* $3/31$	*Late* $5/15$ $5/27$			*Early* $10/12$	*Mid* $11/5$	*Late* $12/5$
			Acer buergeranum				
			Acer ginnala				
			Acer negundo				
			Acer negundo 'Variegatum'				
			Acer opalus var. *obtusatum*				
			Acer platanoides 'Drummondii'				
			Acer platanoides 'Faassen's Black'				
			Acer platanoides 'Schwedleri'				
			Acer pseudoplatanus				
			Acer rubrum 'Autumn Flame'				
			Acer rubrum 'Bowhall'				

[a]Species in the landscape tree trails at the North Willamette Experiment Station, Aurora, Oregon (45° N latitude).

[b]Date of the appearance of the first true leaf.

[c]Date of complete defoliation.

Source: Ticknor 1981

			Dates of Foliation[b]								Dates of Defoliation[c]		
Early	Mid	Late									Early	Mid	Late
$^2/_{26}$	$^3/_{31}$	$^5/_{15}$ $^5/_{27}$									$^{10}/_{12}$	$^{11}/_{5}$	$^{12}/_{5}$

Acer rubrum 'Scanlon'

Acer rufinerve

Acer saccharum 'Green Mountain'

Acer saccharum 'Sweet Shadow'

Albizia julibrissin

Asimina triloba

Betula maximowicziana

Betula papyrifera

Betula papyrifera var. *occidentalis*

Betula pendula

Betula pendula 'Gracilis'

Betula platyphylla var. *japonica*

Carpinus betulus 'Fastigiata'

Carpinus orientalis

Celtis sinensis

Cercidiphyllum japonicum

Cercis siliquastrum

Chionanthus virginicus

Cornus florida 'Fastigiata'

Cornus florida 'Rubra'

Cornus florida 'Welchii'

Cornus nuttallii 'Eddiei'

Corylus colurna

Crataegus 'Autumn Glory'

C. laevigata 'Crimson Cloud'

C. laevigata 'Coccinea Flore Pleno'

C. laevigata 'Winter King'

Crataegus × *lavallei*

Diospyros kaki

Diospyros virginiana

Euodia danielli

641

Early	Mid	Late			Early	Mid	Late
2/26	3/31	5/15	5/27		10/12	11/5	12/5

Euodia henryi

Euodia hupehensis

Fagus sylvatica 'Atropunicea'

Fraxinus pennsylvanica 'Summit'

Gleditsia triacanthos var. *inermis*

G. t. var. *inermis* 'Rubylace'

G. t. var. *inermis* 'Shademaster'

G. t. var. *inermis* 'Skyline'

G. t. var. *inermis* 'Sunburst'

Halesia monticola

Koelreuteria paniculata

Laburnocytisus adamii

Laburnum alpinum f. *pendulum*

Laburnum × *watereri* 'Vossii'

Lagerstroemia indica

Liquidambar formosana

Liquidambar formosana 'Afterglow'

Liquidambar orientalis

Liquidambar styraciflua

Liquidambar styraciflua 'Burgundy'

Liquidambar styraciflua 'Festival'

Liquidambar styraciflua 'Gumball'

Liquidambar styraciflua 'Palo Alto'

Liriodendron tulipifera

Magnolia fraseri

Magnolia × *soulangiana*

Malus floribunda

Nyssa sylvatica

Ostrya carpinifolia

Dates of Foliation[b]					Dates of Defoliation[c]		
Early $\frac{2}{26}$	Mid $\frac{3}{31}$	Late $\frac{5}{15}$ $\frac{5}{27}$			Early $\frac{10}{12}$	Mid $\frac{11}{5}$	Late $\frac{12}{5}$

▦▦▦▦▦	*Parrotia persica*	▦▦▦▦▦	
▦▦▦▦	*Phellodendron amurense*	▦▦▦	
▦▦▦▦	*Prunus cerasifera* 'Thundercloud'	▦▦▦▦	
▦▦▦▦	*Prunus sargentii* 'Columnaris'	▦▦▦▦	
▦▦	*Prunus subhirtella* 'Autumnalis'	▦▦	
▦▦▦▦	*Pterostyrax corymbosus*	▦▦▦▦	
▦▦▦▦	*Pyrus calleryana* 'Aristocrat'	▦▦▦▦	
▦▦▦▦	*Pyrus calleryana* 'Bradford'	▦▦▦▦	
▦▦	*Quercus aliena*	▦▦▦▦	
▦	*Quercus coccinea*	▦▦▦▦▦	
▦▦	*Quercus douglasii*	▦▦▦▦▦	
▦▦	*Quercus lobata*	▦▦▦▦▦	
▦	*Quercus palustris*	▦▦▦▦▦	
▦	*Quercus phellos*	▦▦▦▦	
▦	*Quercus robur* 'Fastigiata'	▦▦▦▦▦	
▦▦	*Quercus shumardii*	▦▦▦▦▦	
▦	*Rhus typhina*	▦▦▦	
▦▦	*Robinia* × *ambigua* 'Idahoensis'	▦▦▦▦	
▦▦▦▦▦	*Salix alba* var. *tristis*	▦▦▦	
▦▦▦▦▦	*Salix babylonica*	▦▦▦▦▦	
▦▦▦▦	*Sorbus alnifolia*	▦▦▦▦	
▦▦▦▦	*Sorbus aucuparia*	▦▦▦	
▦▦▦▦	*Sorbus aucuparia* 'Cardinal Royal'	▦▦▦▦	
▦▦▦▦	*Stewartia pseudo-camellia*	▦▦▦	
▦▦▦▦	*Styrax japonicus* 'Kusan'	▦▦▦▦	
▦▦▦▦▦	*Syringa reticulata*	▦▦▦▦	
▦▦	*Tilia americana*	▦▦▦	
▦	*Tilia cordata*	▦▦▦▦	
▦	*Tilia cordata* 'Greenspire'	▦▦▦	
▦▦▦▦▦	*Ulmus pumila* var. *arborea*	▦▦▦▦	
▦▦	*Zelkova serrata* 'Village Green'	▦▦▦▦▦	

Flood-Tolerant Woody Plants in the Contiguous United States

The species listed withstood 180 or more days of water covering the soil under trees. Observations were made in the ten U.S. Corps of Engineers Divisions in the contiguous United States (Whitlow and Harris 1979).

Botanical Name	Common Name
Acer negundo	Box Elder
Acer rubrum	Red Maple
Acer saccharinum	Silver Maple
Carya aquatica	Water Hickory
Carya illinoensis	Pecan
Carya ovata	Shagbark Hickory
Cephalanthus occidentalis	Buttonbush
Cornus stolonifera	Red-osier Dogwood
Crataegus mollis	Red Hawthorn
Diospyros virginiana	Persimmon
Eucalyptus camaldulensis	Red Gum
Forestiera acuminata	Swamp Privet
Fraxinus pennsylvanica	Green Ash
Gleditsia aquatica	Water Locust
Gleditsia triacanthos	Honey Locust
Ilex decidua	Deciduous Holly
Liquidambar styraciflua	Liquidambar
Nyssa aquatica	Water Tupelo
Planera aquatica	Water Elm
Platanus X acerifolia	London Plane
Platanus occidentalis	Sycamore
Populus deltoides	Eastern Cottonwood
Quercus bicolor	Swamp White Oak
Quercus lyrata	Overcup Oak
Quercus macrocarpa	Bur Oak
Quercus nuttallii	Nuttall's Oak
Quercus palustris	Pin Oak
Salix spp.	Willow
Salix alba var. *tristis*	Golden Weeping Willow
Salix exigua	Narrow-leaf Willow
Salix hookeriana	Hooker Willow
Salix lasiandra	Pacific Willow
Salix nigra	Black Willow
Taxodium distichum	Bald Cypress
Ulmus americana	American Elm
Washingtonia robusta	Mexican Fan Palm

APPENDIX SIX

Tolerance of Woody Plants to Landfill Conditions

Species Tolerant of Landfill Conditions

Botanical Name	Common Name
Acacia latifolia	Broadleaf Acacia
Acer rubra	Red Maple
Arbutus unedo	Strawberry Tree
Eucalyptus lehmannii	Bushy Yate
Fraxinus pennsylvania	Green Ash
Ginkgo biloba	Ginkgo
Gleditsia triacanthos	Honey Locust
Grevillea robusta	Silk Oak
Liquidambar styraciflua	Liquidambar
Melaleuca quinquenervia	Cajeput Tree
Myoporum laetum	False Sandalwood
Myrica pensylvanica	Bayberry
Nyssa sylvatica	Black Gum
Picea abies	Norway Spruce
Pinus halepensis	Aleppo Pine
Pinus pinea	Italian Stone Pine
Pinus strobus	White Pine
Pinus thunbergii	Japanese Black Pine
Pittosporum undulatum	Victorian Box
Platanus occidentalis	American Sycamore
Populus hybrid	Hybrid Poplar (rooted cuttings)
Quercus palustris	Pin Oak
Schinus molle	California Pepper Tree
Taxus cuspidata	Japanese Yew
Ulmus parvifolia	Siberian Elm

Species Sensitive to Landfill Conditions

Botanical Name	Common Name
Abies concolor	White Fir
Acer saccharum	Sugar Maple
Carya ovata	Shagbark Hickory
Cornus florida	Dogwood
Malus	Apple
Picea glauca	White Spruce
Picea pungens	Colorado Blue Spruce
Pinus resinosa	Red Pine
Populus nigra 'Italica'	Lombardy Poplar
Prunus	Black Cherry
Pseudotsuga menziesii	Douglas Fir
Quercus velutina	Black Oak
Salix sp.	Willow
Sorbus aucuparia	Mountain Ash
Taxus cuspidata	Japanese Yew
Thuja	Arborvitae
Tilia americana	American Basswood
Ulmus americana	American Elm
Ulmus fulva	Slippery Elm

REFERENCES

FLOWER, F. B., E. F. GILMAN, and I. A. LEONE. 1981. Landfill Gas: What It Does to Trees and How Its Injurious Effects May Be Prevented. *J. Arboriculture* 7(2):43–52.

LEONE, I. A., and others. 1977. Damage to Woody Species by Anaerobic Landfill Gases, *J. Arboriculture* 3(12):221–225.

VAN HEUIT, R. E. 1979. *Sanitation Districts of Los Angeles County: Generation of Methane Gas in Sanitary Landfills and Its Effect on Growth of Plant Materials*. Unpublished. Delivered at 1979 meetings of Amer. Soc. Consulting Arborists. (Abstract in *Horizons*. 1980. Washington, D.C.: Amer. Assoc. Nurserymen, Sept. 1.)

Sources of Other Plant Lists

Plant Characteristics	Wyman (1971)	Reader's Digest (Mullins 1979)	Ortho Books (Walheim 1977)	McMinn and Maino (1963)	Maino and Howard (1955)	Lord (1979)	Lane Magazine and Book Co. (Clark 1979)	Hoyt (1978)	Hillier (1973)	Harris, Leiser, and Mathe:y (1982)	Clouston (1977)
Adaptable to acid soil	X	X						X	X		
Adaptable to alkaline soil		X		X				X	X		
Adaptable to wet soil	X	X		X		X	X	X	X		
Adaptable to dry soil	X	X		X	X	X	X				X
Adaptable to clay soil		X							X		
Adaptable to shallow soil								X	X		
Adaptable to saline soil								X		X	
Tolerant of smoke, dust, and abuse			X					X	X		X
Tolerant of air pollution										X	
Tolerant of street conditions	X	X	X	X	X		X	X			X
Tolerant of seacoast conditions	X	X	X	X		X	X	X	X	X	
Tolerant of shade	X	X			X		X	X			
Tolerant of heat		X	X	X		X	X				
Tolerant of frost					X		X	X	X		
Pest resistant	X		X	X				X			
Lawn and shade plantings				X			X				
Erosion controlling				X				X	X		X
Sand holding				X							
Windbreaks	X			X	X		X	X			X
Rapid growth				X		X	X		X		
Slow growth								X			
Landscape and aesthetic effects	X	X	X	X	X	X	X	X	X		X

REFERENCES

CLARK, D. E., ed. 1979. *Sunset New Western Garden Book*. Menlo Park, Calif.: Lane.

CLOUSTON, B., ed. 1977. *Landscape Design with Plants*. London: Heinemann.

HARRIS, R. W., A. T. LEISER, and N. MATHENY. 1982. *Urban Tree Manual*. U.S. Department of Agriculture Forest Service.

HILLIER, H. C. 1973. *Hilliers' Manual of Trees and Shrubs*. Newton Abbot, England: David & Charles.

HOYT, R. S. 1978. *Ornamental Plants for Subtropical Regions*. Anaheim, Calif.: Livingston.

LORD, E. E. 1979. *Shrubs and Trees for Australian Gardens*. 4th ed. Melbourne: Lothian.

MAINO, E., and F. HOWARD. 1955. *Ornamental Trees*. Berkeley: University of California Press.

McMINN, H. E., and E. MAINO. 1963. *An Illustrated Manual of Pacific Coast Trees*. Berkeley: University of California Press.

MULLINS, M. G. 1979. *Reader's Digest Illustrated Guide to Gardening*. Sydney: Reader's Digest Services.

WALHEIM, L. 1977. *The World of Trees*. San Francisco: Ortho Books.

WYMAN, D. 1971. *Wyman's Gardening Encyclopedia*. New York: Macmillan.

Bibliography

AGRIOS, G. N. 1969. *Plant Pathology*. New York: Academic Press.

——. 1975. Virus and Mycoplasma Diseases of Shade and Ornamental Trees. *J. Arboriculture* 1:41–47.

AIZUMI, S., and S. KEIPER. n.d. *Plants Suitable for Industrial Areas*. Unpublished paper. Davis: University of California.

ALJIBURY, F. K., J. L. MEYER, and W. E. WILDMAN. 1979. *Managing Compacted and Layered Soils*. Univ. Calif. Agr. Sci. Leaflet 2635.

ALVAREZ R. G. 1977. Growth Regulators. *J. Arboriculture* 3(5):94–97.

AMERICAN HORTICULTURAL SOCIETY. 1980. Are Wood Ashes Good for the Soil? *Area Nursery and Garden Store Newsletter* (Ohio State University) 22(4):8.

AMERICAN NATIONAL STANDARDS INSTITUTE (ANSI Z60.1). 1980. *American Standard for Nursery Stock*. Washington, D.C.: American National Standards Institute.

AMERICAN NURSERYMAN. 1970. Steel-Clad Balled Trees. *American Nurseryman* 131(1):14.

ANDERSON, R. F. 1960. *Forest and Shade Tree Entomology*. New York: John Wiley.

ANDRESEN, J. W. 1977. Survey of Growth and Survival of Trees Indicated No Detrimental Effects Caused by High-Pressure Sodium Street Lighting. *Utility Arborist Assn. Newsletter* 7(4):1–2.

ARBORICULTURAL ASSOCIATION. 1976. *Trees: A Guide to Pruning*. Romsey, Hampshire, England: Arboricultural Association.

ARMSTRONG, N. 1925. Open Cavities. *Proc. Natl. Shade Tree Conf.* 1:104–8.

ASHBAUGH, F. A. 1968. Edison Electrical Institute Tree Growth Control Project, 1968. *Proc. Intl. Shade Tree Conf.* 44:169–72.

AYERS, R. S. 1977. Quality of Water for Irrigation. *J. Irrigation and Drainage Div., Amer. Soc. Civil Engr.* 103(IR2):135–54.

BAILEY, L. H. 1916. *The Pruning Manual*. London: Macmillan.

——. 1949. *Manual of Cultivated Plants*. New York: Macmillan.

BAILEY, L. H., E. Z. BAILEY, and STAFF OF L. H. BAILEY HORTORIUM. 1976. *Hortus Third.* New York: Macmillan.

BAKER, K. F. 1957. *The U. C. System for Producing Healthy Container-Grown Plants.* Calif. Agr. Exp. Sta. Manual 23.

BANCROFT-WHITNEY. 1967. *Deering's California Codes: Agricultural Code Annotated.* Div. 18, Ch. 5, Art 6. San Francisco: Bancroft-Whitney.

BARFIELD, B. V., and V. F. GERBER, eds. 1979. *Modification of the Aerial Environment of Plants.* St. Joseph, Mich.: Amer. Soc. of Agr. Engineers.

BARROWS, W. 1970. Some Practical Steps to Quality Control of New Street Plantings. *Int. Shade Tree Conf. West Chapter Newsletter,* Sept., pp. 12–15.

BARTLETT, F. A. 1958. Pruning Trees. In *Plants & Gardens Pruning Handbook* (Brooklyn Botanic Garden) 14(3):199–203.

BAUMGARDT, J. P. 1968. *How to Prune Almost Everything.* New York: Morrow.

BEAN, W. J. 1950. *Trees and Shrubs Hardy in the British Isles.* 3 vols. London: Butler and Tanner.

BEAUCHAMP, K. H. 1955. Tile Drainage: Its Installation and Upkeep. In *Water: Yearbook of Agriculture,* ed. A. Stefferud. Washington, D.C.: U.S. Printing Office, pp. 508–20.

BECKENBACH, J., and J. H. GOURLEY. 1932. Some Effects of Different Cultural Methods upon Root Distribution of Apple Trees. *Proc. Amer. Soc. Hort. Sci.* 29:202–4.

BECKER, W. B. 1938. *Leaf-Feeding Insects of Shade Trees.* Mass. Agric. Exp. Sta. Bull. 353.

BEN-JAACOV, J., G. NATANSON, and A. HAGILADI. 1980. The Use of a Wind-Controlled Overhead Irrigation System to Prevent Damage by Wind-Borne Salts. *J. Amer. Soc. Hort. Sci.* 105(6):833–35.

BENNETT, J. H., and A. C. HILL. 1975. Interactions of Air Pollutants with Canopies of Vegetation. In *Responses of Plants to Air Pollution,* ed. J. B. Mudd and T. T. Kozlowski. New York: Academic Press, pp. 273–306.

BENNETT, J. P. 1931. *The Treatment of Lime-Induced Chlorosis with Iron Salts.* Univ. Calif. Agr. Exp. Sta. Circ. 321.

BENSON, L., and R. A. DARROW. 1954. *Trees and Shrubs of the Southwestern Deserts.* Tucson: University of Arizona Press.

BERDAHL, P. and others. 1978. *California Solar Data Manual.* Berkeley: Lawrence Berkeley Lab., University of California.

BERNATZKY, A. 1978. *Tree Ecology and Preservation.* New York: Elsevier Publishing.

BERNSTEIN, L. 1964. Salt Tolerance of Plants. *USDA Inf. Bull.* 283.

BERNSTEIN, L., and L. E. FRANCOIS. 1973. Sensitivity of Alfalfa to Salinity of Irrigation and Drainage Water. *Soil Sci. Soc. Amer. Proc.* 37:931–43.

BERNSTEIN, L., L. E. FRANCOIS, and R. A. CLARK. 1972. Salt Tolerance of Ornamental Shrubs and Ground Covers. *J. Amer. Soc. Hort. Sci.* 97(4):550–56.

BEY, C. F. 1980. Growth Gains from Moving Black Walnut Provenances Northward. *J. Forestry* 78(10):640–41.

BIDDLE, P. G. 1979. Tree Root Damage to Buildings: An Arboriculturist's Experience. *Arboricultural J.* 3(6):397–412.

BINGHAM, C. 1968. *Trees in the City.* Chicago: Amer. Soc. Planning Off. Report 236.

BIRAN, I., and A. M. KOFRANEK. 1976. Evaluation of Fluorescent Lamps As an Energy Source for Plant Growth. *J. Amer. Soc. Hort. Sci.* 101(6):9–17.

BLACK, M. 1978. Reducing Vandalism. *Western Chapter News* (Intl. Soc. Arboriculture), July-August, p. 42.

BONNER, F. T. 1974. Seed Testing. In *Seeds of Woody Plants in the United States.* Washington, D.C.: USDA Agr. Handbook 450:136–52.

BONNER, I., and A. GALSTON. 1952. *Principles of Plant Physiology.* San Francisco: W. H. Freeman.

BOOHER, L. J., and C. E. HOUSTON. 1958. Water Holding Characteristics of Some California Soils. *Calif. Agr. Ext. Serv. File* 12.

BORST, H. L., and R. WOODBURN. 1942. The Effect of Mulching and Methods of Cultivation on Runoff and Erosion from Muskingham Silt Loam. *Agr. Engineering* 23:19–22.

BOWEN, G. D. 1973. Mineral Nutrition of Ectomycorrhizae. In *Ectomycorrhizae: Their Ecology and Physiology,* ed. G. C. Marks and T. T. Kozlowski. New York: Academic Press, pp. 151–207.

BOYCE, S. G. 1954. The Salt Spray Community. *Ecology Monographs* 24:29–67.

BOYNTON, D., and G. H. *Oberly.* 1966. Apple Nutrition. In *Nutrition of Fruit Crops: Temperature, Subtropical, Tropical,* ed. N. F. Childers. New Brunswick, N.J.: Horticultural Pub., Rutgers University, pp. 1–50.

BRADFORD, G. R. 1966. Boron. In *Diagnostic Criteria for Plants and Soil,* ed. H. D. Chapman. Riverside, Calif.: Chapman, pp. 33–61.

BRANSON, R. L. 1980. Wood Ashes as a Soil Additive. *Soil and Water.* Univ. Calif. Coop. Ext. 45:2–3.

BRIDGEMAN, P. H. 1976. *Tree Surgery.* London: David & Charles.

BRITISH BROADCASTING CORPORATION. 1979. The Effect of Trees on Television Reception. *Arboriculture Res. Note* 14:1–2.

BRITISH STANDARDS INSTITUTION. 1966. *Recommendations for Tree Work.* London: British Standards Inst. 3998.

BROECKER, W. S. 1970. Enough Air. *Environment* 12(7):28–31.

BROOKLYN BOTANIC GARDEN. 1957. Directory of Mulches. In Handbook on Mulches. *Plants & Gardens* 13(1):4–5.

———. 1981. Handbook on Pruning. *Plants & Gardens* 37(2).

BROWN, C. L. 1971. Primary Growth. In *Trees: Structure and Function,* ed. M. Zimmerman and C. L. Brown. New York: Springer-Verlag, pp. 1–66.

BROWN, C. L., R. G. McALPINE, and P. P. KORMANIK. 1967. Apical Dominance and Form in Woody Plants: A Reappraisal. *Amer. J. Bot.* 54(2):153–62.

BROWN, G. E. 1972. *The Pruning of Trees, Shrubs, and Conifers.* London: Faber and Faber.

BROWN, L. R., and C. O. EADS. 1966. *A Technical Study of Insects Affecting the Elm Tree in Southern California.* Calif. Agr. Exp. Sta. Bull. 821.

BUTIN, H., and A. SHIGO. 1981. *Radial Shakes and "Frost Cracks" in Living Oak Trees.* USDA For. Serv. Res. Paper NE-478.

CABORN, J. M. 1965. *Shelterbelts and Windbreaks.* London: Faber and Faber.

CAMPANA, R. J. 1978. Comparative Aspects of Dutch Elm Disease in Eastern North America and California. *California Plant Pathology* (University of California) 41:1–4.

CARPENTER, E. D. 1970. Salt Tolerance of Ornamental Plants. *American Nurseryman* 131(2):12, 54–71.

CARTER, J. C. 1975. *Diseases of Midwest Trees*. Urbana: University of Illinois.

CATHEY, H. M. 1975. Comparative Plant Growth-Retarding Activities of Ancymidol with APC, Phosfon, Chlormequat and SADH on Ornamental Plant Species. *HortSci.* 10:204–16.

CATHEY, H. M., and L. E. CAMPBELL. 1975a. Effectiveness of Five Vision-Lighting Sources on Photo-Regulation of 22 Species of Ornamental Plants. *Amer. Soc. Hort. Sci.* 100(1):65–71.

CATHEY, H. M., and L. E. CAMPBELL. 1975b. Security Lighting and Its Impact on the Landscape. *J. Arboriculture* 1:181–87.

CHADWICK, L. C. 1941. *Fertilization of Ornamental Trees, Shrubs, and Evergreens.* Ohio Agr. Exp. Sta. Bull. 620.

———. 1971. 3000 Years of Arboriculture: Past, Present and Future. *Arborist's News* 36(6):73a–78a.

———. 1980. Review of Guide for Establishing Values of Trees and Other Plants. *J. Arboriculture* 6(2):48–50.

CHAN, F. J., R. W. HARRIS, and A. T. LEISER. 1977. *Direct Seeding Woody Plants in the Landscape*. Univ. Calif. Agr. Sci. Leaflet 2577.

CHANCELLOR, W. J. 1976. *Compaction of Soil by Agricultural Equipment*. Univ. Calif. Agr. Sci. Bull. 1881.

CHANDLER, W. H., and D. S. BROWN. 1951. *Deciduous Orchards in California Winters*. Calif. Agr. Ext. Ser. Circ. 179.

CHANDLER, W. H., and R. D. CORNELL. 1952. *Pruning Ornamental Trees, Shrubs and Vines*. Calif. Agr. Ext. Circ. 183.

CHANG, J. H. 1968. *Climate and Agriculture*. Chicago: Aldine.

CHAPMAN, H. D. 1960. *Leaf and Soil Analyses in Citrus Orchards*. Calif. Agr. Ext. Serv. Manual 25.

———. 1966a. Calcium. In *Diagnostic Criteria for Plants and Soils*, ed. H. D. Chapman. Riverside, Calif.: H. D. Chapman, pp. 65–92.

———. 1966b. Zinc. In *Diagnostic Criteria for Plants and Soils*, ed. H. D. Chapman. Riverside, Calif.: H. D. Chapman, pp. 484–99.

CHITTENDON, F. J., ed. 1951. *Dictionary of Gardening*. Oxford: Clarendon Press.

CLARK, D. E., ed. 1979. *Sunset New Western Garden Book*. Menlo Park, Calif.: Lane.

CLARK, F. E. 1977. Living Organisms in the Soil. In *Soil: The 1957 Yearbook of Agriculture*. Washington, D.C.: U.S. Government Printing Office.

CLOUSTON, B., ed. 1977. *Landscape Design with Plants*. London: Heinemann.

COLE, L. C. 1968. Can the World Be Saved? *Biol. Sci.* 18:679–84.

———. 1970. Gas Exchange. *Environment* 12(10):39–42.

COLLINS, J. F. 1920. *Tree Surgery*. USDA Farmers' Bull. 1178.

CONKLIN, E. L. 1978. Interior Landscaping. *J. Arboriculture* 4(4):73–79.

CONNELL, J. H., and others. 1979. Gaseous Ammonia Losses Following Nitrogen Fertilization. *Calif. Agr.* 33(1):11–12.

CONOVER, C. A. 1978. Plant Conditioning: It's Here. *New Horizons*. Washington, D.C.: Hort. Res. Inst., pp. 6–9.

CONOVER, C. A., and R. T. POOLE. n.d. *Fertilization of Indoor Foliage Plants*. Apopka, Florida: Agr. Res. Ctr., pp. 130–33.

CONOVER, C. A., R. T. POOLE, and R. W. HENLEY. 1975. Growing Acclimatized Foliage Plants. *Florida Foliage Grower* 12(9):1–7.

COOK, D. I., and D. F. VAN HAVERBEKE. 1971. *Trees and Shrubs for Noise Abatement*. U.S. Forest Serv. Res. Bull. 246.

COOL, R. A. 1976. Tree Spade vs. Bare Root Planting. *J. Arboriculture* 2(5):92–95.

COPELAND, E. B. 1906. On the Water Relations of the Coconut Palm. *Philippine J. Sci.* 1:6–57.

COSTELLO, L., and J. L. PAUL. 1975. Moisture Relations in Transplanted Container Plants, *HortSci.* 10:371–72.

COUNCIL ON ENVIRONMENTAL QUALITY. 1972. *Integrated Pest Management*. Washington, D.C.: U.S. Government Printing Office.

CREECH, J. L., and W. HAWLEY. 1960. Effects of Mulching on Growth and Winter Injury of Evergreen Azaleas. *Proc. Amer. Soc. Hort. Sci.* 75:650–57.

CRIPE, R. E. 1978. Lighting Protection for Trees. Reported by J. A. Weidhaas, Jr. *Amer. Nurseryman* 148(9):58–60.

DAVENPORT, D. C., R. M. HAGAN, and P. E. MARTIN. 1969. Antitranspirants: Uses and Effects on Plant Life. *Calif. Turfgrass Culture* 19(4):25–27.

DAVEY, K. L. 1967. Pruning Mature Trees. *Western Landscaping News* 7(12):14.

DAVIS, S. H., JR. 1971. *What's Wrong with My Tree?* N.J. Coop. Ext. Serv. Leaflet 156-A.

———. 1975. Getting Rid of Tree Roots in Sewer Lines. *Arboricultural Consultant*, Jan.-Feb., p. 1.

DAVIS, W. B., and others. 1964. *Landscaping in Containers without Natural Drainage*. Univ. Calif. Agr. Sci. Leaflet 2577.

DEERING, R. B. 1955. Effective Use of Living Shade. *Calif. Agr.* 9(9):10–11.

DEERING, R. B., and F. A. BROOKS. 1953. Landscaping for Summer Shade. *Calif. Agr.* 7(5):11, 16.

DEL MORAL, R., and C. H. MULLER. 1970. Alleopathic Effects of *Eucalyptus camaldulensis*. *Amer. Midland Naturist* 83:254–82.

DEWALLE, D. R., and E. P. FARRAND. 1978. *Windbreaks and Shade Trees: Their Use in Home Energy Conservation*. Penn. State Univ. Agr. Special Circular 245.

DICKEY, R. D. 1977. *Nutritional Deficiencies of Woody Ornamental Plants Used in Florida Landscapes*. Fla. Agr. Exp. Bull. 791.

DOCHINGER, L. S. 1974. *Air Pollution Problems in Nurseries*. Delaware, Ohio: USDA Forest Serv. N.E. Forest Exp. Sta.: New Horizons, pp. 1–5.

DOMIR S. C. 1978. Chemical Control of Tree Height. *J. Arboriculture* 4(7):145–53.

DOULL, J., C. D. CLAASSEN, and M. O. AMDUR. 1980. *Casarett and Doull's Toxicology: The Basic Science of Poisons*. 2nd ed. New York: Macmillan.

DULING, J. Z. 1969. Recommendations for Treatment of Soil Fills around Trees. *Arborist's News* 34(6):1–4.

DUNN, N. P. 1975. *The Mechanics of Shade Tree Evaluation*. Memphis, Tenn.: Private publication.

EATON, F. M. 1966. Chlorine. In *Diagnostic Criteria for Plants and Soils*, ed. H. D. Chapman. Riverside, Calif.: H. D. Chapman, pp. 98–135.

ENVIRONMENTAL PROTECTION AGENCY. 1977. *Economic Analysis, Root Control, and Backwater Flow Control as Related to Infiltration/Inflow Control*. Cincinnati: Environmental Protection Technology Series EPA 600/2-77-017a.

EPSTEIN, E. 1972. *Mineral Nutrition of Plants: Principles and Perspectives*. New York: John Wiley.

ESSIG, E. D. 1958. *Insects and Mites of Western North America*. New York: Macmillan.

ESTEVEZ, M. T., ed. 1976. *From the Plants' Point of View*. Proc. Symposium of the Use of Living Plants in the Interior Environment. Alexandria, Va.: Soc. Amer. Florists.

FAYLE, D. C. F. 1968. *Radial Growth in Tree Roots: Distribution, Timing, Anatomy*. University of Toronto, Forestry Tech. Rep. 9.

———. 1976. Stem Sway Affects Ring Width and Compression Wood Formation in Exposed Root Bases. *Forest Sci.* 22(2):193–94.

FEDERER, C. A. 1971. *Effects of Trees in Modifying Urban Microclimate*. Proc. Symp. on the Role of Trees in the South's Urban Environment (USDA Forest Service).

FELIX, R., and A. L. SHIGO. 1977. Rots and Rods. *J. Arboriculture* 3(10):187–90.

FELT, E. P. 1940. *Plant Galls and Gall-Makers*. Ithaca, N.Y.: Comstock.

FISHER, M. E., E. SATCHELL, and J. M. WATKINS. 1975. *Gardening with New Zealand Plants, Shrubs, and Trees*. Auckland and London: William Collins.

FLINT, M. L., and R. VAN DEN BOSCH. 1981. *Introduction to Integrated Pest Management*. New York: Plenum Press.

FLOWER, F. B., E. F. GILMAN, and I. A. LEONE. 1981. Landfill Gas: What It Does to Trees and How Its Injurious Effects May Be Prevented. *J. Arboriculture* 7(2):43–52.

FOSTER, R. S. 1978. *Landscaping That Saves Energy Dollars*. New York: McKay.

FRANCIS, H. L., J. R. BREECE, and R. L. BALDWIN. 1974. *Christmas Tree Farming with Monterey Pines in Southern California*. Univ. Calif. Agr. Ext. AXT-n 200.

FRANCOIS, L. E. 1980. *Salt Injury to Ornamental Shrubs and Ground Covers*. USDA Home and Garden Bull. 213.

FRANCOIS, L. E., and R. A. CLARK, 1978. Salt Tolerance of Ornamental Shrubs, Trees, and Iceplant. *J. Amer. Soc. Hort. Sci.* 103(2):280–83.

FRANCOIS, L. E., and R. A. CLARK. 1979. Boron Tolerance of Twenty-five Ornamental Shrub Speices. *J. Amer. Soc. Hort. Sci.* 104(3):319–22.

FREE, M. 1961. *Plant Pruning in Pictures*. Garden City, N.Y.: Doubleday.

FRYDENLUND, M. M. 1977. *Lightning Protection for Home, Farm and Family*. Harvard, Ill.: Lightning Protection Inst.

FULLER, R. G., D. E. BELL, and H. E. KAZMAIER. 1965. Tree Wound Dressing for Sprout Control in Pole Line Clearing. *Edison Electric Inst. Bull.* 33:290–94.

GALE, J., and R. M. HAGAN. 1966. Plant Antitranspirants. *Ann. Rev. Plant Physiol.* 17:269–82.

GANS, H. J. 1967. *The Levittowners*. New York: Pantheon Books.

GARDNER, V. R., F. C. BRADFORD, and H. D. HOOKER, JR. 1939. *The Fundamentals of Fruit Production*. New York: McGraw-Hill.

GARDNER, W. R., and M. FIREMAN. 1958. Laboratory Studies of Evaporation from Soil Columns in the Presence of a Water Table. *Soil Sci.* 85:244–49.

GARNER, J. H. B. 1974. The Death of Woody Ornamental Plants Associated with Leaking Natural Gas. *Arborist's News* 39(12):13–17.

GEIGER, R. 1961. *The Climate Near the Ground.* Cambridge: Harvard University Press.

GEORGE, E. J. 1956. *Cultural Practices for Growing Shelterbelt Trees on the Northern Great Plains.* USDA Tech. Bull. 1138.

———. 1966. *Shelterbelts for the Northern Great Plains.* USDA Farmers Bull. 2109.

GILL, L. S., and F. G. HAWKSWORTH. 1961. *The Mistletoes: A Literature Review.* USDA Forest Serv. Pest Leaflet 147.

GLASS, E. H., ed. 1975. *Integrated Pest Management: Rationale, Potential, Needs, and Implementation.* Washington, D.C.: Entomological Soc. of Amer.

GOODIN, J. R., and V. T. STOUTEMYER. 1962. Carob Tree Growth Stimulated with Gibberellin. *Calif. Agr.* 16(9):4–5.

GOUIN, F. R. 1979. Anti-Desiccants for Protecting Evergreens. *News and Views* (Amer. Hort. Soc.) 21(1):6.

GOWANS, K. D. 1970. Water Pollution Considerations in Landscape Management. *Proc. Calif. Park and Recreation Administrators Inst.* 11(20):1–5.

GOWANS, K. D., and H. L. HALL. 1971. *Providing Drainage for Ornamental Trees in Layered Soils.* Univ. Calif. Agr. Sci. Leaflet 2575.

GRANT, J. A., and C. L. GRANT. 1955. *Trees and Shrubs for Pacific Northwest Gardens.* San Carlos, Calif.: Brown & Nourse.

GREEN, J. L. 1978. Plant Injury Caused by Sawdusts and Barks. *Ornamentals Northwest Newsletter* (Oregon State University) Dec.-Jan. 1977–78:8–12.

GREER, M., and T. A. GASKIN. 1978. Thrive or Survive: Reaction of Certain Plants to Air Pollution in California. *Western Landscaping News* 18(8):14–15.

GRESS, D. R. 1979. Western Arborists Discuss Need to Be Aware and More Involved. *American Nurseryman* 150(3):70–75.

GREY, G. W., and F. J. DENEKE. 1978. *Urban Forestry.* New York: John Wiley.

GROUNDS, R. 1973. *The Complete Handbook of Pruning.* New York: Macmillan.

HACKETT, W. P. 1976. Control of Phase Change in Woody Plants. *Acta Horticultural* 56:143–54.

HAISE, H. R. 1955. How to Measure the Moisture in Soil. In *Water: the Yearbook of Agriculture,* ed. A. Stefferod. Washington, D.C.: U.S. Printing Office, pp. 362–71.

HALFACRE, R. G., and J. A. BARDEN. 1979. *Horticulture.* New York: McGraw-Hill.

HALLIWELL, B., J. TURPIN, and J. WRIGHT. 1979. *The Complete Handbook of Pruning.* 2nd ed. London: Ward Lock.

HAMILTON, W. D. 1976. Street Tree Root Problem Survey. *Landscape Supervisors Forum Newsletter, May 11,* Hayward: Calif. Coop. Ext., pp. 1–3.

———. 1977. *Preventing Formation of Ginkgo Fruit.* Hayward: Calif. Coop. Ext., p. 1.

———. 1978. Drip Irrigation I, II, & III. *Growing Points.* Hayward: Calif. Coop. Ext., April, pp. 3–4; May, pp. 2–4; June, pp. 1–4.

———. 1979. *Olive Flower and Fruit Prevention Trials.* Unpublished report. Hayward: Calif. Coop. Ext., p. 4.

HAMILTON, W. D., and D. MARLING. 1981. Large Tree Cavity Work and Cabling. *J. Arboriculture* 7(7):180–82.

HARRIS, R. W. 1962. Water: Hazardous Necessity. *Proc. Intl. Shade Tree Conf.* 38: 182–88.

———. 1966. Influence of Turfgrass on Young Landscape Trees. *Proc. XVII Intl. Hort. Cong.* 1:81.

———. 1967. Factors Influencing Root Development of Container-Grown Trees. *Proc. Intl. Shade Tree Conf.* 43:304–14.

———. 1972. High-Temperature Limb Breakage. *Proc. Intl. Shade Tree Conf.* 48: 133–34.

HARRIS, R. W., and L. BALICS. 1963. Strong Branch Structure for Modesto Ash. *Calif. Agr.* 17(2):10–11.

HARRIS, R. W., and R. H. COPPOCK. 1977. *Saving Water in Landscape Irrigation.* Univ. Calif. Agr. Sci. Leaflet 2976.

HARRIS, R. W., and W. B. DAVIS. 1976. *Planting Landscape Trees.* Univ. Calif. Agr. Sci. Leaflet 2583.

HARRIS, R. W. and W. D. HAMILTON. 1969. Staking and Pruning Young *Myoporum laetum* Trees. *J. Amer. Soc. Hort. Sci.* 94:359–61.

HARRIS, R. W., A. T. LEISER, and W. B. DAVIS. 1976. *Staking Landscape Trees.* Univ. Calif. Agr. Ext. Leaflet 2576.

HARRIS, R. W., A. T. LEISER, and N. M. MATHENY. 1982. *Urban Tree Manual.* U.S. Department of Agriculture Forest Service

HARRIS, R. W., D. LONG, and W. B. DAVIS. 1967. *Root Problems in Nursery Liner Production.* Univ. Calif. Agr. Sci. Leaflet 2563.

HARRIS, R. W., J. L. PAUL, and A. T. LEISER. 1977. *Fertilizing Woody Plants.* Univ. Calif. Agr. Sci. Leaflet 2958.

HARRIS, R. W., R. M. SACHS, and R. E. FISSELL. 1971. Control of Trunk Sprouts with Growth Regulators. *Calif. Agr.* 25:1–3.

HARRIS, R. W., and others. 1969. *Pruning Landscape Trees.* Univ. Calif. Agr. Sci. Leaflet 2574.

HARRIS, R. W., and others. 1971. Root Pruning Improves Nursery Tree Quality. *J. Amer. Soc. Hort. Sci.* 96:105–8.

HARRIS, R. W., and others. 1972. Spacing of Container-Grown Trees in the Nursery. *J. Amer. Soc. Hort. Sci.* 97(4)503–6.

HARTIG, R. 1878. *Die Zersetzungserscheinungen des Holzes der Nadelbäume und der Eiche* [*Decay Phenomena in the Wood of Conifers and Oaks*]. Berlin: Springer.

HARTMANN, H. T. 1967. 'Swan Hill': A New Ornamental Fruitless Olive for California. *Calif. Agr.* 21(1):4–5.

HARTMANN, H. T., and D. E. KESTER. 1983. *Plant Propagation: Principles and Practices.* 4th ed. Englewood Cliffs, N.J.: Prentice-Hall.

HAUPT, E. H. 1980. The Private Tree Worker and Energized Lines. *J. Arboriculture* 6(4):93–95.

HAUSENBUILLER, R. L. 1978. *Soil Science: Principles and Practices.* 2nd ed. New York: Macmillan.

HENDRICKSON, A. H. 1918. What Size Nursery Trees? *Calif. State Dept. of Agr. Monthly Bull.* 7:171–74.

HEPTING, G. H. 1935. Decay Following Fire in Young Mississippi Delta Hardwoods. USDA Tech. Bull. 494.

———— . 1968. Diseases of Forest and Tree Crops Caused by Air Pollutants. *Phytopathology* 58:1098–101.

HERRICK, G. W. 1935. *Insect Enemies of Shade Trees.* Ithaca, N.Y.: Comstock.

HERRINGTON, L. P. 1974. Trees and Acoustics in Urban Areas. *J. Forestry* 72(8):462–65.

HERRINGTON, L. P., G. E. BERTOLIN, and R. E. LEONARD. 1972. Microclimate of a Suburban Park. In *Proc. Conf. on Urban Environment and 2nd Conf. on Biometeorology:* 43–44 (Amer. Meteorological Soc., Worcester, Mass.).

HEWITT, W. B., and F. L. JENSEN. 1979. *Powdery Mildew Disease of Grapevines.* Univ. Calif. Div. Agr. Sci. Leaflet 2212.

HIELD, H. 1979. Trunk Bark Banding with Chlorflurenol for Growth Control. *J. Arboriculture* 5(3):59–61.

HIELD, H., R. M. SACHS, and S. HEMSTREET. 1978. Foliar Spray and Bark Banding with Dikegulac for Ornamental Tree Growth Inhibition. *HortSci.* 13(4):440–42.

HILLEL, D. 1980. *Applications of Soil Physics.* New York: Academic Press.

HILLIER, H. C. 1973. *Hilliers' Manual of Trees and Shrubs.* Newton Abbot, England: David & Charles.

HIMELICK, E. B. 1969. *Tree and Shrub Hosts of Verticillium albo-atrum.* Ill. Natural History Survey Biol. Notes 66.

———— . 1970. Frost Cracks on London Plane Tree an Important Outdoor Winter Thermometer. *Arborist's News* 35(1):5–6.

———— . 1981. *Transplanting Manual for Trees and Shrubs.* Urbana, Ill.: Intl. Soc. Arboriculture.

HIMELICK, E. B., and D. W. CEPLECHA. 1976. Dutch Elm Disease Eradication by Pruning. *J. Arboriculture* 2(5):81–84.

HITCHCOCK, A. E., W. CROCKER, and P. W. ZIMMERMAN. 1932. Toxicity of Illuminating Gas in Soils. *Proc. Natl. Shade Tree Conf.* 9:34–36.

HOCKS, J. 1972. Changes in Composition of Soil Air Near Leaks in Natural Gas Mains. *Soil Sci.* 13:46–54.

HOITINK, H. A. J., and H. A. POOLE. 1977. Composted Bark Media for Control of Soil Borne Plant Pathogens. *Ohio Florists' Assn. Bull.* 567, pp. 10–11.

HOUSER, J. S. 1937. Borer Control Experiments. *Proc. Natl. Shade Tree Conf.* 13:159–68.

HOUSTON, D. R. 1980. Effects of Defoliation on Trees and Shrubs. In *Gypsy Moth Compendium*, ed. P. M. Wargo. USDA Tech. Bull. 1584:217–18.

HOYT, R. S. 1978. *Ornamental Plants for Subtropical Regions.* Anaheim, Calif.: Livingston.

HUDLER, G. 1981. Salt Injury to Roadside Plants. *Grounds Maintenance* 16(2):80–84.

HUDLER, G. W., and M. A. BEALE. 1981. Anatomical Features of Girdling Root Injury. *J. Arboriculture* 7(2):29–32.

HUDSON, R. L. 1972. *The Pruning Handbook.* Englewood Cliffs, N.J.: Prentice-Hall.

HUFFAKER, C. B., ed. 1973. *Biological Control.* New York: Plenum.

HULL, J., JR., and L. N. LEWIS. 1959. Response of One-Year-Old Cherry and Mature Bearing Cherry, Peach and Apple Trees to Gibberellin. *Proc. Amer. Soc. Hort. Sci.* 74:93–100.

HUTCHINSON, F. E. 1968. The Relationship of Road Salt Applications to Sodium and Chloride Ion Levels in the Soil Bordering Major Highways. In *Proc. Symp. Pollutants in the Roadside Environment* (Univ. Conn.), ed. E. D. Carpenter, pp. 24–35.

INTERAGENCY AGRICULTURAL INFORMATION TASK FORCE. n.d. (after 1974). *Drought Tips: Common Irrigation Problems: Some Solutions*. Davis: Univ. Calif. Land, Air, and Water Resources Ext.

IRVINE, 1975. *Comparative Costs of Park Maintenance*. Unpublished report. City of Irvine, Calif.

IYER, J. G., R. B. CAVEY, and S. A. WILDE. 1980. Mycorrhizae: Facts and Fallacies. *J. Arboriculture* 6(8):213-20.

JACKS, G. V., W. D. BRIND, and R. SMITH. 1955. *Mulching*. Commonwealth Bureau of Soil Sci. Tech. Comm. 49.

JACOBS, M. R. 1939. *A Study of the Effect of Sway on Trees*. Australian Commonwealth Forest Bull. 26.

JACOBSON, J. S., and A. C. HILL, eds. 1970. *Recognition of Air Pollution Injury to Vegetation: A Pictorial Atlas*. Pittsburgh: Agr. Committee, Air Pollution Control Assn. Inf. Report 1, TR-7.

JEFFERS, W. A., and R. ABBOTT. 1979. New System of Guying Trees. *J. Arboriculture* 5(6):121-23.

JENSEN, I. B., and R. L. HODDER. 1979. *Tubelings, Condensation Traps, Mature Tree Transplanting and Root Sprigging Techniques for Tree and Shrub Establishment in Semiarid Areas*. Mont. Agr. Exp. Sta. Res. Report 141, Vol. 2.

JOHNSON, C. 1977. *Mistletoe Control in Shade Trees*. Univ. Calif. Agr. Sci. Leaflet 2571.

JOHNSON, C. M. 1966. Molybdenum. In *Diagnostic Criteria for Plants and Soils*, ed. H. D. Chapman. Riverside, Calif.: H. D. Chapman, pp. 286-301.

JOHNSON, W. T. 1982. The Scale Insect, a Paragon of Confusion. *J. Arboriculture* 8(5):113-23.

JOHNSON, W. T., and H. H. LYON. 1976. *Insects That Feed on Trees and Shrubs*. Ithaca, N.Y.: Comstock.

JOINER, J. N., ed. 1981. *Foliage Plant Production*. Englewood Cliffs, N.J.: Prentice-Hall.

KELLOGG, A. 1882. *Forest Trees of California*. Sacramento, Calif.: State Printing Office.

KELLY, S. 1969. *Eucalypts*. Melbourne: Thomas Nelson.

KENWORTHY, A. L. 1953. Depletion of Soil Moisture in a Mature Apple Orchard with a Sod-Mulch System of Soil Management: *Mich. Agr. Sta. Quarterly Bull.* 36:39-45.

KENWORTHY, A. L., and J. E. MOTES. 1975. *Trickle Irrigation*. Michigan State Res. Report 285.

KEPNER, R. A. 1950. *The Principles of Orchard Heating*. Calif. Agr. Exp. Sta. Cir. 400.

KIMBALL, M. H., and F. A. BROOKS. 1959. Plant Climates of California. *Calif. Agr.* 13(5):2-7.

KING, G. C., C. BEATLY, and M. McKENZIE. 1970. *Polyurethane for Filling Tree Cavities*. Univ. Mass. Pub. 58.

KOEHLER, C. S. 1980. *Galls on Plants Caused by Insects*. Univ. Calif. Agr. Sci. Leaflet 21198.

KOZLOWSKI, T. T., and C. H. WINGET. 1964. Diurnal and Seasonal Variation in Radii of Tree Stems. *Ecology* 45:149-55.

KRAEBEL, C. J. 1936. *Erosion Control of Mountain Roads*. USDA Cir. 380.

KRAMER, P. J., and T. T. KOZLOWSKI. 1979. *Physiology of Woody Plants*. New York: Academic Press.

LABANAUSKAS, C. K. 1966. Manganese. In *Diagnostic Criteria for Plants and Soils*, ed. H. D. Chapman. Riverside, Calif.: H. D. Chapman, pp. 264–85.

LANDSBERG, H. E. 1970. Climates and Urban Planning. In *Urban Climates*. Geneva, Switzerland: World Meteorological Org. Tech. Note 108, pp. 364–74.

LANGE, A. H., C. L. ELMORE, and A. B. SAGHIR. 1973. *Diagnosis of Phytotoxicity from Herbicides in Soils*. Univ. Calif. Agr. Ext. Leaflet TA-69.

LANPHEAR, F. O. 1971. Urban Vegetation: Values and Stresses. *HortSci.* 6:332–34.

LARSON, P. R. 1965. Stem Form of Young Larix as Influenced by Wind and Pruning. *Forest Sci.* 11(4):412–24.

LATHROP, J. K., and R. A. MECKLENBURG. 1971. Root Regeneration and Root Dormancy in *Taxus. J. Amer. Soc. Hort. Sci.* 96(1):111–14.

LEAF, A. L. 1968. K, Mg, and S Deficiencies in Forest Trees. In *Forest Fertilization: Theory and Practice*. Knoxville: Tennessee Valley Authority, pp. 88–122.

LEAR, B., and D. E. JOHNSON. 1975. *Controlling Nematodes in the Home Garden*. Univ. Calif. Agr. Sci. Leaflet 2112.

LEDIN, R. B. 1961. Pruning Palms., *Amer. Hort. Magazine* 40:142–43.

LEIGHTON, G. M., R. D. HARTER, and G. R. CROMBIE. 1978. *Sewage Sludge Composting in Small Towns*. Univ. N.H. Sta. Bull. 508.

LEISER, A. T., and J. D. KEMPER. 1968. A Theoretical Analysis of a Critical Height of Staking Landscape Trees. *Proc. Amer. Soc. Hort. Sci.* 92:713–20.

LEISER, A. T., and J. D. KEMPER. 1973. Analysis of Stress Distribution in the Sapling Tree Trunk. *J. Amer. Soc. Hort. Sci.* 98(2):164–70.

LEISER, A. T., and G. NYLAND. 1972. Unpublished report.

LEISER, A. T., and others. 1972. Staking and Pruning Influence Trunk Development of Young Trees. *J. Amer. Soc. Hort. Sci.* 97(4):498–503.

LEISER, A. T., and others. 1974. *Revegetation of Disturbed Soils in the Tahoe Basin*. Sacramento: Calif. Dept. of Transportation Final Report.

LEISER, A. T., and others. 1980. *Highway Operation and Plant Damage*. Univ. Calif., Davis, Dept. of Env. Hort. Report for Calif. Dept. of Transportation.

LEIUS, K. 1967. Influence of Wild Flowers on Parasitism of Tent Caterpillar and Codling Moth. *Canadian Entomology* 99:444–46.

LEONARD, D. A., D. E. BAYER, and R. K. GLENN. 1974. Control of Tree Roots. *Weed Sci.* 22:516–20.

LEONARD, O. A., and W. A. HARVEY. 1965. *Chemical Control of Woody Plants*. Calif. Agr. Exp. Sta. Bull. 812.

LEONARD, O. A., and J. A. PINKARD. 1946. The Effect of O_2 and CO_2 Levels on Cotton Root Development. *Plant Physiol.* 21:18–36.

LEONARD, O. A., and N. R. TOWNLEY. 1971. Control of Tree Roots in Sewers and Drains. *Calif. Agr.* 25(11):13–15.

LEONE, I. A., and others. 1977. Damage to Woody Species by Anaerobic Landfill Gases. *J. Arboriculture* 3(12):221–25.

LERMAN, S. L., and E. F. DARLEY. 1975. Particulates. In *Responses of Plants to Air Pollution*, ed. J. B. Mudd and T. T. Kozlowski. New York: Academic Press, pp. 141–58.

LINDOW, S. E. 1980. New Method of Frost Control Through Control of Epiphytic Ice Nucleation Active Bacteria. *Calif. Plant Pathology* 48:1–5.

LOCKE, L. F., and H. V. ECK. 1965. *Iron Deficiency in Plants*. USDA Home and Garden Bull. 102.

LONG, D. 1961. Developing and Maintaining Street Trees. *Proc. Intl. Shade Tree Conf.* 37:172.

LOOMIS, R. C., and W. H. PADGETT. n.d. *Air Pollution and Trees in the East*. Upper Darby, Pa.: USDA Forest Serv., Northeastern Area. Adapted from *Air Pollution Damages Trees*. 1973. USDA Forest Serv.

LORD, E. E. 1979. *Shrubs and Trees for Australian Gardens*. 4th ed. Melbourne: Lothian.

LOVELADY, S. M. 1965. Handsome Factories Yield Unexpected Joys. *Wall Street Journal*, Dec. 1, p. 1.

LULL, H. W. 1959. *Soil Compaction on Forest and Range Lands*. USDA Forest Serv. Misc. Pub. 768.

LUMIS, G. P., G. HOFSTRA, and R. HALL. 1973. Sensitivity of Roadside Trees and Shrubs to Aerial Drift of Deicing Salt. *HortSci.* 8(6):475–77.

McCAIN, A. H. 1975. *Fire Blight of Fruits and Ornamentals*. Univ. Calif. Agr. Sci. Leaflet 2715.

————. 1979a. *Sycamore Anthracnose*. Univ. Calif. Div. Agr. Sci. 2618.

————. 1979b. *Verticillium Wilt*. Univ. Calif. Agr. Sci. Leaflet 2592.

McCAIN, A. H., and R. D. RAABE. 1972. *Armillaria Root Rot*. Univ. Calif. Agr. Ext. OSA 80.

McCLINTOCK, E., and A. T. LEISER. 1979. *An Annotated Checklist of Woody Ornamental Plants of California, Oregon, and Washington*. Univ. Calif. Agr. Sci. Pub. 4091.

MacDANIELS, L. H. 1932. Factors Affecting the Breaking Strength of Apple Tree Crotches. *Proc. Amer. Soc. Hort. Sci.* 29:44.

MacDONALD, J. D. 1982. Effect of Salinity Stress on Development of Phytophthora Root Rot of Chrysanthemum. *Phytopathology* 72:214–19.

MacGREGOR J. 1971. Why the Wind Howls around Those Plazas Close to Skyscrapers. *Wall Street Journal*, Feb. 18, 1971, p. 1.

McMINN, H. E. 1939. *An Illustrated Manual of California Shrubs*. San Francisco: J. W. Stacy.

McMINN, H. E., and E. MAINO. 1963. *An Illustrated Manual of Pacific Coast Trees*. Berkeley: University of California Press.

McQUILKIN, W. E. 1950. Effects of Some Growth Regulators and Dressings on the Healing of Tree Wounds. *J. Forestry* 48:423–28.

MACE, J. 1978. Toxicity of Caffeine. *J. Pediatrics* 92:345.

MAFTOUN, M., and W. L. PRITCHETT. 1970. Effects of Added Nitrogen on the Availability of Phosphorus to Slash Pine on Two Lower Coastal Plain Soils. *Soil Sci. Soc. Amer. Proc.* 34(4):685–90.

MAGILL, A. W. 1970. *Five California Campgrounds: Conditions Improve after 5 Years of Recreational Use*. USDA Forest Serv. Res. Paper PSW-62.

MAINO, E., and F. HOWARD. 1955. *Ornamental Trees*. Berkeley: University of California Press.

MAIRE, R. G., and R. L. BRANSON. 1972. Salinity Tolerance of Landscape Plants. Los Angeles County Agr. Ext. Serv. Progress Report, May 11.

MANAKER, G. H. 1981. *Interior Plantscapes: Installation, Maintenance, and Management.* Englewood Cliffs, N.J.: Prentice-Hall.

MANION, P. D. 1981. *Tree Disease Concepts.* Englewood Cliffs, N.J.: Prentice-Hall.

MAROTZ, G. A., and J. C. COINER. 1973. Acquisition and Characterization of Surface Material Data for Urban Climatological Studies. *J. Applied Meteorology* 12: 919-23.

MARSH, A. W. 1975. *Questions and Answers about Tensiometers.* Univ. Calif. Coop. Ext. Leaflet 2264.

———. 1977. *Drip Irrigation.* Univ. Calif. Agr. Sci. Leaflet 2740.

MARSH, R. E., and M. W. CUMMINGS. 1976. *Pocket Gopher Control with Mechanical Bait Applicator.* Univ. Calif. Div. Agr. Sci. Leaflet 2699.

MARSHALL, R. P. 1931. *The Relation of Season of Wounding and Shellacking to Callus Formation in Tree Wounds.* USDA Tech. Bull. 246.

MARTIN, W. E., J. VLAMIS, and N. W. STICE. 1953. Field Correction of Calcium Deficiency on a Serpentine Soil. *Agronomy J.* 45:204-8.

MARX, D. H. 1973. Mycorrhizae and Feeder Root Disease. In *Ectomycorrhizae,* ed. G. C. Marks and T. T. Kozlowski. New York: Academic Press, pp. 351-82.

MAUGH, T. H., II, 1979. SO$_2$ Pollution May Be Good for Plants. *Science* 205:383.

MAYNE, L. S. 1975. Cabling and Bracing. *J. Arboriculture* 1(6):101-6.

MEE, T. R., and J. F. BARTHOLIC. 1979. Man-Made Fog. In *Modification of the Aerial Environment of Crops,* ed. B. J. Barfield and J. F. Gerber. St. Joseph, Mich.: Amer. Soc. Agr. Engineers, pp. 334-52.

MEISTER, E. L., JR. 1982. *Farm Chemicals Handbook.* Willoughby, Ohio: Meister.

MENNINGER, E. A. 1964. *Seaside Plants of the World.* New York: Hearthside Press.

MERCER, P. C. 1979. Attitudes to Pruning Wounds. *Arboricultural J.* 3:457-65.

MERRILL, J., and M. SOLOMONSON. 1977. Mycorrhizae and Their Horticultural Importance. *Univ. Wash. Arboretum Bull.* 40(2):11-15.

MERRILL, W., D. H. LAMBERT, and W. LIESE. 1975. In *Phytopathological Classics* 12. [Translation of *Wichtige Krankheiten der Waldbäume* or *Important Diseases of Forest Trees.* R. Hartig. 1874. Berlin: Springer.] St. Paul, Minn.: Amer. Phytopath. Soc., unnumbered introduction.

MESKIMEN, G. 1970. Combating Grass Competition for Eucalyptus Planted in Turf. *Tree Planter Notes* 21(4):3-5.

METCALF, L. J. 1975. *The Cultivation of New Zealand Trees and Shrubs.* Wellington, N.Z.: A. H. and A. W. Reed.

MIDDLETON, J. T., E. F. DARLEY, and R. F. BREWER. 1958. Damage to Vegetation from Polluted Atmosphere. *J. Air Pollution Control Assn.* 8:7-15.

MIKOLA, P. 1973. Application of Mycorrhizal Symbiosis in Forestry Practice. In *Ectomycorrhizae: Their Ecology and Physiology,* ed. G. C. Marks and T. T. Kozlowski. New York: Academic Press, pp. 383-411.

MILLER, H. C., S. B. SILVERBORG, and R. J. CAMPANA. 1969. Dutch Elm Disease: Relation of Spread and Intensification to Control by Sanitation in Syracuse, New York. *Plant Disease Reporter* 53(7):551-55.

MILLER, P. R., and A. A. MILLECAN. 1971. Extent of Oxidant Air Pollution Damage to Some Pines and Other Conifers in California. *Plant Disease Reporter* 55(6): 555-59.

MILLER, V. J. 1959. Crotch Influence on Strength and Breaking Point of Apple Tree Branches. *Proc. Amer. Soc. Hort. Sci.* 73:27-32.

MONCK, J. W. 1976. Get to the Root of Sewer Problems. *Amer. City and County Magazine*, Nov., pp. 64-65.

MONTGOMERY, K. R. 1974. *Green Belts for Brush Fire Protection and Soil Erosion Control in Hillside Residential Areas.* Los Angeles: Los Angeles Arboreta and Botanic Gardens.

MOORE, L. S. 1979. Research Update on Crown Gall and Hairy Root Diseases in the Northwest. *Ornamentals Northwest Newsletter* 3(5):17-19.

MOORE, W. S., C. S. DAVIS, and C. S. KOEHLER. 1979. *Scale Insects and Their Control.* Univ. Calif. Agr. Sci. Leaflet 2237.

MORALES, D. J. 1975. *The Contribution of Trees to Residential Property Value.* Manchester: University of Connecticut M.S. Thesis.

MORGAN, W. C. 1965. New Irrigation and Aerification Methods. *Calif.Turfgrass Culture* 15(2):11-15.

MORLING, R. J. 1963. *Trees: Including Preservation, Planting, Law, Highways.* London: Estates Gazette.

MUIRHEAD, D. 1961. *Palms.* Globe, Ariz.: D. S. King.

MULLINS, M. G. 1979. *Reader's Digest Illustrated Guide to Gardening.* Sydney: Reader's Digest Services.

MURPHY, R. C., and W. E. MEYER. 1969. *The Care and Feeding of Trees.* New York: Crown.

NATIONAL ARBORIST ASSOCIATION. 1979. *National Arborist Association Standards.* Wantagh, N.Y.: Natl. Arborist Assn.

NATIONAL FIRE PROTECTION ASSOCIATION. 1980. *Lightning Protection Code 1980.* Boston: Natl. Fire Protection Assn.

NEEL, P. L. 1967. Factors Influencing Trunk Development of Landscape Trees. *Proc. Intl. Shade Tree Conf.* 43:293-303.

————. 1971. Experimental Manipulation of Trunk Growth in Young Trees. *Arborist's News* 36(3):25a-31a.

NEEL, P. L., and R. W. HARRIS. 1971. Motion-Induced Inhibition of Elongation and Induction of Dormancy in Liquidambar. *Science* 173:58-59.

NEELY, D. 1970. Healing of Wounds on Trees. *J. Amer. Soc. Hort. Sci.* 95(5):536-40.

————. 1972. Hints on Diagnosis of Tree Problems. *Proc. Intl. Shade Tree Conf.* 48:33-37.

————. 1976. Iron Deficiency Chlorosis of Shade Trees. *J. Arboriculture* 2(7):128-30.

————. 1978. Municipal Control of Dutch Elm Disease in Illinois. *Plant Disease Reporter* 62(2):130-31.

————. 1979a. *Guide for Establishing Values of Trees and Other Plants.* Urbana, Ill.: Intl. Soc. Arboriculture.

————. 1979b. Tree Wounds and Wound Closure. *J. Arboriculture* 5(6):135-40.

NEELY, D., and E. B. HIMELICK. 1968. Fertilization and Watering Trees. *Ill. Natural History Survey Bull.* 52:1-20.

NEELY, D., E. B. HIMELICK, and W. R. CROWLEY, JR. 1970. Fertilization of Established Trees: A Report of Field Studies. *Ill. Natural History Survey Bull.* 30(4):235-66.

NEW, P. B., and A. KERR. 1972. Biological Control of Crown Gall: Field Measurements and Glasshouse Experiments. *J. Applied Bacteriology* 35:279-87.

NEWMAN, C. J. 1963. Transplanting Semi-Mature Trees. In *Trees*, ed. R. J. Morling. London: Estates Gazette, pp. 29–39.

NOBLE, W., and W. TERRY. 1969. *A Partial List of Smog Resistant Plants*. Arcadia, Calif.: Los Angeles State and County Arboretum.

OKE, T. R. 1972. *Evapotranspiration in Urban Areas and Its Implications for Urban Climate Planning*. WNO-CIB Intl. Colloq. on Building Climatology Proc.

OLKOWSKI, W. C. 1973. *A Model Ecosystem Management Program*. Proc. Tall Timbers Conf. on Animal Control by Habitat Management 5:103–17.

OPITZ, K. W. 1970. *Spray to Prevent Fruit Set on Ornamental Olive Trees*. Univ. Calif. Agr. Sci. Leaflet 2479, Rev.

OSBORNE, R. 1975. *Garden Trees*. Menlo Park, Calif.: Lane.

OTTOSON, R. 1976. The Effect of Soil Residual Herbicides on Landscape Planting and Maintenance. *Proc. Calif. Weed Conf.* 28:153–56.

OUTCALT, S. I. 1972. A Reconnaissance Experiment in Mapping and Modeling the Effect of Land Use on Urban Thermal Regimes. *J. Applied Meteorology* 11: 1369–73.

PARKER, E. R., and R. W. SOUTHWICK. 1941. Manganese Deficiency in Citrus. *Proc. Amer. Soc. Hort. Sci.* 39:51–58.

PAYNE, B. R. 1973. The Twenty-Nine Tree Home Improvement Plan. *Natl. History* 82(9):74–75.

PELLETT, H. 1971. Effects of Soil Amendments on Growth of Landscape Plants. *Amer. Nurseryman* 134(10):12, 103–6.

———. 1982. Minimizing Winter Injury of Shade Trees. *J. Arboriculture* 7(12): 309–12.

PETERSON, J. T. 1969. *The Climate of Cities: A Survey of Recent Literature*. Raleigh, N.C.: Natl. Air Pollution Control Adm., U.S. Public Health Serv.

PHILIP, J. R., and D. A. deVRIES. 1957. Moisture Movement in Porous Material Under Thermal Gradients. *Trans. Amer. Geophys. Union* 38:222–32.

PIESTER, E. A. 1958. Rose Pruning. *Plants & Gardens Pruning Handbook* (Brooklyn Botanic Garden) 14(3):220–24.

PINNOCK, D. E., and others. 1978. Integrated Pest Management in Highway Landscapes. *Calif. Agr.* 32(2):33–34.

PIRONE, P. P. 1978a. *Tree Maintenance*. 5th ed. New York: Oxford University Press.

———. 1978b. *Diseases and Pests of Ornamental Plants*. 5th ed. New York: John Wiley.

POWERS, R. F. 1979. Nutrient Deficiency Symptoms in Conifers. In *Principles of Silviculture*, by T. W. Daniel, J. A. Helms, and F. S. Baker. 2nd ed. New York: McGraw-Hill.

———. 1981a. *Nutritional Characteristics of Ponderosa Pine and Associated Species*. Ph.D. Dissertation, University of California, Berkeley.

———. 1981b. Response of California True Fir to Fertilization. In *Forest Fertilization Conference*, ed. S. P. Gessel, R. M. Kennedy, and W. A. Atkinson. University of Washington, Institute of Forest Resources Contrib. 50:95–101.

PRIDHAM, A. M. S. 1938. Growth of Pin Oak (*Quercus palustris*): Report of Seven Years' Observations. *Proc. Amer. Soc. Hort. Sci.* 35:739–41.

PRITCHETT, W. L. 1979. *Properties and Management of Forest Soils*. New York: John Wiley.

PROEBSTING, E. L. 1935. Field and Laboratory Studies on the Behavior of NH₄ Fertilizer with Special Reference to the Almond. *Proc. Amer. Soc. Hort. Sci.* 33:46–50.

———. 1958. *Fertilizers and Cover Crops for California Orchards.* Calif. Agr. Ext. Serv. Cir. 466.

PRUITT, W. O. 1971. *Factors Affecting Potential Evapotranspiration.* West Lafayette, Ind. (Purdue University): Proc. Third Intl. Sem. Hydrology Professors, pp. 82–102.

PRYKE, J. F. S. 1979. Trees and Buildings. *Arboricultural J.* 3(6):388–96.

QUICK, J., and J. M. RIBLE. 1967. *Soil Analysis.* Calif. Agr. Ext. OSA 98.

RAABE, R. D. 1974. A Look at Rapid Composting. *Calif. Hort. J.* 35(1):17–18.

———. 1979. *Resistance or Susceptibility of Certain Plants to Armillaria Root Rot.* Univ. Calif. Agr. Sci. Leaflet 2591.

RAE, W. A. 1969. Large Tree Moving by Frozen Root Balls. *Trees,* Jan.-Feb., pp. 8–9.

RASMUSSEN, R. A. 1972. What Do Hydrocarbons from Trees Contribute to Air Pollution? *J. Air Pollut. Contr. Assn.* 22:537–43.

REECE, R. A. 1979. Trees and Insurance. *Arboricultural J.* 3(7):492–99.

REHDER, A. 1940. *Manual of Cultivated Trees and Shrubs Hardy in North America.* 2nd ed. New York: Macmillan.

REIL, W. O. 1979. Pressure-Injecting Chemicals into Trees. *Calif. Agr.* 6:16–19.

REUTHER, W., and C. K. LABANAUSKAS. 1966. Copper. In *Diagnostic Criteria for Plant and Soils,* ed. H. D. Chapman. Riverside, Calif.: H. D. Chapman, pp. 157–79.

RIBLE, J. M., and J. L. MEYER. 1979. Cleaning Drip Irrigation Systems. *Soil and Water* (Univ. Calif. Ext.) 42:1–4.

RICH, A. E. 1971. Salt Injury to Roadside Trees. *Proc. Intl. Shade Tree Conf.* 47: 77a–79a.

RICH, S. 1975. Air Pollution and Agricultural Practices. In *Responses of Plants to Air Pollution,* ed. J. B. Mudd and T. T. Kozlowski. New York: Academic Press, pp. 335–60.

RICHARDS, S. J. and A. W. MARSH. 1961. Irrigation Based on Soil Suction Measurements. *Soil Sci. Soc. Amer. Proc.* 25:65–69.

RICHARDSON, S. D. 1958. Bud Dormancy and Root Development in *Acer saccharinum.* In *The Physiology of Forest Trees,* ed. K. V. Thimann. New York: Ronald Press, pp. 409–25.

RICKMAN, R. W. 1979. *Bad with the Good.* USDA Agr. Res. 28(3):15.

RISHBETH, J. 1970. The Role of Basidiospores in Stump Infection by *Armillaria mellea.* In *Root Diseases and Soil-Borne Pathogens,* ed. T. A. Toussoun and others. Berkeley: University of California Press, 141–46.

ROBBINS, W. W., T. E. WEIER, and C. R. STOCKING. 1950. *Botany: An Introduction to Plant Science.* 2nd ed. New York: John Wiley.

ROBERTS, B. R., G. K. BROWN, and C. L. WILSON. 1979. *New Methods and Chemicals to Control Regrowth in Trees.* USDA Nursery Corps Res. Lab. EL-1112 Res. Project 214.

ROBINETTE, G. O. 1972. *Plants/People/and Environmental Quality.* Washington, D.C.: U.S. Natl. Park Serv.

ROSS, N., and others. 1970. *Reducing Loss from Crown Gall Disease.* Calif. Agr. Exp. Sta. Bull. 845.

RUSHFORTH, K. D. 1979. Summer Branch Drop. *Arboriculture Res. Note* (British Dept. of the Environment), Dec.: 1-2.

RUSSEL, J. C. 1939. The Effect of Surface Cover on Soil Moisture Losses by Evaporation. *Proc. Soil Sci. Soc. Amer.* 4:65-70.

RUTH, W. A., and V. W. KELLEY. 1932. *A Study of the Framework of the Apple Tree and Its Relation to Longevity.* Ill. Agr. Exp. Sta. Bull. 376:509-637.

SACHS, R. M. 1970. Chemical Control of Plant Growth. Univ. Calif. Ext., Davis. *Proc. Park and Recreation Administrators Institute* (9):1-16.

⸻. 1975. Unpublished information.

SACHS, R. M., and W. P. HACKETT. 1972. Chemical Inhibition of Plant Height. *Hort-Sci.* 7:440-47.

SACHS, R. M., T. KRETCHUN, and T. MOCK. 1975. Minimum Irrigation Requirements for Landscape Plants. *J. Amer. Soc. Hort. Sci.* 100(5):499-502.

SACHS, R. M., and others. 1970. *Chemical Control of Plant Growth in Landscapes.* Calif. Agr. Exp. Sta. Bull. 844.

SANTAMOUR, F. S., JR. 1972. Shade-Tree Improvement Research at the U.S. National Arboretum. *Proc. Intl. Shade Tree Conf.* 48:132-33.

⸻. 1979a. Inheritance of Wound Compartmentalization of Soft Maples. *J. Arboriculture* 5(10):220-25.

⸻. 1979b. Root Hardiness of Green Ash Seedlings from Different Proveniences. *J. Arboriculture* 5(12):276-79.

SCARLETT, A. L., and C. L. WAGENER. 1973. *Managing and Marketing California Forest-Grown Christmas Trees.* Calif. Agr. Ext. Serv. AXT-182.

SCHIECHTL, H. M. 1973. *Sicherungsarbeiten im landschaftsbau* [Stabilization Methods in Landscape Construction]. Munich: Verlag D. W. Callwey.

⸻. 1978. Umweltsverträgliche Abhangsbefestigung [Environmentally Compatible Slope Stabilization]. Organ der Deutschen Gesellschaft für Erd und Grundbau [German Natl. Soc. Soil Mechanics and Foundation Engineering]. *Geotechnik* 1:10-21.

SCHMID, J. A. 1975. *Urban Vegetation.* Univ. Chicago Dept. of Geography Res. Pap. 161.

SCHOENEWEISS, D. F. 1973. Diagnosis of Physiological Disorders of Woody Ornamentals. *Proc. Intl. Shade Tree Conf.* 49:33a-38a.

⸻. 1978. The Influence of Stress on Diseases of Nursery and Landscape Plants. *J. Arboriculture* 4(10):217-25.

SCHREIBER, L. R., and J. W. PEACOCK. 1974. *Dutch Elm Disease and Its Control.* USDA Forest Serv. Inf. Bull. 193.

SCHUDER, D. L. 1978. Identifying and Controlling Insect Galls. *American Nurseryman* 148(6):14, 115-19.

SCHULTZ, H. B., and J. V. LIDER. 1968. *Frost Protection with Overhead Sprinklers.* Calif. Agr. Ext. Serv. Leaflet 201.

SCHÜTTE, K. H. 1966. Trace Element Deficiencies in Cape Vegetation. *J. S. African Bot.* 26:145-49.

SCOTT, D. H. 1973. *Air Pollution Injury to Plant Life.* Washington, D.C.: Natl. Landscape Assn.

SCROGGINS, T. 1971. Using Leached Cedar Tow for Packing? Be Careful! *Oregon Ornamental and Nursery Digest* 15(2):4.

SEMONIN, R. G. 1978. Severe Weather Climatology in the Midwest and Arboriculture. *J. Arboriculture* 4(6):128-36.

SHARON, E. M. 1973. Some Histological Features of *Acer saccharum* Wood Formed after Wounding. *Canadian Jour. Forest Res.* 3:83-89.

SHAW, E. J., ed. 1980. *Western Fertilizer Handbook.* 6th ed. Sacramento, Calif.: Calif. Fertilizer Assoc.

SHEARMAN, R. C., and others. 1979. A Comparison of Turfgrass Clippings, Oat Straw, and Alfalfa as Mulching Material. *J. Amer. Soc. Hort. Sci.* 104(4):461-63.

SHIGO, A. L. 1969. The Death and Decay of Trees. *Natural History* 78(3):43-47.

———. 1977. Superior Tree Production Fights against Wound Fatalities. *Amer. Nurseryman* 146(12):10-11.

———. 1979a. Tree Care. *J. Arboriculture* 5(9):vi.

———. 1979b. *Tree Decay: An Expanded Concept.* USDA Forest Serv. Agr. Inf. Bull. 419.

———. 1981. To Paint or Not to Paint. In *Handbook of Pruning.* Brooklyn: Brooklyn Botanic Garden, *Plants & Gardens* 37(2):20-23.

SHIGO, A. L., and R. CAMPANA. 1977. Discolored and Decayed Wood Associated with Injection Wounds in American Elm. *J. Arboriculture* 3(12):230-35.

SHIGO, A. L., and R. FELIX. 1980. Cabling and Bracing. *J. Arboriculture* 6(1):5-9.

SHIGO, A. L., and E. vH. LARSON. 1969. *Photo Guide to the Patterns of Discoloration and Decay in Living Northern Hardwood Trees.* USDA Forest Serv. Res. Pap. NE-127.

SHIGO, A. L., and H. G. MARX. 1977. *Compartmentalization of Decay in Trees.* USDA Forest Serv. Agr. Inf. Bull. 405.

SHIGO, A. L., W. E. MONEY, and D. I. DODDS. 1977. Some Internal Effects of Mauget Tree Injections. *J. Arboriculture* 3(11):213-20.

SHIGO, A. L., and A. SHIGO. 1974. *Detection and Decay in Living Trees and Utility Poles.* USDA Forestry Serv. Res. Pap. NE-294.

SHIGO, A. L., W. C. SHORTLE, and P. W. GARRETT. 1977. Genetic Control Suggested in Compartmentalization of Discolored Wood Associated with Tree Wounds. *Forest Sci.* 23:179-82.

SHIGO, A. L., and C. L. WILSON. 1977. Wound Dressings on Red Maple and American Elm: Effectiveness after Five Years. *J. Arboriculture* 3(5):81-87.

SHIGO, A. L., and others. 1979. *Internal Defects Associated with Pruned and Non-pruned Branch Stubs in Black Walnut.* USDA Forest Serv. Res. Pap. 440.

SHORTLE, W. C. 1979. New Look at Tree Care. *J. Arboriculture* 5(12):281-84.

SHURTLEFF, M. C. 1980. The Search for Disease-Resistant Trees. *J. Arboriculture* 6(9):238-44.

SINCLAIR, W. A. 1978. Range, Suscepts, Losses. In *Dutch Elm Disease: Perspectives after 60 Years,* ed. W. A. Sinclair and R. J. Campana. Cornell Univ. Agr. Exp. Sta. *Search Agr.* 8(5):6-8.

SINCLAIR, W. A., E. J. BAUN, and A. O. LARSEN. 1976. Update on Phloem Necrosis of Elms. *J. Arboriculture* 2:106-13.

SINCLAIR, W. A., and R. J. CAMPANA. 1978. Development and Status of Dutch Elm Disease. In *Dutch Elm Disease: Perspectives after 60 Years,* ed. W. A. Sinclair and R. J. Campana. Cornell Univ. Agr. Exp. Sta. *Search Agr.* 8(5):5-6.

SINCLAIR, W. A., and T. H. FILER, JR. 1974. Diagnostic Features of Elm Phloem Necrosis. *Arborist's News* 39:145-49.

SINCLAIR, W. A., and W. T. JOHNSON. 1975. *Verticillium Wilt*. Cornell Tree Pest Leaflet A-3.

SKUTT, H. R., A. L. SHIGO, and R. A. LESSARD. 1972. Detection of Discolored and Decayed Wood in Living Trees Using a Pulsed Electric Current. *Canadian J. Forest Res.* 2:54–56.

SMALLEY, E. B. 1978. Systemic Chemical Treatments of Trees for Protection and Therapy. In *Dutch Elm Disease: Perspectives after 60 Years*, ed. W. A. Sinclair and R. J. Campana. Cornell Univ. Agr. Exp. Sta. *Search Agr.* 8(5):34–39.

SMITH, E. M. 1976. Pin Oak Chlorosis: A Serious Landscape Problem. *Amer. Nurseryman* 143(3):15, 44.

SMITH, E. M., and T. A. FRETZ. 1979. *Chemical Weed Control in Commercial Nursery and Landscape Plantings*. Ohio Coop. Ext. Serv. MM-297.

SMITH, E. M., and K. W. REISCH. 1975. Fertilizing Trees in the Landscape: Progress Report. Abstract. In *J. Arboriculture* 1(4):77.

SMITH, R. C. 1977. Planting Trees with a Power Auger. *Grounds Maintenance*, May, p. 84.

SMITH, W. H. 1970. *Tree Pathology: A Short Introduction*. New York: Academic Press.

STASIUK, W. N., and P. E. COFFEY. 1974. Rural and Urban Ozone Relationships. *J. Air Pollution* 24(9):819.

STIMMANN, M. W. 1977. *Pesticide Application and Safety Training*. Univ. Calif. Agr. Sci. Pub. 4070.

STONE, E. L. 1968. Microelement Nutrition of Forest Trees: A Review. In *Forest Fertilization: Theory and Practice*. Knoxville: Tennessee Valley Authority, pp. 132–75.

STORIE, R. E. 1932. *An Index for Rating the Agricultural Value of Soils*. Calif. Agr. Exp. Sta. Bull. 556.

STREETS, R. B. 1969. *Diseases of the Cultivated Plants of the Southwest*. Tucson: University of Arizona Press.

STRIBLING'S NURSERIES. 1966. New System in Tree Plantings. *Stribling's Fruit and Grape Growers Newsletter*, Jan.

STROMBERG, L. K. 1975. *Water Quality for Irrigation*. Fresno County: Calif. Coop. Ext. Ser. Pub., Nov. 18.

TALBOTT, J. A., and others. 1976. Flowering Plants as Therapeutic/Environmental Agent in a Psychiatric Hospital. *HortSci.* 11(4):365–66.

TATTAR, T. A. 1978. *Diseases of Shade Trees*. New York: Academic Press.

TEMPLE, A. R. 1981. Acute and Chronic Aspirin Toxicity and Its Treatment. *Archives Internal Medicine* 141:364.

TENNESSEE VALLEY AUTHORITY. 1968. *Forest Fertilization: Theory and Practice*. Symposium on Forest Fertilization. Muscle Shoals, Ala.: Tennessee Valley Authority.

THAYER, R. L., JR. 1981. *Solar Access: It's the Law*. Univ. Calif., Davis, Inst. Govt. Affairs and Ecology, Environmental Quality Series 34.

THOMPSON, A. R. 1940. *Transplanting Trees and Other Woody Plants*. U.S. Natl. Park Serv. Tree Preservation Bull. 9.

――――. 1959. *Tree Bracing*. U.S. Natl. Park Serv. Tree Preservation Bull. 3.

――――. 1961. *Shade Tree Pruning*. Rpt. U.S. Natl. Park Serv. Tree Preservation Bull. 4.

THORTON, P. L. 1971. Managing Urban and Suburban Trees and Woodland for Timber Products. In *Trees and Forests in an Urbanizing Environment*. Univ.

Mass., Amherst, Coop. Ext. Serv. Planning and Resource Devel. Monograph 17: 129-32.

TICKNOR, R. L. 1981. Deciduous Trees Modify Temperature of Buildings. *Weeds, Trees, and Turf* 20(2):22-23, 25-26.

TISDALE, S. L., and W. L. NELSON. 1975. *Soil Fertility and Fertilizers.* 3rd ed. New York: Macmillan.

TODHUNTER, M. N., and W. F. BEINEKE. 1979. Effect of Fescue on Black Walnut Growth. *Tree Planter's Notes,* Summer, 20-23.

TORNGREN, T. S., and F. CHAN. 1978. *Mistletoe Control Progress Report: Sacramento County.* Sacramento County: Univ. Calif. Coop. Ext. Ser., Jan. 16.

TORNGREN, T. S., E. J. PERRY, and C. L. ELMORE. 1980. *Mistletoe Control in Shade Trees.* Univ. Calif. Agr. Sci. Leaflet 2571.

TOWNSEND, A. M. 1977. Improving the Adaptation of Maples and Elms to the Urban Environment. In *Proc. 16th Meeting Canadian Improvement Assoc.: Part 2.* Winnipeg: University of Manitoba Press, pp. 27-31.

TROUSE, A. C., JR., and R. P. HUMBERT. 1959. Deep Tillage in Hawaii: Subsoiling. *Soil Sci.* 88:150-58.

TURNER, N. C., and H. C. DeRoo. 1974. Hydration of Eastern Hemlock as Influenced by Waxing and Weather. *Forest Sci.* 20:19-24.

UNITED STATES DEPARTMENT OF AGRICULTURE. 1957. *Soil: The 1957 Yearbook of Agriculture.* Washington, D.C.: U.S. Superintendent of Documents.

————. 1960. *Plant Hardiness Zone Map.* USDA Misc. Pub. 814.

————. 1969. *Pruning Ornamental Shrubs and Vines.* USDA Home and Garden Bull. 165.

————. 1973a. *Air Pollution Damages Trees.* Upper Darby, Pa.: USDA Forest Serv., Northeast Area.

————. 1973b. *Controlling the Japanese Beetle.* USDA Home and Garden Bull. 159.

————. 1973c. *Trees for Polluted Air.* Washington, D.C.: USDA Forest Serv., Misc. Pub. 1230.

————. 1975. *Protecting Shade Trees During Home Construction.* USDA Home and Garden Bull. 104.

————. 1979. *Seed and Planting Stock Dealers.* Washington, D.C.: USDA Forest Serv., FS-331.

VANALFEN, N. K., and W. E. MACHARDY. 1978. Symptoms and Host-Pathogen Interactions. In *Dutch Elm Disease: Perspectives after 60 Years,* ed. W. A. Sinclair and R. J. Campana. Cornell Univ. Agr. Exp. Sta. *Search Agr.* 8(5):20-25.

VAN DEN BOSCH, R., and P. S. MESSENGER. 1973. *Biological Control.* New York: Harper & Row.

VAN DE WERKEN, H. 1981. Fertilization and Other Factors Enhancing the Growth Rate of Young Shade Trees. *J. Arboriculture* 7(2):33-37.

VAN EIMERN, J. 1964. *Windbreaks and Shelterbelts.* Geneva: World Meteorological Org. Tech. Note 59.

VAN HEUIT, R. E. 1979. *Sanitation Districts of Los Angeles County: Generation of Methane Gas in Sanitary Landfills and Its Effect on Growth of Plant Materials.* Unpublished paper delivered at 1979 meeting of Amer. Soc. of Consulting Arborists. (Abstract in *Horizons* 1980. Washington, D.C.: Amer. Assoc. Nurserymen, Sept.)

VEIHMEYER, F. J., and A. H. HENDRICKSON. 1955. Does Transpiration Decrease as the Soil Moisture Decreases? *Transactions, Amer. Geophysical Union* 36:425-48.

VERNER, L. 1955. *Hormone Relations in the Growth and Training of Apple Trees.* Univ. Idaho, College of Agr. Res. Bull. 28.

VIETS, F. G., JR. 1966. *Zinc Deficiency in the Silver Oak.* Ann. Admin. Rept. Sci. Dept. (Tea Sect.) United Planters Assoc. S. India 1962/63, pp. 70-71. (For Abstr. 26:2320).

VON RUMKER, R. V., and others. 1975. *A Study of the Efficiency of the Use of Pesticides in Agriculture.* Washington, D.C.: Office of Pesticide Programs, Office of Water and Hazardous Materials, Environmental Protection Agency Report 540/9-75-025.

WALHEIM, L. 1977. *The World of Trees.* San Francisco: Ortho Books.

WALLER, J. 1965. *Cooperative Enterprises-Activities of Civic Groups, Organizations and Individuals in Beautification Programs.* Proc. Intl. Shade Tree Conf. 41: 195-201.

WALLIHAN, E. F. 1966. Iron. In *Diagnostic Criteria for Plants and Soils,* ed. H. D. Chapman. Riverside, Calif.: H. D. Chapman, pp. 203-12.

WALTERS, D. T., and A. R. GILMORE. 1976. Allelopathic Effects of Fescue on the Growth of Sweetgum. *J. Chem. Ecol.* 2:469-79.

WARD, J. C., and W. Y. PONG. 1981. *Wetwood in Trees: A Timber Resource Problem.* Pacific NW Sta., U.S. Forest Serv. Gen. Tech. Report PNW-112.

WARE, G. W. 1978. *The Pesticide Book.* San Francisco: W. H. Freeman.

WARGO, P. M. 1980. *Armillaria mellea:* An Opportunist. *J. Arboriculture* 6(10):276-78.

WATKINS, J. V. 1961. *Your Guide to Florida Landscape Plants.* Gainesville: University of Florida Press.

WEAVER, R. J. 1972. *Plant Growth Substances in Agriculture.* San Francisco: W. H. Freeman.

Webster's Seventh New Collegiate Dictionary. 1976. Springfield, Mass.: Merriam.

WEIER, T. E., C. R. STOCKING, and M. G. BARBOUR. 1974, *Botany: An Introduction to Plant Biology.* 5th ed. New York: John Wiley.

WEISER, C. J. 1970a. Cold Resistance and Acclimation in Woody Plants. *HortSci.* 5(5):403-10.

———. 1970b. Cold Resistance and Injury in Woody Plants. *Science* 169:1269-78.

WELCH, D. S. 1949. The Cause and Treatment of Cavities in Trees. *Proc. Natl. Shade Tree Conf.* 25:126-31.

WHITCOMB, C. E. 1979a. Amendments and Tree Establishment. *J. Arboriculture* 5(7):167.

———. 1979b. Factors Affecting the Establishment of Urban Trees. *J. Arboriculture* 5(10):217-19.

———. 1980. Effects of Black Plastic and Mulches on Growth and Survival of Landscape Plants. *J. Arboriculture* 6(10):10-12.

WHITE, D. P. 1956. Aerial Application of Potash Fertilizer to Coniferous Plantations. *J. Forestry* 54:762-68.

WHITE, R. F. 1945. *Effects of Landscape Development on the Natural Ventilation of Buildings and Their Adjacent Areas.* College Station: Texas Engr. Exp. Sta. Res. Report 45.

WHITEHEAD, F. H. 1963. The Effects of Exposure on Growth and Development. In *Water Relations of Plants*, ed. A. J. Rutter and F. H. Whitehead. New York: John Wiley.

WHITLOW, T. H., and R. W. HARRIS. 1979. *Flood Tolerance in Plants: A State-of-the-Art Review*. Vicksburg, Miss.: U.S. Army Engineer Waterways Exp. Sta. Tech. Report E-79-2.

WHITTAKER, R. H. 1954. The Ecology of Serpentine Soils: A Symposium. *Ecology* 35(2):258–59.

WILDMAN, W. E. 1969. Water Penetration: Where Is the Restricting Layer. *Soil and Water* (Univ. Calif. Coop. Ext. Serv.) 7:1–4.

——— . 1976. *Diagnosing Soil Physical Problems*. Univ. Calif. Ext. Leaflet 2664.

WILDMAN, W. E., and K. D. GOWANS. 1975. *Soil: Physical Environment and How It Affects Plant Growth*. Univ. Calif. Ext. Leaflet 2280.

WILDMAN, W. E., J. L. MEYER, and R. A. NEJA. 1975. *Managing and Modifying Problem Soils*. Univ. Calif. Ext. Leaflet 2791.

WILLIAMSON, J. F. 1972. *Sunset Pruning Handbook*. Menlo Park, Calif.: Lane.

WILSON, B. F. 1970. *The Growing Tree*. Amherst: University of Massachusetts Press.

WILSON, C. L., and A. L. SHIGO. 1973. Dispelling Myths in Arboriculture Today. *Amer. Nurseryman* 127:24–28.

WITTROCK, G. I. 1971. *The Pruning Book*, Emmaus, Pa.: Rodale Press.

WONG, T. L., R. W. HARRIS, and R. E. FISSELL. 1971. Influence of High Soil Temperatures of Five Woody Species. *J. Amer. Soc. Hort. Sci.* 96:80–83.

WORLD METEOROLOGICAL ORGANIZATION. 1971. *Meteorological Services of the World*. Geneva: World Meteo. Org. 2.

WYMAN, D. P. 1936. *Growth Experiments with Pin Oaks Which Are Growing under Lawn Conditions*. Cornell Univ. Agr. Exp. Sta. Bull. 646.

——— . 1957. Mulching Practices at the Arnold Arboretum. *Plants & Gardens* (Brooklyn Botanic Garden) 13(1):27–30.

——— . 1965. *Trees for American Gardens*. New York: Macmillan.

——— . 1969. *Shrubs and Vines for American Gardens*. New York: Macmillan.

——— . 1971. *Wyman's Gardening Encyclopedia*. New York: Macmillan.

ZANETTO, J. 1978. Trees for Solar Neighborhoods. *Landscape Architecture*, Nov., pp. 514–19.

ZELAZNY, L. W. 1968. Salt Tolerance of Roadside Vegetation in Proceedings *Symposium: Pollutants in the Roadside Environment*, ed. E. D. Carpenter. Storrs, Conn.: Plant Science Dept., University of Connecticut, pp. 50–56.

ZIMMERMAN, M. H., and C. L. BROWN. 1971. *Trees: Structure and Function*. New York: Springer-Verlag.

Index
of Plants Mentioned
in the Text

Plants listed in the appendices are not included.

General Index

Illustrations and tables are indicated in boldface type. Cultivated plants are listed in the Plant Index.